AF389516

Mechanisms of ER Protein Import

Mechanisms of ER Protein Import

Editors

Richard Zimmermann
Sven Lang

MDPI • Basel • Beijing • Wuhan • Barcelona • Belgrade • Manchester • Tokyo • Cluj • Tianjin

Editors
Richard Zimmermann
Medical Biochemistry and
Molecular Biology
Saarland University
Homburg
Germany

Sven Lang
Medical Biochemistry and
Molecular Biology
Saarland University
Homburg
Germany

Editorial Office
MDPI
St. Alban-Anlage 66
4052 Basel, Switzerland

This is a reprint of articles from the Special Issue published online in the open access journal *International Journal of Molecular Sciences* (ISSN 1422-0067) (available at: www.mdpi.com/journal/ijms/special_issues/mechanisms_ER_protein_import).

For citation purposes, cite each article independently as indicated on the article page online and as indicated below:

LastName, A.A.; LastName, B.B.; LastName, C.C. Article Title. *Journal Name* **Year**, *Volume Number*, Page Range.

ISBN 978-3-0365-4094-8 (Hbk)
ISBN 978-3-0365-4093-1 (PDF)

Contents

About the Editors

Richard Zimmermann

Prof. Dr. Richard Zimmermann is a professor emeritus at the Department of Medical Biochemistry and Molecular Biology of Saarland University in Homburg, Germany. His scientific interests are human endoplasmic reticulum, in particular: i) the components and mechanisms of protein import into the ER; ii) the components and physiological role of calcium leakage from the ER; iii) the components and mechanisms of ATP/ADP exchange between the ER and the cytosol; iv) the molecular chaperones of the ER; v) the components and mechanisms of Sec61 channel gating; vi) Sec61-channelopathies; and vii) chaperonopathies. He has earned various awards, including the *Heinz-Maier-Leibnitz* Award 1984 from the DFG for research on biological membranes, was elected as a member of the DFG study section for Biochemistry, Biophysics, Clinical Chemistry, and Pathobiochemistry in 2002, and as member of the DFG study section for Biochemistry, Biophysics, Structural Biology, Bioinformatics, and Theoretical Biology in 2008 for two terms. Richard Zimmermann was a graduate student with Walter Neupert at the University of Göttingen, Germany, between 1977 and 1980, a postdoctoral fellow with William Wicker at UCLA in Los Angeles, USA, in 1981 and 1982, a research associate at the Ludwig-Maximilians-University Munich, Germany, from 1983 to 1987, an assistant professor at the Ludwig-Maximilians-University Munich from 1988 to 1990, an associate professor for Molecular Cell Biology at the Biochemistry Center, Georg-August-University Göttingen between 1991 and 1995, full professor at the Institute for Medical Biochemistry and Molecular Biology of Saarland University between 1995 and 2017, senior professor at the Institute for Medical Biochemistry and Molecular Biology of Saarland University in 2018 and 2019, and Head of Competence Center for Molecular Medicine of Saarland University between 2005 and 2017.

Sven Lang

Dr. Sven Lang is an independent research group leader at the Department of Medical Biochemistry and Molecular Biology of Saarland University in Homburg, Germany. The scientific focus of his team is on aspects of protein and energy homeostasis of the endoplasmic reticulum (ER). His current projects address the dynamic assembly of the Sec61 complex and ER protein translocase, the structure–function relationship of the Sec61 complex with an emphasis on disease-associated variants, the regulation and signaling capacity of ER calcium leakage, the maintenance and components of the ER energy homeostasis, and elements of different protein targeting networks. Dr. Lang started as graduate student in the lab of Prof. Dr. Zimmermann at Saarland University (2008-2012). He was a postdoctoral fellow in the labs of Prof. Dr. Pankaj Kapahi at the Buck Institute for Research on Aging (2012-2015) and Prof. Dr. Chi at the University of California, San Francisco (2015-2016). In 2016, he started as a junior research group leader at his alma mater Saarland University.

Preface to "Mechanisms of ER Protein Import"

Protein import into the endoplasmic reticulum (ER) is the first step in the biogenesis of approximately 10,000 different soluble and membrane proteins of human cells, which amounts to about 30% of the proteome. Most of these proteins fulfill their functions either in the membrane or lumen of the ER plus the nuclear envelope, in one of the organelles of the pathways for endo- and exocytosis (ERGIC, Golgi apparatus, endosome, lysosome, and trafficking vesicles), or at the cell surface as plasma membrane or secreted proteins. In addition, an increasing number of membrane proteins destined to lipid droplets, peroxisomes or mitochondria are first targeted to and inserted into the ER membrane prior to their integration into budding lipid droplets or peroxisomes or prior to their delivery to mitochondria via the ER-SURF pathway. ER protein import involves two stages, ER targeting, which guarantees membrane specificity, and the insertion of nascent membrane proteins into or translocation of soluble precursor polypeptides across the ER membrane. In most cases, both processes depend on amino-terminal signal peptides or transmembrane helices, which serve as signal peptide equivalents. However, the targeting reaction can also involve the ER targeting of specific mRNAs or ribosome–nascent chain complexes. In addition, both processes may occur co- or post-translationally and are facilitated by various sophisticated machineries, which reside in the cytosol and the ER membrane, respectively. Except for resident ER and mitochondrial membrane proteins, the mature proteins are delivered to their functional locations by vesicular transport. In this Special Issue, international experts in this area of cell biology report on their structural and mechanistic insights into various aspects of targeting, insertion, and translocation machineries, such as the signal recognition particle (SRP), its corresponding receptor (SR) and the Sec61 complex. Furthermore, small-molecule inhibitors and toxins that interfere with ER protein import are discussed, thereby providing a link to human medicine, specifically to the so-called Sec61-channelopathies.

Richard Zimmermann and Sven Lang

Editors

International Journal of
Molecular Sciences

Review

Targeting of Proteins for Translocation at the Endoplasmic Reticulum

Martin R. Pool

School of Biological Science, Faculty of Biology, Medicine and Health, University of Manchester, Manchester M13 9PL, UK; martin.r.pool@manchester.ac.uk; Tel.: +44-161-275-5392

Abstract: The endoplasmic reticulum represents the gateway to the secretory pathway. Here, proteins destined for secretion, as well as soluble and membrane proteins that reside in the endomembrane system and plasma membrane, are triaged from proteins that will remain in the cytosol or be targeted to other cellular organelles. This process requires the faithful recognition of specific targeting signals and subsequent delivery mechanisms to then target them to the translocases present at the ER membrane, which can either translocate them into the ER lumen or insert them into the lipid bilayer. This review focuses on the current understanding of the first step in this process representing the targeting phase. Targeting is typically mediated by cleavable N-terminal hydrophobic signal sequences or internal membrane anchor sequences; these can either be captured co-translationally at the ribosome or recognised post-translationally and then delivered to the ER translocases. Location and features of the targeting sequence dictate which of several overlapping targeting pathway substrates will be used. Mutations in the targeting machinery or targeting signals can be linked to diseases.

Keywords: endoplasmic reticulum; ribosome; signal sequence; signal recognition particle; protein targeting; GET; SND; Sec61 translocase; NAC

Citation: Pool, M.R. Targeting of Proteins for Translocation at the Endoplasmic Reticulum. *Int. J. Mol. Sci.* **2022**, *23*, 3773. https://doi.org/10.3390/ijms23073773

Academic Editor: Alexandre Mironov

Received: 24 December 2021
Accepted: 24 March 2022
Published: 29 March 2022

Publisher's Note: MDPI stays neutral with regard to jurisdictional claims in published maps and institutional affiliations.

1. Targeting Signals

Secretory and soluble proteins that reside in the endomembrane system are typically synthesised with an N-terminal signal sequence, which directs them to the ER [1,2]. Once translocated to the ER lumen, the signal is then cleaved off (Figure 1A). Signal sequences are highly variable in sequence but typically are 12–30 residues in length and composed of an N-terminal region, often positively charged, a core of eight or more hydrophobic residues, and a short polar C-terminal domain which often contains helix-breaking glycine and proline residues, as well as amino acids with short-side chains at the -1 and -2 positions, the consensus site for the cleavage by the signal peptidase [2,3].

Integral membrane proteins can be classified based on their topology in the membrane. Single spanning membrane proteins (with just one trans-membrane (TM) domain) that have their N-termini in the ER lumen and C-terminus in the cytoplasm are termed type I, whereas those with their N-termini in the cytoplasm and C-terminus in the lumen are termed type II. In addition to this classification, a subclass of type I membrane proteins that possess very short luminal domains are called type III membrane proteins. Similarly, type II proteins with very short luminal domains are called tail-anchored proteins, reflecting the location of the TM domain close to the C-terminus (Figure 1A). Membrane proteins with multiple TM domains are termed polytopic membrane proteins.

Type I membrane proteins, similar to secretory proteins, are also targeted by cleavable signal sequences. In contrast, type II membrane proteins, including tail-anchored proteins, use their TM domain as a non-cleavable targeting sequence (Figure 1A) [4,5]. The same mechanism is also used by Type III membrane proteins. Targeting of polytopic membrane proteins is typically linked to the topology of the first TM domain; thus, where there is a large N-terminal luminal domain, targeting typically takes place via an N-terminal signal sequence. In contrast, where there is a short N-terminal luminal domain or where the

N-terminus is in the cytosol, then, the first TM domain serves as a non-cleavable targeting signal as with type II membrane proteins. A key feature of all these targeting signals is their hydrophobic nature, and hence, a vital role of cellular targeting pathways is to prevent their aggregation while they are being targeted.

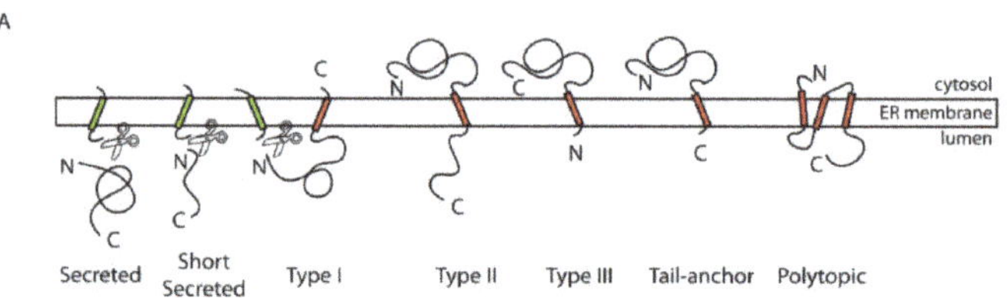

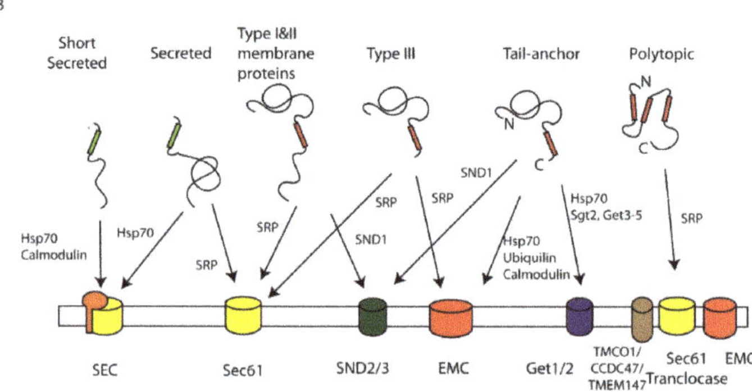

Figure 1. Targeting signals, pathways, and translocases: (**A**) Soluble secretory and luminal proteins as well as type I membrane proteins are targeted by cleavable N-terminally located signal sequences (green). Type II, type III and tail-anchor membrane proteins as well as multi-spanning polytopic membrane proteins utilise internal TM domains (red) for targeting. (**B**) These different classes of secretory and membrane proteins are translocated across or inserted into the membrane by a number of translocases with overlapping specificity. Short secretory proteins are translocated by the SEC translocase and maintained in a translocation-competent state by Hsp70 chaperones or calmodulin. Longer secretory proteins can use the same pathway as well as delivery to the Sec61 translocase by SRP. Type I and II membrane proteins can also be delivered to Sec61 by SRP. Those with more central and C-terminal TM domains can also be targeted by the SND pathway. Type III membrane proteins can be delivered by SRP to either Sec61 or the EMC translocase. Tail-anchored membrane proteins can utilise the SND pathway and the GET pathway as well as delivery to the EMC complex involving Hsp70, calmodulin, or ubiquilins. Finally, polytopic membrane proteins are targeted to the Sec61 translocon by SRP and may recruit the EMC or TMCO1 to assist in their integration.

2. ER Protein Translocases and Insertases

At the ER membrane, a number of protein translocases are able to either translocate proteins into the ER lumen or insert them into the lipid bilayer (Figure 1B). The canonical Sec61 translocase can perform both these functions with a large range of substrates. The more recently discovered GET insertase and EMC translocase can function independently of Sec61 but are limited to substrates with short luminal domains, while the recently discovered TMCO1 translocon functions in collaboration with Sec61 during polytopic membrane protein biogenesis [5–7]. Here, we shall focus on the targeting mechanisms which recognise substrates and deliver them to these translocases. The details of structure, function, and insertion mechanisms of these translocases are covered by a number of excellent recent reviews [5,6,8,9].

3. SRP-Dependent Targeting

The signal recognition particle (SRP) is a highly-conserved targeting machine present in all domains of life [10]. SRP is able to bind to translating ribosomes and can scan the emerging N-terminus of the nascent chain for the presence of signal sequences [11] (Figure 2A). Signal sequence recognition by SRP leads targeting of the ribosome together with the nascent chain to the ER membrane via the action of its cognate receptor (SRP receptor SR), an integral ER membrane protein which then facilitates the transfer of the ribosome and nascent chain to the Sec61 protein-conducting channel [12–14] (Figure 2A). SRP also induces a transient slowdown in translation, termed elongation arrest, which extends the time window where the ribosome-nascent chain complex remains competent to be targeted to Sec61 by the action of SR [15]. Once the nascent chain is released from the SRP–SR complex, the two can dissociate to allow further rounds of targeting.

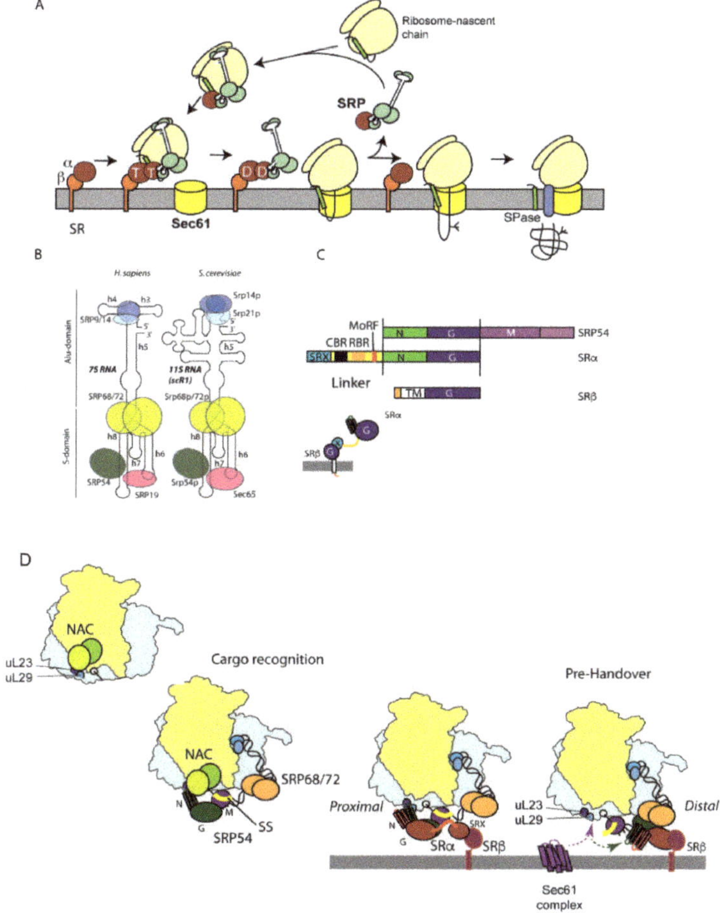

Figure 2. SRP-dependent targeting pathway: (**A**). Overview of the SRP targeting pathway: SRP can bind to ribosomes and scan for the presence of a signal sequence in the nascent chain as the N-terminus emerges from the ribosome exit tunnel. Engagement with the nascent chain with the SRP54 subunit (red) leads to a transient retardation in translation elongation and targeting to the ER membrane via an interaction with SRP receptor (SR). Complex formation between SRP and SR is driven by GTP (T) binding to SRP54 and SRα. The Sec61 translocase triggers release of signal sequence

from SRP concomitant with hydrolysis of GTP to GDP (D). The ribosome binds Sec61 such that as translation resumes the nascent chain is threaded through the channel into the ER lumen where signal peptidase (SPase) can then remove the signal sequence. Finally, SRP and SR can dissociate, and nucleotide is released, permitting further rounds of targeting. (**B**). Domain organisation of yeast and mammalian SRP: Mammalian SRP is composed of a 7SL SRP RNA and six SRP proteins. SRP54, SRP19, SRP68, and SRP72 bind the RNA to form the S-domain whilst SRP9 and SRP14 form the Alu domain. Yeast SRP comprises a larger 11S RNA and homologues of SRP54, SRP19 (Sec65), SRP68, and SRP72 in the S domain and Srp14 and a dimer of Srp21 in the Alu domain. (**C**). Domain organisation of SRP54 and SRP receptor: SRP54 is comprised of a composite NG domain containing a four α-helical N- and GTPase (G) domain as well as a M-domain that comprises the RNA and signal sequence binding site. SRα has an N-terminal SRX domain that interacts with SRβ, a flexible linker with conserved CBR, RBR, and MoRF motifs and an NG domain closely related to that of SRP54. SRβ comprises an N-terminal TM domain and GTPase domain that interacts with SRX. (**D**). All ribosomes are associated with NAC, a nascent chain chaperone, which is bound at the exit site; SRP can transiently bind to the ribosome, positioning SRP54 at the exit site via an interaction with uL23 and uL29, allowing it to scan the emerging nascent chain for signal sequences. NAC aids the specificity of cargo loading of SRP54 with the signal sequence, which then stabilises ribosome binding. Initial interaction with SR at the ER membrane involves interaction of the SRP54 and SRα NG domains adjacent to uL23/uL29 (the proximal site) and is accelerated by the SRα linker MoRF motif binding to the SRP RNA when SRP54 is bound to a signal sequence. Subsequent dissociation and the NG domains from uL23/L29 and SR compaction move the NG domains to the distal site, where interactions involving SRP68, SR CBR motif, SRX, and SRβ stabilise its binding to form the 'prehandover complex'. An interaction of the GTPase domains with SRP72 blocks GTP hydrolysis until Sec61 arrives. Movement to the distal site allows access to uL23/uL29 which is a key binding site for the Sec61 translocase and exposes the signal sequence and M-domain. This allows efficient transfer of the ribosome and nascent chain to Sec61 coordinated by concerted GTP hydrolysis by the NG domains.

The conserved core of the SRP targeting system, present in all domains of life, comprises the signal-sequence-binding protein of SRP, SRP54, associated with a conserved SRP RNA together with the SRα subunit of SR. SRP54 and SRα are both GTPases and possess closely related GTPase domains. In higher eukaryotes, SRP comprises a 300 nt 7S SRP RNA and six polypeptides organised in two domains: the S-domain which includes the conserved core as well as proteins SRP19, SRP68, and SRP72, and the Alu domain comprising proteins SRP9 and SRP14 [16] (Figure 2B). Yeast SRP possesses a slightly larger 11S RNA, a larger homologue of SRP19 (Sec65), and SRP9 is replaced with the related Srp21, and Srp14 is present as a homodimer [17] (Figure 2B). The S-domain of SRP is primarily involved with signal sequence recognition and accurate handover to the Sec61 translocation channel at the ER membrane facilitated by the interaction with SR. In contrast, the Alu domain is responsible for inducing the elongation arrest activity [18].

The SRP receptor comprises the conserved SRα subunit anchored at the ER membrane by single spanning-membrane protein SRβ, which also possesses a GTPase domain but which is more closely related to ARF and Sar1 than SRP54 and SRα [19] (Figure 2C).

Structural insight into SRP and its interaction with signal sequences and the ribosome have been provided by both X-ray crystallography and cryo-EM studies [11,20,21]. SRP54 is composed of two domains, the NG and M-domains, separated by a flexible linker which allows communication between the two domains [22] (Figure 2C). The N-domain comprises a four α-helical bundle which folds onto the GTPase domain [23]. The M-domain contains a helix-loop-helix structure, which allows SRP54 to bind to the SRP RNA and a hydrophobic groove which forms the binding site for signal sequences [24,25]. SRP19 does not contact SRP54 directly but also binds the SRP RNA and modifies its structure allowing SRP54 to bind [26].

Initial, low-affinity binding of SRP to ribosomes is independent of the signal sequence allowing it to dynamically scan ribosomes for the presence of signal sequences in the emerging nascent chain prior to high affinity binding upon recognition of the signal sequence [27,28]. Consistent with this observation, ribosome profiling experiments indicate that SRP can also be recruited to ribosomes prior to emergence of the signal sequence [29].

The S-domain of SRP binds to the ribosome on the 60S subunit at the exit site where the nascent chain emerges from the exit tunnel which conveys it from the peptidyl-transferase centre to the ribosome surface (Figure 2D) [20]. The N-domain of SRP54 contacts ribosomal proteins uL23 and L29 at one side of the exit tunnel, as well as three additional contact sites involving the SRP54 M-domain and SRP68/72 [11,20,30]. This positioning of SRP54 permits efficient scanning and capture of the signal sequence by the M-domain as it emerges [11,20,30,31]. Blocking access of SRP to uL23/L29 is known to lead to targeting defects in vivo [32].

The Alu domain of SRP contacts the interface of the large and small subunits at the translation elongation factor-binding site, thus rationalising the slowdown in translation elongation by antagonising factor binding [11,20] (Figure 2D). The C-terminus of SRP14 represents one of the contact sites, and its removal leads to targeting defects, which can be ameliorated by elevating SR concentration [20,33,34].

Slowdown of translation is not solely mediated by SRP; analysis of codon-optimality downstream from the targeting sequences revealed a prevalence of poorly translated codons distal to the signal sequence, which also contributes to slowed translation and replacement with more synonymous optimal codons, reducing targeting efficiency [35].

Careful kinetic and structural analysis of the SRP targeting cycle has provided key insight into the roles played by the two GTPases, SRP54 and SRα, in regulating targeting. SRP54 and SRα are both members of the SIMIBI family of GTPases which are characterised by a relatively low affinity for GTP compared to other small GTPases, such as Ras [36]. This is explained structurally by the presence of an additional insertion domain called the I-box which stabilises the nucleotide-free state [37].

SRα is organised into two domains, an N-terminal SRX domain, which folds together with the SRβ GTPase [38–40], and the SRP54-related NG domain, separated by a conserved flexible linker rich in positive charge (Figure 2C). SRP54 and SRα NG domains dimerise in a GTP-dependent manner, whereby the two bound nucleotides sit at the interface contacting one another in an anti-parallel organisation [41,42].

Once bound to a signal sequence, SRP54 can then engage the SRP receptor. This occurs in two steps, an initial low affinity binding where the SRP54 NG domain is still adjacent to ribosomal proteins, uL23 and uL29, termed the proximal site, followed by rearrangement to a high-affinity bound state where the NG-domains together move away from the ribosome surface together towards SRP68/72, termed the distal site, to form a 'prehandover' complex [43,44]. Initial complex formation at the proximal site is facilitated by GTP-dependent NG–NG domain interaction and is dramatically accelerated by the conserved molecular recognition feature (MoRF) in the unstructured SRα linker region, which contacts the SRP RNA [45]. This kinetic acceleration is dependent on cargo loading to SRP54, thereby strongly favouring complex formation upon recognition of the signal sequence [45]. Moreover, this is directly analogous to the role played by the SRP RNA in bacterial SRP–SR complex formation [46,47].

Following the initial complex formation by the two NG domains, there is destabilisation of the first α-helix in each of the two N domains and their associated loops which contact the ribosome [44,48]. This rationalises their dissociation from uL23/uL29 at the proximal site. Once the NG domains are dissociated, a compaction of SR brings the NG domain close to the SRX domain at the distal site [49]. This is stabilised by interactions between SRP68, elements of SRα NG, and X domains, as well as with SRβ [44]. The charged CBR motif in the SRα linker also makes an interaction with the distal site elements of the SRP RNA [44] (Figure 2D).

This prehandover complex is now primed to interact with the Sec61 translocon as the uL23/L29 site, which Sec61 also contacts, and is now accessible, as well as the signal sequence in the M-domain [30,43,44,50]. Whilst at the distal site, an interaction of the

GTPase domain interface with the C-terminus of SRP72 delays hydrolysis of GTP by SRP54 and SRα until Sec61 is present [44,51].

The arrival of Sec61 leads to rearrangement of the prehandover complex which induces concerted GTP hydrolysis by SRP54 and SRα, thereby releasing the nascent chain from SRP and allowing its transfer together with the ribosome to the Sec61 complex [14,43,52]. Following GTP hydrolysis, SRP and SR can dissociate to undertake further rounds of targeting [53].

Recent studies have shown that SRP alone is unable to discriminate with high precision cognate and near-cognate (non-functional) signal sequences. Another ribosome-associated biogenesis factor, NAC, also needs to collaborate with SRP to enhance fidelity [54] (Figure 2D). NAC binds to all ribosomes and contacts overlapping regions at the exit site to where SRP binds [20,55–57]. It is also able to initially insert a domain deep into the exit tunnel which would prohibit interactions of factors with non-translating ribosomes; subsequent displacement of this domain by the nascent chain has been proposed to lead to dynamic rearrangement of NAC which appears important for the correct triage of nascent chains to SRP in the case of secretory and membrane proteins and RAC and Hsp70 in the case of cytosolic proteins [57,58]. Hence, the increased fidelity in signal sequence recognition in the presence of NAC is entirely consistent with this model.

Proteomic approaches in yeast have used ribosome profiling to identify substrates and binding sites for SRP on translating polysomes [29]. SRP is associated mainly with ribosomes translating integral membrane proteins and to a lesser extent proteins with N-terminal signal sequences [29]. This parallels similar studies in bacteria which also show SRP is mainly involved in membrane protein biogenesis [59]. These findings are also supported by SRP-substrate dependency as measured proteome-wide by loss of ER mRNA association upon rapid depletion of SRP [60]. This analysis revealed a similar SRP-dependency profile that is again biased towards integral membrane proteins [60]. The ribosome profiling experiments also revealed that whilst for some substrates SRP recruitment occurs as the targeting sequence emerges from the ribosome as expected in the canonical model of SRP function, in many cases SRP is recruited earlier, often well before the signal sequence/anchor has been synthesised [29]. This non-canonical binding of SRP is still dependent upon the ribosome and in some cases requires features of the $3'$ UTR, but it is not presently known how preferential pre-delivery of SRP to these substrates occurs and if it requires specific mRNA binding proteins [29]. Interestingly, it was recently shown that the ribosome-associated factor, Hel2, which binds to collided ribosomes that arise where stalling occurs, was shown to bind to membrane protein polysomes in a pattern that strongly overlaps with that of SRP [61].

In the absence of SRP, substrates that are usually delivered to the ER can become aggregated [62] or are instead mistargeted to mitochondria causing disruption to mitochondrial function [56,60], a phenotype also observed if Hel2 is disrupted [61]. Hence, another key function of SRP is to prevent promiscuous misbehaviour of hydrophobic targeting signals.

4. SND-Targeting Pathway

SRP is not the only co-translational ER-targeting pathway. Studies in yeast investigating targeting of GPI-anchored proteins identified the SND pathway, comprising the cytosolic factor SND1 associated with the ribosome and ER-membrane associated SND2/3 proteins [63]. Exploitation of this pathway is not exclusive to GPI-anchored proteins but rather membrane proteins with centrally located TM domains. The SND pathway also operates in mammalian cells, although to date, only the homologue of SND2 has been identified [64]. As seen in yeast, human GPI-anchored proteins also use SND2 for their targeting to the ER [65]. Detailed molecular details of the SND-targeting pathway await further investigation.

5. Sec62-Dependent Targeting

Work in yeast showed that ER targeting of many secretory precursors is independent of SRP and SR but rather requires the ER membrane protein, Sec62 [66], which forms a larger SEC complex along with the core Sec61 translocase as well as Sec63, Sec71, and Sec72 [67]. Sec62 is dispensable for targeting of substrates that use SRP and SR [68,69]. Also, in contrast to the SRP pathway, targeting can be uncoupled from translation [70]. Indeed, the SEC complex cannot bind ribosomes directly [71,72], rationalised by recent structures of the SEC complex which show the ribosome-binding sites on the core Sec61 components are all occupied by the additional components of the SEC complex [73].

Cytosolic Hsp70s and Hsp40 are required to maintain substrates in a translocation-competent state [74,75], and Sec71/72 have been shown to be receptors for Hsp70 family chaperones [76]. Properties of the signal sequence largely determine which targeting pathway a substrate will use, with more hydrophobic sequences exclusively using the SRP-pathway and less hydrophobic ones using Sec62, while intermediate hydrophobic ones can access both [69].

Although Sec62-dependent translocation can be uncoupled from translation, proteome-wide proximity labelling experiments in yeast have shown that most substrates are translocated co-translationally, likely driven by engagement of a pioneer ribosome-associated signal-sequence-bearing protein with the SEC complex before its translation has been completed [77]. This will bring the polysome close to the membrane such that subsequent nascent chains are also highly likely to engage the SEC translocon before their synthesis is complete [77].

Homologues of Sec62 as well as Sec63 (but not Sec71/72) also exist in higher eukaryotes [72,78]. Unlike its yeast counterpart, mammalian Sec62 has an additional ribosome-binding domain [79]. They have mainly been implicated in the targeting and translocation of short-secretory proteins (<100 amino acids), such as the insect proteins, preprocecropin A and prepromelittin [80–83], and mammalian proteins, such as insulin, apelin, and statherin [81,84]. These proteins are so short that they are released from the ribosome before SRP can effectively engage with them and so are targeted post-translationally. In the case of preprocecropin A, an interaction with calmodulin is important to maintain the protein in an insertion-competent state [85]. This is a calcium-dependent phenomenon that can occur already at resting cytosolic calcium levels [85]. Proteomic approaches in human cells have identified 199 proteins whose biogenesis is negatively affected by loss of Sec62 [86]. Not all of these are short secretory proteins, but as in yeast, they typically possess signal peptide or trans-membrane domains with lower hydrophobicity [86]. This may reflect a role of Sec62 in their SRP-independent targeting and/or a requirement of Sec62 at the translocon at the later stages of translocation following targeting via SRP and SR [87,88]. Overall, targeting sequences in mammalian cells have higher hydrophobicity than in yeast, which likely reflects the difference in bias in SRP versus Sec62-dependence in the two systems [85].

6. GET-Targeting Pathway

Proteins which possess a C-terminal TM anchor (tail anchor [TA]) are unable to access the classical co-translational SRP-targeting machinery as they are released from the ribosomes before the TA sequence has emerged from the exit tunnel [89]. Rather they use the distinct GET (guided-entry of tail-anchor proteins) machinery for their delivery to the ER [4,90,91]. In yeast, they are first bound by the cytosolic targeting factor, Sgt2, which acts as a pre-loading complex and are then transferred to Get3, facilitated by its accessory proteins, Get4 and Get5 [92,93] (Figure 3A). The Get3-targeting factor can then deliver the proteins to the Get1/Get2 ER membrane heterodimer, which acts as both a receptor for Get3 and also a membrane insertase [92,94]. In higher eukaryotes, homologous of all these key components are present: SGTA (Sgt2), TRC40 (Get3), TRC35 (Get4), Ubl4A (Get5), and CAML/WRB (Get1/Get2) [95–98] (Figure 3B). In addition, an extra factor, BAG6, is present in a complex with TRC35 and Ubl4A (called the BAG6 complex) and acts to triage hydrophobic clients between TRC40 and the ubiquitin-proteasome system [99,100]. While *bona fide* ER TA proteins are delivered faithfully to TRC40, mis-localised membrane and secretory proteins are also bound by the BAG6 complex and directed for ubiquitination and disposal via the proteasome [99]. Hence, BAG6 has a role in both targeting fidelity and proteostasis.

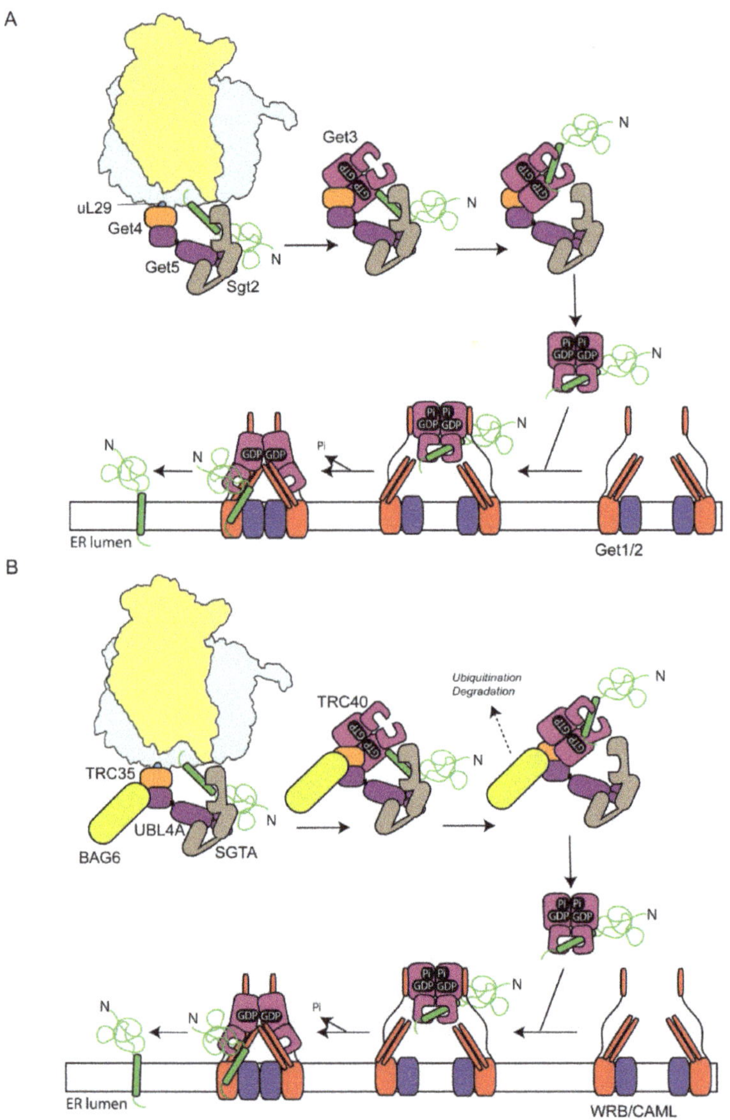

Figure 3. GET-targeting pathways for tail-anchored proteins in yeast and mammals. (**A**) In yeast, the Get4/5 complex is able to bind to the ribosome via an interaction of Get4, close to the exit site and uL29. Get5 can then recruit Sgt2, which is able to capture a tail-anchor TM domain as it emerges from the exit tunnel. The Sgt2-Get4/5-TA protein can be released from the ribosome and then the TA protein can be directly transferred to Get3, a homodimeric ATPase, which is recruited in the ATP bound state. Release from Get4/5 leads to ATP hydrolysis to ADP and P$_i$, which remain bound, altering the conformation of the TA-binding domain such that the TA remains bound but can now

bind the membrane receptor Get1/2. Get1/2 is from a dimeric complex embedded in the ER membrane that can assemble to a tetramer. Cytosolic elements of Get1 bind Get3 and allow insertion of a coiled coil, like a wedge into the ATPase domain interface of Get3, which triggers phosphate release and reorganisation of the TA-binding regions such that the TA is now released into a hydrophilic cavity between Get3 and the inter-membrane regions. This directs the TA to the insertase formed by the membrane-spanning regions of Get1/2. (**B**) In mammals, initial TA capture involves SGTA and the BAG complex (formed by Bag6, Ubl4A and TRC35). Again, this occurs co-translationally at the ribosome. TA proteins are then efficiently transferred from SGTA to TRC40; BAG6 can additionally triage hydrophobic, non-TA proteins to the ubiquitin proteasome system for degradation. As with the yeast TRC40 homologue, Get3, the TRC40 ATPase cycle controls recruitment and transfer of the TA protein from SGTA and subsequent delivery to the membrane insertase formed by the Get1/2 homologues, WRB/CAML.

Initial capture of TA sequences by the SGTA pre-loading complex occurs at the ribosome [97] (Figure 3B). Furthermore, a recent in vitro study indicated that SGTA can be pre-recruited to the translating ribosome, prior to the emergence of a hydrophobic sequence from the exit tunnel, thereby allowing co-translational capture of TA segments for subsequent handover to TRC40 [101]. BAG6, TRC40, and TRC35 are also ribosome associated, indicating this step also occurs at the ribosome [97]. In the case of the Sec61β, delayed termination of translation due to pausing at the stop codon also likely enhances TA capture by TRC40 [97,102].

Interestingly, SRP-dependent targeting sequences could be similarly engaged by SGTA, this may reflect a holdase function prior to SRP binding, or a mechanism for targeting signal sequences that SRP fails to recognise for degradation via BAG6 [101].

Studies in yeast have shown that Get4 and Get5 bind to non-programmed ribosomes with high affinity mediated by an interaction of Get4 with the ribosome close to ribosomal proteins uL29 and uL26 at the exit site and that this enhances recruitment of Sgt2 [103]. Consistent with the observations with SGTA, the presence of a TM inside the ribosome exit tunnel enhances ribosome binding of Get4/5 and Sgt2 as has also been seen previously with SRP [103–105]. The ribosome-binding site of SRP and Get5 overlap [20,30,103], and once SRP binds to a SA/SS that has emerged from the exit tunnel, Get4/5 binding is inhibited [103]. The Hsp70 ATPase Ssa1 has also been shown to be important for efficient transfer of TA proteins to Sgt2 in the absence of the ribosome; thus, it may rescue any TA clients that fail to be captured by Sgt2 at the ribosome [106].

Get3/TRC40, the central player of the GET pathway is a homodimeric ATPase, composed of a P-type ATPase domain and alpha-helical domain, which is involved in TA binding [107,108]. The ATPase cycle of Get3 controls the conformation of the TA-binding region, such that in the ATP-bound state the homodimer forms a fully closed state that forms a composite hydrophobic groove that can accommodate a 20 amino acid long hydrophobic helix [107,108]. Hydrolysis of ATP to the ADP bound state partially disrupts the groove but still allows the TA to remain bound, while in the nucleotide-free state the composite groove is completely disrupted [90].

Get3 is recruited to Get4 in the ATP-bound state, which inhibits ATP hydrolysis and promotes direct transfer of the substrate from Sgt2 to the hydrophobic groove in Get3 [109,110]. A dynamic α-helix in Get3 that forms a lid facilitates transfer and prevents delivery of hydrophobic helices that were not pre-bound to Sgt2 [111]. Binding of the substrate to Get3 stimulates ATP hydrolysis releasing it from the preloading complex and allowing it to interact with the Get1/2 insertase at the ER membrane in the ADP-bound conformation [110,112,113].

Get1 and Get2 both possess cytoplasmic domains that extend from the membrane and engage Get3. Upon binding, a coiled-coil domain within Get1 inserts between the subunits of Get3 and thereby stabilises the nucleotide-free conformation, triggering substrate release [113–115].

Recent structures of the complete Get1/2/3 complex, including the TM regions of Get1/2, show that Get1/2 forms a hetero-tetramer with symmetric recruitment of Get3, where both its subunits contact separate Get2 subunits [116]. Furthermore, mutations at the hetero-tetramer interface decrease the efficiency of membrane insertion, suggesting the tetramer is the active form. However, it may still be possible for insertion to occur via a Get1/2 heterodimer as has also been proposed [112]. A hydrophilic cavity is formed between the trans-membrane region and Get3. Furthermore, a hydrophobic helix (a3′) in Get1 forms a gate adjacent to the TA-binding region of Get3 and the cavity such that as the cargo is released, the a3′ helix is displaced forming a wall to direct the cargo into the cavity [116].

Despite the essential function of many tail-anchor proteins, loss of the GET machinery is not lethal to yeast cells, indicating functional redundancy [94]. In particular, the alternate co-translational SND pathway can also accommodate TA proteins in the absence of GET components [63]. This is also reflected in mammalian cells; TA proteins with moderately hydrophobic TMs can be recognised by calmodulin and then targeted to the EMC complex [85,117,118]. Hsp70s and ubiquilin family proteins can similarly maintain TA proteins in a soluble state in the cytosol competent for insertion [4], whilst the more hydrophobic TA protein, Sec61β, can engage both SRP and TRC40 as well as the mammalian SND complex [64,97,119].

7. EMC Translocase

The alternative EMC translocase has been implicated in the biogenesis of both tail-anchor membrane proteins, as mentioned above, and those with a type III orientation. In both cases, the substrates possess a targeting membrane domain that is positioned close to the C- and N-termini, respectively, such that only a small hydrophilic domain has to traverse the bilayer [117,118,120,121]. It can also collaborate with Sec61 during the biogenesis of some polytopic membrane proteins, particularly those with a short N-terminal luminal domain, akin to type III proteins, such as many GPCRs [122,123]. Targeting of type III membrane proteins in both yeast and mammalian cells involves SRP and SR [60,69,121,124], yet the manner in which they are handed over from the ribosome–SRP–SR complex to EMC remains unknown. Intriguingly, insertion of type III proteins in mammalian cells is insensitive to Sec61-inihbitors, including Ipomoeassin-F, which blocks translocation of all other known substrates via the Sec61 channel [121]. In contrast, depletion of Sec61 does impact type III insertion [121], suggesting Sec61 might play a non-canonical role, perhaps facilitating the release from the SRP–SR complex and handover to EMC.

8. Defective ER Targeting and Human Disease

A number of patient mutations have been identified which map to components of the ER protein-targeting machinery. Point mutations in SRP54 are associated with severe neutropenia and Shwachmann–Diamond syndrome-like symptoms, which affect development and function of tissues with high secretory activity, such as the pancreas, as well as skeletal and neurodevelopmental defects [125,126]. While some of the mutations are attributable to loss of function and haploinsufficiency, others, including T115A, T117Δ, and G226E, have a dominant phenotype. All mutations to date map to the G-domain where they are implicated to impact either nucleotide binding or overall structure. Several of the mutants have been analysed in detail (T115A, T117Δ, and G226E) revealing structural changes to the core GTPase, which impair GTP binding and prevent complex formation of isolated SRP54 and SRα NG domains [127]. A more detailed analysis of the G226E mutant when assembled in SRP in the context of SR and the RNC reveals that while initial SRP–SR complex assembly can occur, it becomes locked in an RNC–SRP–SR intermediate that cannot relocate the NG domains from the proximal to distal position, thereby rationalising the dominant negative phenotype associated with the mutation [49].

Using a zebrafish model, the severe neutropenia-associated phenotypes associated with autosomal dominant mutations (T115A, T117Δ, and G226E) are phenocopied along with pancreatic dysfunction [128]. Furthermore, the neutropenia phenotype has been linked to impaired splicing of the *XBP1* transcription factor required for the unfolded protein response and which is spliced in an unconventional manner by *IRE1* following its membrane targeting by SRP [129]. Loss of Xbp1 in the zebrafish also shows a similar neutropenia phenotype, consistent with this being a key driver of the disease phenotype [128].

Disease mutations are not limited to SRP54; a recently identified biallelic mutant of SRP68 that leads to loss of exon1 is also associated with neutropenia and Shwachmann–Diamond-like symptoms [130], whilst SRP72 mutants have been linked to aplastic anemia (AA) and myelodysplasia (MDS) [131].

As well as mutations in SRP, disease mutations that impact protein targeting are also linked to mutations in targeting sequences in client proteins [132]. First shown with synthetic mutations in the well-studied model signal sequence from bovine preprolactin, a reduction in the length of the hydrophobic core led to loss of SRP binding and instead the nascent chains interacted with Argonaut2 (Ago2). As well as blocking translocation in vitro, this also promoted rapid degradation of the mRNA in vivo in a quality control pathway termed RAPP (Regulation of Aberrant Protein Production) [133]. A number of disease-linked mutations in signal sequence mutations also trigger this pathway in response [132,134], for example in granulin linked to fronto-temporal lobal degeneration [134,135], aspartyl-glucosaminidase in aspartylglucosaminuria [134,136], UDP-glucuronosyltransferase in Crigler–Najjar disease [134,137], and cathepsin K in pycnodysostosis [134,138].

Future Outlook

Despite more than 40 years of research into ER protein-targeting, questions clearly remained unanswered. In particular, the SND complex remains only very basically characterised in terms of mechanism and structural characterisation, and the identity of the mammalian components beyond hSnd2 remain elusive. While the canonical mode of action of SRP recruitment has been studied in much detail, the observed recruitment of SRP to polysomes translating membrane proteins prior to synthesis of the signal sequence/anchor remains poorly understood, likewise the recent link between SRP binding and the Hel2 protein. Mechanistic understanding of the role NAC plays at the ribosome has been hampered by lack of structural images until very recently, and there is much scope to understand how NAC functions to enhance SRP targeting. Proteomic profiling of substrate and targeting factor interaction in yeast has proved highly informative, and extending this to the more complex mammalian system should shed new light on the interplay between targeting pathways. Hence, there is still much to find out for the future.

Funding: Work in the author's laboratory was funded by the BBSRC (BB/V015109/1) and the Biochemical Society.

Institutional Review Board Statement: Not applicable.

Informed Consent Statement: Not applicable.

Data Availability Statement: Not applicable.

Conflicts of Interest: The author declares no conflict of interest.

References

1. Blobel, G.; Dobberstein, B. Transfer of proteins across membranes I. Presence of proteolytically processed and unprocessed nascent immunoglobulin light chains on membrane-bound ribosomes of murine myeloma. *J. Cell Biol.* **1975**, *67*, 835–851. [CrossRef]
2. Martoglio, B.; Dobberstein, B. Signal sequences: More than just greasy peptides. *Trends Cell Biol.* **1998**, *8*, 410–415. [CrossRef]
3. Von Heijne, G. Signal sequences. The limits of variation. *J. Mol. Biol.* **1985**, *184*, 99–105. [CrossRef]
4. Hegde, R.S.; Keenan, R.J. The mechanisms of integral membrane protein biogenesis. *Nat. Rev. Mol. Cell Biol.* **2021**, *23*, 107–124. [CrossRef]
5. O'Keefe, S.; Pool, M.R.; High, S. Membrane protein biogenesis at the ER: The highways and byways. *FEBS J.* **2021**. [CrossRef]
6. Hegde, R.S.; Keenan, R.J. Tail-anchored membrane protein insertion into the endoplasmic reticulum. *Nat. Rev. Mol. Cell Biol.* **2011**, *12*, 787–798. [CrossRef]

7. McGilvray, P.T.; Anghel, S.A.; Sundaram, A.; Zhong, F.; Trnka, M.J.; Fuller, J.R.; Hu, H.; Burlingame, A.L.; Keenan, R.J. An ER translocon for multi-pass membrane protein biogenesis. *eLife* **2020**, *9*, e56889. [CrossRef]

8. Mateja, A.; Keenan, R.J. A structural perspective on tail-anchored protein biogenesis by the GET pathway. *Curr. Opin. Struct. Biol.* **2018**, *51*, 195–202. [CrossRef]

9. Chitwood, P.J.; Hegde, R.S. The Role of EMC during Membrane Protein Biogenesis. *Trends Cell Biol.* **2019**, *29*, 371–384. [CrossRef] [PubMed]

10. Pool, M.R. Signal recognition particles in chloroplasts, bacteria, yeast and mammals (review). *Mol. Membr. Biol.* **2005**, *22*, 3–15. [CrossRef] [PubMed]

11. Voorhees, R.M.; Hegde, R.S. Structures of the scanning and engaged states of the mammalian SRP-ribosome complex. *eLife* **2015**, *4*, e07975. [CrossRef]

12. Walter, P.; Blobel, G. Purification of a membrane-associated protein complex required for protein translocation across the endoplasmic reticulum. *Proc. Natl. Acad. Sci. USA* **1980**, *77*, 7112–7116. [CrossRef]

13. Meyer, D.I.; Dobberstein, B. Identification and characterization of a membrane component essential for the translocation of nascent secretory proteins across the membrane of the endoplasmic reticulum. *J. Cell Biol.* **1980**, *87*, 503–508. [CrossRef]

14. Song, W.; Raden, D.; Mandon, E.C.; Gilmore, R. Role of Sec61a in the regulated transfer of the ribosome-nascent chain complex from the signal recognition particle to the translocation channel. *Cell* **2000**, *100*, 333–343. [CrossRef]

15. Siegel, V.; Walter, P. Elongation arrest is not a prerequisite for secretory protein translocation across the microsomal membrane. *J. Cell Biol.* **1985**, *100*, 1913–1921. [CrossRef]

16. Gundelfinger, E.D.; Krause, E.; Melli, M.; Dobberstein, B. The organization of the 7SL RNA in the signal recognition particle. *Nucleic Acids Res.* **1983**, *11*, 7363–7374. [CrossRef]

17. Brown, J.D.; Hann, B.C.; Medzihradszky, K.F.; Niwa, M.; Burlingame, A.L.; Walter, P. Subunits of the *Saccharomyces cerevisiae* signal recognition particle required for its functional expression. *EMBO J.* **1994**, *13*, 4390–4400. [CrossRef]

18. Siegel, V.; Walter, P. Each of the activities of signal recognition particle (SRP) is contained within a distinct domain: Analysis of biochemical mutants of SRP. *Cell* **1988**, *52*, 39–49. [CrossRef]

19. Ogg, S.C.; Barz, W.P.; Walter, P. A functional GTPase domain, but not its transmembrane domain, is required for function of the SRP receptor beta-subunit. *J. Cell Biol.* **1998**, *142*, 341–354. [CrossRef]

20. Halic, M.; Becker, T.; Pool, M.R.; Spahn, C.M.; Grassucci, R.A.; Frank, J.; Beckmann, R. Structure of the signal recognition particle interacting with the elongation-arrested ribosome. *Nature* **2004**, *427*, 808–814. [CrossRef]

21. Wild, K.; Becker, M.M.M.; Kempf, G.; Sinning, I. Structure, dynamics and interactions of large SRP variants. *Biol. Chem.* **2019**, *401*, 63–80. [CrossRef]

22. Hainzl, T.; Sauer-Eriksson, A.E. Signal-sequence induced conformational changes in the signal recognition particle. *Nat. Commun.* **2015**, *6*, 7163. [CrossRef]

23. Freymann, D.M.; Keenan, R.J.; Stroud, R.N.; Walter, P. Structure of the conserved GTPase domain of the signal recognition particle. *Nature* **1997**, *385*, 361–364. [CrossRef]

24. Janda, C.Y.; Li, J.; Oubridge, C.; Hernandez, H.; Robinson, C.V.; Nagai, K. Recognition of a signal peptide by the signal recognition particle. *Nature* **2010**, *465*, 507–510. [CrossRef]

25. Keenan, R.J.; Freymann, D.M.; Walter, P.; Stroud, R.M. Crystal structure of the signal sequence binding protein of the signal recognition particle. *Cell* **1998**, *94*, 181–191. [CrossRef]

26. Hainzl, T.; Huang, S.; Sauer-Eriksson, A.E. Structure of the SRP19 RNA complex and implications for signal recognition particle assembly. *Nature* **2002**, *417*, 767–771. [CrossRef]

27. Ogg, S.C.; Walter, P. SRP samples nascent chains for the presence of signal sequences by interacting with the ribosome at a discrete step during translation elongation. *Cell* **1995**, *81*, 1075–1084. [CrossRef]

28. Flanagan, J.J.; Chen, J.C.; Miao, Y.; Shao, Y.; Lin, J.; Bock, P.E.; Johnson, A.E. SRP binds to ribosome-bound signal sequences with fluorescence-detected subnanomolar affinity that does not diminish as the nascent chain lengthens. *J. Biol. Chem.* **2003**, *278*, 18628–18637. [CrossRef]

29. Chartron, J.W.; Hunt, K.C.; Frydman, J. Cotranslational signal-independent SRP preloading during membrane targeting. *Nature* **2016**, *536*, 224–228. [CrossRef]

30. Pool, M.R.; Stumm, J.; Fulga, T.A.; Sinning, I.; Dobberstein, B. Distinct Modes of Signal Recognition Particle Interaction with the Ribosome. *Science* **2002**, *297*, 1345–1348. [CrossRef]

31. Hauser, S.; Bacher, G.; Dobberstein, B.; Lütcke, H. A complex comprising the signal sequence binding protein and the SRP RNA promotes translocation of nascent proteins. *EMBO J.* **1995**, *14*, 5485–5493. [CrossRef]

32. Dalley, J.A.; Selkirk, A.; Pool, M.R. Access to Ribosomal Protein Rpl25p by the Signal Recognition Particle Is Required for Efficient Cotranslational Translocation. *Mol. Biol. Cell* **2008**, *19*, 2876–2884. [CrossRef] [PubMed]

33. Mason, N.; Ciufo, L.F.; Brown, J.D. Elongation arrest is a physiologically important function of the signal recognition particle. *EMBO J.* **2000**, *19*, 4164–4174. [CrossRef]

34. Lakkaraju, A.K.; Mary, C.; Scherrer, A.; Johnson, A.E.; Strub, K. SRP keeps polypeptides translocation-competent by slowing translation to match limiting ER-targeting sites. *Cell* **2008**, *133*, 440–451. [CrossRef]

35. Pechmann, S.; Chartron, J.W.; Frydman, J. Local slowdown of translation by nonoptimal codons promotes nascent-chain recognition by SRP in vivo. *Nat. Struct. Mol. Biol.* **2014**, *21*, 1100–1105. [CrossRef]

36. Bange, G.; Sinning, I. SIMIBI twins in protein targeting and localization. *Nat. Struct. Mol. Biol.* **2013**, *20*, 776–780. [CrossRef]
37. Montoya, G.; Svensson, C.; Luirink, J.; Sinning, I. Crytsal structure of the NG domain from the signal recognition particle receptor FtsY. *Nature* **1997**, *385*, 365–368. [CrossRef]
38. Young, J.C.; Ursini, J.; Legate, K.R.; Miller, J.D.; Walter, P.; Andrews, D.W. An amino-terminal domain containing hydrophobic and hydrophilic sequences binds the signal recognition particle receptor a subunit to the b subunit on the endoplasmic reticulum membrane. *J. Biol. Chem.* **1995**, *270*, 15650–15657. [CrossRef]
39. Schwartz, T.; Blobel, G. Structural basis for the function of the beta subunit of the eukaryotic signal recognition particle receptor. *Cell* **2003**, *112*, 793–803. [CrossRef]
40. Schlenker, O.; Hendricks, A.; Sinning, I.; Wild, K. The structure of the mammalian signal recognition particle (SRP) receptor as prototype for the interaction of small GTPases with Longin domains. *J. Biol. Chem.* **2006**, *281*, 8898–8906. [CrossRef]
41. Egea, P.F.; Shan, S.O.; Napetschnig, J.; Savage, D.F.; Walter, P.; Stroud, R.M. Substrate twinning activates the signal recognition particle and its receptor. *Nature* **2004**, *427*, 215–221. [CrossRef]
42. Focia, P.J.; Shepotinovskaya, I.V.; Seidler, J.A.; Freymann, D.M. Heterodimeric GTPase core of the SRP targeting complex. *Science* **2004**, *303*, 373–377. [CrossRef]
43. Kobayashi, K.; Jomaa, A.; Lee, J.H.; Chandrasekar, S.; Boehringer, D.; Shan, S.O.; Ban, N. Structure of a prehandover mammalian ribosomal SRP.SRP receptor targeting complex. *Science* **2018**, *360*, 323–327. [CrossRef]
44. Jomaa, A.; Eitzinger, S.; Zhu, Z.; Chandrasekar, S.; Kobayashi, K.; Shan, S.O.; Ban, N. Molecular mechanism of cargo recognition and handover by the mammalian signal recognition particle. *Cell Rep.* **2021**, *36*, 109350. [CrossRef]
45. Hwang Fu, Y.H.; Chandrasekar, S.; Lee, J.H.; Shan, S.O. A molecular recognition feature mediates ribosome-induced SRP-receptor assembly during protein targeting. *J. Cell Biol.* **2019**, *218*, 3307–3319. [CrossRef]
46. Peluso, P.; Herschlag, D.; Nock, S.; Freymann, D.; Johnson, A.E.; Walter, P. Role of the 4.5S RNA in Assembly of the Bacterial Signal Recognition Particle with its Receptor. *Science* **2000**, *288*, 1640–1643. [CrossRef]
47. Bradshaw, N.; Neher, S.B.; Booth, D.S.; Walter, P. Signal sequences activate the catalytic switch of SRP RNA. *Science* **2009**, *323*, 127–130. [CrossRef]
48. Neher, S.B.; Bradshaw, N.; Floor, S.N.; Gross, J.D.; Walter, P. SRP RNA controls a conformational switch regulating the SRP-SRP receptor interaction. *Nat. Struct. Mol. Biol.* **2008**, *15*, 916–923. [CrossRef]
49. Lee, J.H.; Jomaa, A.; Chung, S.; Hwang Fu, Y.H.; Qian, R.; Sun, X.; Hsieh, H.H.; Chandrasekar, S.; Bi, X.; Mattei, S.; et al. Receptor compaction and GTPase rearrangement drive SRP-mediated cotranslational protein translocation into the ER. *Sci. Adv.* **2021**, *7*, eabg0942. [CrossRef]
50. Halic, M.; Gartmann, M.; Schlenker, O.; Mielke, T.; Pool, M.R.; Sinning, I.; Beckmann, R. Signal Recognition Particle Receptor Exposes the Ribosomal Translocon Binding Site. *Science* **2006**, *312*, 745–747. [CrossRef]
51. Lee, J.H.; Chandrasekar, S.; Chung, S.; Hwang Fu, Y.H.; Liu, D.; Weiss, S.; Shan, S.O. Sequential activation of human signal recognition particle by the ribosome and signal sequence drives efficient protein targeting. *Proc. Natl. Acad. Sci. USA* **2018**, *115*, E5487–E5496. [CrossRef]
52. Jomaa, A.; Fu, Y.H.; Boehringer, D.; Leibundgut, M.; Shan, S.O.; Ban, N. Structure of the quaternary complex between SRP, SR, and translocon bound to the translating ribosome. *Nat. Commun.* **2017**, *8*, 15470. [CrossRef] [PubMed]
53. Connolly, T.; Rapiejko, P.J.; Gilmore, R. Requirement of GTP hydrolysis for dissociation of the signal recognition particle from its receptor. *Science* **1991**, *252*, 1171–1173. [CrossRef] [PubMed]
54. Hsieh, H.H.; Lee, J.H.; Chandrasekar, S.; Shan, S.O. A ribosome-associated chaperone enables substrate triage in a cotranslational protein targeting complex. *Nat. Commun.* **2020**, *11*, 5840. [CrossRef] [PubMed]
55. Wiedmann, B.; Sakai, H.; Davis, T.A.; Wiedmann, M. A protein complex required for signal-sequence-specific sorting and translocation. *Nature* **1994**, *370*, 434–440. [CrossRef]
56. Nyathi, Y.; Pool, M.R. Analysis of the interplay of protein biogenesis factors at the ribosome exit site reveals new role for NAC. *J. Cell Biol.* **2015**, *210*, 287–301. [CrossRef] [PubMed]
57. Gamerdinger, M.; Kobayashi, K.; Wallisch, A.; Kreft, S.G.; Sailer, C.; Schlomer, R.; Sachs, N.; Jomaa, A.; Stengel, F.; Ban, N.; et al. Early Scanning of Nascent Polypeptides inside the Ribosomal Tunnel by NAC. *Mol. Cell* **2019**, *75*, 996–1006. [CrossRef]
58. Del Alamo, M.; Hogan, D.J.; Pechmann, S.; Albanese, V.; Brown, P.O.; Frydman, J. Defining the specificity of cotranslationally acting chaperones by systematic analysis of mRNAs associated with ribosome-nascent chain complexes. *PLoS Biol.* **2011**, *9*, e1001100. [CrossRef]
59. Schibich, D.; Gloge, F.; Pohner, I.; Bjorkholm, P.; Wade, R.C.; von Heijne, G.; Bukau, B.; Kramer, G. Global profiling of SRP interaction with nascent polypeptides. *Nature* **2016**, *536*, 219–223. [CrossRef]
60. Costa, E.A.; Subramanian, K.; Nunnari, J.; Weissman, J.S. Defining the physiological role of SRP in protein-targeting efficiency and specificity. *Science* **2018**, *359*, 689–692. [CrossRef]
61. Matsuo, Y.; Inada, T. The ribosome collision sensor Hel2 functions as preventive quality control in the secretory pathway. *Cell Rep.* **2021**, *34*, 108877. [CrossRef] [PubMed]
62. Hann, B.C.; Walter, P. The Signal Recognition Particle in *S. cerevisiae*. *Cell* **1991**, *67*, 131–144. [CrossRef]
63. Aviram, N.; Ast, T.; Costa, E.A.; Arakel, E.C.; Chuartzman, S.G.; Jan, C.H.; Hassdenteufel, S.; Dudek, J.; Jung, M.; Schorr, S.; et al. The SND proteins constitute an alternative targeting route to the endoplasmic reticulum. *Nature* **2016**, *540*, 134–138. [CrossRef] [PubMed]

64. Hassdenteufel, S.; Sicking, M.; Schorr, S.; Aviram, N.; Fecher-Trost, C.; Schuldiner, M.; Jung, M.; Zimmermann, R.; Lang, S. hSnd2 protein represents an alternative targeting factor to the endoplasmic reticulum in human cells. *FEBS Lett.* **2017**, *591*, 3211–3224. [CrossRef]
65. Yang, J.; Hirata, T.; Liu, Y.S.; Guo, X.Y.; Gao, X.D.; Kinoshita, T.; Fujita, M. Human SND2 mediates ER targeting of GPI-anchored proteins with low hydrophobic GPI attachment signals. *FEBS Lett.* **2021**, *595*, 1542–1558. [CrossRef]
66. Deshaies, R.J.; Schekman, R. *SEC62* encodes a putative membrane protein required for protein translocation into the yeast endoplasmic reticulum. *J. Cell Biol.* **1989**, *109*, 2644–2653. [CrossRef]
67. Deshaies, R.J.; Sanders, S.L.; Feldheim, D.A.; Schekman, R. Assembly of yeast Sec proteins involved in translocation into the endoplasmic reticulum into a membrane-bound multisubunit complex. *Nature* **1991**, *349*, 806–808. [CrossRef]
68. Stirling, C.J.; Rothblatt, J.; Hosobuchi, M.; Deshaies, R.; Schekman, R. Protein translocation mutants defective in the insertion of integral membrane proteins into the endoplasmic reticulum. *Mol. Biol. Cell* **1992**, *3*, 129–142. [CrossRef]
69. Ng, D.T.; Brown, J.D.; Walter, P. Signal sequences specify the targeting route to the endoplasmic reticulum membrane. *J. Cell Biol.* **1996**, *134*, 269–278. [CrossRef]
70. Waters, M.G.; Blobel, G. Secretory protein translocation in a yeast cell-free system can occur post-translationally and requires ATP-hydrolysis. *J. Cell Biol.* **1986**, *102*, 1543–1550. [CrossRef]
71. Prinz, A.; Behrens, C.; Rapoport, T.A.; Hartmann, E.; Kalies, K.U. Evolutionarily conserved binding of ribosomes to the translocation channel via the large ribosomal RNA. *EMBO J.* **2000**, *19*, 1900–1906. [CrossRef] [PubMed]
72. Meyer, H.A.; Grau, H.; Kraft, R.; Kostka, S.; Prehn, S.; Kalies, K.U.; Hartmann, E. Mammalian Sec61 is associated with Sec62 and Sec63. *J. Biol. Chem.* **2000**, *275*, 14550–14557. [CrossRef] [PubMed]
73. Weng, T.H.; Steinchen, W.; Beatrix, B.; Berninghausen, O.; Becker, T.; Bange, G.; Cheng, J.; Beckmann, R. Architecture of the active post-translational Sec translocon. *EMBO J.* **2021**, *40*, e105643. [CrossRef]
74. Chirico, W.J.; Waters, M.G.; Blobel, G. 70 K heat shock related proteins stimulate protein translocation into microsomes. *Nature* **1988**, *332*, 805–810. [CrossRef]
75. Ngosuwan, J.; Wang, N.M.; Fung, K.L.; Chirico, W.J. Roles of cytosolic Hsp70 and Hsp40 molecular chaperones in post-translational translocation of presecretory proteins into the endoplasmic reticulum. *J. Biol. Chem.* **2003**, *278*, 7034–7042. [CrossRef]
76. Tripathi, A.; Mandon, E.C.; Gilmore, R.; Rapoport, T.T.A. Two alternative binding mechanisms connect the protein translocation Sec71-Sec72 complex with heat shock proteins. *J. Biol. Chem.* **2017**, *292*, 8007–8018. [CrossRef] [PubMed]
77. Jan, C.H.; Williams, C.C.; Weissman, J.S. Principles of ER cotranslational translocation revealed by proximity-specific ribosome profiling. *Science* **2014**, *346*, 1257521. [CrossRef] [PubMed]
78. Tyedmers, J.; Lerner, M.; Bies, C.; Dudek, J.; Skowronek, M.H.; Haas, I.G.; Heim, N.; Nastainczyk, W.; Volkmer, J.; Zimmermann, R. Homologs of the yeast Sec complex subunits Sec62p and Sec63p are abundant proteins in dog pancreas microsomes. *Proc. Natl. Acad. Sci. USA* **2000**, *97*, 7214–7219. [CrossRef]
79. Muller, L.; de Escauriaza, M.D.; Lajoie, P.; Theis, M.; Jung, M.; Muller, A.; Burgard, C.; Greiner, M.; Snapp, E.L.; Dudek, J.; et al. Evolutionary gain of function for the ER membrane protein Sec62 from yeast to humans. *Mol. Biol. Cell* **2010**, *21*, 691–703. [CrossRef]
80. Müller, G.; Zimmermann, R. Import of honeybee prepromelittin into the endoplasmic reticulum: Structural basis for independence of SRP and docking protein. *EMBO J.* **1987**, *6*, 2099–2107. [CrossRef]
81. Lakkaraju, A.K.; Thankappan, R.; Mary, C.; Garrison, J.L.; Taunton, J.; Strub, K. Efficient secretion of small proteins in mammalian cells relies on Sec62-dependent posttranslational translocation. *Mol. Biol. Cell* **2012**, *23*, 2712–2722. [CrossRef]
82. Lang, S.; Benedix, J.; Fedeles, S.V.; Schorr, S.; Schirra, C.; Schauble, N.; Jalal, C.; Greiner, M.; Hassdenteufel, S.; Tatzelt, J.; et al. Different effects of Sec61alpha, Sec62 and Sec63 depletion on transport of polypeptides into the endoplasmic reticulum of mammalian cells. *J. Cell Sci.* **2012**, *125*, 1958–1969. [CrossRef] [PubMed]
83. Johnson, N.; Hassdenteufel, S.; Theis, M.; Paton, A.W.; Paton, J.C.; Zimmermann, R.; High, S. The signal sequence influences post-translational ER translocation at distinct stages. *PLoS ONE* **2013**, *8*, e75394. [CrossRef] [PubMed]
84. Johnson, N.; Vilardi, F.; Lang, S.; Leznicki, P.; Zimmermann, R.; High, S. TRC40 can deliver short secretory proteins to the Sec61 translocon. *J. Cell Sci.* **2012**, *125*, 3612–3620. [CrossRef] [PubMed]
85. Shao, S.; Hegde, R.S. A calmodulin-dependent translocation pathway for small secretory proteins. *Cell* **2011**, *147*, 1576–1588. [CrossRef]
86. Schorr, S.; Nguyen, D.; Hassdenteufel, S.; Nagaraj, N.; Cavalie, A.; Greiner, M.; Weissgerber, P.; Loi, M.; Paton, A.W.; Paton, J.C.; et al. Identification of signal peptide features for substrate specificity in human Sec62/Sec63-dependent ER protein import. *FEBS J.* **2020**, *287*, 4612–4640. [CrossRef]
87. Reithinger, J.H.; Kim, J.E.; Kim, H. Sec62 protein mediates membrane insertion and orientation of moderately hydrophobic signal anchor proteins in the endoplasmic reticulum (ER). *J. Biol. Chem.* **2013**, *288*, 18058–18067. [CrossRef]
88. Jung, S.J.; Kim, J.E.; Reithinger, J.H.; Kim, H. The Sec62-Sec63 translocon facilitates translocation of the C-terminus of membrane proteins. *J. Cell Sci.* **2014**, *127*, 4270–4278. [CrossRef]
89. Kutay, U.; Ahnert-Hilger, G.; Hartmann, E.; Wiedenmann, B.; Rapoport, T.A. Transport route for synaptobrevin via a novel transport pathway of insertion into the endoplasmic reticulum membrane. *EMBO J.* **1995**, *14*, 217–223. [CrossRef]
90. Shan, S.O. Guiding tail-anchored membrane proteins to the endoplasmic reticulum in a chaperone cascade. *J. Biol. Chem.* **2019**, *294*, 16577–16586. [CrossRef]

91. Borgese, N.; Coy-Vergara, J.; Colombo, S.F.; Schwappach, B. The Ways of Tails: The GET Pathway and more. *Protein J.* **2019**, *38*, 289–305. [CrossRef] [PubMed]
92. Jonikas, M.C.; Collins, S.R.; Denic, V.; Oh, E.; Quan, E.M.; Schmid, V.; Weibezahn, J.; Schwappach, B.; Walter, P.; Weissman, J.S.; et al. Comprehensive characterization of genes required for protein folding in the endoplasmic reticulum. *Science* **2009**, *323*, 1693–1697. [CrossRef] [PubMed]
93. Wang, F.; Brown, E.C.; Mak, G.; Zhuang, J.; Denic, V. A chaperone cascade sorts proteins for posttranslational membrane insertion into the endoplasmic reticulum. *Mol. Cell* **2010**, *40*, 159–171. [CrossRef] [PubMed]
94. Schuldiner, M.; Metz, J.; Schmid, V.; Denic, V.; Rakwalska, M.; Schmitt, H.D.; Schwappach, B.; Weissman, J.S. The GET complex mediates insertion of tail-anchored proteins into the ER membrane. *Cell* **2008**, *134*, 634–645. [CrossRef]
95. Stefanovic, S.; Hegde, R.S. Identification of a targeting factor for posttranslational membrane protein insertion into the ER. *Cell* **2007**, *128*, 1147–1159. [CrossRef]
96. Favaloro, V.; Spasic, M.; Schwappach, B.; Dobberstein, B. Distinct targeting pathways for the membrane insertion of tail-anchored (TA) proteins. *J. Cell Sci.* **2008**, *121*, 1832–1840. [CrossRef]
97. Mariappan, M.; Li, X.; Stefanovic, S.; Sharma, A.; Mateja, A.; Keenan, R.J.; Hegde, R.S. A ribosome-associating factor chaperones tail-anchored membrane proteins. *Nature* **2010**, *466*, 1120–1124. [CrossRef]
98. Yamamoto, Y.; Sakisaka, T. Molecular machinery for insertion of tail-anchored membrane proteins into the endoplasmic reticulum membrane in mammalian cells. *Mol. Cell* **2012**, *48*, 387–397. [CrossRef]
99. Shao, S.; Rodrigo-Brenni, M.C.; Kivlen, M.H.; Hegde, R.S. Mechanistic basis for a molecular triage reaction. *Science* **2017**, *355*, 298–302. [CrossRef]
100. Leznicki, P.; High, S. SGTA antagonizes BAG6-mediated protein triage. *Proc. Natl. Acad. Sci. USA* **2012**, *109*, 19214–19219. [CrossRef]
101. Leznicki, P.; High, S. SGTA associates with nascent membrane protein precursors. *EMBO Rep.* **2020**, *21*, e48835. [CrossRef] [PubMed]
102. Han, P.; Shichino, Y.; Schneider-Poetsch, T.; Mito, M.; Hashimoto, S.; Udagawa, T.; Kohno, K.; Yoshida, M.; Mishima, Y.; Inada, T.; et al. Genome-wide Survey of Ribosome Collision. *Cell Rep.* **2020**, *31*, 107610. [CrossRef] [PubMed]
103. Zhang, Y.; De Laurentiis, E.; Bohnsack, K.E.; Wahlig, M.; Ranjan, N.; Gruseck, S.; Hackert, P.; Wolfle, T.; Rodnina, M.V.; Schwappach, B.; et al. Ribosome-bound Get4/5 facilitates the capture of tail-anchored proteins by Sgt2 in yeast. *Nat. Commun.* **2021**, *12*, 782. [CrossRef] [PubMed]
104. Zhang, Y.; Berndt, U.; Golz, H.; Tais, A.; Oellerer, S.; Wolfle, T.; Fitzke, E.; Rospert, S. NAC functions as a modulator of SRP during the early steps of protein targeting to the endoplasmic reticulum. *Mol. Biol. Cell* **2012**, *23*, 3027–3040. [CrossRef]
105. Berndt, U.; Oellerer, S.; Zhang, Y.; Johnson, A.E.; Rospert, S. A signal-anchor sequence stimulates signal recognition particle binding to ribosomes from inside the exit tunnel. *Proc. Natl. Acad. Sci. USA* **2009**, *106*, 1398–1403. [CrossRef]
106. Cho, H.; Shan, S.O. Substrate relay in an Hsp70-cochaperone cascade safeguards tail-anchored membrane protein targeting. *EMBO J.* **2018**, *37*, e99264. [CrossRef]
107. Mateja, A.; Szlachcic, A.; Downing, M.E.; Dobosz, M.; Mariappan, M.; Hegde, R.S.; Keenan, R.J. The structural basis of tail-anchored membrane protein recognition by Get3. *Nature* **2009**, *461*, 361–366. [CrossRef]
108. Bozkurt, G.; Stjepanovic, G.; Vilardi, F.; Amlacher, S.; Wild, K.; Bange, G.; Favaloro, V.; Rippe, K.; Hurt, E.; Dobberstein, B.; et al. Structural insights into tail-anchored protein binding and membrane insertion by Get3. *Proc. Natl. Acad. Sci. USA* **2009**, *106*, 21131–21136. [CrossRef]
109. Chartron, J.W.; Suloway, C.J.; Zaslaver, M.; Clemons, W.M., Jr. Structural characterization of the Get4/Get5 complex and its interaction with Get3. *Proc. Natl. Acad. Sci. USA* **2010**, *107*, 12127–12132. [CrossRef]
110. Rome, M.E.; Rao, M.; Clemons, W.M.; Shan, S.O. Precise timing of ATPase activation drives targeting of tail-anchored proteins. *Proc. Natl. Acad. Sci. USA* **2013**, *110*, 7666–7671. [CrossRef]
111. Chio, U.S.; Chung, S.; Weiss, S.; Shan, S.O. A Chaperone Lid Ensures Efficient and Privileged Client Transfer during Tail-Anchored Protein Targeting. *Cell Rep.* **2019**, *26*, 37–44. [CrossRef]
112. Zalisko, B.E.; Chan, C.; Denic, V.; Rock, R.S.; Keenan, R.J. Tail-Anchored Protein Insertion by a Single Get1/2 Heterodimer. *Cell Rep.* **2017**, *20*, 2287–2293. [CrossRef] [PubMed]
113. Stefer, S.; Reitz, S.; Wang, F.; Wild, K.; Pang, Y.Y.; Schwarz, D.; Bomke, J.; Hein, C.; Lohr, F.; Bernhard, F.; et al. Structural basis for tail-anchored membrane protein biogenesis by the Get3-receptor complex. *Science* **2011**, *333*, 758–762. [CrossRef] [PubMed]
114. Wang, F.; Whynot, A.; Tung, M.; Denic, V. The mechanism of tail-anchored protein insertion into the ER membrane. *Mol. Cell* **2011**, *43*, 738–750. [CrossRef] [PubMed]
115. Mariappan, M.; Mateja, A.; Dobosz, M.; Bove, E.; Hegde, R.S.; Keenan, R.J. The mechanism of membrane-associated steps in tail-anchored protein insertion. *Nature* **2011**, *477*, 61–66. [CrossRef]
116. McDowell, M.A.; Heimes, M.; Fiorentino, F.; Mehmood, S.; Farkas, A.; Coy-Vergara, J.; Wu, D.; Bolla, J.R.; Schmid, V.; Heinze, R.; et al. Structural Basis of Tail-Anchored Membrane Protein Biogenesis by the GET Insertase Complex. *Mol. Cell* **2020**, *80*, 72–86. [CrossRef]
117. Guna, A.; Volkmar, N.; Christianson, J.C.; Hegde, R.S. The ER membrane protein complex is a transmembrane domain insertase. *Science* **2018**, *359*, 470–473. [CrossRef]

118. Volkmar, N.; Thezenas, M.L.; Louie, S.M.; Juszkiewicz, S.; Nomura, D.K.; Hegde, R.S.; Kessler, B.M.; Christianson, J.C. The ER membrane protein complex promotes biogenesis of sterol-related enzymes maintaining cholesterol homeostasis. *J. Cell Sci.* **2019**, *132*, jcs223453. [CrossRef]
119. Abell, B.M.; Pool, M.R.; Schlenker, O.; Sinning, I.; High, S. Signal recognition particle mediates post-translational targeting in eukaryotes. *EMBO J.* **2004**, *23*, 2755–2764. [CrossRef]
120. Shurtleff, M.J.; Itzhak, D.N.; Hussmann, J.A.; Schirle Oakdale, N.T.; Costa, E.A.; Jonikas, M.; Weibezahn, J.; Popova, K.D.; Jan, C.H.; Sinitcyn, P.; et al. The ER membrane protein complex interacts cotranslationally to enable biogenesis of multipass membrane proteins. *eLife* **2018**, *7*, e37018. [CrossRef]
121. O'Keefe, S.; Zong, G.; Duah, K.B.; Andrews, L.E.; Shi, W.Q.; High, S. An alternative pathway for membrane protein biogenesis at the endoplasmic reticulum. *Commun. Biol.* **2021**, *4*, 828. [CrossRef] [PubMed]
122. Chitwood, P.J.; Juszkiewicz, S.; Guna, A.; Shao, S.; Hegde, R.S. EMC Is Required to Initiate Accurate Membrane Protein Topogenesis. *Cell* **2018**, *175*, 1507–1519.e16. [CrossRef] [PubMed]
123. Miller-Vedam, L.E.; Brauning, B.; Popova, K.D.; Schirle Oakdale, N.T.; Bonnar, J.L.; Prabu, J.R.; Boydston, E.A.; Sevillano, N.; Shurtleff, M.J.; Stroud, R.M.; et al. Structural and mechanistic basis of the EMC-dependent biogenesis of distinct transmembrane clients. *eLife* **2020**, *9*, e62611. [CrossRef] [PubMed]
124. Spiller, M.P.; Stirling, C.J. Preferential targeting of a signal recognition particle-dependent precursor to the Ssh1p translocon in yeast. *J. Biol. Chem.* **2011**, *286*, 21953–21960. [CrossRef] [PubMed]
125. Carapito, R.; Konantz, M.; Paillard, C.; Miao, Z.; Pichot, A.; Leduc, M.S.; Yang, Y.; Bergstrom, K.L.; Mahoney, D.H.; Shardy, D.L.; et al. Mutations in signal recognition particle SRP54 cause syndromic neutropenia with Shwachman-Diamond-like features. *J. Clin. Investig.* **2017**, *127*, 4090–4103. [CrossRef]
126. Bellanne-Chantelot, C.; Schmaltz-Panneau, B.; Marty, C.; Fenneteau, O.; Callebaut, I.; Clauin, S.; Docet, A.; Damaj, G.L.; Leblanc, T.; Pellier, I.; et al. Mutations in the SRP54 gene cause severe congenital neutropenia as well as Shwachman-Diamond-like syndrome. *Blood* **2018**, *132*, 1318–1331. [CrossRef]
127. Juaire, K.D.; Lapouge, K.; Becker, M.M.M.; Kotova, I.; Michelhans, M.; Carapito, R.; Wild, K.; Bahram, S.; Sinning, I. Structural and Functional Impact of SRP54 Mutations Causing Severe Congenital Neutropenia. *Structure* **2021**, *29*, 15–28.e7. [CrossRef]
128. Schurch, C.; Schaefer, T.; Muller, J.S.; Hanns, P.; Arnone, M.; Dumlin, A.; Scharer, J.; Sinning, I.; Wild, K.; Skokowa, J.; et al. SRP54 mutations induce congenital neutropenia via dominant-negative effects on XBP1 splicing. *Blood* **2021**, *137*, 1340–1352. [CrossRef]
129. Kanda, S.; Yanagitani, K.; Yokota, Y.; Esaki, Y.; Kohno, K. Autonomous translational pausing is required for XBP1u mRNA recruitment to the ER via the SRP pathway. *Proc. Natl. Acad. Sci. USA* **2016**, *113*, E5886–E5895. [CrossRef]
130. Schmaltz-Panneau, B.; Pagnier, A.; Clauin, S.; Buratti, J.; Marty, C.; Fenneteau, O.; Dieterich, K.; Beaupain, B.; Donadieu, J.; Plo, I.; et al. Identification of biallelic germline variants of SRP68 in a sporadic case with severe congenital neutropenia. *Haematologica* **2021**, *106*, 1216–1219. [CrossRef]
131. Kirwan, M.; Walne, A.J.; Plagnol, V.; Velangi, M.; Ho, A.; Hossain, U.; Vulliamy, T.; Dokal, I. Exome sequencing identifies autosomal-dominant SRP72 mutations associated with familial aplasia and myelodysplasia. *Am. J. Hum. Genet.* **2012**, *90*, 888–892. [CrossRef] [PubMed]
132. Karamyshev, A.L.; Tikhonova, E.B.; Karamysheva, Z.N. Translational Control of Secretory Proteins in Health and Disease. *Int. J. Mol. Sci.* **2020**, *21*, 2538. [CrossRef] [PubMed]
133. Karamyshev, A.L.; Patrick, A.E.; Karamysheva, Z.N.; Griesemer, D.S.; Hudson, H.; Tjon-Kon-Sang, S.; Nilsson, I.; Otto, H.; Liu, Q.; Rospert, S.; et al. Inefficient SRP interaction with a nascent chain triggers a mRNA quality control pathway. *Cell* **2014**, *156*, 146–157. [CrossRef]
134. Tikhonova, E.B.; Karamysheva, Z.N.; von Heijne, G.; Karamyshev, A.L. Silencing of Aberrant Secretory Protein Expression by Disease-Associated Mutations. *J. Mol. Biol.* **2019**, *431*, 2567–2580. [CrossRef]
135. Pinarbasi, E.S.; Karamyshev, A.L.; Tikhonova, E.B.; Wu, I.H.; Hudson, H.; Thomas, P.J. Pathogenic Signal Sequence Mutations in Progranulin Disrupt SRP Interactions Required for mRNA Stability. *Cell Rep.* **2018**, *23*, 2844–2851. [CrossRef]
136. Saarela, J.; von Schantz, C.; Peltonen, L.; Jalanko, A. A novel aspartylglucosaminuria mutation affects translocation of aspartylglucosaminidase. *Hum. Mutat.* **2004**, *24*, 350–351. [CrossRef] [PubMed]
137. Seppen, J.; Steenken, E.; Lindhout, D.; Bosma, P.J.; Elferink, R.P. A mutation which disrupts the hydrophobic core of the signal peptide of bilirubin UDP-glucuronosyltransferase, an endoplasmic reticulum membrane protein, causes Crigler-Najjar type II. *FEBS Lett.* **1996**, *390*, 294–298. [CrossRef]
138. Donnarumma, M.; Regis, S.; Tappino, B.; Rosano, C.; Assereto, S.; Corsolini, F.; Di Rocco, M.; Filocamo, M. Molecular analysis and characterization of nine novel CTSK mutations in twelve patients affected by pycnodysostosis. Mutation in brief #961. Online. *Hum. Mutat.* **2007**, *28*, 524. [CrossRef]

International Journal of
Molecular Sciences

MDPI

Fidelity of Cotranslational Protein Targeting to the Endoplasmic Reticulum

Hao-Hsuan Hsieh and Shu-ou Shan *

Division of Chemistry and Chemical Engineering, California Institute of Technology, Pasadena, CA 91125, USA;
hhsieh@caltech.edu
* Correspondence: sshan@caltech.edu

Abstract: Fidelity of protein targeting is essential for the proper biogenesis and functioning of organelles. Unlike replication, transcription and translation processes, in which multiple mechanisms to recognize and reject noncognate substrates are established in energetic and molecular detail, the mechanisms by which cells achieve a high fidelity in protein localization remain incompletely understood. Signal recognition particle (SRP), a conserved pathway to mediate the localization of membrane and secretory proteins to the appropriate cellular membrane, provides a paradigm to understand the molecular basis of protein localization in the cell. In this chapter, we review recent progress in deciphering the molecular mechanisms and substrate selection of the mammalian SRP pathway, with an emphasis on the key role of the cotranslational chaperone NAC in preventing protein mistargeting to the ER and in ensuring the organelle specificity of protein localization.

Keywords: protein targeting; signal recognition particle; nascent polypeptide-associated complex; ribosome; endoplasmic reticulum; membrane proteins; fidelity

Citation: Hsieh, H.-H.; Shan, S.-o. Fidelity of Cotranslational Protein Targeting to the Endoplasmic Reticulum. *Int. J. Mol. Sci.* **2022**, *23*, 281. https://doi.org/10.3390/ijms23010281

Academic Editors: Richard Zimmermann and Sven Lang

Received: 4 December 2021
Accepted: 18 December 2021
Published: 28 December 2021

Publisher's Note: MDPI stays neutral with regard to jurisdictional claims in published maps and institutional affiliations.

1. Introduction

Over ~30% of the newly synthesized proteins in eukaryotic cells are initially delivered to the endoplasmic reticulum (ER) membrane, where they initiate their journeys through the endomembrane system including the ER, the Golgi apparatus, secretory vesicles, and the plasma membrane [1,2]. These membrane and organellar proteins are prone to misfolding, aggregation, and consequent degradation in the cytosol where their biosynthesis begins [3]. For this reason, proteins destined to the endomembrane system predominantly use a cotranslational pathway of targeting and translocation mediated by signal recognition particle (SRP), giving rise to the ribosome-studded morphology of the rough ER (Figure 1A). By coupling the synthesis of proteins to their localization, the SRP pathway minimizes the off-pathway interactions of nascent membrane and organellar proteins in the cytosol and provides the most efficient mechanism for membrane protein biogenesis, a process that is kinetically demanding and energetically costly [3,4].

How fidelity is achieved in protein localization has been a long-standing question that is challenging to address conceptually and experimentally. SRP-dependent proteins contain a transmembrane domain (TMD) on integral membrane proteins or an ER signal sequence, characterized by a contiguous stretch of hydrophobic amino acids, on secretory and organellar proteins. However, signal sequences and TMDs are divergent in length, sequence and amino acid composition [5,6]. The degenerate nature of these targeting signals demands that protein targeting machineries, such as SRP, distinguish between the correct and incorrect substrates based on minor differences in the molecular features of signal sequences. In addition, eukaryotic cells contain multiple membrane-enclosed compartments, such as mitochondria and peroxisomes, to which a nascent protein could be targeted (Figure 1A). The recent observation that SRP depletion leads to the mis-localization of proteins to mitochondria [7] provides a salient example of the promiscuity of the targeting signals and pathways. Finally, translation termination effectively abolishes the SRP pathway. In addition, it has been reported that SRP loses targeting competence after the nascent chain reaches a critical length of ~130 amino acids (aa) [8,9]. These effects impose a limited time window for SRP to complete the targeting reaction (Figure 1A). The significantly slower translation elongation rate for eukaryotic (3–6 aa/s) than bacterial (10–20 aa/s) ribosomes implies that this time window is significantly longer in eukaryotic cells, and could increase the probability of mis-targeting.

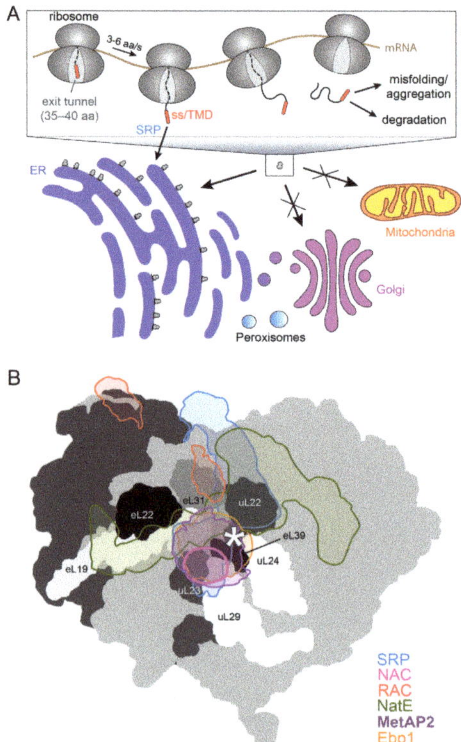

Figure 1. Overview of cotranslational ER targeting and ribosome-associated protein biogenesis factors (RPBs). (**A**) Overview of cotranslational protein targeting in eukaryotic cells. Proteins destined to the endomembrane system are initially targeted to the ER during translation, based on recognition of highly hydrophobic signal sequences (SS) or TMDs on the nascent chain by the SRP pathway. Cotranslational protein targeting is likely in kinetic competition with translation elongation, and failure to complete the targeting reaction within the duration of protein synthesis can lead to the misfolding, aggregation and downstream degradation of nascent secretory and membrane proteins. (**B**) Overlay of known RPB structures onto the surface facing the nascent polypeptide exit site (marked by '*') of the 80S eukaryotic ribosome (PDB-4UG0; *light grey* and *dark grey* indicate the 60S and 40S subunits, respectively). Electron densities of ribosomes bound with SRP (EMD-3037), NatE (EMD-4745), NAC (EMD-4938), RAC (EMD-6105), Ebp1 (EMDB-10321), or MetAP2 (PDB-1BN5) are aligned according to the 60S density. The density of MetAP2 is derived from homology modeling with Arx1-ribosome structure (PDB-5APN). The densities of RAC and NAC contain only part of the complex due to low resolution of the EM density map. The silhouettes of the individual RPBs are shown in the indicated colors. Ribosomal proteins in the vicinity of the tunnel exit are colored in different shades of grey and indicated following the nomenclature proposed in Ban et al. [10], with 'L' indicating the large ribosomal subunit, 'u' indicating universally conserved ribosomal proteins, and 'e' indicating eukaryote-specific ribosomal protein subunits.

SRP also provides a salient example of an emerging concept: protein biogenesis begins early on translating ribosomes, long before synthesis of the nascent polypeptide is completed [11–13]. Indeed, the vicinity of the nascent polypeptide exit tunnel of the ribosome provides a platform to recruit multiple ribosome-associated protein biogenesis factors (RPBs), including cotranslational chaperones (nascent polypeptide associated complex (NAC) and ribosome-associated complex (RAC)), protein targeting and translocation machineries (SRP and Sec61p), nascent protein modification enzymes (methionine aminopeptidase (MetAP) and N-acetyl transferase E (NatE)), and quality control factors (Figure 1B) [13]. The RPBs dock at conserved and overlapping sites near the nascent polypeptide exit tunnel on the ribosome, and their engagement with the nascent chain directs the newly synthesized protein to distinct biogenesis pathways. How a nascent protein recruits the correct set of RPBs and thus commits to the proper biogenesis pathway in a timely manner is an emerging question at the heart of accurate protein biogenesis.

In this article, we review recent progress in understanding the molecular mechanism and substrate selection of the eukaryotic SRP pathway, with an emphasis on results demonstrating how regulation of SRP by the cotranslational chaperone NAC enhances the fidelity of protein targeting to the ER. Based on these and recent work on related pathways, we suggest that cells evolved multiple mechanisms to overcome the physicochemical challenges in recognizing degenerate targeting signals. These include allosteric regulation by macromolecular crowding at the ribosome exit site, kinetic competition with translation elongation, rivalry of opposing targeting pathways with overlapping but distinct substrate preferences, and surveillance and error correction mechanisms at the organelle membrane. It is likely that each individual mechanism generates a modest degree of specificity, but collectively, the combination of these mechanisms ensures the accuracy of membrane protein localization and organelle biogenesis.

2. SRP-Dependent Cotranslational Protein Targeting

SRP is a universally conserved ribonucleoprotein particle comprised of the 7SL SRP RNA on which six protein subunits (SRP19, SRP54, SRP68, SRP72, SRP9, SRP14) are assembled (Figure 2A). SRP is responsible for the targeted delivery of newly synthesized membrane and secretory proteins to the SecYEG translocase at the bacterial plasma membrane, or the Sec61p translocase at the eukaryotic ER membrane. The universally conserved core of SRP is a GTPase, SRP54, with two structural and functional domains: a methionine-rich M-domain, which binds the SRP RNA and provides the docking site for ER signal sequences (Figure 2A) [14–19]. The M-domain is connected via a flexible linker to a special GTPase domain, termed NG, which can interact with ribosomal protein uL23 near the exit site [20–23]. SRP54-NG assembles a stable, GTP-dependent dimer with a highly homologous NG-domain in the SRP receptor (SR; Figure 2A) [24–28]. The two NG-domains undergo cooperative conformational rearrangements in their heterodimer that culminates in their reciprocal GTPase activation, followed by GTP hydrolysis that drives the disassembly and recycling of SRP and SR [2,25,29–36]. Extensive work on this simplest SRP system in bacteria showed how this dimerization-activated GTPase cycle ensures the fidelity of the prokaryotic SRP pathway: ribosomes bearing an SRP-dependent signal sequence not only bind SRP more strongly, but also mediate SRP–SR assembly at rates that are 100–1000 fold faster than those on signal-less ribosomes or ribosomes with suboptimal signal sequences (Figure 3A,B, *E. coli*) [37,38]. Furthermore, ribosomes bearing an SRP-dependent substrate effectively delays GTP hydrolysis in the SRP•SR complex until the arrival of the SecYEG translocase, and thus effectively couples the SRP/SR GTPase cycle to productive protein translocation [33,37–39]. In contrast, SRP•SR complex assembled on signal-less ribosomes prematurely hydrolyzes GTP, aborting the targeting reactions to help reject nascent proteins that lack an ER targeting signal [37,38].

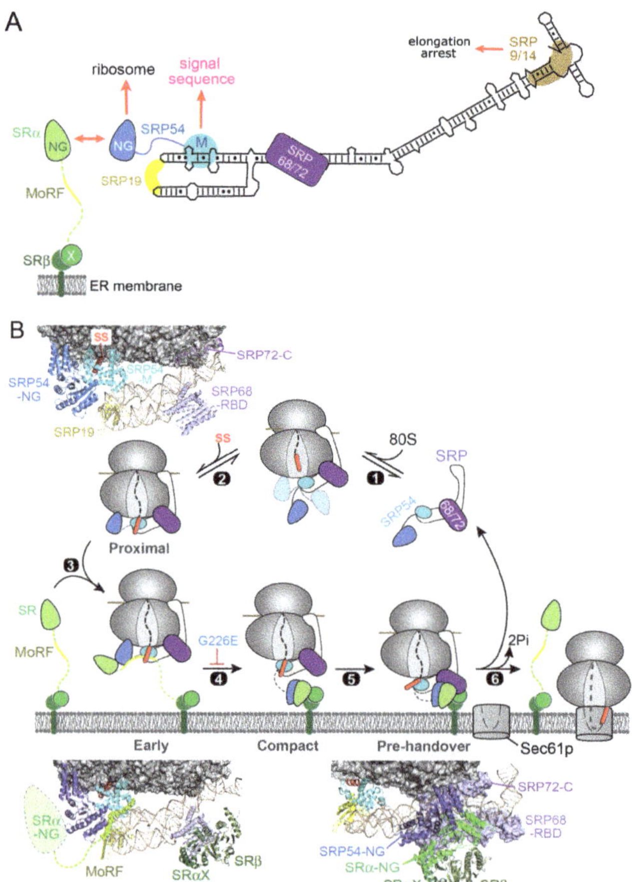

Figure 2. Model of the mammalian SRP pathway. (**A**) Schematic of the composition and interactions of the mammalian SRP and SRP receptor (SR). The individual subunits, domains, and important sequence motifs are defined in the text and indicated. MoRF, molecular recognition feature. (**B**) Current molecular model of the mammalian SRP pathway. Step 1, SRP binds to the translating ribosomes, on which it samples multiple conformations. Step 2, emergence of an ER signal sequence (*red*) drives SRP into the Proximal conformation. Step 3, early stage of SRP–SR assembly, mediated by dynamic interactions between the SRP and SR NG domains and by the SR MoRF (*lime*) interaction with SRP54. Step 4, a stable SRP/SR NG-heterodimer detaches from the ribosome exit site and docks onto the X/β domain of SR. Step 5, the NG•X/β complex docks onto the distal site of SRP to form the Pre-handover conformation, in which the translating ribosome is primed for handover to the Sec61p complex. Step 6, cargo is loaded on Sec61p to initiate protein translocation, and GTP hydrolysis drives the detachment of SRP from SR. The insets show the structural models of the RNC-SRP complex (PDB: 7OBR, upper left), the early RNC-SRP–SR complex (PDB: 7NFX, lower left), and RNC-SRP–SR pre-handover complex (PDB: 6FRK, lower right). Dashed outline in the early complex structure depicts SRα-NG that dynamically interacts with SRP54-NG at this stage and was not resolved in the structure. The M- and NG-domains of SRP54 are in *cyan* and *dark blue*, respectively; signal sequence (ss) is in *red*; SRP19 is in *yellow*, SRP68/72 is in *purple*, SRP RNA is in *tan*, SRαNG, SRαX, and SRβ are in *light green*, *dark green*, and *mustard*, respectively; MoRF in the SR linker is highlighted in spacefill model in *lime*.

While these core GTPases in SRP and SR are highly conserved across species, both SRP and SR undergo extensive expansions in size and complexity during evolution. While bacterial SRP is a complex of the 4.5S RNA with the SRP54 homologue Ffh, eukaryotic SRP contains a larger 7SL RNA on which five additional protein subunits (SRP19, SRP68/72, SRP9/14) are assembled (Figure 2A) [40,41]. While the bacterial SRP receptor is a single protein FtsY in which the NG-domain is preceded by two amphiphilic lipid-binding helices [42–46], eukaryotic SR is a heterodimer of SRα and SRβ subunits (Figure 2A). SRβ is a single-pass transmembrane protein anchored at the ER. SRα binds tightly to SRβ via its N-terminal X-domain [47,48], which is connected to the NG-domain through a ~200-residue intrinsically disordered linker that contains sites for ribosome interaction and sensing [49,50]. Extensive progress has been made in elucidating the function of many of the eukaryote-specific SRP components and deciphering the molecular mechanism of the mammalian SRP pathway in recent years, owing in large part to the ability to reconstitute human SRP and SR with recombinant components [51]. This enabled detailed biochemical and biophysical analyses of the molecular events in the pathway, the identification of new targeting intermediates and the elucidation of their structures, together generating a molecular model for the pathway that incorporates structural, dynamic, and kinetic information.

To summarize, free SRP appears to be locked in a latent conformation that is inactive in its interaction with SR (Figure 2B) [51]. The particle is activated upon binding to the ribosome, on which it can sample a variety of conformations with its NG domain positioned differently relative to the proximal end of SRP (Figure 2B, step 1) [51]. The emergence of an ER signal sequence drives SRP into the 'Proximal' conformation, in which SRP54 NG docks at uL23 in close proximity to the ribosome exit site (step 2) [20,21,51]. In this conformation, SRP initiates assembly with SR via the interaction between their NG domains (step 3). Early SRP–SR association is assisted by a molecular recognition feature (MoRF) in the SR linker, which contacts both the M- and NG-domains of SRP54 to stabilize the earliest stage of targeting (Figure 2B, 'Early') [49,52]. Formation of a stable NG dimer drives a series of conformational rearrangements, leading to the detachment of the NG-dimer from the ribosome exit site and its docking onto the membrane-proximal X and β-domains of SR, resulting in a global compaction of the SR (Figure 2B, 'Compact') [52]. A new molecular surface is generated in the resulting NG•Xβ complex, allowing it to dock onto the distal end of SRP where SRP68/72 is located (Figure 2B, step 5) [21,52,53]. In this 'pre-handover' conformation of the targeting complex, the ribosome is brought close to the membrane surface, and the ribosome exit site is vacated and thus primed to initiate interaction with the Sec61p translocation machinery (step 6).

Thus, eukaryotic cotranslational protein targeting requires multiple largescale conformational rearrangements in both SRP and SR, which allow this targeting machine to transition successively through the cargo recognition, targeting, and cargo handover stages in the targeting cycle. These structural and functional transitions are driven by the dimerization-activated GTPase cycle of SRP/SR, the translating ribosome, and possibly other components of the pathway. Notably, multiple mutations in SRP54 NG are linked to severe syndromic neutropenia with Shwachman–Diamond-like features; these mutations block either the assembly of the NG heterodimer [54,55] or the conformational rearrangements that lead to the pre-handover complex (Figure 2B, '⊥') [52], demonstrating the critical role of the GTPase-driven conformational rearrangements in the proper functioning of SRP. As described in the next section, these conformational rearrangements also provide multiple opportunities for allosteric regulation of this targeting machine, for example by additional RPBs at the ribosome exit site.

The ability to quantitatively measure the individual molecular events in the mammalian SRP pathway also enabled a comparison of the molecular interactions of the mammalian and bacterial SRP, which raised intriguing questions as to how high fidelity is achieved during cotranslational protein targeting (Figure 3A,B). As described earlier, bacterial SRP and SR form a self-sufficient system that can generate a high level of targeting specificity by using a combination of differential binding, induced fit, and kinetic proof-reading mechanisms (Figure 3A,B, *E. coli*). In contrast, ribosomes with and without an ER targeting signal differ only ~four-fold in the binding of human SRP, and ~two-fold in activating the assembly between SRP and SR (Figure 3A,B, human) [56,57]. These results suggest that, unexpectedly, mammalian SRP and SR by themselves are insufficient to generate the specificity required for high fidelity protein targeting to the ER. As described in the next Section, SRP requires a cotranslational chaperone, the nascent polypeptide associated complex (NAC), to act as a triage factor during substrate selection in eukaryotic cells.

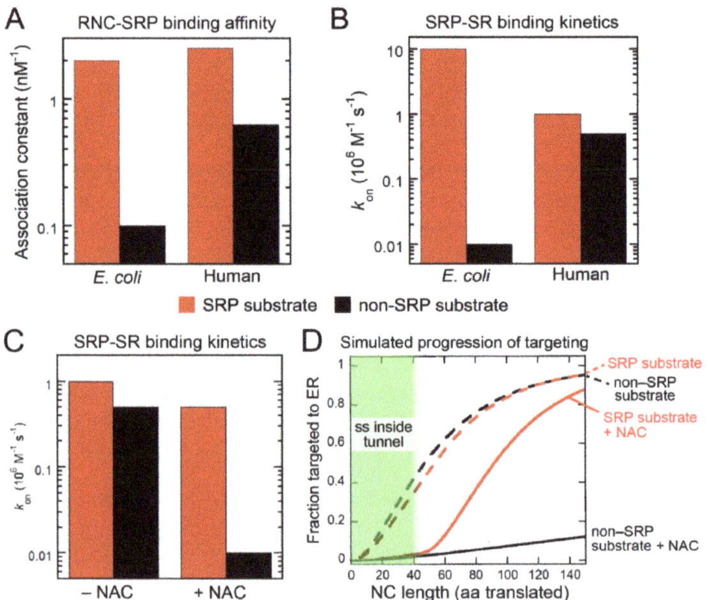

Figure 3. NAC regulates the activity of mammalian SRP to help reject signal-less ribosomes from ER targeting. Comparison of the specificity during the cargo recognition (**A**) and targeting steps (**B**) between the *E. coli* and human SRPs. The rate and equilibrium constants were measured with purified RNC, SRP and SR in vitro and show the promiscuity of human SRP in both steps. Adapted with modifications from the data in [56,57]. (**C**) NAC enhances the specificity of human SRP during the SR recruitment step from <2-fold to ~50-fold. Adapted from [56]. (**D**) Simulation of the progression of cotranslational protein targeting to the ER with and without NAC present. The simulation is based on a kinetic model of SRP-dependent protein targeting using experimentally determined rate and equilibrium constants of the cargo binding and SRP–SR assembly steps, followed by a commitment step in which the RNC-SRP–SR complex loads the translating ribosome onto Sec61p. The enhanced selectivity of ER targeting in the presence of NAC arise from two effects: suppression of pre-mature targeting before a signal sequence emerges from the ribosome, and inhibition of non-specific SRP–SR association on ribosomes that do not expose an ER targeting signal. Adapted from [56].

3. NAC: A Triage Factor during Cotranslational Protein Targeting

NAC is an abundant cotranslational chaperone expressed at equimolar concentrations relative to the ribosome in eukaryotic organisms [11,58]. Given its abundance and high ribosome binding affinity (K_d ~1 nM; [56]), NAC can bind to virtually every ribosome in eukaryotic cells. How NAC interacts with the ribosome is incompletely understood. NAC is a heterodimer of α and β subunits, both containing a central NAC domain that dimerizes into a β-barrel structure (Figure 4A, pink and magenta) [59]. In addition, NACβ contains an N-terminal extension harboring a conserved basic motif crucial for its ribosome binding (Figure 4A, '++') [60]. Crosslinking data suggested multiple interaction sites of NAC on the ribosome, including uL23, uL29, eL31 and eL39, all of which are located near the exit tunnel but on opposite sides [61]. A recent cryoEM structure revealed an unexpected mode of NAC interaction: the N-terminal tail of NACβ inserts deeply into the nascent polypeptide exit tunnel of the 60S ribosomal subunit [62]. In support of the structural observation, the N-terminus of NACβ can crosslink to nascent chains on ribosome-nascent chain complexes (RNCs) as short as 10 amino acids (Figure 4B, (1) and (2)), suggesting that NAC acts at the earliest stage of protein synthesis [62]. The crosslink to the NACβ N-terminus becomes weaker when the nascent chain exceeds 30 amino acids in length, suggesting that the inserted tail of NACβ is pushed out of the exit tunnel during translation elongation, and that NAC switches to a distinct mode(s) of interaction once the nascent chain emerges from the tunnel exit (Figure 4B, (3) and (4)).

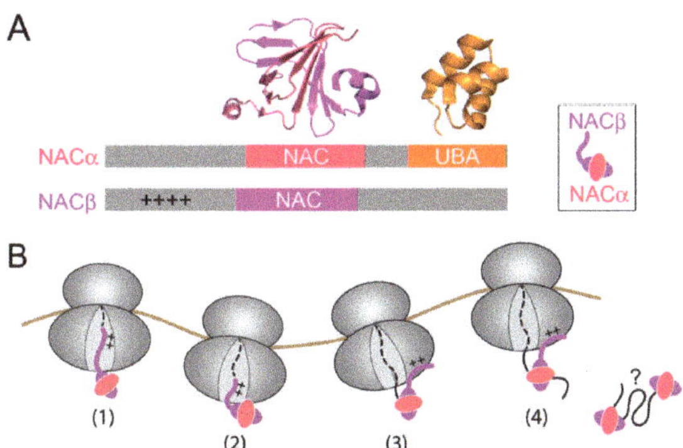

Figure 4. Proposed structure and interactions of NAC. (**A**) Overview of the domain composition and available structural information of NAC. Grey depicts unstructured extensions from the folded NAC and UBA domains. '++++' denotes the basic RRKKK motif in the N-terminal extension of NACβ crucial for its ribosome binding. The crystal structures are shown for the C-terminal UBA domain of NACα (PDB: 1TR8) as well as the α/β NAC domain, which dimerizes into a β-barrel like structure (PDB: 3MCB). (**B**) Summary of models of NAC interaction with ribosome and the nascent chain during protein synthesis. NAC engages ribosome early during translation, with the N-terminal tail of NACβ inserting deeply into the exit tunnel of the ribosome (stage (1)). The inserted N-terminal tail of NACβ begins to retract in the tunnel as the nascent chain elongates to ~30 amino acids (stage (2)), and switches to interact with the ribosome surface upon the emergence of the nascent protein from the exit tunnel (stage (3)). NAC is likely anchored by the N-terminal NACβ tail on the ribosome, and the central NAC β-barrel could continue to contact portions of the nascent protein that just emerge from the tunnel exit as the nascent chain further elongates (stage (4)). Finally, NAC could also associate with aggregation prone proteins when released from the ribosome, although the molecular mechanism of NAC interactions off the ribosome awaits further investigation (?). '++++' denotes the basic RRKKK motif at the N-terminus of NACβ.

NAC is generally described as a chaperone-like molecule that assists in the maturation of newly synthesized proteins [63,64]. The embryonic lethality of NAC mutants in *C. elegans*, *Drosphila melanogaster* and mice demonstrates an essential function of this chaperone in higher eukaryotic organisms [65–68]. However, the precise cellular roles and biochemical activities of NAC still await to be clearly defined. A variety of functions have been ascribed to NAC, including the de novo folding of nascent proteins and the biogenesis/maturation of ribosomes [69,70]. NAC weakly interacts with a variety of proteins and helps maintain the solubility of aggregation-prone proteins, such as alpha-synuclein and polyQ [71,72]. NAC is also proposed to be a proteostasis sensor that relocalizes from the ribosome to aggregated puncta in the presence of proteostasis stress [73]. These observations potentially reflect a small heat shock protein-like activity of NAC off the ribosome. Other suggested roles for NAC include protein import to mitochondria [74–77], NACα as a transcription activator [78,79], and suppression of apoptosis [67]. Many of these proposed roles, including the direct or indirect involvement of NAC in these processes, remain to be explored.

The best studied function of NAC is its role in the regulation of protein targeting to the ER. NAC was initially identified in rabbit reticulocyte lysate as a factor that prevents the nonspecific engagement of SRP with nascent chains that lack an ER signal sequence, and whose depletion leads to the mistargeting of cytosolic and mitochondrial proteins to ER microsomes [80]. Many ensuing studies corroborated the involvement of NAC in regulating protein sorting to the ER [81–88] and further showed that NAC binds to ribosomes with short nascent chains and forms a protective environment for regions of the nascent polypeptide just emerging from the tunnel exit [58,89]. Significantly, knockdown of NAC in *C. elegans* led to ER stress and the mislocalization of reporter proteins with a mitochondrial signal sequence to the ER [65,90], providing strong support for the role of NAC as a specificity factor during ER targeting.

Despite these earlier works, the mechanism by which NAC prevents protein mistargeting remained controversial. Nevertheless, a globular domain of NAC is located at the ribosome tunnel exit in its 'inserted' conformation described above (Figure 4B, (1)), in a position that can block the binding of SRP or Sec61p to the ribosome [62]. This and the observed antagonistic effect of NAC on the binding of signal-less ribosomes to SRP and to the ER membrane [81–88] gave rise to a primarily competitive model, in which NAC excludes SRP and Sec61p from binding to ribosomes without an ER signal sequence. However, recent quantitative measurements suggested otherwise. While NAC weakened the binding affinity of SRP for RNCs, in agreement with earlier observations, the binding antagonism saturated at 4–6 fold; this saturation behavior is in contrast to expectations from a strictly competitive model, which predicts that the observed binding affinity will continue to decrease with increasing concentrations of the competitor [56]. In addition, the effects of NAC on SRP binding affinity were similar between ribosomes with and without an ER signal sequence and insufficient to explain the ability of NAC to specifically suppress the targeting of signal-less nascent chains [56]. Finally, co-binding of SRP and NAC on the same RNC can be observed in single-molecule colocalization experiments, and efficient FRET was also observed between the two factors on the RNC [56], indicating that they are positioned in close proximity to each other on the same ribosome.

These observations indicate that NAC does not act solely by excluding SRP from ribosome binding but instead, exerts regulation via an allosteric mechanism. Indeed, under conditions where SRP and NAC are co-bound on the ribosome, NAC specifically reduced the kinetics of SRP–SR assembly on ribosomes without an ER signal sequence, increasing the discrimination against signal-less ribosomes to ~50-fold in this membrane-targeting step (Figure 3C; [56]). In addition, significant promiscuous SRP–SR association were observed on ribosomes with nascent chains shorter than 35 amino acids, when the targeting signal is still buried inside the nascent polypeptide exit tunnel (Figure 3D, area shaded in green) [56]. NAC also strongly suppresses these pre-mature targeting events and thus delays the onset of targeting [56]. Kinetic modeling of the SRP pathway, based on these experimentally measured parameters, showed that the combination of these regulatory effects of NAC are necessary and sufficient to generate a high degree of specificity during cotranslational protein targeting to the ER (Figure 3D), whereas in the absence of NAC, both ribosomes with and without an ER signal sequence are delivered to the ER within ~100 amino acids of their synthesis (Figure 3D, dashed lines) [56].

The allosteric regulation of SRP by NAC was directly probed in single molecule FRET measurements (Figure 5A) [56]. A pair of FRET dyes, engineered between SRP54-NG and SRP19, was used to specifically monitor the formation of the 'Proximal' conformation of SRP that is most active for SR recruitment [51]. On ribosomes exposing an ER targeting signal, SRP is dominated by the high FRET population corresponding to the Proximal conformation, and its conformational distribution is not substantially affected by NAC (Figure 5A, right panel). On ribosomes exposing a mutated signal sequence, in contrast, SRP is conformationally much more dynamic and heterogeneous, sampling low-, medium-, and high-FRET states all with substantial frequency [51,56]. Significantly, ~30% of SRP samples in the Proximal conformation were conducive to SR binding even when bound to signal-less ribosomes (Figure 5A), which may explain the low substrate specificity during SRP–SR assembly. However, NAC largely eliminated the population of SRP that resides in the Proximal conformation on signal-less ribosomes, forcing SRP into low- and medium-FRET states that are presumably inactive in binding with SR. (Figure 5A) [56].

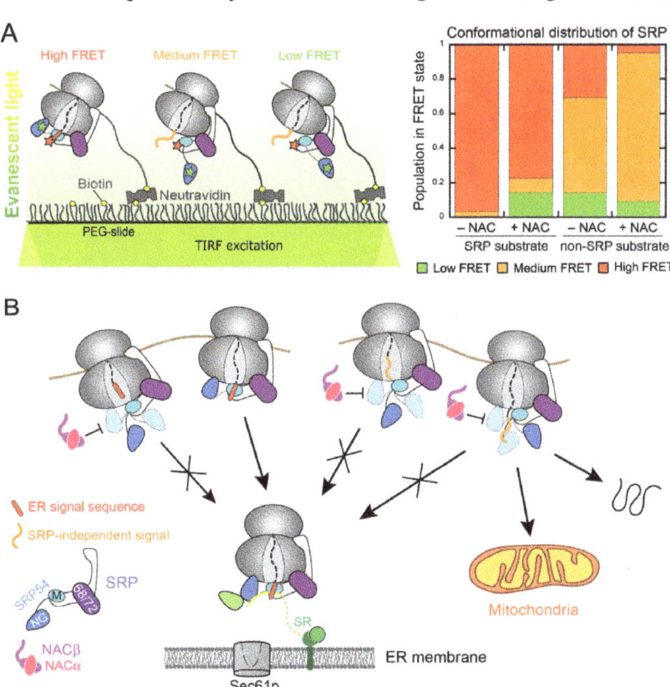

Figure 5. NAC enhances the specificity of ER targeting by regulating the conformation and activity of SRP. (**A**) Single-molecule FRET (smFRET) measurements show how NAC remodels the conformation of SRP on the ribosome. Left panel: schematic of the smFRET experiment. A FRET dye pair was incorporated on SRP54-NG and SRP19 to detect the Proximal conformation of SRP most active in SR recruitment. SRP was recruited to RNCs immobilized on the microscope slide surface, on which it can sample distinct conformational states that generate different FRET efficiencies between the dye pair. Right: summary of the conformational distribution of SRP on signal sequence-containing and signal-less ribosomes. NAC reduces the population of SRP in the high FRET state on signal-less ribosomes, directly demonstrating NAC allosteric regulation of SRP. Adapted from [56]. (**B**) Model of the mechanism by which NAC improves the targeting specificity of SRP. NAC (magenta) inhibits SRP from adopting the Proximal conformation on both short-chain RNCs and on ribosomes that expose a non-ER targeting signal, thus delaying the onset of ER targeting and preventing promiscuous targeting to the ER.

Collectively, the recent results provide strong evidence that NAC acts as a triage factor that enforces the correct timing and specificity of SRP-dependent protein targeting (Figure 5B). In addition to the binding antagonism proposed previously, NAC further exerts its regulation allosterically, by remodeling the conformational landscape of SRP on the ribosome and preventing SRP from adopting the targeting-active conformation in the absence of an exposed ER signal sequence (Figure 5B). This ensures that SRP is activated to initiate targeting only upon the emergence of a correct signal sequence from the ribosome exit tunnel, thus preventing the promiscuous ER localization of ribosomes translating cytosolic proteins, or proteins destined to other organelles such as mitochondria (Figure 5B).

4. Perspectives and Open Questions

Emerging data show that the observation with NAC is not an isolated example, but rather, represents a general mechanism whereby the fidelity of individual protein biogenesis pathways can be reshaped by macromolecular crowding at the ribosome exit site. For example, the abundant cotranslational chaperone trigger factor (TF) in bacteria can help the bacterial SRP reject borderline secretory protein substrates with weakly hydrophobic signal sequences [91]. Analogously to NAC, TF co-binds with SRP on the ribosome and regulates the activity of SRP via a multi-layered mechanism: it selectively reduces SRP–SR assembly rates on ribosomes displaying weakly hydrophobic signal sequences [91]. TF also restricts SRP-dependent targeting after the nascent polypeptide exceeds a critical length of ~130 amino acids, imposing a limited time window during translation for SRP to complete the targeting reaction [91]. This combination of allosteric and timing mechanisms allows TF to suppress the leaky cotranslational targeting of secretory proteins that can otherwise use the SecB/A post-translational translocation pathway. In another recent example, the specificity of an essential nascent protein modification enzyme in bacteria, methionine amino peptidase (MAP), was shown to be critically dependent on RPBs on the ribosome [92]. Cotranslational excision of the initiator methionine by MAP is rapid and diffusion-limited, with the irreversible chemical step significantly faster than the dissociation of MAP from the ribosome. As such, ribosome-bound MAP displays limited discrimination against suboptimal substrates with large side chains at the second amino acid [92]. A combination of RPBs, SRP and TF, selectively reduces the reaction rate of MAP at nascent chain lengths below 67 aa and beyond 82 aa [92]. This effectively restricts the action of MAP to a limited time window during translation elongation and thus re-establishes the sequence specificity of MAP during cotranslational processing of the nascent protein [92]. These and other recent work highlight the rich and dynamic mechanisms of molecular coordination between protein biogenesis factors on the ribosome and show that this coordination plays a vital role in ensuring the fidelity of nascent protein selection into their appropriate biogenesis pathways.

Recent findings in both co- and post-translational protein targeting pathways further emphasize the principle that the appropriate sorting of nascent proteins to cellular organelles is a result of the balanced action of multiple protein biogenesis factors and pathways. While mitochondrial proteins are mislocalized to the ER in the absence of NAC, acute depletion of SRP in yeast leads to the mistargeting of ribosomes translating normally ER-destined proteins to mitochondria, triggering rapid mitochondria fragmentation and dysfunction [7]. These results are reminiscent of the observations during the post-translational targeting of tail-anchored membrane proteins (TAs), in which deletion of components of the guided-entry-of-tail-anchored protein (GET) pathway resulted in the mistargeting of some ER-destined TAs to mitochondria [93]. These observations likely reflect the general tendency of hydrophobic, aggregation-prone membrane proteins to be mislocalized to membrane-enclosed organelles. They also suggest that any individual protein targeting pathway does not generate sufficient specificity of protein localization in the cell. Instead, organelle specificity of protein localization relies critically on the proper functioning of a combination of pathways and factors with opposing activities. These pathways likely possess overlapping yet distinct substrate preferences, which could enable more effective differentiation of degenerate targeting signals that share many physicochemical features.

While NAC provides a triage factor that facilitates the correct selection of translating ribosomes at early stages of ER targeting, it is likely that additional mechanisms are in place to ensure the fidelity of protein targeting and translocation to cellular membranes. In the case of bacterial SRP and MAP enzymes, kinetic rivalry with translation elongation proved to be an effective strategy to reject suboptimal substrates [91,92]; whether this principle operates in the mammalian SRP pathway to tune substrate selection remains to be determined. Intriguingly, the SRP9/14 subunits of the mammalian SRP competes with eEF1 and slows translation elongation (Figure 2A); whether and how this activity plays a role in the efficiency and substrate selection of SRP remain open questions [94–101]. In addition, early work showed that the Sec61p translocase provides a post-targeting mechanism to reject ribosomes with mutated signal sequences [102]. Subsequent biochemical and structural work revealed a lateral gate formed by TM2 and TM7 in the SecYEG/Sec61p complex that forms a docking site for TMDs and signal sequences [103,104], providing a molecular basis for the ability of this translocation machinery to recognize the targeting signal. Furthermore, surveillance and quality control pathways have been identified on both mitochondria and the ER that provide mechanisms for clearance of mislocalized membrane proteins. The conserved AAA-ATPase, Msp1 in yeast or ATAD1 in mammalian cells, localizes to the outer membrane of mitochondria and extracts ER-destined TAs that are mislocalized to mitochondria in the absence of a functioning GET pathway, as well as mislocalized peroxisomal TAs in the absence of the peroxisome targeting factor Pex19 [105–109]. Msp1 facilitates the transfer of mistargeted TAs from mitochondria to the ER, where the TA is recognized and degraded by the ubiquitin ligase Doa10 [110]. Reciprocally, an ER-resident P5A AAA-ATPase, ATP13A1 (Spf1 in yeast), recognizes and mediates the extraction of mitochondrial TAs mislocalized at the ER membrane [111,112]. While the observation of error-correction mechanisms has thus far used TAs as model substrates, whether analogous quality control machineries exist to correct mistakes in cotranslational protein targeting and to handle topologically more complex membrane proteins remain an outstanding question. The molecular mechanism by which these quality control machineries detect errors in protein localization also remain to be determined.

Author Contributions: Writing—original draft preparation, S.-o.S.; writing—review and editing, H.-H.H.; visualization, H.-H.H. and S.-o.S. All authors have read and agreed to the published version of the manuscript.

Funding: This work is funded by grants NSF-1929452 from the National Science Foundation and R01 GM078024 and R35 GM136321 from the National Institute of Health to S.S.

Institutional Review Board Statement: Not applicable.

Informed Consent Statement: Not applicable.

Conflicts of Interest: The authors declare no conflict of interest.

References

1. Walter, P.; Johnson, A.E. Signal sequence recognition and protein targeting to the endoplasmic reticulum membrane. *Ann. Rev. Cell Biol.* **1994**, *10*, 87–119. [CrossRef]
2. Akopian, D.; Shen, K.; Zhang, X.; Shan, S. Signal recognition particle: An essential protein-targeting machine. *Annu. Rev. Biochem.* **2013**, *82*, 693–721. [CrossRef] [PubMed]
3. Hegde, R.S.; Keenan, R.J. The mechanisms of integral membrane protein biogenesis. *Nat. Rev. Mol. Cell Biol.* **2021**, 1–18. [CrossRef]
4. Spiess, M.; Junne, T.; Janoschke, M. Membrane Protein Integration and Topogenesis at the ER. *Protein J.* **2019**, *38*, 306–316. [CrossRef] [PubMed]
5. Von Heijne, G. Signal sequences: The limits of variation. *J. Mol. Biol.* **1985**, *184*, 99–105. [CrossRef]
6. Gierasch, L.M. Signal sequences. *Biochemistry* **1989**, *28*, 923–930. [CrossRef] [PubMed]
7. Costa, E.A.; Subramanian, K.; Nunnari, J.; Weissman, J.S. Defining the physiological role of SRP in protein-targeting efficiency and specificity. *Science* **2018**, *359*, 689–692. [CrossRef] [PubMed]
8. Flanagan, J.J.; Chen, J.C.; Miao, Y.; Shao, Y.; Lin, J.; Bock, P.E.; Johnson, A.E. Signal recognition particle binds to ribo-some-bound signal sequences with fluorescence-detected subnanomolar affinity that does not diminish as the nascent chain lengthens. *J. Biol. Chem.* **2003**, *278*, 18628–18637. [CrossRef] [PubMed]
9. Siegel, V.; Walter, P. The affinity of signal recognition particle for presecretory proteins is dependent on nascent chain length. *EMBO J.* **1988**, *7*, 1769–1775. [CrossRef]
10. Ban, N.; Beckmann, R.; Cate, J.H.; Dinman, J.D.; Dragon, F.; Ellis, S.R.; Lafontaine, D.L.J.; Lindahl, L.; Liljas, A.; Lipton, J.M.; et al. A new system for naming ribosomal proteins. *Curr. Opin. Struct. Biol.* **2014**, *24*, 165–169. [CrossRef] [PubMed]
11. Deuerling, E.; Gamerdinger, M.; Kreft, S.G. Chaperone Interactions at the Ribosome. *Cold Spring Harb. Perspect. Biol.* **2019**, *11*, a033977. [CrossRef]

12. Cassaignau, A.M.E.; Cabrita, L.D.; Christodoulou, J. How Does the Ribosome Fold the Proteome? *Annu. Rev. Biochem.* **2020**, *89*, 389–415. [CrossRef] [PubMed]
13. Kramer, G.; Shiber, A.; Bukau, B. Mechanisms of Cotranslational Maturation of Newly Synthesized Proteins. *Annu. Rev. Biochem.* **2019**, *88*, 337–364. [CrossRef] [PubMed]
14. Batey, R.T.; Rambo, R.P.; Lucast, L.; Rha, B.; Doudna, J.A. Crystal structure of the ribonucleoprotein core of the signal recognition particle. *Science* **2000**, *287*, 1232–1239. [CrossRef]
15. Batey, R.T.; Sagar, M.B.; Doudna, J.A. Structural and energetic analysis of RNA recognition by a universally conserved protein from the signal recognition particle. *J. Mol. Biol.* **2001**, *307*, 229–246. [CrossRef]
16. Hainzl, T.; Huang, S.; Merilainen, G.; Brannstrom, K.; Sauer-Eriksson, A.E. Structural basis of signal sequence recognition by the signal recognition particle. *Nat. Struct. Mol. Biol.* **2011**, *18*, 389–391. [CrossRef] [PubMed]
17. Janda, C.Y.; Li, J.; Oubridge, C.; Hernandez, H.; Robinson, C.V.; Nagai, K. Recognition of a signal peptide by the signal recognition particle. *Nature* **2010**, *465*, 507–510. [CrossRef]
18. Krieg, U.C.; Walter, P.; Johnson, A.E. Photocrosslinking of the signal sequence of nascent preprolactin to the 54-kilodalton polypeptide of the signal recognition particle. *Proc. Natl. Acad. Sci. USA* **1986**, *83*, 8604–8608. [CrossRef]
19. Kurzchalia, T.V.; Wiedmann, M.; Girshovich, A.S.; Bochkareva, E.S.; Bielka, H.; Rapoport, T.A. The signal sequence of nascent preprolactin interacts with the 54K polypeptide of the signal recognition particle. *Nature* **1986**, *320*, 634–636. [CrossRef] [PubMed]
20. Voorhees, R.M.; Hegde, R.S. Structures of the scanning and engaged states of the mammalian SRP-ribosome complex. *Elife* **2015**, *4*, e07975. [CrossRef]
21. Jomaa, A.; Eitzinger, S.; Zhu, Z.; Chandrasekar, S.; Kobayashi, K.; Shan, S.O.; Ban, N. Molecular mechanism of cargo recognition and handover by the mammalian signal recognition particle. *Cell Rep.* **2021**, *36*, 109350. [CrossRef]
22. Jomaa, A.; Boehringer, D.; Leibundgut, M.; Ban, N. Structures of the *E. coli* translating ribosome with SRP and its receptor and with the translocon. *Nat. Commun.* **2016**, *7*, 10471. [CrossRef] [PubMed]
23. Schaffitzel, C.; Oswald, M.; Berger, I.; Ishikawa, T.; Abrahams, J.P.; Koerten, H.K.; Koning, R.I.; Ban, N. Structure of the E. coli signal recognition particle bound to a translating ribosome. *Nature* **2006**, *444*, 503–506. [CrossRef] [PubMed]
24. Peluso, P.; Herschlag, D.; Nock, S.; Freymann, D.M.; Johnson, A.E.; Walter, P. Role of 4.5S RNA in assembly of the bacterial signal recognition particle with its receptor. *Science* **2000**, *288*, 1640–1643. [CrossRef] [PubMed]
25. Peluso, P.; Shan, S.; Nock, S.; Herschlag, D.; Walter, P. Role of SRP RNA in the GTPase cycles of Ffh and FtsY. *Biochemistry* **2001**, *40*, 15224–15233. [CrossRef]
26. Egea, P.F.; Shan, S.; Napetschnig, J.; Savage, D.F.; Walter, P.; Stroud, R.M. Substrate twinning activates the signal recog-nition particle and its receptor. *Nature* **2004**, *427*, 215–221. [CrossRef]
27. Focia, P.J.; Shepotinovskaya, I.V.; Seidler, J.A.; Freymann, D.M. Heterodimeric GTPase Core of the SRP Targeting Complex. *Science* **2004**, *303*, 373–377. [CrossRef]
28. Wild, K.; Bange, G.; Motiejunas, D.; Kribelbauer, J.; Hendricks, A.; Segnitz, B.; Wade, R.C.; Sinning, I. Structural Basis for Conserved Regulation and Adaptation of the Signal Recognition Particle Targeting Complex. *J. Mol. Biol.* **2016**, *428*, 2880–2897. [CrossRef]
29. Shan, S.; Stroud, R.; Walter, P. Mechanism of association and reciprocal activation of two GTPases. *PLoS Biol.* **2004**, *2*, e320. [CrossRef]
30. Shan, S.O. ATPase and GTPase Tangos Drive Intracellular Protein Transport. *Trends Biochem. Sci.* **2016**, *41*, 1050–1060. [CrossRef]
31. Saraogi, I.; Shan, S. Molecular mechanism of co-translational protein targeting by the Signal Recognition Particle. *Traffic* **2011**, *12*, 535–542. [CrossRef]
32. Estrozi, L.F.; Boehringer, D.; Shan, S.; Ban, N.; Schaffitzel, C. Cryo-EM structure of the *E. coli* translating ribosome in complex with SRP and its receptor. *Nat. Struct. Mol. Biol.* **2011**, *18*, 88–90. [CrossRef]
33. Zhang, X.; Schaffitzel, C.; Ban, N.; Shan, S. Multiple conformational changes in a GTPase complex regulate protein targeting. *Proc. Natl. Acad. Sci. USA* **2009**, *106*, 1754–1759. [CrossRef]
34. Zhang, X.; Kung, S.; Shan, S. Demonstration of a two-step mechanism for assembly of the SRP–SRP receptor complex: Implications for the catalytic role of SRP RNA. *J. Mol. Biol.* **2008**, *381*, 581–593. [CrossRef] [PubMed]
35. Ataide, S.F.; Schmitz, N.; Shenk, K.; Ke, A.; Shan, S.; Doudna, A.; Ban, N. The crystal structure of the Signal Recognition Particle in complex with its receptor. *Science* **2011**, *381*, 881–886. [CrossRef] [PubMed]
36. Shen, K.; Arslan, S.; Akopian, D.; Ha, T.; Shan, S. Activated GTPase movement on an RNA scaffold drives cotranslational protein targeting. *Nature* **2012**, *492*, 271–275. [CrossRef] [PubMed]
37. Zhang, X.; Shan, S.O. Fidelity of cotranslational protein targeting by the signal recognition particle. *Annu. Rev. Biophys.* **2014**, *43*, 381–408. [CrossRef] [PubMed]
38. Zhang, X.; Rashid, R.; Wang, K.; Shan, S. Sequential checkpoints govern fidelity during co-translational protein targeting. *Science* **2010**, *328*, 757–760. [CrossRef]
39. Akopian, D.; Dalai, K.; Shen, K.; Duong, F.; Shan, S. SecYEG activates GTPases to drive the completion of cotranslational protein targeting. *J. Cell Biol.* **2013**, *200*, 397–405. [CrossRef]
40. Walter, P.; Blobel, G. Disassembly and reconstitution of signal recognition particle. *Cell* **1983**, *34*, 525–533. [CrossRef]
41. Walter, P.; Blobel, G. Signal recognition particle contains a 7S RNA essential for protein translocation across the endo-plasmic reticulum. *Nature* **1982**, *299*, 691–698. [CrossRef]

42. Bahari, L.; Parlitz, R.; Eitan, A.; Stjepanovic, G.; Bochkareva, E.S.; Sinning, I.; Bibi, E. Membrane targeting of ribosomes and their release require distinct and separable functions of FtsY. *J. Biol. Chem.* **2007**, *282*, 32168–32175. [CrossRef]
43. Parlitz, R.; Eitan, A.; Stjepanovic, G.; Bahari, L.; Bange, G.; Bibi, E.; Sinning, I. Escherichia coli signal recognition particle receptor FtsY contains an essential and autonomous membrane-binding amphipathic helix. *J. Biol. Chem.* **2007**, *282*, 32176–32184. [CrossRef]
44. De Leeuw, E.; Poland, D.; Mol, O.; Sinning, I.; ten Hagen-Jongman, C.M.; Oudega, B.; Luirink, J. Membrane association of FtsY, the *E. coli* SRP receptor. *FEBS Lett.* **1997**, *416*, 225–229. [CrossRef]
45. de Leeuw, E.; te Kaat, K.; Moser, C.; Memestrina, G.; Demel, R.; de Kruijff, B.; Oudegam, B.; Luirink, J.; Sinning, I. Anionic phospholipids are involved in membrane association of FtsY and stimulate its GTPase activity. *EMBO J.* **2000**, *19*, 531–541. [CrossRef] [PubMed]
46. Hwang Fu, Y.H.; Huang, W.Y.C.; Shen, K.; Groves, J.T.; Miller, T.; Shan, S.O. Two-step membrane binding by the bacterial SRP receptor enable efficient and accurate Co-translational protein targeting. *eLife* **2017**, *6*, e25885. [CrossRef] [PubMed]
47. Jadhav, B.; Wild, K.; Pool, M.R.; Sinning, I. Structure and Switch Cycle of SRbeta as Ancestral Eukaryotic GTPase Associated with Secretory Membranes. *Structure* **2015**, *23*, 1838–1847. [CrossRef] [PubMed]
48. Schlenker, O.; Hendricks, A.; Sinning, I.; Wild, K. The structure of the mammalian signal recognition particle (SRP) receptor as prototype for the interaction of small GTPases with Longin domains. *J. Biol. Chem.* **2006**, *281*, 8898–8906. [CrossRef]
49. Hwang Fu, Y.H.; Chandrasekar, S.; Lee, J.H.; Shan, S.O. A molecular recognition feature mediates ribosome-induced SRP-receptor assembly during protein targeting. *J. Cell Biol.* **2019**, *218*, 3307–3319. [CrossRef]
50. Jadhav, B.; McKenna, M.; Johnson, N.; High, S.; Sinning, I. Pool MR. Mammalian SRP receptor switches the Sec61 translocase from Sec62 to SRP-dependent translocation. *Nat. Commun.* **2015**, *6*, 10133. [CrossRef]
51. Lee, J.H.; Chandrasekar, S.; Chung, S.; Hwang Fu, Y.H.; Liu, D.; Weiss, S.; Shan, S. Sequential activation of human signal recognition particle by the ribosome and signal sequence drives efficient protein targeting. *Proc. Natl. Acad. Sci. USA* **2018**, *115*, E5487–E5496. [CrossRef] [PubMed]
52. Lee, J.H.; Jomaa, A.; Chung, S.; Hwang Fu, Y.H.; Qian, R.; Sun, X.; Hsieh, H.H.; Chandrasekar, S.; Bi, X.; Mattei, S.; et al. Receptor compaction and GTPase rearrangement drive SRP-mediated cotranslational protein translocation into the ER. *Sci. Adv.* **2021**, *7*, eabg0942. [CrossRef] [PubMed]
53. Kobayashi, K.; Jomaa, A.; Lee, J.H.; Chandrasekar, S.; Boehringer, D.; Shan, S.O.; Ban, N. Structure of a prehandover mammalian ribo-somal SRP.SRP receptor targeting complex. *Science* **2018**, *360*, 323–327. [CrossRef] [PubMed]
54. Schurch, C.; Schaefer, T.; Muller, J.S.; Hanns, P.; Arnone, M.; Dumlin, A.; Schärer, J.; Sinning, I.; Wild, K.; Skokowa, J.; et al. SRP54 mutations induce congenital neutropenia via dominant-negative effects on XBP1 splicing. *Blood J. Am. Soc. Hematol.* **2021**, *137*, 1340–1352. [CrossRef] [PubMed]
55. Juaire, K.D.; Lapouge, K.; Becker, M.M.M.; Kotova, I.; Michelhans, M.; Carapito, R.; Wild, K.; Bahram, S.; Sinning, I. Structural and Functional Impact of SRP54 Mutations Causing Severe Congenital Neutropenia. *Structure* **2021**, *29*, 15–28. [CrossRef] [PubMed]
56. Hsieh, H.H.; Lee, J.H.; Chandrasekar, S.; Shan, S.O. A ribosome-associated chaperone enables substrate triage in a cotranslational protein targeting complex. *Nat. Commun.* **2020**, *11*, 5840. [CrossRef] [PubMed]
57. Shen, K.; Zhang, X.; Shan, S. Synergiestic action between the SRP RNA and translating ribosome allows efficient delivery of correct cargos during co-translational protein targeting. *RNA* **2011**, *17*, 892–902. [CrossRef] [PubMed]
58. Raue, U.; Oellerer, S.; Rospert, S. Association of protein biogenesis factors at the yeast ribosomal tunnel exit is affected by the translational status and nascent polypeptide sequence. *J. Biol. Chem.* **2007**, *282*, 7809–7816. [CrossRef] [PubMed]
59. Wang, L.; Zhang, W.; Wang, L.; Zhang, X.C.; Li, X.; Rao, Z. Crystal structures of NAC domains of human nascent polypep-tide-associated complex (NAC) and its alphaNAC subunit. *Protein. Cell* **2010**, *1*, 406–416. [CrossRef] [PubMed]
60. Wegrzyn, R.D.; Hofmann, D.; Merz, F.; Nikolay, R.; Rauch, T.; Graf, C.; Deuerling, E. A conserved motif is prerequisite for the interaction of NAC with ribosomal protein L23 and nascent chains. *J. Biol. Chem.* **2006**, *281*, 2847–2857. [CrossRef]
61. Pech, M.; Spreter, T.; Beckmann, R.; Beatrix, B. Dual binding mode of the nascent polypeptide-associated complex reveals a novel universal adapter site on the ribosome. *J. Biol. Chem.* **2010**, *285*, 19679–19687. [CrossRef] [PubMed]
62. Gamerdinger, M.; Kobayashi, K.; Wallisch, A.; Kreft, S.G.; Sailer, C.; Schlomer, R.; Sachs, N.; Jomaa, A.; Stengel, F.; Ban, N.; et al. Early Scanning of Nascent Polypeptides inside the Ribosomal Tunnel by NAC. *Mol. Cell* **2019**, *75*, 996–1006. [CrossRef] [PubMed]
63. Wegrzyn, R.D.; Deuerling, E. Molecular guardians for newborn proteins: Ribosome-associated chaperones and their role in protein folding. *Cell. Mol. Life Sci.* **2005**, *62*, 2727–2738. [CrossRef]
64. Balchin, D.; Hayer-Hartl, M.; Hartl, F.U. In Vivo aspects of protein folding and quality control. *Science* **2016**, *353*, aac4354. [CrossRef] [PubMed]
65. Arsenovic, P.T.; Maldonado, A.T.; Colleluori, V.D.; Bloss, T.A. Depletion of the *C. elegans* NAC engages the unfolded protein response, resulting in increased chaperone expression and apoptosis. *PLoS ONE* **2012**, *7*, e44038. [CrossRef]
66. Deng, J.M.; Behringer, R.R. An insertional mutation in the BTF3 transcription factor gene leads to an early postimplantation lethality in mice. *Transgenic Res.* **1995**, *4*, 264–269. [CrossRef]
67. Bloss, T.A.; Witze, E.S.; Rothman, J.H. Suppression of CED-3-independent apoptosis by mitochondrial betaNAC in Caeno-rhabditis elegans. *Nature* **2003**, *424*, 1066–1071. [CrossRef]
68. Markesich, D.C.; Gajewski, K.M.; Nazimiec, M.E.; Beckingham, K. bicaudal encodes the Drosophila beta NAC homolog, a component of the ribosomal translational machinery*. *Development* **2000**, *127*, 559–572. [CrossRef]

69. Koplin, A.; Preissler, S.; Ilina, Y.; Koch, M.; Scior, A.; Erhardt, M.; Deuerling, E. A dual function for chaperones SSB-RAC and the NAC nascent polypeptide-associated complex on ribosomes. *J. Cell Biol.* **2010**, *189*, 57–68. [CrossRef]
70. Ott, A.K.; Locher, L.; Koch, M.; Deuerling, E. Functional Dissection of the Nascent Polypeptide-Associated Complex in Saccharomyces cerevisiae. *PLoS ONE* **2015**, *10*, e0143457. [CrossRef]
71. Shen, K.; Gamerdinger, M.; Chan, R.; Gense, K.; Martin, E.M.; Sachs, N.; Knight, P.D.; Schlömer, R.; Calabrese, A.N.; Stewart, K.L.; et al. Dual Role of Ribosome-Binding Domain of NAC as a Potent Suppressor of Protein Aggregation and Aging-Related Proteinopathies. *Mol. Cell* **2019**, *74*, 729–741. [CrossRef]
72. Martin, E.M.; Jackson, M.P.; Gamerdinger, M.; Gense, K.; Karamonos, T.K.; Humes, J.R.; Deuerling, E.; Ashcroft, A.E.; Radford, S.E. Conformational flexibility within the nascent polypeptide-associated complex enables its interactions with structurally diverse client proteins. *J. Biol. Chem.* **2018**, *293*, 8554–8568. [CrossRef] [PubMed]
73. Kirstein-Miles, J.; Scior, A.; Deuerling, E.; Morimoto, R.I. The nascent polypeptide-associated complex is a key regulator of pro-teostasis. *EMBO J.* **2013**, *32*, 1451–1468. [CrossRef] [PubMed]
74. Funfschilling, U.; Rospert, S. Nascent polypeptide-associated complex stimulates protein import into yeast mitochondria. *Mol. Biol. Cell* **1999**, *10*, 3289–3299. [CrossRef] [PubMed]
75. Beddoe, T.; Lithgow, T. Delivery of nascent polypeptides to the mitochondrial surface. *Biochim. Biophys. Acta.* **2002**, *1592*, 35–49. [CrossRef]
76. Yogev, O.; Karniely, S.; Pines, O. Translation-coupled translocation of yeast fumarase into mitochondria in Vivo. *J. Biol. Chem.* **2007**, *282*, 29222–29229. [CrossRef]
77. Ponce-Rojas, J.C.; Avendano-Monsalve, M.C.; Yanez-Falcon, A.R.; Jaimes-Miranda, F.; Garay, E.; Torres-Quiroz, F.; DeLuna, A.; Funes, S. αβ′-NAC cooperates with Sam37 to mediate early stages of mitochondrial protein import. *FEBS J.* **2017**, *284*, 814–830. [CrossRef]
78. Yotov, W.V.; Moreau, A.; St-Arnaud, R. The alpha chain of the nascent polypeptide-associated complex functions as a transcriptional coactivator. *Mol. Cell. Biol.* **1998**, *18*, 1303–1311. [CrossRef]
79. Quelo, I.; Hurtubise, M.; St-Arnaud, R. alphaNAC requires an interaction with c-Jun to exert its transcriptional coactivation. *Gene Expr.* **2002**, *10*, 255–262. [CrossRef]
80. Wiedmann, B.; Sakai, H.; Davis, T.A.; Wiedmann, M. A protein complex required for signal-sequence-specific sorting and translocation. *Nature* **1994**, *370*, 434–440. [CrossRef]
81. Lauring, B.; Sakai, H.; Kreibich, G.; Wiedmann, M. Nascent polypeptide-associated complex protein prevents mistargeting of nascent chains to the endoplasmic reticulum. *Proc. Natl. Acad. Sci. USA* **1995**, *92*, 5411–5415. [CrossRef]
82. Powers, T.; Walter, P. The nascent polypeptide-associated complex modulates interactions between the signal recognition particle and the ribosome. *Curr. Biol.* **1996**, *6*, 331–338. [CrossRef]
83. Wiedmann, B.; Prehn, S. The nascent polypeptide-associated complex (NAC) of yeast functions in the targeting process of ribosomes to the ER membrane. *FEBS Lett.* **1999**, *458*, 51–54. [CrossRef]
84. Raden, D.; Gilmore, R. Signal recognition particle-dependent targeting of ribosomes to the rough endoplasmic reticulum in the absence and presence of the nascent polypeptide-associated complex. *Mol. Biol. Cell* **1998**, *9*, 117–130. [CrossRef]
85. Neuhof, A.; Rolls, M.M.; Jungnickel, B.; Kalies, K.U.; Rapoport, T.A. Binding of signal recognition particle gives ribosome/nascent chain complexes a competitive advantage in endoplasmic reticulum membrane interaction. *Mol. Biol. Cell* **1998**, *9*, 103–115. [CrossRef]
86. Lauring, B.; Kreibich, G.; Wiedmann, M. The intrinsic ability of ribosomes to bind to endoplasmic reticulum membranes is regulated by signal recognition particle and nascent-polypeptide-associated complex. *Proc. Natl. Acad. Sci. USA* **1995**, *92*, 9435–9439. [CrossRef]
87. Zhang, Y.; Berndt, U.; Golz, H.; Tais, A.; Oellerer, S.; Wolfle, T.; Fitzke, E.; Rospert, S. NAC functions as a modulator of SRP during the early steps of protein targeting to the endoplasmic reticulum. *Mol. Biol. Cell* **2012**, *16*, 3027–3040. [CrossRef]
88. Del Alamo, M.; Hogan, D.J.; Pechmann, S.; Albanese, V.; Brown, P.O.; Frydman, J. Defining the specificity of cotranslationally acting chaperones by systematic analysis of mRNAs associated with ribosome-nascent chain complexes. *PLoS Biol.* **2011**, *9*, e1001100.
89. Wang, S.; Sakai, H.; Wiedmann, M. NAC covers ribosome-associated nascent chains thereby forming a protective envi-ronment for regions of nascent chains just emerging from the peptidyl transferase center. *J. Cell Biol.* **1995**, *130*, 519–528. [CrossRef] [PubMed]
90. Gamerdinger, M.; Hanebuth, M.A.; Frickey, T.; Deuerling, E. The principle of antagonism ensures protein targeting specificity at the endoplasmic reticulum. *Science* **2015**, *348*, 201–207. [CrossRef] [PubMed]
91. Ariosa, A.; Lee, J.H.; Wang, S.; Saraogi, I.; Shan, S.O. Regulation by a chaperone improves substrate selectivity during cotranslational protein targeting. *Proc. Natl. Acad. Sci. USA* **2015**, *112*, E3169–E3178. [CrossRef]
92. Yang, C.I.; Hsieh, H.H.; Shan, S.O. Timing and specificity of cotranslational nascent protein modification in bacteria. *Proc. Natl. Acad. Sci. USA* **2019**, *116*, 23050–23060. [CrossRef] [PubMed]
93. Schuldiner, M.; Metz, J.; Schmid, V.; Denic, V.; Rakwalska, M.; Schmitt, H.D.; Schwappach, B.; Weissman, J.S. The GET complex mediates insertion of tail-anchored proteins into the ER membrane. *Cell* **2008**, *134*, 634–645. [CrossRef]

94. Walter, P.; Blobel, G. Translocation of proteins across the endoplasmic reticulum III. Signal recognition protein (SRP) causes signal sequence-dependent and site-specific arrest of chain elongation that is released by microsomal membranes. *J. Cell Biol.* **1981**, *91*, 557–561. [CrossRef]
95. Siegel, V.; Walter, P. Elongation arrest is not a prerequisite for secretory protein translocation across the microsomal membrane. *J. Cell Biol.* **1985**, *100*, 1913–1921. [CrossRef] [PubMed]
96. Wolin, S.L.; Walter, P. Signal recognition particle mediates a transient elongation arrest of preprolactin in reticulocyte lysate. *J. Cell Biol.* **1989**, *109*, 2617–2622. [CrossRef] [PubMed]
97. Halic, M.; Becker, T.; Pool, M.R.; Spahn, C.M.T.; Grassucci, R.A.; Frank, J.; Beckmann, R. Structure of the signal recognition particle interacting with the elongation-arrested ribosome. *Nature* **2004**, *427*, 808–814. [CrossRef] [PubMed]
98. Mary, C.; Scherrer, A.; Huck, L.; Lakkaraju, A.K.; Thomas, Y.; Johnson, A.E.; Strub, K. Residues in SRP9/14 essential for elongation arrest activity of the signal recognition particle define a positively charged functional domain on one side of the protein. *RNA* **2010**, *16*, 969–979. [CrossRef]
99. Lakkaraju, A.K.K.; Mary, C.; Scherrer, A.; Johnson, A.E.; Strub, K. SRP keeps polypeptides translocation-competent by slowing translation to match limiting ER-targeting sites. *Cell* **2008**, *133*, 440–451. [CrossRef] [PubMed]
100. Pechmann, S.; Chartron, J.W.; Frydman, J. Local slowdown of translation by nonoptimal codons promotes nascent-chain recognition by SRP In Vivo. *Nat. Struct. Mol. Biol.* **2014**, *21*, 1100–1105. [CrossRef]
101. Chartron, J.W.; Hunt, K.C.; Frydman, J. Cotranslational signal-independent SRP preloading during membrane targeting. *Nature* **2016**, *536*, 224–248. [CrossRef] [PubMed]
102. Jungnickel, B.; Rapoport, T.A. A posttargeting signal sequence recognition event in the endoplasmic reticulum membrane. *Cell* **1995**, *82*, 261–270. [CrossRef]
103. Gorlich, D.; Rapoport, T.A. Protein translocation into proteoliposomes reconstituted from purified compoennts of the endoplasmic reticulum membrane. *Cell* **1993**, *75*, 615–630. [CrossRef]
104. Voorhees, R.M.; Hegde, R.S. Structure of the Sec61 channel opened by a signal sequence. *Science* **2016**, *351*, 88–91. [CrossRef] [PubMed]
105. Okreglak, V.; Walter, P. The conserved AAA-ATPase Msp1 confers organelle specificity to tail-anchored proteins. *Proc. Natl. Acad. Sci. USA* **2014**, *111*, 8019–8024. [CrossRef]
106. Li, L.; Zheng, J.; Wu, X.; Jiang, H. Mitochondrial AAA-ATPase Msp1 detects mislocalized tail-anchored proteins through a du-al-recognition mechanism. *EMBO Rep.* **2019**, *20*, e46989. [CrossRef]
107. Chen, Y.C.; Umanah, G.K.; Dephoure, N.; Andrabi, S.A.; Gygi, S.P.; Dawson, T.M.; Dawson, V.L.; Rutter, J. Msp1/ATAD1 maintains mitochondrial function by facilitating the degradation of mislocalized tail-anchored proteins. *EMBO J.* **2014**, *33*, 1548–1564. [CrossRef] [PubMed]
108. Wang, L.; Walter, P. Msp1/ATAD1 in Protein Quality Control and Regulation of Synaptic Activities. *Annu. Rev. Cell Dev. Biol.* **2020**, *36*, 141–164. [CrossRef]
109. Wang, L.; Myasnikov, A.; Pan, X.; Walter, P. Structure of the AAA protein Msp1 reveals mechanism of mislocalized membrane protein extraction. *eLife* **2020**, *9*, e54031. [CrossRef]
110. Matsumoto, S.; Nakatsukasa, K.; Kakuta, C.; Tamura, Y.; Esaki, M.; Endo, T. Msp1 Clears Mistargeted Proteins by Facilitating Their Transfer from Mitochondria to the ER. *Mol. Cell* **2019**, *76*, 191–205.e10. [CrossRef]
111. McKenna, M.J.; Sim, S.I.; Ordureau, A.; Wei, L.; Harper, J.W.; Shao, S.; Park, E. The endoplasmic reticulum P5A-ATPase is a trans-membrane helix dislocase. *Science* **2020**, *369*. [CrossRef] [PubMed]
112. Dederer, V.; Lemberg, M.K. Transmembrane dislocases: A second chance for protein targeting. *Trends Cell Biol.* **2021**, *31*, 898–911. [CrossRef] [PubMed]

International Journal of
Molecular Sciences

MDPI

Review

ER-SURF: Riding the Endoplasmic Reticulum Surface to Mitochondria

Christian Koch [1], Maya Schuldiner [2] and Johannes M. Herrmann [1,*]

[1] Cell Biology, University of Kaiserslautern, 67663 Kaiserslautern, Germany; c_koch@rhrk.uni-kl.de
[2] Department of Molecular Genetics, Weizmann Institute of Science, Rehovot 7610001, Israel; maya.schuldiner@weizmann.ac.il
* Correspondence: hannes.herrmann@biologie.uni-kl.de

Abstract: Most mitochondrial proteins are synthesized in the cytosol and targeted to the mitochondrial surface in a post-translational manner. The surface of the endoplasmic reticulum (ER) plays an active role in this targeting reaction. ER-associated chaperones interact with certain mitochondrial membrane protein precursors and transfer them onto receptor proteins of the mitochondrial surface in a process termed ER-SURF. ATP-driven proteins in the membranes of mitochondria (Msp1, ATAD1) and the ER (Spf1, P5A-ATPase) serve as extractors for the removal of mislocalized proteins. If the re-routing to mitochondria fails, precursors can be degraded by ER or mitochondria-associated degradation (ERAD or MAD respectively) in a proteasome-mediated reaction. This review summarizes the current knowledge about the cooperation of the ER and mitochondria in the targeting and quality control of mitochondrial precursor proteins.

Keywords: chaperones; contact sites; endoplasmic reticulum; ER-SURF; membrane extraction; mitochondria; protein targeting

Citation: Koch, C.; Schuldiner, M.; Herrmann, J.M. ER-SURF: Riding the Endoplasmic Reticulum Surface to Mitochondria. *Int. J. Mol. Sci.* **2021**, 22, 9655. https://doi.org/10.3390/ijms22179655

Academic Editor: Alessandra Ferramosca

Received: 25 August 2021
Accepted: 29 August 2021
Published: 6 September 2021

Publisher's Note: MDPI stays neutral with regard to jurisdictional claims in published maps and institutional affiliations.

1. Introduction

It is the hallmark of eukaryotic cells that intracellular membranes define multiple functionally different compartments. As a consequence, many, in some cell types even most, proteins that are initially synthesized in the cytosol have to leave the cytosol to reach another cellular compartment [1–3]. Thus, eukaryotic cells face the challenge to specifically direct thousands of proteins to their respective position and, equally important, to remove those proteins that get stranded at foreign and inappropriate locations. While localization signals, targeting factors, receptors, and translocases for many of the residents of the different organelles were identified, we are only starting to unravel how chaperones, proteases, retro-translocases, extractors, and other "correction factors" marshal and proofread the sorting of proteins to ensure well-defined proteomes and, hence, functional cellular compartments. As if this disorder was not complicated enough, recent studies suggest that the surfaces of different organelles actively cooperate in the sorting, the targeting and the clean-up of translocation intermediates on the passage to their final residence. In particular, the relevance of the endoplasmic reticulum (ER) as the professional cellular sorting station is not restricted to proteins that enter the secretory pathway, but also supports nascent proteins destined to mitochondria, peroxisomes, lipid droplets, and chloroplasts [4–10]. In this review, we will provide an overview of the role of the ER in targeting and degradation of mitochondrial precursor proteins.

2. Targeting and Translocation of Mitochondrial Proteins

Mitochondria contain a small genome coding for a handful of proteins, most of which represent hydrophobic core subunits of the respiratory chain; these proteins are presumably difficult to import from the cytosol, and their expression in mitochondria allows organelle-controlled synthesis [11–13]. The vast majority of mitochondrial proteins, many hundreds to thousands, are encoded in the nucleus, synthesized in the cytosol and subsequently targeted and imported into the organelle (for review see [14]).

Most of these proteins are synthesized with N-terminal presequences that serve as a matrix targeting signal (MTS) [15,16]. MTSs are amphipathic helices with one positively charged and one hydrophobic surface [17]. They are recognized at the mitochondrial outer membrane by three receptor proteins, Tom20, Tom22, and Tom70 that are part of the TOM (translocase of the outer membrane) complex. However, the association of Tom70 with the TOM complex is presumably dynamic and short-lived, potentially to gather the substrates for the import pore [18–20]. The three receptors, which differ in their exact substrate preference, pass precursor proteins in a cooperative manner to the protein-conducting channel formed by the β-barrel protein Tom40. Tom40 serves as a general entry gate for all proteins destined to the matrix, the inner membrane, the intermembrane space (IMS), and for many outer membrane proteins. After translocation through Tom40, precursor proteins are sorted to their respective submitochondrial localizations (Figure 1): Matrix and many inner membrane proteins are directed to the TIM23 (translocase of the inner membrane) complex which threads presequence-containing proteins through the inner membrane. The transfer into the matrix is promoted by the PAM (presequence translocase-associated motor) machinery by hydrolysis of matrix ATP [21] and is followed by the proteolytic removal of the presequence mediated by the mitochondrial processing peptidase, MPP.

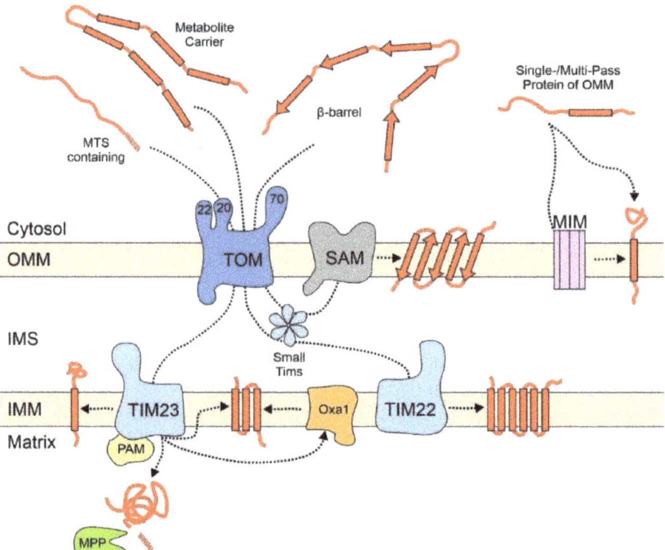

Figure 1. Different groups of mitochondrial proteins embark on different import pathways. Proteins of the matrix and many inner membrane proteins are synthesized as precursor proteins with N-terminal matrix targeting signals (MTSs) and imported via the TOM and TIM23 complexes. The PAM complex serves as motor for their translocation reaction. The mitochondrial processing peptidase (MPP) removes the MTS of most of these proteins. Metabolite carriers lack presequences and are integrated into the inner membrane by the TIM22 complex. The SAM complex integrates β-barrel proteins into the outer membrane. Many outer membrane proteins with helical transmembrane domains bypass the TOM complex but can be dependent on the MIM complex. IMM, inner mitochondrial membrane; IMS, intermembrane space; OMM, outer mitochondrial membrane.

Carrier proteins (also referred to as the SLC25A or metabolite carrier family) are highly abundant proteins of the mitochondrial inner membrane that mediate the exchange of ATP and metabolites between mitochondria and the cytosol. They usually contain six transmembrane domains and do not contain an MTS but use internal targeting signals that are scattered across their sequence [22]. These types of proteins are first recognized on the mitochondrial surface by Tom70, which also tightly cooperates with the chaperone system of the cytosol [23,24]. Soluble chaperone complexes in the IMS, formed by small TIM proteins, and the TIM22 complex then integrate carrier proteins into the inner membrane [25,26].

Many proteins of the IMS lack presequences and their import relies on cysteine motifs [27]. The oxidoreductase Mia40 (CHCHD4 in humans) and the sulfhydryl oxidase Erv1 (ALR in human) drive the import reaction of these proteins [28,29].

In the outer membrane, β-barrel proteins form large pores that allow the facilitated diffusion of molecules up to a mass of several kDa. Precursors of β-barrel proteins are first recognized at the TOM complex by their β-hairpin element. These β-barrel proteins are imported through the TOM pore and inserted into the outer membrane by the SAM (sorting and assembly) machinery in a reaction that is conserved between bacteria and mitochondria [30,31].

Tail-anchored (TA) outer membrane proteins, that harbor only a single transmembrane domain (TMD) in their very C terminus, bypass the pore in the TOM complex. The mechanisms and factors that promote their insertion into the outer membrane are still poorly defined [32].

The individual steps of these import reactions were elucidated by use of very powerful in vitro import assays for which radiolabeled precursor proteins were mixed with isolated mitochondria. Whereas this approach is very well suited to study protein translocation across the mitochondrial membranes, it does not reveal the initial targeting process that occurs before precursors reach the mitochondrial surface receptors.

3. Productive Targeting via the ER Surface: ER-SURF

In the context of protein biogenesis, protein targeting (the passage of nascent precursors from the ribosome to the mitochondrial surface) has to be distinguished from protein translocation (which refers to the threading of precursors through mitochondrial protein translocases) [33]. Whereas mitochondrial protein translocation was studied extensively over the last two decades, mitochondrial protein targeting is by far less understood. Many recent studies documented the general importance of the cytosolic chaperone network and the ubiquitin-proteasome-system (UPS). However, we know very little about which chaperones and which ubiquitin ligases interact with which types of precursors and how their directional movement to and subsequent release from the mitochondrial membrane is mediated. Furthermore, it is unclear whether these quality control systems usher every single precursor protein along its way to the mitochondrial surface or whether they only deal with the fraction of stranded or structurally compromised precursor proteins. Thus, the early reactions of mitochondrial protein biogenesis still await to be discovered. Several recent reviews discussed these issues in depth [33–35]. In this article, we therefore specifically focus on the relevance of the ER surface for mitochondrial protein biogenesis.

A large fraction of all ribosomes is bound to the surface of the ER. Ribosome-bound nascent chains that expose highly hydrophobic signal sequences or transmembrane domains are recognized by the signal recognition particle (SRP) and recruited to the ER surface. Many mitochondrial proteins contain transmembrane domains, and therefore, it is not surprising that mitochondrial membrane proteins were also identified among the SRP clients [36]. For example, transmembrane domains of the inner membrane proteins Oxa1 and Psd1 were found to be recognized by the SRP. Both proteins are synthesized with presequences and use the TIM23-mediated import pathway [37,38]. However, presumably due to the lower hydrophobicity of transmembrane domains in mitochondrial membrane proteins [39,40], the SRP is able to discriminate between secretory proteins and most mitochondrial proteins. A recent study even proposed that the major mission of the SRP system is the reliable distinction of these two large groups of cellular proteins [41]. Cotranslational binders of the nascent chain, such as Ssb1/2 chaperones and the nascent chain-associated complex (NAC), fine-tune the SRP-mediated discrimination process [42,43], which might contribute to their observed relevance for mitochondrial protein biogenesis [44–48]. Interestingly, the factors of the guided entry of tail-anchored proteins (GET) pathway [49] also play a role in mitochondrial protein targeting. Get3, a cytosolic chaperone and targeting factor, was found to directly interact with some mitochondrial precursors, in addition to its ER-destined clients [50]. Moreover, if mitochondrial precursor proteins accumulate in the cytosol, they can be "rescued" by the GET pathway, which directs them onto the ER surface from where they finally reach the mitochondrial import machinery; this GET-mediated detour might be particularly relevant for carrier proteins and prevents their incorporation into non-productive protein aggregates [51].

Thus, mitochondrial proteins might initially find themselves stranded at the ER due to "mis-localization" if the SRP and cellular quality control components are not effective enough in recognizing them. However, alternatively and not mutually exclusive, some mitochondrial proteins might deliberately associate with the ER surface or be synthesized by ER-bound ribosomes, in line with a considerable number of nascent mitochondrial proteins that were observed on the ER by ribosome profiling experiments [52,53]. Regardless of whether ER targeting of nascent mitochondrial proteins is an active, intentional mechanism or a cellular mistake, cells have evolved a pathway to help such precursors make it to their final destination. This stopover-mediated targeting route to mitochondria via the ER surface was termed the ER-SURF pathway (Figure 2) and the ER-bound J protein Djp1 was identified as a component that increased the efficiency of this import route [5,54]. The ER-SURF model proposes that the ER surface acts as an antenna that helps to funnel precursors to mitochondria. Consistent with this idea, in in vitro experiments the addition of ER fractions increases the import rate of proteins into isolated mitochondria [5,55], an observation that can hardly be reconciled with the idea that ER binding is synonymous with non-productive mis-localization.

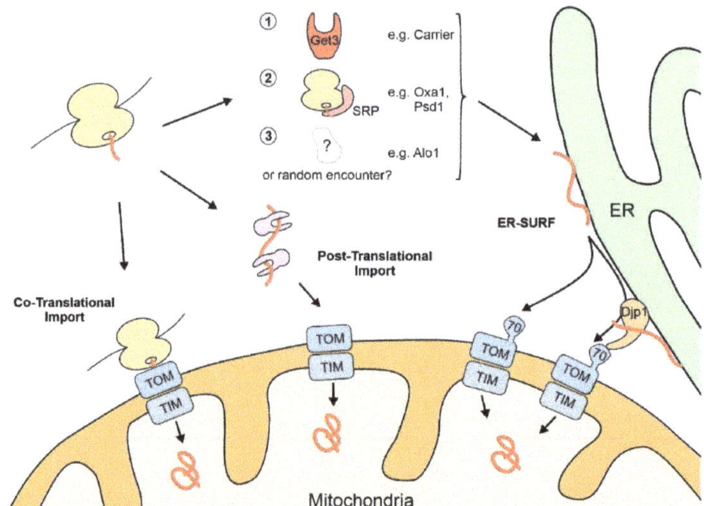

Figure 2. The surface of the ER facilitates mitochondrial targeting of proteins that use the ER-SURF pathway. Precursor proteins can reach mitochondria during, or after, their synthesis on cytosolic ribosomes. Some mitochondrial precursors are directed to the ER surface. For example, metabolite carriers were observed to be bound by Get3, a chaperone that facilitates ER-targeting of TA proteins. Some mitochondrial membrane proteins, such as Oxa1 and Psd1, are recognized by the SRP. For other mitochondrial proteins, such as Alo1, targeting factors were not identified so far. Djp1 is an ER-associated protein that facilitates the transfer of ER-bound proteins to mitochondria.

When mitochondrial import sites are limiting so that precursor proteins accumulate outside of mitochondria, a large number of precursors of mitochondrial membrane proteins were found to associate with the ER surface [56] and to induce the unfolded protein response pathway of the ER [57]. Since these precursor proteins, in particular those of the carriers, have a highly toxic potential [24,58,59], ER binding might serve as a safeguard mechanism [51]. This is because the ER surface is coated by a number of cellular chaperones. For example, Ydj1, the most abundant DnaJ-type co-chaperone of yeast cells, is tethered to the ER surface by a farnesyl anchor [60]. The relevance of Ydj1 for mitochondrial protein biogenesis is well documented [61–63], but it is not known at which intracellular localization Ydj1 "meets" mitochondrial precursors.

Djp1 forms a complex with Tom70 [64]. It is not clear whether the Djp1 species that partners Tom70 is bound to the ER. However, the ability of Tom70 to form mitochondria-ER contact sites is well established [20,65,66]. It still needs to be elucidated whether Djp1 hands over precursors from the ER to the mitochondrial surface via direct contacts or, alternatively, releases them to the cytosolic chaperone system.

A recent study in mammalian cells showed that Bcl2 can be transferred from the ER to the mitochondrial surface in a Tom20-mediated reaction that occurs at ER-mitochondria contact sites dubbed mitochondria-associated membranes (MAMs) [67]. This indicates that the ER-SURF pathway also exists in mammalian cells and that the close collaboration of the ER and mitochondria in protein sorting is a conserved feature of eukaryotic cells.

4. Contact Sites of the ER and Mitochondria

A function for the close cooperation of ER and mitochondria via membrane contact sites was identified three decades ago when MAMs were first purified [68,69]. The MAM was described as a membranous fraction that was inseparable even from highly purified mitochondria and found to play a role in phospholipid biosynthesis. However, molecular tethers that connect mitochondria to the ER remained unclear until the ER-MES (ER mitochondria encounter structure) complex was identified in budding yeast in 2009 [70]. This complex consists of the four structural components Mmm1, Mdm12, Mdm34, and Mdm10, which form a chain-like bridge holding the two organelles in close proximity [70–73]. ERMES is mainly involved in the transfer of phophatidylserine from the ER to mitochondria where it is converted to phosphatidylethanolamine. Deletion of any ERMES component leads to a decrease in the rate of phosphatidylethanolamine synthesis and the overall levels of cardiolipin in mitochondria and to a collapse of the mitochondrial network [74–78]. Moreover, ERMES promotes the formation of mitochondria-derived compartments (MDCs) [6], defines the position of intra-mitochondrial complexes such as nucleoids, the MICOS (mitochondrial contact site and cristae organizing center), and the coenzyme Q synthome [78–80].

In addition to the ERMES complex, at least two further tethering complexes connect the ER with mitochondria in yeast: One tether is formed by the ER-resident sterol transporter Lam6/Ltc1 and Tom70 [65,66], the other was proposed to form by the ER membrane complex (EMC) and Tom5 [81].

In mammalian cells, multiple tethering molecules have been suggested to act at contact sites between the ER and mitochondria [82]. For example, a single recent split proximity labeling approach proposed that 30 proteins are enriched in ER mitochondrial contact zones [83], but the individual functions of these factors still need to be elucidated; however, there is no doubt that these contacts are highly important for cellular functionality [84,85]. Thus, the ER and mitochondria form entangled intracellular networks that are connected by several specific tethering complexes. These contact sites ensure a close proximity of the ER and mitochondrial membranes and thereby support the biogenesis of (membrane) proteins and lipids.

5. Destructive Targeting via ERAD and MAD

Proteasomal degradation of ER and mitochondrial proteins (Figure 3) is often summarized under the umbrella terms ERAD (for ER-associated degradation) and MAD (for mitochondria-associated degradation) [86–89]. The components and underlying mechanisms of ERAD are rather well understood, and even structures of the ERAD machinery were recently published [90]. In contrast, the puzzle pieces of MAD are only in the process of being collected and therefore a comprehensive, generally accepted picture still has to emerge.

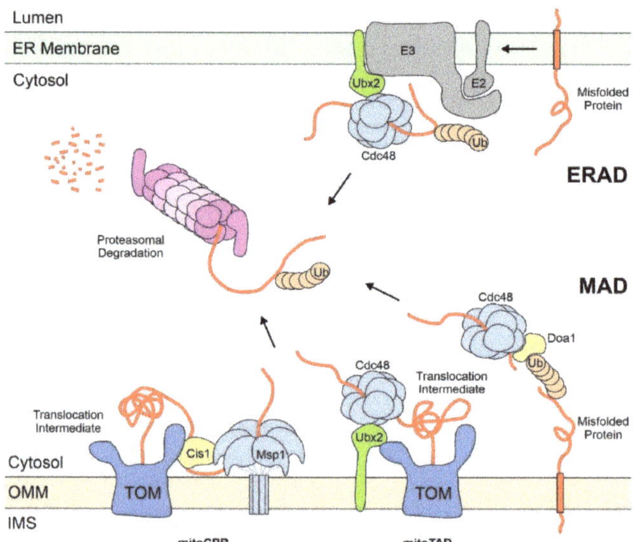

Figure 3. Proteasomal degradation of proteins at the ER and the mitochondrial surface. ER and mitochondrial proteins are released by the AAA proteins Cdc48/p97 (on both ER and mitochondria) or Msp1 (on mitochondria) into the cytosol to be degraded by the proteasome. Poly-ubiquitin chains serve as degradation signals and as handles on the proteins for unfolding and insertion into the proteasome. Adaptor proteins such as Ubx2 or Doa1 play crucial role in the substrate binding of AAA proteins. Stalled translocation intermediates induce the recruitment of Msp1 to the TOM complex by Cis1 in a process called mitochondrial compromised protein import response (mitoCPR). MAD of translocation intermediates is also referred to as mitochondrial protein translocation-associated degradation (mitoTAD).

The UPS plays a general role in the degradation of non-functional or damaged outer membrane proteins. Mitofusins (Fzo1 in yeast) were among the first mitochondrial proteasome substrates that were identified and the proteasomal turnover of these fusion factors is crucial for mitochondrial morphogenesis [91–93]. Surprisingly, many of the factors found to carry out MAD were initially established as ERAD factors. Both processes share the AAA unfoldase Cdc48/p97 and the adaptor proteins Ubx2 and Doa1/Ufd3 [94–97]. For example, Ubx2, a well-established ER-embedded ERAD factor, also forms a pool on the outer membrane of mitochondria. It monitors the TOM complex and in case of accumulating translocation intermediates, Ubx2 recruits Cdc48 to prevent clogging of mitochondrial import sites. This Ubx2-mediated subcategory of MAD, which is important to cleanse stalled translocation intermediates from the TOM complex, was called mitoTAD for mitochondrial protein translocation-associated degradation [98]. Doa1/Ufd3 might play an Ubx2-equivalent role in the degradation of outer membrane proteins [94–97].

At least in mammalian cells, ubiquitination might even regulate and fine-tune the import through the TOM complex: precursors are ubiquinated by the ubiquitin ligase

March5 (also called MITOL) before ubiquitin is removed by the deubiquitinase USP30. If deubiquitination does not keep pace with ubiquitination, translocation intermediates are arrested in the TOM complex, resulting in their degradation [99]. The USP30-mediated deubiquitination of translocation intermediates is also relevant to prevent mitophagy via activation of the ubiquitin ligase Parkin and the protein kinase PINK1 [100].

6. The Role of Membrane Extractors

The degradation of membrane proteins and translocation intermediates by ERAD and MAD requires their extraction and presentation to the proteasome. The cytosolic AAA protein Cdc48/p97 serves as such an extractor: this hexamer consists of two stacked ATPase rings that pull substrate proteins through the central cavity, thereby generating the force required to dislodge membrane proteins [101]. As mentioned before, different adaptor proteins, such as Ubx2 and Doa1/Ufd3, facilitate the binding of ubiquitinated substrates to Cdc48/p97.

The outer membrane contains an additional AAA protein for Cdc48/p97-independent extraction. This protein complex is called Msp1 in yeast and ATAD1 in mammalian cells, shares the overall hexameric organization with Cdc48/p97 and uses a comparable mechanism for protein extraction (Figure 4). Msp1 was initially found as the dislocase for peroxisomal and ER TA proteins that were aberrantly integrated into the mitochondrial outer membrane [102,103]. After extraction these TA proteins are either sent for degradation or passed on to their cognate target membrane [8,104]. Msp1 also extracts translocation intermediates that get stuck in the import tunnel, a function for which it requires being specifically recruited to the TOM complex by the adaptor Cis1 [105,106]. Interestingly, proteins extracted from mitochondria by Msp1 can be targeted to the ER surface where they are turned over by ERAD. Once again, this emphasizes the close alliance of the ER and mitochondria during protein biogenesis.

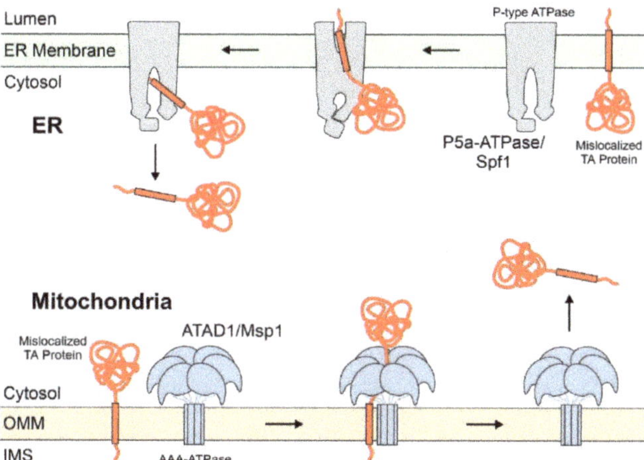

Figure 4. ATP-driven membrane extractors give mis-localized TA proteins a second chance to find their correct location. The ATAD1/Msp1 complex is a hexameric AAA dislocase on the outer membrane of mitochondria that recognizes non-mitochondrial TA proteins as well as translocation intermediates stalled in the TOM complex. It removes these proteins, which then either find their respective target membrane or are degraded, for example, via ERAD. An analogous extraction system exists on the ER membrane, where the P-type ATPase P5a-ATPase (also called ATP13A1 or CATP-8) recognizes and dislocates membrane proteins destined to mitochondria.

A membrane-bound extractor is also found on the ER membrane; this protein is called Spf1 in yeast or P5A-ATPase in mammals [107,108]. It safeguards the ER by dislocation of mis-localized mitochondrial proteins, analogous to the function of Msp1/ATAD1 on mitochondria. A potential role of Spf1/P5A-ATPase in ER-SURF still has to be elucidated.

7. ER-Mitochondria Contact Zones as Protein Nurseries

While the ER has been known for 30 years to serve as a general sorting station for proteins of the secretory pathway, the biogenesis of proteins of mitochondria, chloroplasts, and, to some degree, peroxisomes and lipid droplets was traditionally regarded as processes that occur independently from the ER. This assumption was fueled by the observation that, in vitro, isolated proteins can be efficiently imported into mitochondria, chloroplasts, and peroxisomes in a post-translational reaction. The success of high-resolution light microscopy and cryo-electron microscopy revealed fascinating insights into the interplay of the different organelles, and many recent studies discovered the close collaborations between the intracellular networks formed by the ER and by mitochondria. Examples for such cooperations of the ER and mitochondria include the control of mitochondrial fusion and fission [109–112], the positioning of genomes within mitochondria [78], the formation of isolation membranes for autophagy [113–115], or the transfer of lipids [70,72]. It therefore is no surprise that ER mitochondria contact sites are also relevant for protein targeting reactions and that the surfaces of the different membranes closely collaborate for protein sorting.

We are only beginning to appreciate the dynamic interplay of organellar surfaces in the context of protein biogenesis. These inter-organellar interactions are presumably particularly relevant in the context of the large number of dually localized proteins [116], such as Psd1 [117], DAKAP1 [118] or NADH-cytochrome b_5 reductase [119].

Thus, the contact zone between mitochondria and the ER apparently serves as a nursery where nascent proteins, under surveillance of cytosolic chaperones, and the quality control factors of MAD and ERAD, find their appropriate destination membrane. It will be exciting to further explore how eukaryotic cells orchestrate these biogenesis hotspots to avoid an unproductive chaotic jumble.

Author Contributions: C.K., M.S., J.M.H. jointly wrote the manuscript. All authors have read and agreed to the published version of the manuscript.

Funding: We are grateful to the Deutsche Forschungsgemeinschaft for funding (DIP MitoBalance to M.S. and J.M.H., grant number 323127228). The work in this manuscript is part of a project in the M.S. lab that has received funding from the European Research Council (ERC) under the European Union's Horizon 2020 research and innovation programme (EU-H2020-ERC-CoG; grant name OnTarget, grant number 864068). M.S. is an incumbent of the Dr. Gilbert Omenn and Martha Darling Professorial Chair in Molecular Genetics.

References

1. Weill, U.; Yofe, I.; Sass, E.; Stynen, B.; Davidi, D.; Natarajan, J.; Ben-Menachem, R.; Avihou, Z.; Goldman, O.; Harpaz, N.; et al. Genome-wide SWAp-Tag yeast libraries for proteome exploration. *Nat. Methods* **2018**, *15*, 617–622. [CrossRef]
2. Foster, L.J.; de Hoog, C.L.; Zhang, Y.; Zhang, Y.; Xie, X.; Mootha, V.K.; Mann, M. A mammalian organelle map by protein correlation profiling. *Cell* **2006**, *125*, 187–199. [CrossRef]
3. Huh, W.K.; Falvo, J.V.; Gerke, L.C.; Carroll, A.S.; Howson, R.W.; Weissman, J.S.; O'Shea, E.K. Global analysis of protein localization in budding yeast. *Nature* **2003**, *425*, 686–691. [CrossRef] [PubMed]
4. Herker, E.; Vieyres, G.; Beller, M.; Krahmer, N.; Bohnert, M. Lipid Droplet Contact Sites in Health and Disease. *Trends Cell Biol.* **2021**, *31*, 345–358. [CrossRef] [PubMed]
5. Hansen, K.G.; Aviram, N.; Laborenz, J.; Bibi, C.; Meyer, M.; Spang, A.; Schuldiner, M.; Herrmann, J.M. An ER surface retrieval pathway safeguards the import of mitochondrial membrane proteins in yeast. *Science* **2018**, *361*, 1118–1122. [CrossRef] [PubMed]
6. English, A.M.; Schuler, M.H.; Xiao, T.; Kornmann, B.; Shaw, J.M.; Hughes, A.L. ER-mitochondria contacts promote mitochondrial-derived compartment biogenesis. *J. Cell Biol.* **2020**, *219*. [CrossRef]
7. Baune, M.C.; Lansing, H.; Fischer, K.; Meyer, T.; Charton, L.; Linka, N.; von Schaewen, A. The Arabidopsis Plastidial Glucose-6-Phosphate Transporter GPT1 is Dually Targeted to Peroxisomes via the Endoplasmic Reticulum. *Plant Cell* **2020**, *32*, 1703–1726. [CrossRef] [PubMed]

8. Dederer, V.; Khmelinskii, A.; Huhn, A.G.; Okreglak, V.; Knop, M.; Lemberg, M.K. Cooperation of mitochondrial and ER factors in quality control of tail-anchored proteins. *Elife* **2019**, *8*. [CrossRef]

9. Villarejo, A.; Buren, S.; Larsson, S.; Dejardin, A.; Monne, M.; Rudhe, C.; Karlsson, J.; Jansson, S.; Lerouge, P.; Rolland, N.; et al. Evidence for a protein transported through the secretory pathway en route to the higher plant chloroplast. *Nat. Cell Biol.* **2005**, *7*, 1224–1231. [CrossRef]

10. Radhamony, R.N.; Theg, S.M. Evidence for an ER to Golgi to chloroplast protein transport pathway. *Trends Cell Biol.* **2006**, *16*, 385–387. [CrossRef]

11. Bjorkholm, P.; Harish, A.; Hagstrom, E.; Ernst, A.M.; Andersson, S.G. Mitochondrial genomes are retained by selective constraints on protein targeting. *Proc. Natl. Acad. Sci. USA* **2015**, *112*, 10154–10161. [CrossRef] [PubMed]

12. Bertgen, L.; Muhlhaus, T.; Herrmann, J.M. Clingy genes: Why were genes for ribosomal proteins retained in many mitochondrial genomes? *Biochim. Biophys. Acta-Bioenerg.* **2020**, *1861*, 148275. [CrossRef] [PubMed]

13. Allen, J.F. Why chloroplasts and mitochondria retain their own genomes and genetic systems: Colocation for redox regulation of gene expression. *Proc. Natl. Acad. Sci. USA* **2015**, *112*, 10231–10238. [CrossRef] [PubMed]

14. Chacinska, A.; Koehler, C.M.; Milenkovic, D.; Lithgow, T.; Pfanner, N. Importing mitochondrial proteins: Machineries and mechanisms. *Cell* **2009**, *138*, 628–644. [CrossRef]

15. Vögtle, F.N.; Wortelkamp, S.; Zahedi, R.P.; Becker, D.; Leidhold, C.; Gevaert, K.; Kellermann, J.; Voos, W.; Sickmann, A.; Pfanner, N.; et al. Global analysis of the mitochondrial N-proteome identifies a processing peptidase critical for protein stability. *Cell* **2009**, *139*, 428–439. [CrossRef] [PubMed]

16. Garg, S.G.; Gould, S.B. The Role of Charge in Protein Targeting Evolution. *Trends Cell Biol.* **2016**, *26*, 894–905. [CrossRef]

17. von Heijne, G. Mitochondrial targeting sequences may form amphiphilic helices. *EMBO J.* **1986**, *5*, 1335–1342. [CrossRef]

18. Araiso, Y.; Tsutsumi, A.; Qiu, J.; Imai, K.; Shiota, T.; Song, J.; Lindau, C.; Wenz, L.S.; Sakaue, H.; Yunoki, K.; et al. Structure of the mitochondrial import gate reveals distinct preprotein paths. *Nature* **2019**, *575*, 395–401. [CrossRef]

19. Shiota, T.; Imai, K.; Qiu, J.; Hewitt, V.L.; Tan, K.; Shen, H.H.; Sakiyama, N.; Fukasawa, Y.; Hayat, S.; Kamiya, M.; et al. Molecular architecture of the active mitochondrial protein gate. *Science* **2015**, *349*, 1544–1548. [CrossRef]

20. Kreimendahl, S.; Rassow, J. The Mitochondrial Outer Membrane Protein Tom70-Mediator in Protein Traffic, Membrane Contact Sites and Innate Immunity. *Int. J. Mol. Sci.* **2020**, *21*, 7262. [CrossRef]

21. Mokranjac, D. How to get to the other side of the mitochondrial inner membrane—The protein import motor. *Biol. Chem.* **2020**, *401*, 723–736. [CrossRef]

22. Horten, P.; Colina-Tenorio, L.; Rampelt, H. Biogenesis of Mitochondrial Metabolite Carriers. *Biomolecules* **2020**, *10*, 1008. [CrossRef] [PubMed]

23. Young, J.C.; Hoogenraad, N.J.; Hartl, F.U. Molecular chaperones Hsp90 and Hsp70 deliver preproteins to the mitochondrial import receptor Tom70. *Cell* **2003**, *112*, 41–50. [CrossRef]

24. Backes, S.; Bykov, Y.S.; Flohr, T.; Raschle, M.; Zhou, J.; Lenhard, S.; Kramer, L.; Muhlhaus, T.; Bibi, C.; Jann, C.; et al. The chaperone-binding activity of the mitochondrial surface receptor Tom70 protects the cytosol against mitoprotein-induced stress. *Cell Rep.* **2021**, *35*, 108936. [CrossRef]

25. Sirrenberg, C.; Bauer, M.F.; Guiard, B.; Neupert, W.; Brunner, M. Import of carrier proteins into the mitochondrial inner membrane mediated by Tim22. *Nature* **1996**, *384*, 582–585. [CrossRef] [PubMed]

26. Koehler, C.M.; Jarosch, E.; Tokatlidis, K.; Schmid, K.; Schweyen, R.J.; Schatz, G. Import of mitochondrial carrier proteins mediated by essential proteins of the intermembrane space. *Science* **1998**, *279*, 369–373. [CrossRef]

27. Mesecke, N.; Terziyska, N.; Kozany, C.; Baumann, F.; Neupert, W.; Hell, K.; Herrmann, J.M. A disulfide relay system in the intermembrane space of mitochondria that mediates protein import. *Cell* **2005**, *121*, 1059–1069. [CrossRef]

28. Finger, Y.; Riemer, J. Protein import by the mitochondrial disulfide relay in higher eukaryotes. *Biol. Chem.* **2020**, *401*, 749–763. [CrossRef]

29. Edwards, R.; Gerlich, S.; Tokatlidis, K. The biogenesis of mitochondrial intermembrane space proteins. *Biol. Chem.* **2020**, *401*, 737–747. [CrossRef]

30. Takeda, H.; Tsutsumi, A.; Nishizawa, T.; Lindau, C.; Busto, J.V.; Wenz, L.S.; Ellenrieder, L.; Imai, K.; Straub, S.P.; Mossmann, W.; et al. Mitochondrial sorting and assembly machinery operates by beta-barrel switching. *Nature* **2021**, *590*, 163–169. [CrossRef]

31. Diederichs, K.A.; Ni, X.; Rollauer, S.E.; Botos, I.; Tan, X.; King, M.S.; Kunji, E.R.S.; Jiang, J.; Buchanan, S.K. Structural insight into mitochondrial beta-barrel outer membrane protein biogenesis. *Nat. Commun.* **2020**, *11*, 3290. [CrossRef]

32. Drwesh, L.; Rapaport, D. Biogenesis pathways of alpha-helical mitochondrial outer membrane proteins. *Biol. Chem.* **2020**, *401*, 677–686. [CrossRef] [PubMed]

33. Bykov, Y.S.; Rapaport, D.; Herrmann, J.M.; Schuldiner, M. Cytosolic Events in the Biogenesis of Mitochondrial Proteins. *Trends Biochem. Sci.* **2020**, *45*, 650–667. [CrossRef]

34. Song, J.; Herrmann, J.M.; Becker, T. Quality control of the mitochondrial proteome. *Nat. Rev. Mol. Cell Biol.* **2020**. [CrossRef] [PubMed]

35. Avendano-Monsalve, M.C.; Ponce-Rojas, J.C.; Funes, S. From cytosol to mitochondria: The beginning of a protein journey. *Biol. Chem.* **2020**, *401*, 645–661. [CrossRef] [PubMed]

36. Chartron, J.W.; Hunt, K.C.; Frydman, J. Cotranslational signal-independent SRP preloading during membrane targeting. *Nature* **2016**, *536*, 224–228. [CrossRef] [PubMed]
37. Herrmann, J.M.; Neupert, W.; Stuart, R.A. Insertion into the mitochondrial inner membrane of a polytopic protein, the nuclear encoded Oxa1p. *EMBO J.* **1997**, *16*, 2217–2226. [CrossRef]
38. Horvath, S.E.; Bottinger, L.; Vogtle, F.N.; Wiedemann, N.; Meisinger, C.; Becker, T.; Daum, G. Processing and topology of the yeast mitochondrial phosphatidylserine decarboxylase 1. *J. Biol. Chem.* **2012**, *287*, 36744–36755. [CrossRef]
39. Pogozheva, I.D.; Lomize, A.L. Evolution and adaptation of single-pass transmembrane proteins. *Biochim. Biophys. Acta Biomembr.* **2018**, *1860*, 364–377. [CrossRef]
40. Perez-Martinez, X.; Vazquez-Acevedo, M.; Tolkunova, E.; Funes, S.; Claros, M.G.; Davidson, E.; King, M.P.; Gonzalez-Halphen, D. Unusual location of a mitochondrial gene. Subunit III of cytochrome C oxidase is encoded in the nucleus of Chlamydomonad algae. *J. Biol. Chem.* **2000**, *275*, 30144–30152. [CrossRef]
41. Costa, E.A.; Subramanian, K.; Nunnari, J.; Weissman, J.S. Defining the physiological role of SRP in protein-targeting efficiency and specificity. *Science* **2018**, *359*, 689–692. [CrossRef]
42. Doring, K.; Ahmed, N.; Riemer, T.; Suresh, H.G.; Vainshtein, Y.; Habich, M.; Riemer, J.; Mayer, M.P.; O'Brien, E.P.; Kramer, G.; et al. Profiling Ssb-Nascent Chain Interactions Reveals Principles of Hsp70-Assisted Folding. *Cell* **2017**, *170*, 298–311.e20. [CrossRef]
43. Hsieh, H.H.; Lee, J.H.; Chandrasekar, S.; Shan, S.O. A ribosome-associated chaperone enables substrate triage in a cotranslational protein targeting complex. *Nat. Commun.* **2020**, *11*, 5840. [CrossRef] [PubMed]
44. Ponce-Rojas, J.C.; Avendano-Monsalve, M.C.; Yanez-Falcon, A.R.; Jaimes-Miranda, F.; Garay, E.; Torres-Quiroz, F.; DeLuna, A.; Funes, S. alphabeta'-NAC cooperates with Sam37 to mediate early stages of mitochondrial protein import. *FEBS J.* **2017**, *284*, 814–830. [CrossRef]
45. Fünfschilling, U.; Rospert, S. Nascent polypeptide-associated complex stimulates protein import into yeast mitochondria. *Mol. Biol. Cell* **1999**, *10*, 3289–3299. [CrossRef] [PubMed]
46. Gamerdinger, M.; Hanebuth, M.A.; Frickey, T.; Deuerling, E. The principle of antagonism ensures protein targeting specificity at the endoplasmic reticulum. *Science* **2015**, *348*, 201–207. [CrossRef]
47. Dunn, C.D.; Jensen, R.E. Suppression of a defect in mitochondrial protein import identifies cytosolic proteins required for viability of yeast cells lacking mitochondrial DNA. *Genetics* **2003**, *165*, 35–45. [CrossRef] [PubMed]
48. Lesnik, C.; Cohen, Y.; Atir-Lande, A.; Schuldiner, M.; Arava, Y. OM14 is a mitochondrial receptor for cytosolic ribosomes that supports co-translational import into mitochondria. *Nat. Commun.* **2014**, *5*, 5711. [CrossRef] [PubMed]
49. Schuldiner, M.; Metz, J.; Schmid, V.; Denic, V.; Rakwalska, M.; Schmitt, H.D.; Schwappach, B.; Weissman, J.S. The GET complex mediates insertion of tail-anchored proteins into the ER membrane. *Cell* **2008**, *134*, 634–645. [CrossRef]
50. Vitali, D.G.; Sinzel, M.; Bulthuis, E.P.; Kolb, A.; Zabel, S.; Mehlhorn, D.G.; Figueiredo Costa, B.; Farkas, A.; Clancy, A.; Schuldiner, M.; et al. The GET pathway can increase the risk of mitochondrial outer membrane proteins to be mistargeted to the ER. *J. Cell Sci.* **2018**, *131*. [CrossRef]
51. Xiao, T.; Shakya, V.P.S.; Hughes, A.L. The GET pathway safeguards against non-imported mitochondrial protein stress. *bioRxiv* **2020**. [CrossRef]
52. Williams, C.C.; Jan, C.H.; Weissman, J.S. Targeting and plasticity of mitochondrial proteins revealed by proximity-specific ribosome profiling. *Science* **2014**, *346*, 748–751. [CrossRef] [PubMed]
53. Jan, C.H.; Williams, C.C.; Weissman, J.S. Principles of ER cotranslational translocation revealed by proximity-specific ribosome profiling. *Science* **2014**, *346*, 1257521. [CrossRef] [PubMed]
54. Papic, D.; Elbaz-Alon, Y.; Koerdt, S.N.; Leopold, K.; Worm, D.; Jung, M.; Schuldiner, M.; Rapaport, D. The role of Djp1 in import of the mitochondrial protein Mim1 demonstrates specificity between a cochaperone and its substrate protein. *Mol. Cell Biol.* **2013**, *33*, 4083–4094. [CrossRef]
55. Laborenz, J.; Bykov, Y.S.; Knoringer, K.; Raschle, M.; Filker, S.; Prescianotto-Baschong, C.; Spang, A.; Tatsuta, T.; Langer, T.; Storchova, Z.; et al. The ER protein Ema19 facilitates the degradation of nonimported mitochondrial precursor proteins. *Mol. Biol. Cell* **2021**, *32*, 664–674. [CrossRef] [PubMed]
56. Shakya, V.P.; Barbeau, W.A.; Xiao, T.; Knutson, C.S.; Schuler, M.H.; Hughes, A.L. A nuclear-based quality control pathway for non-imported mitochondrial proteins. *Elife* **2021**, *10*. [CrossRef]
57. Knöringer, K.; Groh, C.; Krämer, L.; Stein, K.C.; Hansen, K.G.; Herrmann, J.M.; Frydman, J.; Boos, F. The unfolded protein response of the endoplasmic reticulum supports mitochondrial biogenesis by buffering non-imported proteins. *bioRxiv* **2021**. [CrossRef]
58. Schlagowski, A.M.; Knoringer, K.; Morlot, S.; Sanchez Vicente, A.; Flohr, T.; Kramer, L.; Boos, F.; Khalid, N.; Ahmed, S.; Schramm, J.; et al. Increased levels of mitochondrial import factor Mia40 prevent the aggregation of polyQ proteins in the cytosol. *EMBO J.* **2021**, e107913. [CrossRef]
59. Wang, X.; Chen, X.J. A cytosolic network suppressing mitochondria-mediated proteostatic stress and cell death. *Nature* **2015**, *524*, 481–484. [CrossRef]
60. Caplan, A.J.; Tsai, J.; Casey, P.J.; Douglas, M.G. Farnesylation of YDJ1p is required for function at elevated growth temperatures in Saccharomcyes cerevisiae. *J. Biol. Chem.* **1992**, *267*, 18890–18895. [CrossRef]
61. Caplan, A.J.; Cyr, D.M.; Douglas, M.G. YDJ1p facilitates polypeptide translocation across different intracellular membranes by a conserved mechanism. *Cell* **1992**, *71*, 1143–1155. [CrossRef]

62. Jores, T.; Lawatscheck, J.; Beke, V.; Franz-Wachtel, M.; Yunoki, K.; Fitzgerald, J.C.; Macek, B.; Endo, T.; Kalbacher, H.; Buchner, J.; et al. Cytosolic Hsp70 and Hsp40 chaperones enable the biogenesis of mitochondrial beta-barrel proteins. *J. Cell Biol.* **2018**, *217*, 3091–3108. [CrossRef] [PubMed]

63. Becker, J.; Walter, W.; Yan, W.; Craig, E.A. Functional interaction of cytosolic hsp70 and a DnaJ-related protein, Ydj1p, in protein translocation in vivo. *Mol. Cell Biol.* **1996**, *16*, 4378–4386. [CrossRef] [PubMed]

64. Opalinski, L.; Song, J.; Priesnitz, C.; Wenz, L.S.; Oeljeklaus, S.; Warscheid, B.; Pfanner, N.; Becker, T. Recruitment of Cytosolic J-Proteins by TOM Receptors Promotes Mitochondrial Protein Biogenesis. *Cell Rep.* **2018**, *25*, 2036–2043.e5. [CrossRef]

65. Murley, A.; Sarsam, R.D.; Toulmay, A.; Yamada, J.; Prinz, W.A.; Nunnari, J. Ltc1 is an ER-localized sterol transporter and a component of ER-mitochondria and ER-vacuole contacts. *J. Cell Biol.* **2015**, *209*, 539–548. [CrossRef] [PubMed]

66. Elbaz-Alon, Y.; Eisenberg-Bord, M.; Shinder, V.; Stiller, S.B.; Shimoni, E.; Wiedemann, N.; Geiger, T.; Schuldiner, M. Lam6 Regulates the Extent of Contacts between Organelles. *Cell Rep.* **2015**, *12*, 7–14. [CrossRef]

67. Lalier, L.; Mignard, V.; Joalland, M.P.; Lanoe, D.; Cartron, P.F.; Manon, S.; Vallette, F.M. TOM20-mediated transfer of Bcl2 from ER to MAM and mitochondria upon induction of apoptosis. *Cell Death Dis.* **2021**, *12*, 182. [CrossRef]

68. Rusinol, A.E.; Cui, Z.; Chen, M.H.; Vance, J.E. A unique mitochondria-associated membrane fraction from rat liver has a high capacity for lipid synthesis and contains pre-Golgi secretory proteins including nascent lipoproteins. *J. Biol. Chem.* **1994**, *269*, 27494–27502. [CrossRef]

69. Vance, J.E. Phospholipid synthesis in a membrane fraction associated with mitochondria. *J. Biol. Chem.* **1990**, *265*, 7248–7256. [CrossRef]

70. Kornmann, B.; Currie, E.; Collins, S.R.; Schuldiner, M.; Nunnari, J.; Weissman, J.S.; Walter, P. An ER-mitochondria tethering complex revealed by a synthetic biology screen. *Science* **2009**, *325*, 477–481. [CrossRef]

71. Stroud, D.A.; Oeljeklaus, S.; Wiese, S.; Bohnert, M.; Lewandrowski, U.; Sickmann, A.; Guiard, B.; van der Laan, M.; Warscheid, B.; Wiedemann, N. Composition and topology of the endoplasmic reticulum-mitochondria encounter structure. *J. Mol. Biol.* **2011**, *413*, 743–750. [CrossRef]

72. Kawano, S.; Tamura, Y.; Kojima, R.; Bala, S.; Asai, E.; Michel, A.H.; Kornmann, B.; Riezman, I.; Riezman, H.; Sakae, Y.; et al. Structure-function insights into direct lipid transfer between membranes by Mmm1-Mdm12 of ERMES. *J. Cell Biol.* **2018**, *217*, 959–974. [CrossRef]

73. Jeong, H.; Park, J.; Jun, Y.; Lee, C. Crystal structures of Mmm1 and Mdm12-Mmm1 reveal mechanistic insight into phospholipid trafficking at ER-mitochondria contact sites. *Proc. Natl. Acad. Sci. USA* **2017**, *114*, E9502–E9511. [CrossRef]

74. Sogo, L.F.; Yaffe, M.P. Regulation of mitochondrial morphology and inheritance by Mdm10p, a protein of the mitochondrial outer membrane. *J. Cell Biol.* **1994**, *130*, 1361–1373. [CrossRef] [PubMed]

75. Burgess, S.M.; Delannoy, M.; Jensen, R.E. MMM1 encodes a mitochondrial outer membrane protein essential for establishing and maintaining the structure of yeast mitochondria. *J. Cell Biol.* **1994**, *126*, 1375–1391. [CrossRef] [PubMed]

76. Berger, K.L.; Sogo, L.F.; Yaffe, M.P. Mdm12p, a component required for mitochondrial inheritance that is conserved between budding and fission yeast. *J. Cell Biol.* **1997**, *136*, 545–553. [CrossRef] [PubMed]

77. Dimmer, K.S.; Fritz, S.; Fuchs, F.; Messerschmitt, M.; Weinbach, N.; Neupert, W.; Westermann, B. Genetic basis of mitochondrial function and morphology in Saccharomyces cerevisiae. *Mol. Biol. Cell* **2002**, *13*, 847–853. [CrossRef] [PubMed]

78. Murley, A.; Lackner, L.L.; Osman, C.; West, M.; Voeltz, G.K.; Walter, P.; Nunnari, J. ER-associated mitochondrial division links the distribution of mitochondria and mitochondrial DNA in yeast. *Elife* **2013**, *2*, e00422. [CrossRef] [PubMed]

79. Eisenberg-Bord, M.; Tsui, H.S.; Antunes, D.; Fernandez-Del-Rio, L.; Bradley, M.C.; Dunn, C.D.; Nguyen, T.P.T.; Rapaport, D.; Clarke, C.F.; Schuldiner, M. The Endoplasmic Reticulum-Mitochondria Encounter Structure Complex Coordinates Coenzyme Q Biosynthesis. *Contact* **2019**, *2*. [CrossRef] [PubMed]

80. Tirrell, P.S.; Nguyen, K.N.; Luby-Phelps, K.; Friedman, J.R. MICOS subcomplexes assemble independently on the mitochondrial inner membrane in proximity to ER contact sites. *J. Cell Biol.* **2020**, *219*. [CrossRef]

81. Lahiri, S.; Chao, J.T.; Tavassoli, S.; Wong, A.K.; Choudhary, V.; Young, B.P.; Loewen, C.J.; Prinz, W.A. A conserved endoplasmic reticulum membrane protein complex (EMC) facilitates phospholipid transfer from the ER to mitochondria. *PLoS Biol.* **2014**, *12*, e1001969. [CrossRef]

82. Zung, N.; Schuldiner, M. New horizons in mitochondrial contact site research. *Biol. Chem.* **2020**, *401*, 793–809. [CrossRef]

83. Cho, K.F.; Branon, T.C.; Rajeev, S.; Svinkina, T.; Udeshi, N.D.; Thoudam, T.; Kwak, C.; Rhee, H.W.; Lee, I.K.; Carr, S.A.; et al. Split-TurboID enables contact-dependent proximity labeling in cells. *Proc. Natl. Acad. Sci. USA* **2020**, *117*, 12143–12154. [CrossRef] [PubMed]

84. Wilson, E.L.; Metzakopian, E. ER-mitochondria contact sites in neurodegeneration: Genetic screening approaches to investigate novel disease mechanisms. *Cell Death Differ.* **2021**, *28*, 1804–1821. [CrossRef] [PubMed]

85. Xu, L.; Wang, X.; Tong, C. Endoplasmic Reticulum-Mitochondria Contact Sites and Neurodegeneration. *Front. Cell Dev. Biol.* **2020**, *8*, 428. [CrossRef] [PubMed]

86. Berner, N.; Reutter, K.R.; Wolf, D.H. Protein Quality Control of the Endoplasmic Reticulum and Ubiquitin-Proteasome-Triggered Degradation of Aberrant Proteins: Yeast Pioneers the Path. *Annu. Rev. Biochem.* **2018**, *87*, 751–782. [CrossRef]

87. Krämer, L.; Groh, C.; Herrmann, J.M. The proteasome: Friend and foe of mitochondrial biogenesis. *FEBS Lett.* **2020**. [CrossRef]

88. Ravanelli, S.; den Brave, F.; Hoppe, T. Mitochondrial Quality Control Governed by Ubiquitin. *Front. Cell Dev. Biol.* **2020**, *8*, 270. [CrossRef]

89. Mohanraj, K.; Nowicka, U.; Chacinska, A. Mitochondrial control of cellular protein homeostasis. *Biochem. J.* **2020**, *477*, 3033–3054. [CrossRef] [PubMed]

90. Wu, X.; Cabanos, C.; Rapoport, T.A. Structure of the post-translational protein translocation machinery of the ER membrane. *Nature* **2019**, *566*, 136–139. [CrossRef]

91. Fritz, S.; Weinbach, N.; Westermann, B. Mdm30 is an F-box protein required for maintenance of fusion-competent mitochondria in yeast. *Mol. Biol. Cell* **2003**, *14*, 2303–2313. [CrossRef] [PubMed]

92. Escobar-Henriques, M.; Westermann, B.; Langer, T. Regulation of mitochondrial fusion by the F-box protein Mdm30 involves proteasome-independent turnover of Fzo1. *J. Cell Biol.* **2006**, *173*, 645–650. [CrossRef] [PubMed]

93. Anton, F.; Dittmar, G.; Langer, T.; Escobar-Henriques, M. Two deubiquitylases act on mitofusin and regulate mitochondrial fusion along independent pathways. *Mol. Cell* **2013**, *49*, 487–498. [CrossRef] [PubMed]

94. Saladi, S.; Boos, F.; Poglitsch, M.; Meyer, H.; Sommer, F.; Muhlhaus, T.; Schroda, M.; Schuldiner, M.; Madeo, F.; Herrmann, J.M. The NADH Dehydrogenase Nde1 Executes Cell Death after Integrating Signals from Metabolism and Proteostasis on the Mitochondrial Surface. *Mol. Cell* **2020**, *77*, 189–202.e6. [CrossRef]

95. Nahar, S.; Chowdhury, A.; Ogura, T.; Esaki, M. A AAA ATPase Cdc48 with a cofactor Ubx2 facilitates ubiquitylation of a mitochondrial fusion-promoting factor Fzo1 for proteasomal degradation. *J. Biochem.* **2020**, *167*, 279–286. [CrossRef]

96. Neuber, O.; Jarosch, E.; Volkwein, C.; Walter, J.; Sommer, T. Ubx2 links the Cdc48 complex to ER-associated protein degradation. *Nat. Cell Biol.* **2005**, *7*, 993–998. [CrossRef]

97. Wu, X.; Li, L.; Jiang, H. Doa1 targets ubiquitinated substrates for mitochondria-associated degradation. *J. Cell Biol.* **2016**, *213*, 49–63. [CrossRef]

98. Mårtensson, C.U.; Priesnitz, C.; Song, J.; Ellenrieder, L.; Doan, K.N.; Boos, F.; Floerchinger, A.; Zufall, N.; Oeljeklaus, S.; Warscheid, B.; et al. Mitochondrial protein translocation-associated degradation. *Nature* **2019**, *569*, 679–683. [CrossRef]

99. Phu, L.; Rose, C.M.; Tea, J.S.; Wall, C.E.; Verschueren, E.; Cheung, T.K.; Kirkpatrick, D.S.; Bingol, B. Dynamic Regulation of Mitochondrial Import by the Ubiquitin System. *Mol. Cell* **2020**, *77*, 1107–1123.e10. [CrossRef]

100. Ordureau, A.; Paulo, J.A.; Zhang, J.; An, H.; Swatek, K.N.; Cannon, J.R.; Wan, Q.; Komander, D.; Harper, J.W. Global Landscape and Dynamics of Parkin and USP30-Dependent Ubiquitylomes in iNeurons during Mitophagic Signaling. *Mol. Cell* **2020**, *77*, 1124–1142.e10. [CrossRef]

101. Bodnar, N.O.; Rapoport, T.A. Molecular Mechanism of Substrate Processing by the Cdc48 ATPase Complex. *Cell* **2017**, *169*, 722–735.e9. [CrossRef]

102. Okreglak, V.; Walter, P. The conserved AAA-ATPase Msp1 confers organelle specificity to tail-anchored proteins. *Proc. Natl. Acad. Sci. USA* **2014**, *111*, 8019–8024. [CrossRef]

103. Chen, Y.C.; Umanah, G.K.; Dephoure, N.; Andrabi, S.A.; Gygi, S.P.; Dawson, T.M.; Dawson, V.L.; Rutter, J. Msp1/ATAD1 maintains mitochondrial function by facilitating the degradation of mislocalized tail-anchored proteins. *EMBO J.* **2014**, *33*, 1548–1564. [CrossRef]

104. Matsumoto, S.; Nakatsukasa, K.; Kakuta, C.; Tamura, Y.; Esaki, M.; Endo, T. Msp1 Clears Mistargeted Proteins by Facilitating Their Transfer from Mitochondria to the ER. *Mol. Cell* **2019**, *76*, 191–205.e10. [CrossRef] [PubMed]

105. Weidberg, H.; Amon, A. MitoCPR-A surveillance pathway that protects mitochondria in response to protein import stress. *Science* **2018**, *360*, eaan4146. [CrossRef] [PubMed]

106. Basch, M.; Wagner, M.; Rolland, S.; Carbonell, A.; Zeng, R.; Khosravi, S.; Schmidt, A.; Aftab, W.; Imhof, A.; Wagener, J.; et al. Msp1 cooperates with the proteasome for extraction of arrested mitochondrial import intermediates. *Mol. Biol. Cell* **2020**, *31*, 753–767. [CrossRef] [PubMed]

107. McKenna, M.J.; Sim, S.I.; Ordureau, A.; Wei, L.; Harper, J.W.; Shao, S.; Park, E. The endoplasmic reticulum P5A-ATPase is a transmembrane helix dislocase. *Science* **2020**, *369*. [CrossRef]

108. Qin, Q.; Zhao, T.; Zou, W.; Shen, K.; Wang, X. An Endoplasmic Reticulum ATPase Safeguards Endoplasmic Reticulum Identity by Removing Ectopically Localized Mitochondrial Proteins. *Cell Rep.* **2020**, *33*, 108363. [CrossRef] [PubMed]

109. Chai, P.; Cheng, Y.; Hou, C.; Yin, L.; Zhang, D.; Hu, Y.; Chen, Q.; Zheng, P.; Teng, J.; Chen, J. USP19 promotes hypoxia-induced mitochondrial division via FUNDC1 at ER-mitochondria contact sites. *J. Cell Biol.* **2021**, *220*. [CrossRef]

110. Friedman, J.R.; Lackner, L.L.; West, M.; DiBenedetto, J.R.; Nunnari, J.; Voeltz, G.K. ER tubules mark sites of mitochondrial division. *Science* **2011**, *334*, 358–362. [CrossRef]

111. Lackner, L.L.; Ping, H.; Graef, M.; Murley, A.; Nunnari, J. Endoplasmic reticulum-associated mitochondria-cortex tether functions in the distribution and inheritance of mitochondria. *Proc. Natl. Acad. Sci. USA* **2013**, *110*, E458–E467. [CrossRef]

112. Zhou, Z.; Torres, M.; Sha, H.; Halbrook, C.J.; Van den Bergh, F.; Reinert, R.B.; Yamada, T.; Wang, S.; Luo, Y.; Hunter, A.H.; et al. Endoplasmic reticulum-associated degradation regulates mitochondrial dynamics in brown adipocytes. *Science* **2020**, *368*, 54–60. [CrossRef] [PubMed]

113. Böckler, S.; Westermann, B. ER-mitochondria contacts as sites of mitophagosome formation. *Autophagy* **2014**, *10*, 1346–1347. [CrossRef]

114. Böckler, S.; Westermann, B. Mitochondrial ER contacts are crucial for mitophagy in yeast. *Dev. Cell* **2014**, *28*, 450–458. [CrossRef]

115. Hamasaki, M.; Furuta, N.; Matsuda, A.; Nezu, A.; Yamamoto, A.; Fujita, N.; Oomori, H.; Noda, T.; Haraguchi, T.; Hiraoka, Y.; et al. Autophagosomes form at ER-mitochondria contact sites. *Nature* **2013**, *495*, 389–393. [CrossRef]

116. Ben-Menachem, R.; Tal, M.; Shadur, T.; Pines, O. A third of the yeast mitochondrial proteome is dual localized: A question of evolution. *Proteomics* **2011**, *11*, 4468–4476. [CrossRef]
117. Friedman, J.R.; Kannan, M.; Toulmay, A.; Jan, C.H.; Weissman, J.S.; Prinz, W.A.; Nunnari, J. Lipid Homeostasis Is Maintained by Dual Targeting of the Mitochondrial PE Biosynthesis Enzyme to the ER. *Dev. Cell* **2018**, *44*, 261–270.e6. [CrossRef] [PubMed]
118. Ma, Y.; Taylor, S.S. A molecular switch for targeting between endoplasmic reticulum (ER) and mitochondria: Conversion of a mitochondria-targeting element into an ER-targeting signal in DAKAP1. *J. Biol. Chem.* **2008**, *283*, 11743–11751. [CrossRef] [PubMed]
119. Colombo, S.; Longhi, R.; Alcaro, S.; Ortuso, F.; Sprocati, T.; Flora, A.; Borgese, N. N-myristoylation determines dual targeting of mammalian NADH-cytochrome b5 reductase to ER and mitochondrial outer membranes by a mechanism of kinetic partitioning. *J. Cell Biol.* **2005**, *168*, 735–745. [CrossRef] [PubMed]

International Journal of
Molecular Sciences

Review

The Molecular Biodiversity of Protein Targeting and Protein Transport Related to the Endoplasmic Reticulum

Andrea Tirincsi [1], Mark Sicking [1], Drazena Hadzibeganovic [1], Sarah Haßdenteufel [2],* and Sven Lang [1],*

[1] Department of Medical Biochemistry and Molecular Biology, Saarland University, 66421 Homburg, Germany; s8antiri@stud.uni-saarland.de (A.T.); mark.sicking@uni-saarland.de (M.S.); drazenapervan@yahoo.com (D.H.)
[2] Department of Molecular Genetics, Weizmann Institute of Science, Rehovot 7610001, Israel
* Correspondence: sarah.hassdenteufel@weizmann.ac.il (S.H.); sven.lang@uni-saarland.de (S.L.)

Abstract: Looking at the variety of the thousands of different polypeptides that have been focused on in the research on the endoplasmic reticulum from the last five decades taught us one humble lesson: no one size fits all. Cells use an impressive array of components to enable the safe transport of protein cargo from the cytosolic ribosomes to the endoplasmic reticulum. Safety during the transit is warranted by the interplay of cytosolic chaperones, membrane receptors, and protein translocases that together form functional networks and serve as protein targeting and translocation routes. While two targeting routes to the endoplasmic reticulum, SRP (signal recognition particle) and GET (guided entry of tail-anchored proteins), prefer targeting determinants at the N- and C-terminus of the cargo polypeptide, respectively, the recently discovered SND (SRP-independent) route seems to preferentially cater for cargos with non-generic targeting signals that are less hydrophobic or more distant from the termini. With an emphasis on targeting routes and protein translocases, we will discuss those functional networks that drive efficient protein topogenesis and shed light on their redundant and dynamic nature in health and disease.

Keywords: endoplasmic reticulum; GET; protein targeting; protein transport; SND; SRP; Sec61 complex; EMC; positive-inside rule; hydrophobicity; signal peptide; transmembrane helix

Citation: Tirincsi, A.; Sicking, M.; Hadzibeganovic, D.; Haßdenteufel, S.; Lang, S. The Molecular Biodiversity of Protein Targeting and Protein Transport Related to the Endoplasmic Reticulum. *Int. J. Mol. Sci.* **2022**, *23*, 143. https://doi.org/10.3390/ijms23010143

Academic Editor: Alexandre Mironov

Received: 27 October 2021
Accepted: 20 December 2021
Published: 23 December 2021

Publisher's Note: MDPI stays neutral with regard to jurisdictional claims in published maps and institutional affiliations.

1. Introduction

Eucaryotic cells use the principle of compartmentalization to streamline the flow of information within the crowded intracellular environment. Different subcellular compartments have occurred during evolution as the result of either endosymbiosis, invagination of the plasma membrane, budding off from other previously formed organelles, or, as discussed more recently, from the autogenous fusion of plasma membrane protrusions [1,2]. Irrespective of its inside-out (protrusions of the procaryotic plasma membrane) or outside-in (invagination of the procaryotic plasma membrane) origin, the lumen of the endoplasmic reticulum (ER) was at first similar to the extracellular milieu and therefore different from the cytosol. In mammalian cells, the ER lumen still reflects its extracellular derivation based on the high concentration and storage of calcium [3,4]. Next to the nucleus, the ER is one of the largest organelles in many cell types and is abundantly present in secretory cells such as those found in the endo- and exocrine portions of the pancreas [5,6]. The correlation between protein secretion and abundance of the ER is not coincidental and substantiates the importance of the ER for this process [7,8]. Indeed, the ER represents the entry point for proteins to the secretory pathway including the exocrine zymogens and endocrine hormones released by the pancreas.

About one-third of eucaryotic genes encode for polypeptides that require targeting to the ER membrane [9,10]. Those polypeptides belong to soluble, membrane-associated, or integral membrane proteins that are all handled and distributed by the ER. Although this organelle represents a vast network that spans from the outer nuclear membrane to the periphery, newly synthesized proteins do not find the ER autonomously. Instead, precursor polypeptides rely on specialized targeting mechanisms that direct them to the ER membrane in a co- or post-translational fashion [11]. In the case of co-translational protein targeting, a nascent polypeptide is recognized during the process of ribosomal translation. As soon as a specific amino acid stretch of the precursor, a so-called targeting signal, emerges from the ribosomal exit tunnel, it is recognized by a targeting factor that directs the complex of ribosome and nascent chain to the ER membrane. The signal recognition particle (SRP) was the first co-translationally acting targeting factor that was discovered and shown to target the ribosome-nascent chain complex (RNC) with the help of the cognate SRP receptor (SR) to the ER membrane [12,13]. In contrast, post-translational protein targeting occurs after a precursor is fully synthesized and released from the ribosome into the cytosol. Key features of post-translationally targeted polypeptides can be a short overall precursor length (<100 amino acids), a C-terminal positioning of a transmembrane helix (TMH) that serves as a targeting signal, or, as is the case for yeast, a "weak" N-terminal signal peptide (SP) that is required for targeting [14–16]. These features prevent efficient recognition by the co-translational targeting factor SRP and require the presence of a post-translational targeting factor such as GET3. GET is the acronym for guided entry of tail-anchored proteins. This implies that GET3 is involved in the targeting of tail-anchored (TA) proteins, substrates that carry a single C-terminally located TMH as a targeting signal. In addition, GET3 has been shown to promote the targeting of short precursor polypeptides with a cleavable N-terminal SP [17–20]. Recent evidence has uncovered another SRP-independent (SND) targeting pathway that seems to function as a backup system with some substrate spectra that overlap with the GET and SRP pathway [21–23]. Additionally, abundant cytosolic chaperones support the targeting of fully-synthesized precursor polypeptides, but they are not discussed here in detail [24,25].

After targeting to the ER membrane, co- and post-translationally arriving polypeptides rely on one of the protein translocation machineries residing in this membrane. Similar to the diversity of precursor polypeptides that are handled by multiple targeting pathways, multiple protein translocases have also evolved to support the insertion or translocation of proteins into or across the ER membrane [26,27]. As the ER membrane originated from the procaryotic plasma membrane, some of the translocation machines of the ER resemble their bacterial ancestors [28,29]. The first ER protein translocase that was described and studied in much detail was the Sec61 complex [30]. This translocase can open up a "fenestrated" conduit through the ER membrane for the translocation of SP-carrying proteins as well as the lateral release of TMHs [30,31]. Other precursor polypeptides, such as TA proteins, can make use of alternative translocation machines such as the GET1/2 complex or the ER membrane protein complex (EMC) [32,33]. Interestingly, precursor proteins relying on the GET1/2 complex or EMC require a differentially shaped opening compared to the membrane-spanning pore that is provided by the Sec61 complex. This shows that nature has apparently found more than one solution for polypeptides to traverse a membrane.

It is now half a century since Blobel and Sabatini published their first "tentative scheme" on what they would later coin the signal hypothesis. They speculated about the "role of the ribosome-membrane interaction in the vectorial discharge of proteins into the ER", which "is indicated by (a) the close association of the nascent polypeptide chain with the ER membrane and (b) the close association of the large ribosomal subunit with this membrane" [34]. As no one size fits all, several types of targeting signals, targeting pathways, and protein translocases have since then been described, which demonstrate the diversity of transferring proteins from the cytosol to the secretory pathway. However, in light of the macromolecular crowding and potential off-target destinations, one underlying theme that unifies those functional targeting and translocation networks is "safety first". During their journey from the inside of the ribosome to the ER membrane, the nascent polypeptides are handed over from one protected compartment and cavity to the next to avoid extensive folding and premature, unproductive interactions. After leaving the shielded environment provided by the ribosome, hydrophobic targeting signals are cradled by a binding pocket of one of the targeting factors before they are transferred to a narrow opening of one of the translocation machineries that acts as a membrane-integrated chaperone. We will summarize some of these protective micro-compartments and discuss key players of the functional targeting and translocation routes that chaperone polypeptides into or across the ER membrane.

2. Two Prominent Types of Hydrophobic Targeting Signals

According to the original concept of the signal hypothesis by Günter Blobel and David Sabatini, proteins themselves contain targeting information within their polypeptide sequence that directs them to the ER membrane. Fifty years ago in 1971, they wrote that a "nascent chain provides the structural conditions for its transfer into the membrane-bounded compartment of the ER". That is, intrinsic, proteinaceous targeting signals govern the transport and localization of precursor proteins to the ER membrane where "the segregated chains may undergo the modifications (proteolytic cleavage, e.g., proinsulin; attachment of carbohydrate, e.g., immunoglobulins) required for secretion, storage, or disposal in the various intracellular membrane-bounded compartments" [34]. Upon arrival at the correct membrane, the targeting signals fulfill a second function. They can initiate the gating of a compatible protein translocase that temporarily opens a pore or somehow helps to overcome the energetically unfavorable process of protein insertion/translocation into or across a lipid bilayer. A wealth of data gathered over the past 50 years demonstrates the validity of this oligopeptide-based targeting concept which relies on intrinsic amino acid stretches of cargo proteins. In addition, we would like to refer the reader to compelling data and reviews highlighting the oligonucleotide-based targeting of mRNA, which instead relies on intrinsic nucleotide motifs. In the case of the latter, specialized membrane proteins recruit either mRNA to membrane-bound ribosomes or mRNA that is present in a RNC to the membrane [35–41].

Here, we will focus on two types of oligopeptide-based ER targeting signals, cleavable SPs and TMHs, that are encoded as part of the primary structure of cargo proteins and help to find the ER. Both variants of targeting signals share a hydrophobic core element that upon its emergence from the ribosomal exit tunnel is usually considered to initiate the process of selective targeting, translocation, and/or membrane insertion [42]. Other hydrophobic sequences, such as the C-terminal glycosylphosphatidylinositol (GPI)-attachment signal [43,44], may further shape the choice of a specific targeting pathway but do not act independently of an N-terminal SP. Moreover, adhering to the Latin prefix "trans", we do not discuss the targeting of monotopic membrane proteins that do not fully traverse both halves of the lipid bilayer and instead insert via an intramembrane domain often showing a hairpin or amphipathic helix [45,46].

2.1. The Tripartite Nature of Cleavable Signal Peptides

Soon after describing their insights into the ribosome–membrane interaction in eucaryotic cells, Blobel and colleagues published a series of landmark papers that demonstrated the presence of N-terminal SPs to direct proteins to the ER via the cytosolic SRP, before they are cleaved off upon translocation [12,13,47–55]. Although it was originally assumed that many, if not all, SPs share the same consensus motif, analyses of a growing number of protein sequences over time showed the opposite: a remarkable sequence variation of SPs. However, upon closer inspection, some preserved features shared by all SPs became evident and pointed for example to the importance of the central hydrophobic h-domain, which is usually rich in leucine [56–58]. Leucine residues stabilize the formation of an α-helix, a feature that is relevant at different stages during protein biogenesis [59,60]. Studies addressing the physicochemical parameters of SPs found a typical overall length of about 20–30 residues and validated the h-domain as a hydrophobic α-helical stretch of ~7–15 residues, which is flanked by two shorter domains named according to their relative position the n- and c-domain. While the n-domain (~1–5 residues long) can carry some cationic residues, the c-domain (~3–7 residues long) is polar in nature and harbors a SP cleavage site recognized by a signal peptidase [61–66]. Thus, most SPs are tripartite in their design, which is themed "cationic-hydrophobic-polar" (Figure 1A).

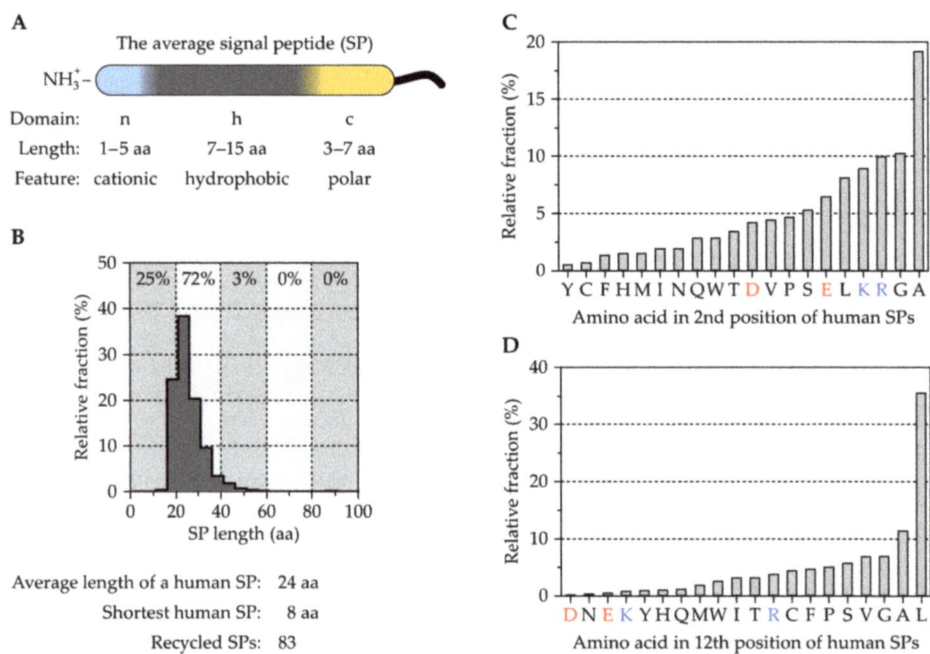

Figure 1. Features of human signal peptides and the amino acid distribution in key positions. (**A**) The tripartite segmentation of a cleavable signal peptide (SP) found at the N-terminus (NH_3^+) of secretory proteins. Characteristic attributes of the three sub-domains are given below. (**B**) A histogram showing the SP length distribution of 3584 human SP annotated at Uniprot. Numbers at the top of the diagram represent frequencies of the highlighted area. Almost three-quarters of SPs are 21–40 amino acids (aa) in length. (**C,D**) Histograms showing the frequencies of the twenty proteinogenic amino acids in the second (**C**) of twelve (**D**) positions of the same human SPs considered in (**B**) ordered from the lowest to the highest fraction. Amino acids are listed on the x-axis according to the one-letter code. The acidic amino acids aspartate (D) and glutamate (E) are labeled in red and the basic amino acids arginine (R) and lysine (K) in blue.

Looking at the ~3500 human SPs and their combined 85,000 amino acids that are annotated at the Uniprot database (www.uniprot.org, accessed 12 June 2020), the same conclusions about physicochemical properties arise. The average length of a human SP is 24 residues and more than 80% are between 15 and 30 residues in length (Figure 1B). The shortest SP of eight residues is found in the chymotrypsin-like elastase family member 1, CELA1. Considering the importance of hydrophobicity and the h-domain, it is not surprising that the most frequent amino acid found in human SPs is leucine with almost 25%. However, the frequency of individual amino acids varies depending on the position, congruent with the tripartite sectioning of SPs. For example, in position #2 right after the starting methionine, the most frequent amino acid is alanine (19%) followed by glycine, arginine, and lysine with ~10% each (Figure 1C). The frequent occurrence of arginine and lysine in this position reflects the observation of a positively charged n-domain. In position #12, which should report on the h-domain of many signal peptides, leucine is the most prevalent amino acid with 35%. As expected, the four charged amino acids, aspartate, glutamate, lysine, and arginine are less frequently found at this position and have a combined frequency of about 5% (Figure 1D). Photo-crosslink studies addressing the targeting and translocation efficiency of different SP variants have shown that the severe transport defects of reporter proteins mainly derive from the shortening of the h-domain [58]. These findings underline previous studies that have tested the functionality of essentially random, 20 amino acid long peptide sequences for secretion [67]. Similarly, the occurrence of charged residues in the h-domain of a preprotein's SP can affect translocation and cause disease, as is the case with mutations identified in the SP of renin or bilirubin UDP-glucuronosyltransferase [68–72]. Equivalent to charged residues, the introduction of a single helix breaking proline in the middle of the h-domain can also suppress the efficient translocation of the preprotein, as was shown for the SP of the yeast invertase [73]. The lack of a universal consensus motif and the detrimental impact on protein targeting in the presence of single mutations in the SP support the conclusion that SPs and the respective downstream mature domain of a protein are specifically matched. Hence, SP diversity does not represent a lack of selective evolutionary pressure. In fact, the opposite is the case and SPs have been found to evolve half as fast as neutral sequences which probably helps to conserve SP functionality [74]. The co-evolution of a SP with the rest of the protein also supports data that show SPs and mature domains are not always readily interchangeable without impacting translocation efficiency [75–77]. Therefore, it is not surprising that only about 80 of the ~3500 human SPs are re-used once or more within the human proteome.

The Correlation between Signal Peptides and Downstream Processes

The broad biophysical similarities plus the individual SP-specific features characterize most SPs as unique cis-acting elements that shape the fate of a protein including its targeting, translocation, SP cleavage, and post-translocation events. These processes involve various trans-acting factors that directly or indirectly associate with the SP. In addition to cytosolic targeting factors and membrane-resident translocases that temporarily harbor the SP, further auxiliary components that fine-tune these steps have been identified over the past years [42]. As will be discussed in more detail in the following sections, the first elements that were shown to accommodate the h-domain of co-translationally targeted SPs were the transiently opening binding pockets of SRP and the Sec61 complex [78–80]. In yeast and mammalian cells, certain secretory proteins can also be directed to the ER membrane SRP-independently in a post-translational fashion. Likewise, the secretion of proteins to the periplasmic space of *E. coli* occurs prevalently post-translation. Interestingly, the hydrophobicity of post-translationally directed SPs seems to differ from those that are recognized co-translationally; however, there are species-specific differences. While in yeast post-translationally targeted SPs show a lower hydrophobicity than the co-translational counterparts, post-translationally transported small secretory proteins of human cells show the opposite: an above-average hydrophobicity [17,81]. Irrespective of the organism, the transport of post-translational substrates requires the Sec61 complex to associate with the auxiliary membrane proteins Sec62/Sec63 that act as allosteric effectors supporting the opening of the translocation pore [82–88]. Other than SPs with a hydrophobicity deviant from the norm, human SPs with an above-average portion of the helix breaking residues glycine and proline require the presence of another auxiliary component, the translocon-associated protein (TRAP) complex [89]. Overall, the plethora of different SPs require the presence of multiple targeting and translocation modalities and provide one reason why the original idea of a unified pathway for the topogenesis of all secretory proteins had to be extended.

2.2. Transmembrane Helices Are Efficient Targeting Signals

In addition to N-terminal SPs, hydrophobic TMHs can also serve as efficient targeting signals and are readily recognized and accommodated by different targeting routes and translocation machineries. The two types of targeting signals are not mutually exclusive. About 40% of the SP-carrying human proteins mentioned above also have one or more TMH(s). Vice versa, of the ~4800 human membrane proteins that are annotated at the Uniprot database, ~1350 proteins carry a SP and ~50 a so-called transit peptide that is destined for the targeting of a nuclear-encoded protein to an organelle other than the ER (www.uniprot.org/help/transit, accessed 14 October 2021). The number of annotated human membrane proteins aligns reasonably well with previous reports stating that membrane proteins correspond to 20–30% of genes encoded in a typical genome [90,91]. On the one hand, TMHs and SPs share (i) the functional equivalence as targeting and gating signals as well as (ii) the extensive sequence variability. On the other hand, they differ with respect to (i) the length of their hydrophobic core and (ii) their positioning within the precursor protein. While SPs are found at the N-terminus and their h-domain comprises ~7–15 residues, TMHs can be localized anywhere within the primary structure and are typically 19–25 residues in size. Although TMHs are on average more hydrophobic than SPs, both types of targeting signals cover a wide range of hydrophobicity, a parameter that equally influences targeting and topogenesis [92,93].

A crude inspection of the ~4800 human membrane proteins annotated at the Uniprot database (accessed 19 May 2020) shows roughly 2100 (44%) bitopic/single-pass and 2700 (56%) polytopic membrane proteins with a totality of ~20,000 TMH (Figure 2A). A total of 88% of all human membrane proteins contain less than 8 TMHs and the shortest one with a total of 31 amino acids and 1 TMH is sarcolipin, a regulator of the ER calcium ATPase SERCA2. Without much deviation, the average length of all first, second, or third TMHs is 21 residues, a number that remains fairly stable even for membrane proteins with many more TMHs (Figure 2B).

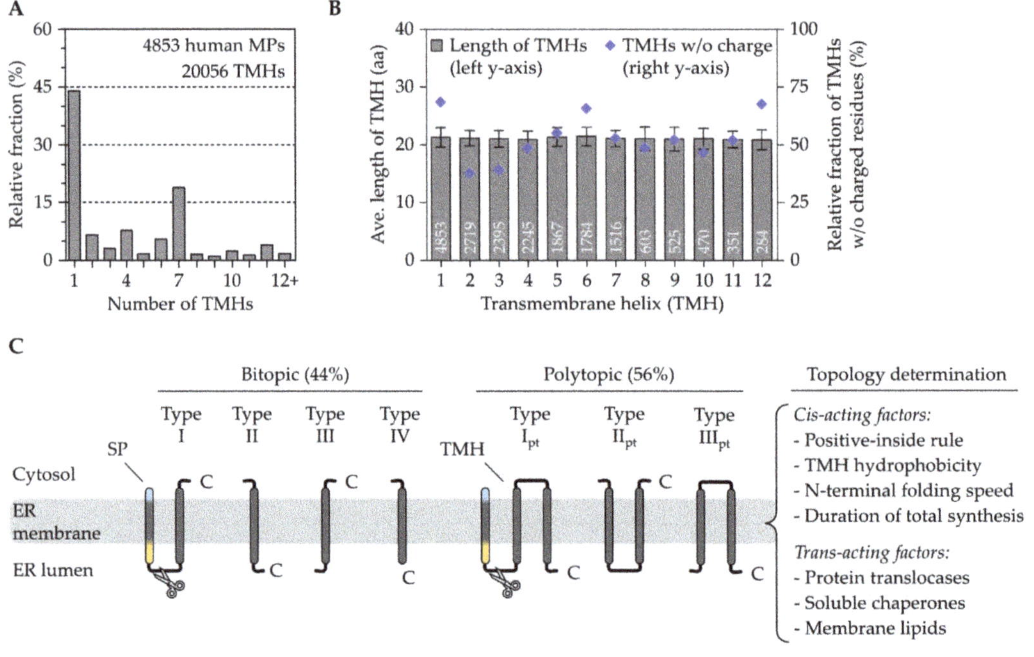

Figure 2. Length, charge, and topology of transmembrane helices of human bi- and polytopic membrane proteins. (**A**) The histogram shows the relative fraction of bitopic membrane proteins (MPs) with one transmembrane helix (TMH) and polytopic MPs with two or more TMHs. The data are based on the 20,056 TMHs of 4853 human MPs annotated at Uniprot. (**B**) Grey bars show the average length (left y-axis) with standard deviation of all first, second, and later TMHs till the twelfth TMH. White numbers at the bottom of the bar indicate the number of TMHs found in the human MPs. The blue diamond represents the relative fraction of the same TMHs that do not carry any charged residues (right y-axis). (**C**) Classification of bi- and polytopic (pt) membrane proteins based

on topology, presence of a signal peptide (SP), and localization of a TMH within the primary structure. SPs are colored according to Figure 1 and their cleavage by the signal peptidase is indicated by scissors. For orientation purposes, the C-terminus (**C**) of each representative is indicated. Black lines indicate domains up- and downstream of a TMH. Cis- and trans-acting factors that influence the final topology of MPs are mentioned on the right and are further described in the text. aa, amino acids; ave., average; *w/o*, without.

2.2.1. The Topological Diversity of Transmembrane Helices

The topological layout is one of many ways to classify membrane proteins. With a focus on TMHs, the following classification excludes β-barrel membrane proteins that are not present in the ER and are limited to the outer membranes of mitochondria, chloroplasts, and bacteria [94,95]. The topology-based grouping (Figure 2C), which is used by many researchers, considers four features including the presence of an SP, the orientation of the first TMH fully traversing the lipid bilayer, the relative position of this first TMH within the primary structure, and the quantity of TMHs giving rise to seven classes of α-helical membrane proteins [96]. Accordingly, single-pass type I membrane proteins start with a cleavable SP dictating the downstream TMH into an N_{exo}/C_{cyto} orientation, whereas the type II version lacks a SP and the single TMH adopts an N_{cyto}/C_{exo} orientation requiring an originally unanticipated flip-turn during insertion [97,98]. Interestingly, as targeting equivalents, the SPs and type II TMHs adopt the same N_{cyto}/C_{exo} topology using a loop-like insertion or flip-turn. As will be discussed later, this is the reason why SP-carrying substrates (including soluble and type I membrane proteins) and type II membrane proteins share the same translocation apparatus, the Sec61 complex. In contrast, single-pass type III membrane proteins lack an SP and the single TMH orients in an N_{exo}/C_{cyto} topology, which, whilst somewhat counter-intuitive, relies preferentially on a different translocation apparatus, the EMC [99,100]. Single pass type IV membrane proteins are more often referred to as TA proteins. They resemble the N_{cyto}/C_{exo} orientation of a type II protein, but their TMH is located close to the protein's C-terminus. This classification system for single-pass membrane proteins is also used at Uniprot (www.uniprot.org/locations/SL-9905 for type I and up to/SL-9908 for type IV). Based on the presence of an SP and the orientation of their first TMH, polytopic membrane proteins can also be classified analogously as type I_{pt}, type II_{pt}, and type III_{pt}. We use the index "pt" to indicate the classification of a first TMH of a "polytopic" membrane protein.

2.2.2. The Crosstalk between Transmembrane Helices and Trans-Acting Factors

The determination of the final topology of a membrane protein is a multifactorial process and depends on both trans-acting factors such as protein translocases, chaperones, and lipids as well as intrinsic parameters encoded within the protein sequence (Figure 2C). The sequence-intrinsic, or cis-acting, elements that influence the orientation of the initial TMH include the positive-inside rule, the folding capacity of preceding N-terminal domains, hydrophobicity, and the length of a protein reflecting on the duration of ribosomal translation [95,101]. The positive-inside rule is a driving factor for the orientation of both targeting signals, TMHs (initial and internal ones) and SPs. This conserved principle reflects on the statistical enrichment of arginine and lysine residues in the cytosolic edge of a TMH as well as in the cytosolic-facing n-domain of SPs [63,93,102,103]. Mutational and protein engineering analyses of membrane protein reporters have provided additional support for the bioinformatic approaches and corroborated the importance of positive charges as a topological determinant [104,105]. Larger, quick folding N-terminal domains that precede a TMH can start to adopt secondary and tertiary structure motifs that prevent their translocation and handling by the limited pore size of a translocase. Therefore, folded domains force the downstream TMH to adopt a type II (N_{cyto}/C_{exo}) topology, irrespective of the positive-inside rule [106]. The hydrophobicity of TMHs varies to a great extent. The TMHs of single-spanning membrane proteins are usually more hydrophobic than those of polytopic membrane proteins [107,108]. Similar to a quick folding N-terminal domain, a very hydrophobic TMH can also overpower the positive-inside rule. However, in this case, the hydrophobicity forces the TMH into a type III (N_{exo}/C_{cyto}) orientation by stimulating the translocation of the N-terminus [98]. Although the experiments by Goder and Spiess were based on an unusually hydrophobic 22 residue long oligo-leucine stretch, they impressively demonstrated two points. One, the total protein length following the TMH influences the topology, most likely as a function of synthesis time/speed. Two, the topology determination is a dynamic equilibrium that can entail TMH inversion starting from a type III insertion and culminating in a type II re-orientation. Furthermore, slowing down translation by low-dose cycloheximide treatments to increase synthesis time also favored TMH inversion from type III to type II. Those data aligned well with experiments using reconstituted conditions that also demonstrated the flip-turn of a canonical type II TMH from aquaporin IV [97,98].

In the case of polytopic membrane proteins, the topology determination of the first TMH dictates, in most cases, the orientation of the following TMHs with each one crossing through the membrane in the alternate orientation as the predecessor. However, the mixture of more and less hydrophobic TMHs sometimes only separated by a short loop can put further steric constraints on the topology determination and additional parameters such as TMH–TMH interactions come into play [109,110]. The cooperative behavior of neighboring TMHs can be deduced from experiments with wild-type and mutant proteins. For example, studies of the bovine heptahelical opsin show that, within the same polypeptide chain, some TMHs tend to insert into the lipid bilayer individually, whereas others bundle up before being released as a small package of TMHs [111,112]. Other studies that have focused on charge inversion mutations in the TMH-flanking regions of Glut1 or single point mutations in TMHs of connexin Cx32 have also demonstrated the cooperativity between TMHs [113–115]. Those findings imply that downstream TMHs can re-initialize correct topology determination and mutations do not necessarily cause an inverted topology of downstream TMHs or of the entire polytopic membrane protein. Less hydrophobic TMHs, sometimes also carrying membrane-aversive charged amino acids, can be found in ion transporters or channels such as the ER calcium ATPase SERCA2 or the voltage-gated potassium channel KCNA1 of humans. Such specialized membrane proteins have hydrophilic residues of different TMHs facing inwards, i.e., away from the hydrocarbon core of the bilayer. In the case of the potassium channel, TMHs and the hydrophilic residues form a tunnel with a selectivity filter and a gated pore as a conduit for potassium ions after shedding the water-shell [116]. With regard to cooperativity, the membrane-insertion of such "unusual" TMHs might require their release into the membrane in small bundles and the assistance of different trans-acting factors.

Although instances are known where TMHs such as the type IV tail-anchor of Cytb5 can insert into protein-free membranes of liposomes [117,118], it is safe to say that other than the local lipid composition, protein translocation machineries also represent relevant trans-acting factors that help to shape topogenesis. As such, they decode the sequence-intrinsic information that influences the orientation of the initial TMH [119]. By analogy, protein translocases act in a similar way to the axon hillock of neuronal cells that collects, considers, and computes incoming inhibitory and activating signals from the dendrites into one unified signal output for the axon. Yet, before reaching a protein translocase at the ER membrane, the crowded intracellular environment requires the hydrophobic targeting signals to be cradled from the start to the end of their journey to prevent aggregation and premature, unproductive interactions. After the emergence of a targeting signal from the protective ribosomal exit tunnel, it is chaperoned by an adequate targeting pathway that safely transfers it to the membrane and a suitable protein translocase. Critical elements of the multiple targeting and translocation pathways that are required to properly accommodate the various targeting signals will be discussed next.

3. Multifactorial Polypeptide Targeting Pathways: The SRP, GET, and SND Systems

For years, the view of ER targeting was dominated by one pathway that would cater to all the polypeptides with the help of the SRP (Figure 3). The discovery of the

GET and SND routes in the last fifteen years has led to the emergence of alternative targeting pathways [18,19,21,22]. Their existence has long been anticipated since the disturbance or absence of the SRP pathway does not result in a complete shutdown of ER targeting [120,121]. Differential energy requirements have been observed in the transport of different cargo [122–124] and the vitality of SRP knock-out cells also argued against its exclusive role in ER targeting [121,125,126]. However, the identification of the GET system as a dedicated pathway for TA proteins was still not sufficient to explain ER targeting of all cargo, hinting at the existence of additional pathways. Accordingly, the knock-out of GET components tends to have rather moderate effects, though heterogeneous phenotypes have been observed for various factors and organisms [127–130]. In line with the notion that GET is not essential in yeast [19], another targeting option was found with SND. Synergistic effects have been observed upon the simultaneous depletion of GET and SND and strikingly, an overexpression of SND could compensate for the loss of SRP function in yeast [21]. The proposed model involves a network of three main targeting pathways that show different, but partially overlapping, client spectra which are in accordance with a non-hierarchical organization where one pathway acts as a back-up in case of the malfunction of another. Specifically, SRP recognizes cleavable SPs and TMHs at the N-terminus, whereas the GET pathway preferentially binds the C-terminal TMHs of TA proteins. While both groups of clients share signals with a similar hydrophobicity but a different localization within the polypeptide, the SND pathway catches less hydrophobic targeting signals which occur internally or at the C-terminus [22,23,43]. These findings have highlighted a previously unanticipated complexity of ER targeting that expands beyond proteins with classical targeting signals and includes the heterogeneous cohorts of TA and GPI-anchored proteins [23,43,131]. Next, the current view of pathway organization and operation will be discussed for SRP, GET, and SND before we continue with a structural comparison of the central cargo binding components to discuss the partially overlapping client spectra.

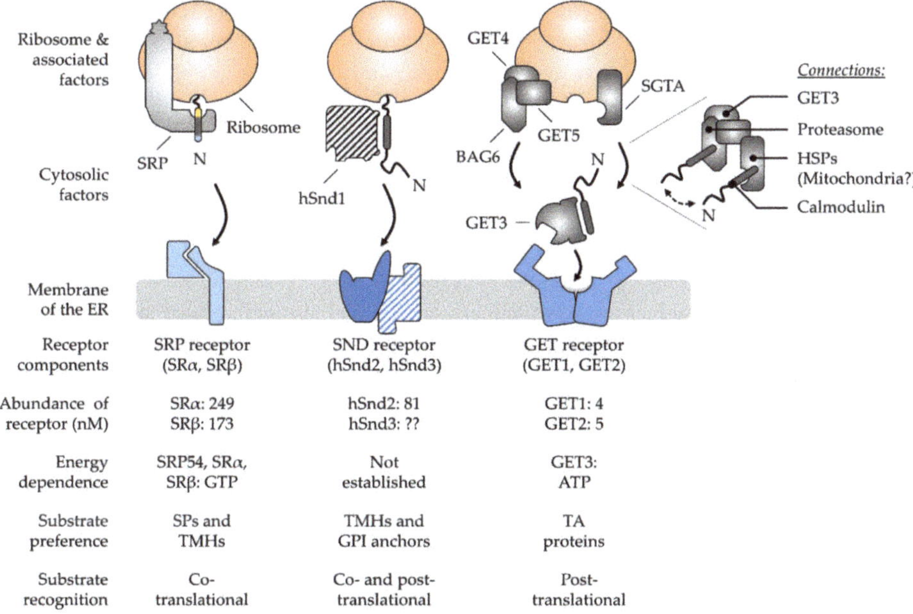

Figure 3. Major components and hallmarks of the mammalian SRP, SND, and GET targeting pathways. The top half shows a graphical output of the major components that shape the three targeting

pathways SRP, SND, and GET. The dotted lines indicate a zoomed-in view of the BAG6 pre-targeting complex cooperating with SGTA and other cellular components. The double-headed arrow suggests the cycling of TA proteins between the two chaperones SGTA and BAG6. The bottom half summarizes some of the key features that differentiate the pathways from each other. Ribosome-associated and cytosolic targeting components are shown in grey colors and the cognate membrane receptors in shades of blue. Components (hSnd1, hSnd3) shown with a hatched color fill have not yet been identified in higher eucaryotes. Their existence is based on findings from yeast and the conserved nature of targeting machineries [21]. Abundance values for the receptor components are based on the quantitative mass spectrometry of mammalian cells [132]. Please note, there is considerable controversy about the abundance of GET1 and GET2 with some sources finding GET2 in four- to sevenfold molar excess over GET1 [133,134]. The N-terminus (N) of newly synthesized polypeptides is shown to accentuate the positioning of signal peptides (SPs) and transmembrane helices (TMHs) of different types of cargos, including tail-anchored (TA) proteins and glycosylphosphatidylinositol (GPI)-anchored proteins. BAG6, BCL2-associated athanogene 6; GET, guided entry of TA proteins; HSPs, heat shock proteins; SGTA, small glutamine rich tetratricopeptide repeat co-chaperone alpha; SND, SRP-independent; SR, SRP receptor; SRP, signal recognition particle.

3.1. The Ancient Silk Road Project, SRP—Direct Way to the ER

In the SRP pathway (Figure 3), two GTPases, the cargo-binding SRP in the cytosol and the corresponding ER-resident receptor SR, ensure efficient and specific targeting to the ER membrane [135]. Energy consumption is coupled to the dissociation of the SRP–SR complex upon the coordinated transfer of cargo to the Sec61 complex [136–138]. SRP is thus recycled into the cytosol for another round of ER targeting only upon the successful termination of the cargo delivery. Underlining its relevance as an important factor in ER targeting, the SRP pathway is conserved in all domains of life [139]. Though, similar to the ribosome, the size and complexity of SRP architecture has increased during evolution [140]. In contrast to the bacterial SRP that only comprises one protein component for cargo recognition (an SRP54 homolog), the mammalian SRP consists of six proteins bound to a universal 7S RNA scaffold. Along with the growth in size and complexity, eucaryotic SRP has developed the ability to arrest translation via the acquired Alu domain (SRP9/SRP14) that binds near the peptidyl transferase center (PTC) and competes with elongation factors [141,142]. It is based on this dogma that ER transport has long been assumed to be strictly co-translational, which entails the coupling of protein synthesis and translocation for maximal shielding of the hydrophobic targeting signal until translocation is completed. In an alternative model, the SRP-mediated slowdown of translation serves to keep the translocation competence of polypeptides to match the limiting targeting sites in the ER membrane, referring to the SR [143]. Today, the view of a strict translational arrest in SRP-mediated targeting has been weakened based on ribosomal profiling experiments in yeast and bacteria [125,144]. Indeed, SRP might not be the only option to locally slow down the translation, as mentioned before. The complete ribosome-independent action of SRP has likewise been proposed based on the binding to certain TA proteins [145], which seemed to hold functional relevance under cellular conditions [23]. In addition, the question of where the ribosomal translation is localized, in the cytosol or at the ER membrane, is an ongoing matter of debate. Accordingly, SRP might be prevalently required for the delivery of the RNC in the first round of ER targeting which is accompanied by local translation initialization at the ER membrane [146–149]. Breaking with another dogma, SRP seems not to be the lonely player it was generally believed to be. Mammalian small glutamine-rich tetratricopeptide repeat co-chaperone alpha (SGTA) has been shown to act co-translationally and downstream of SRP to ensure targeting fidelity [150], yet an overlap between the SRP and GET pathways has been demonstrated.

3.2. The Juggling GET Cascade—Safety First

The GET pathway (Figure 3) is well-established to deal post-translationally with the C-terminal targeting signals of TA proteins or short SP-carrying polypeptides with 100 amino acids or less, both of which are likely to evade co-translational recognition [22]. Despite some compensating mechanisms (see Section 4.1), TA proteins and short secretory proteins are released from the ribosome shortly after their synthesis is terminated without prior exposition of the targeting signal to the cytosol [151]. In such cases, a cascade of trans-acting factors handles the recognition, shielding, and delivery of the cargo [152,153]. Directionality of targeting towards the ER membrane is ensured by the homodimeric ATPase GET3 (Get3 in yeast), the central energy-consuming targeting factor of the GET pathway. Depending on the nucleotide occupancy, it transitions between closed and open conformations that allow cargo binding and release [154,155]. ATP hydrolysis is induced upon capture of the cargo and may represent a proof-reading step leading to full commitment to GET-driven ER targeting [155,156]. Thus, cargo delivery and factor recycling are carefully regulated to prevent premature cargo release before reaching the ER receptor. While such energy-driven control appears as a general theme in protein targeting, recent findings have shown that GET3 is dispensable for the targeting of moderate or low hydrophobic TA proteins. Since they are less at risk for aggregation, a loose binding and re-binding by cytosolic chaperones such as the abundant calmodulin might be sufficient. Calmodulin directs TMHs of lower hydrophobicity to the EMC for insertion [157].

Installment of a Pre-Targeting Complex with Connections to Other Chaperone Systems

In the canonical GET pathway, SGTA (Sgt2 in yeast) transfers cargo to the GET3 targeting factor as part of the so-called pre-targeting complex, together with GET4 and GET5 (Get4 and Get5 in yeast) [158,159]. While GET5 binds SGTA, GET4 recruits ATP-bound GET3 and primes it for capturing cargo from SGTA by the inhibition of ATP hydrolysis. Thus, by regulating the ATPase cycle of GET3, the pre-targeting complex supports the safe and direct loading of cargo from SGTA onto GET3. Such a hand-over is therefore characterized as private, in contrast to a simple release and diffusion of the cargo. Regulation of the GET3 conformation and cargo capture is hence driven by the nucleotide status and the interactions with effectors and cargo. In mammalian cells, the additional protein BAG6 (BCL2-associated athanogene 6) connects GET4 and GET5 to form the pre-targeting BAG6 complex (Figure 3). The C-terminus of BAG6 is involved in the scaffolding and has been shown to be the minimally required domain for TA targeting [160]. Its other domains recruit the E3 ubiquitin ligase RNF126, heat shock proteins (HSPs), and proteasomal subunits or bind misfolded proteins. Thus, BAG6 links the GET pathway to protein degradation [161–165]. Strikingly, BAG6 is antagonized by SGTA, which may involve the cycling of hydrophobic domains between both chaperones until the cargo is either triaged to the ER (via GET3) or the proteasome (Figure 3) [166,167]. The competitive binding between the various factors in building a complex that either drives targeting, or degradation is decisive for its fate. However, SGTA shows the fastest binding kinetic to the cargo, making it a central player of the GET cascade [166–169]. Thus, BAG6 may only have the chance to capture the cargo from SGTA when transfer to GET3 is failing and the cargo undergoes unproductive cycles of release and re-capture by SGTA. Engagement of the cargo is further competed for by calmodulin, an off-pathway chaperone preventing the degradation of the cargo and providing a safeguard of the GET pathway [161].

Likewise, SGTA-mediated sorting may involve the targeting of TA proteins to mitochondria, since the abolished delivery to the ER results in mistargeting to the mitochondria congruent with a default destination [19,170]. Indeed, BAG6 and SGTA are platforms for the recruitment of other off-pathway chaperones, such as Hsp70 and Hsp90, which have been associated with mitochondrial TA protein biogenesis [171]. However, the nature of the mitochondrial targeting pathway is still unclear and alternative implications for the connection of the GET pathway with heat shock proteins have been discussed. A recent yeast model suggested that soluble TA proteins may first be captured by Hsp70 for hand-off to Sgt2 and the subsequent transfer to Get3 [172,173]. Such funneling through a chaperone-guided cascade has been proven to be more efficient and protective against aggregation than the immediate loading of cargo to Sgt2. How this model applies to the mammalian GET pathway remains to be clarified, since whether SGTA and BAG6 directly bind the ribosome with putative implications for initial cargo binding has been discussed (see Section 4.3). Nevertheless, the Hsp70-Sgt2-Get3 triad provides an attractive model for how directionality is maintained in the chaperone cascade [152]. Hsp70 is highly abundant in the cytosol and binds rapidly but loosely to its cargo which subsequently engages stronger interactions with chaperones downstream in the pathway. The initially high binding kinetic might be required to compete with unfavorable misfolding and aggregation, while unidirectionality is later driven by thermodynamic forces ensuring high affinity and stable binding that prevent the reverse transfer to upstream components. This selective hand-over of cargo polypeptides in the targeting cascade from one chaperone to the next reminds us of the directed flow of electrons in the electron transfer chain from the lower to the higher redox potential.

The many factors acting in concert in the GET pathway raise the question of a rationale for such complexity, whereas SRP can fulfil ER targeting in a one-factor show. The multitude of components might support the rather transient and low-affinity interactions in the GET cascade [174]. The characteristic branching into diverse off-road pathways, however, demonstrates the importance of quality control in post-translational targeting. To compensate for misfolding and mistargeting, quality control systems are installed on the road and at the destination [161,175]. The modular organization further provides flexibility and options for internal rearrangement to bypass overloaded components or to allow a targeted regulation of defined cargo pools. On the other hand, SRP and GET might be representations of different targeting concepts that have co-evolved according to the needs of their cargos. They either provide one-to-one guidance for highly hydrophobic cargo in a direct delivery process (SRP) or tolerate the fact that the outcome is a balance between delivery and degradation of the cargo (GET), which may include shuffling between chaperons and moratorium stages in case of malfunctions or stress and starvation [176]. Metazoan GET pathway components are additionally connected to other cellular processes (see Sections 5.2 and 6). GET2, the mammalian subunit of the GET receptor, is described as a signaling molecule and is associated with cell survival in the immune system, independently from TA biogenesis [177–179]. These GET independent functions might explain the controversially discussed four- to sevenfold molar excess of GET2 over GET1, its partner in the receptor complex [133,134,180]. GET5 is also related to other cellular functions and dysfunctions including actin cytoskeletal reorganization, cell migration [181], tumorigenesis [182], and bone development [183]. Those effects occur independently of its action in the GET chaperone cascade.

3.3. The Lonely SND Player—Mysterious All-Rounder

Congruent with the observation of protein transport in the absence of SRP and GET [120,121], a genetic screen in yeast uncovered the SND pathway as an alternative targeting route capable of delivering a broad range of substrates to the ER. So far, three proteins called Snd1, Snd2, and Snd3 have been assigned as components of the SND route [21]. Initial evidence for the involvement of the *SND2* and *SND3* genes in protein targeting arose from another yeast screen showing synthetic lethality when deletions of either component were combined with deletions of the *GET2* or *GET3* gene [184]. Further, SND overexpression has been shown to have the capacity to rescue a SRP growth defect in yeast. This led to the working model of alternative targeting pathways which have overlapping client spectra and compensate for each other. Strikingly, SND as a backup route to the ER seems to be conserved in human cells [21,22]. While a detailed mechanistic description of the pathway is still missing, the currently known client spectrum in mammalian cells comprises co-translational cargo such as membrane proteins with internal TMHs and GPI-anchored precursor proteins with a GPI attachment signal of low hydrophobicity [22,43,185]. TA proteins have likewise been found among its clients when the GET pathway was compromised [22,23]. The three components attributed to the yeast SND pathway include the ribosome-binding factor Snd1 and the heterodimeric ER-resident receptor Snd2/Snd3 [21,186]. TMEM208 was identified as the human Snd2 homolog hSnd2 (Figure 3) [22]. Two sequence paralogs of Snd2 have also been identified in plants, which await functional validation [187]. In contrast to the GET cascade, no cytosolic factor has so far been characterized in mammalian cells either for initial signal recognition or for the chaperoning or directed delivery to the ER membrane. Assuming a similar case as for GET2, where a missing sequence similarity with Get2 complicated its identification, database searches for structural homologues might be instrumental in finding the missing SND components in higher eucaryotes [188]. Eventually, a targeting factor of another pathway, such as SRP, might support SND-mediated targeting at least in some cases. The recent observation of a dual recognition of GPI-anchored proteins at both termini may argue for such a scenario [43], where the C-terminus is captured by an SND component and the N-terminus by SRP for subsequent targeting to the Sec61 complex. Such duality might demonstrate another example of a mixed targeting reaction where modules from different pathways are shuffled together to fulfill a joint targeting activity. In comparison to the model of alternative targeting pathways backing-up each other, single components may also back-up shortcomings in other pathways. Further, Talbot et al. have shown a connection between hSnd2 and the EMC complex, which comes with several implications [185]. First, the SND pathway might target cargo to different ER translocation machineries such as the Sec61 complex or EMC (see Section 5.3). Second, combined with the observation that calmodulin and SGTA act upstream of EMC, the SND receptor component may team up with different targeting factors and translocases depending on the type of cargo. Another explanation for the lack of a cytosolic SND component in higher eucaryotes implies that hSnd2 may initiate a localized translation at the ER membrane which would bypass the challenges of cargo shielding and targeted polypeptide delivery. Lastly, a combination of the above scenarios is possible, assuming multiple roles for hSnd2 or the corresponding SND receptor.

3.4. Comparison of Signal Recognition by the Various Chaperones Involved in Targeting

Looking at the diversity of ER targeting signals (see Section 2), hydrophobicity appears as a central feature. Accordingly, trans-acting factors that mediate targeting to the ER membrane show hydrophobic pockets or binding grooves adapted to the properties of their clients for selective and efficient recognition. In the following, we shortly discuss structural analogies between the various components for cargo binding as well as mechanistic similarities or differences in targeting. Although information about the initial capture of cargo is missing for the SND pathway, understanding the limitations of SRP and GET sheds light on the putative capacity of SND as a back-up system.

3.4.1. Cargo Capture by SRP

SRP binds highly hydrophobic cargo with the methionine-rich M domain of its SRP54 subunit [189]. Its hydrophobic groove is built by five helices and a connecting finger loop which supposedly provides high flexibility for the accommodation of various SPs and TMHs [78,190]. Cryo-EM structures of SRP engaged with a 21-residue TMH have revealed a density in the hydrophobic groove that has been attributed to an α-helix of $\sim$12 residues, corresponding to half of the TMH [191]. Assuming a similar mode of binding for cleavable SPs, the same length would account for the central h-domain as a crucial feature in cargo recognition. The M domain sits on the SRP RNA and is structurally connected via a flexible linker to the GTPase domain of SRP54 with implications for the further regulation of cargo binding [78,192]. Indeed, client specificity is provided by electrostatic interactions of the SRP RNA with charged residues in the targeting signal, such as the N-terminal polar region of cleavable SPs. Communication with the adjacent GTPase domain helps to coordinate between cargo and receptor binding. Mechanistic details revealed by Cryo-EM shed light on the hydrophobic groove which is pre-formed in the absence of a targeting signal and is shielded by amphipathic helices [191]. The binding of a hydrophobic signal displaces amphipathic helix 2 which then covers the targeting signal in a similar way to a protective lid. Intriguingly, this placeholder helix sets at the same time a hydrophobicity threshold for regulating the admission of cargo to the hydrophobic binding groove of SRP.

3.4.2. Cargo Capture by GET3

In the case of GET3, a composite hydrophobic groove is built during dimerization involving extensive interactions between the ATPase domains [193]. Such architecture allows communication between the nucleotide and TMH binding sites for the coordinated capture and release of cargo and the safe delivery from the upstream pre-targeting complex to the downstream ER receptor [33,194]. For instance, ATP binding stabilizes the assembly of hydrophobic helices at the dimer interface required for cargo binding. These helices are rich in methionine residues and those essential for TA binding have been attributed to the so-called "GET3-insert" including helix 8. It shows moderate evolutionary conservation with a varying length and methionine content. In animals, helix 8 has 3 methionine residues among 21 amino acids, while there are only 2 methionine residues among the 15 amino acids of the homologous helix in *S. cerevisiae* [195]. The apparent acquisition of additional methionine residues has been interpreted as a result of the specialization of GET3 in TA biogenesis which evolved from the bacterial ancestor ArsA, an arsenical pump without targeting activity. On the functional level, helix 8 has been described to act in a way similar to a dynamic lid that optionally shields the hydrophobic groove or the hydrophobic targeting signal from the aqueous environment. According to recent mechanistic insights, it engages the TMH first and guides it into the hydrophobic groove while preventing off-pathway chaperones from competing with Get3 for access to the TMH [196]. Strikingly, the lid accelerates the hand-over of cargo from Sgt2 to Get3 in cooperation with the pre-targeting complex. Thus, reminiscent of helix 2 in SRP54 setting the client specificity of the SRP pathway, helix 8 in GET3 may ensure privileged client transfer and pathway specificity in a network of chaperones. Interestingly, it has been debated whether the native targeting complex is comprised of a Get3 tetramer rather than a dimer, in which case a hydrophobic chamber is formed by two dimers allowing the binding of several signals and shielding all-round the TMH [194,197,198]. In this model, the role of helix 8 has been attributed to the stabilization of the tetramer interface. In contrast to these crystallography-based models where a hydrophobic chamber is formed for cargo binding in the closed state, a different model has been derived from single-molecule spectroscopy which has proposed the stable trapping of cargo in a dynamically open conformation of yeast Get3 acting as a "protean clamp" [199]. Here, the different modes of interaction allow the TMH to find the energetically best fit instead of locking into a fixed position in a pre-formed binding site. Moreover, helix 8 does not play a prominent role in this model. However, it solves the mechanistic dilemma of an exclusively closed Get3–cargo complex while allowing the high-affinity binding of the cargo with a dynamic and directed hand-over between factors. For the dimer, however, the overall hydrophobic area measures more than 3000 Å^2 of surface which is twice the size of the binding groove in the M domain of bacterial SRP [200]. The ends of the hydrophobic groove are lined by charged residues, thus defining a length of 30 Å tuned to accommodate sequences of ~20 amino acids, such as the TMH of TA proteins destined for the ER [78,200]. Client hydrophobicity is comparable to that of SRP. Since the TMHs of mitochondrial TA proteins are typically shorter, less hydrophobic, have a lower helicity, and are C-terminally flanked by positively charged residues, GET3 may have some capacity to discriminate between TA signals of different organelles [201–203]. As shown by a systematic study in yeast, Get3 is indeed sensitive to TMHs with variations in hydrophobicity and helicity, thus representing one of multiple filters in the GET pathway for selecting cargo [156]. The physicochemical properties of TA-TMHs from ER, mitochondria, and peroxisomes overlap [153], yet the capacity to detect and select differential C-terminal charges seems to be missing in the GET pathway [156]. Yeast Get3 was indeed found to bind and deliver mitochondrial TA proteins to the ER, which led to the conclusion that an as yet undefined mitochondrial targeting pathway might outcompete the GET pathway under cellular conditions [204].

3.4.3. Cargo Capture by SGTA

As the first selectivity filter in the GET pathway, SGTA has been shown to discriminate signals based on their hydrophobicity and helical content, including a certain distinction between ER and mitochondrial TA proteins [156,165]. Despite similar hydrophobicity thresholds of yeast Sgt2 and Get3, faster binding kinetics were assumed to increase the efficiency of Sgt2 in the rejection of unsuitable cargo [152,156,196]. Looking for features in yeast TMHs that allow better discrimination between different TA proteins, it was recently revealed that a so-called hydrophobic helical face geometry (the clustering of hydrophobic residues on one side of a TMH) was more successful in predicting the destined organelle than the overall hydrophobicity [205]. Strikingly, the finding was reflected by the results of a modeled structure and the binding behavior of yeast Sgt2. For human TA-TMHs, this correlation was less pronounced, hinting at a more complex scenario in the mammalian GET pathway. Here, a sequence of 11 residues within the TMH seems to be crucial for prediction, in combination with the C-terminal charge [206].

In contrast to the array of high-resolution crystal structures for GET3, the structural information of SGTA is limited. Homodimerization solely involves the N-terminus while the TPR domains in the center of the sequence typically interact with heat shock proteins [207]. Exploiting SAXS, NMR, and EPR experiments, the C-terminus has been characterized as functionally important in cargo binding containing one or two α-helical regions and a novel NNP and glutamine-rich domain [174]. In the proposed working model, the transient dimerization of the SGTA C-terminus results in the formation of a hydrophobic binding groove that captures cargo in a tweezer-like motion with all-around shielding of the TMH. How the binding and release of cargo is regulated without nucleotide binding and hydrolysis has yet to be determined. Recent computational modeling supported by biochemical data identified a similar hand structure for the binding of at least 11 residues, preferably containing clusters of leucine residues [205,208]. Since a typical TA-TMH has 18–20 amino acids, it was speculated that the two C-termini in a SGTA homodimer may occupy the full length of the signal, similar to the binding mode of calmodulin. However, a C-terminus is composed of five methionine-rich amphipathic helices forming a hydrophobic groove 15 Å in length. While a long N-terminal loop in yeast Sgt2 is supposed to shield the hydrophobic surface, the corresponding loop in human SGTA is shorter and may not be sufficient to complete the same action. Reminiscent of other chaperones in targeting, cargo binding leads to the stabilization of a rather unordered structure. By contrast, human SGTA appears to be more ordered with a higher content of glutamine residues, which may explain the differences in client binding, despite a similar hydrophobicity of the binding groove. Here, the binding to a more ordered structure causes lower entropic costs which may set a lower hydrophobicity threshold and, indeed, human SGTA was found to be less selective [205]. These data match with other findings indicating that SGTA contributes to the biogenesis of TA proteins with TMHs of high and low hydrophobicity [161]. How mitochondrial TA proteins are excluded from binding in the mammalian system remains to be defined in future studies. In sum, client specificity increases when moving downstream through the GET cascade. A similar working model is known from the HOP co-chaperone family that coordinates cargo transfer between HSP70 and HSP90 and to which an evolutionary connection has been suggested based on the similar domain organization of SGTA [208].

3.4.4. Cargo Capture by BAG6

Signal recognition by BAG6 is widely uncharacterized due to missing structural information. Several domains at the N-terminus, two proline-rich and a novel BUILD domain, have been shown to build a platform for the binding of hydrophobic sequences in different contexts [209]. Poly-ubiquitinated proteins, defective proteins with exposure of short hydrophobic stretches, ERAD substrates, but also GPI-anchored and TA proteins are among its clients, with putative binding of the latter to a more central region of BAG6 [169]. How discrimination with this variety of clients is achieved is currently not known. For TA targeting, however, the BAG6 C-terminus which acts as scaffolding for the GET pre-targeting complex has been shown to be sufficient [160]. Although the capture of cargo by BAG6 seems not to be directly associated with targeting, rather than degradation, its competitive binding to GET clients might still contribute to their targeting fate.

3.4.5. Cargo Capture by a Putative hSnd1

Sequence comparisons of the yeast counterpart have failed to identify the human hSnd1. Assuming that a functional homology may exist, the consideration of structural features could help to identify a cytosolic SND component in mammalian cells. The AlphaFold model of the 100 kDa yeast Snd1 protein shows a structurally complex molecule [210]. In the center of the folded molecule, a layer of seven β-sheets forms a platform that is partially confined by the charged residues of surrounding α-helices. Such a platform might be reminiscent of the hydrophobic groove described in other targeting factors. Yet, none of the 17 methionine residues of the Snd1 primary structure are part of the β-sheet assembly. Instead, the methionine residues appear in regions which are predicted with a low confidence score hinting at disordered domains that may stabilize and reconfigure upon cargo binding. Other characteristic structural motifs predicted for Snd1 comprise a 30 residue long coiled-coil domain in the middle of the sequence followed by an unusually acidic stretch. Structural studies of Snd1, eventually with a bound cargo, will elucidate the substrate binding mechanism and provide a rationale for the broad range of transported substrates.

3.4.6. Common Cargo Capture Principles Used by Some Cytosolic Targeting Factors

Comparing the evolved binding sites in the different targeting components showcases the general mechanistic characteristics and the potential of the client spectrum of each pathway. For instance, weak arrangements of amphipathic helices rich in methionine residues that are stabilized upon client binding and the displacement of a regulatory lid helix appear as common themes. Differences become apparent when comparing the binding kinetics and the energetic aspects of the varying targeting factors, co-chaperones, and off-pathway chaperones. All these have effects on the functional outcome, directionality, and specificity of an interaction. For instance, SRP appears as dominating targeting factor with priority over GET, though both share binding grooves of similar hydrophobicity. Another example is given by SGTA, where the hydrophobicity of the binding site alone seems not to be decisive, rather than its degree of structural order. Since focusing on crystal structures of hydrophobic grooves is not sufficient to explain the client spectrum, novel kinetic approaches (single-particle spectroscopy, molecular modeling), systematic screening for clients (trap mutants), and advances in the bioinformatic determination of characteristic features in the targeting signal complement the current view of how cargo recognition defines the targeting fate. In the native cell, of course, the probability for a putative interaction is equally important in a network of competing chaperone systems. Thus, it is also decisive for the targeting choice which targeting factor is given primary access to the client. Helices of the binding pockets play a regulatory role in providing or preventing interactions. Chances for additional stimuli that can regulate the targeting fate are provided by the different energy requirements of the pathways (GTP versus ATP), with GET3 reversibly switching between a targeting and a holdase activity in accordance with the energy status of the cell [176]. *HSND2* expression is transcriptionally regulated in response to hypoxia [211]. Another option for regulation involves post-translational modification. The phosphorylation or acetylation of a targeting signal or the central NTPase domain could disrupt communication with the cargo binding site for the inhibition of client capture (cf. alkylation of SRP with NEM [189]) and would be reminiscent of the temporary phosphorylation of some transit peptides destined for chloroplasts [212]. Despite all control, mistargeting may occur, for which separate regulation systems have been established by the cell, such as for the removal of proteins from the mitochondrial outer membrane or ER membrane [213,214]. Moreover, the dual targeting, re-targeting, or SURFing of proteins from one organelle to another potentiate the diversity of targeting networks [153,215,216].

4. The Interplay of Targeting Factors at the Ribosomal Exit Tunnel

The biogenesis of proteins destined for entry into the secretory pathway starts at a cytoplasmic ribosome, a megadalton-sized ribonucleoprotein complex composed of core components plus associated factors (Figure 4). Their flexible recruitment reflects the functional connection of protein synthesis with protein folding, quality control, and downstream cellular processes [217]. Variability in the ribosomal composition has likewise been interpreted as a functional specialization of distinct ribosomal populations that selectively translate certain mRNA [218]. The impact of ribosome heterogeneity on localized translation (initiation in the cytosol versus at the membrane of specific ER domains) or protein targeting remains to be further investigated [36,146]. However, it is tempting to speculate that alternative mechanisms of translation initiation influence the choice of a protein targeting pathway. These involve elements in the mRNA, such as internal ribosome entry sites or RNA hypoxia response elements (rHREs), which may prevalently occur in cargos that utilize the same ER targeting route [219–221]. Endowing the corresponding targeting factors with the same mRNA control element (e.g., rHRE) would allow the co-ordinated regulation of both processes, translation and targeting (cf., oxygen-regulated transcription of *HSND2* [211]). Likewise, distinct elements in the mRNA may directly dictate the choice of a targeting pathway by the recruitment of a certain targeting factor. Such capacity has indeed been described for mRNA with certain 3′ untranslated regions and SRP [222]. Specialized ribosomes may thus be primed for a specific targeting pathway and the selective translation of specific mRNAs coding for respective clients. All these aspects are refining our current view of ribosomal protein synthesis towards a more active role of the ribosome in proteostasis.

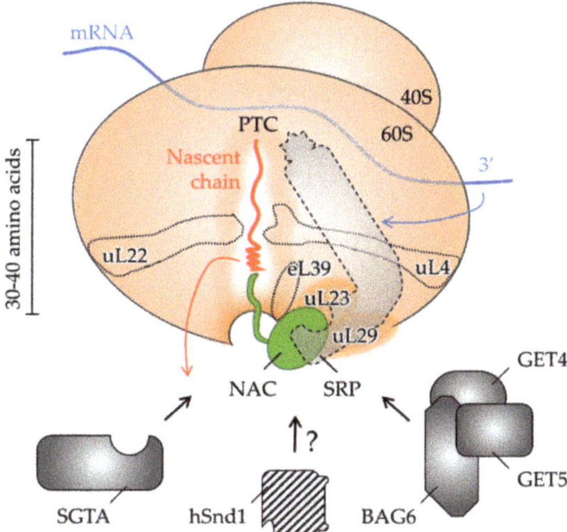

Figure 4. The impact of ribosome-binding proteins and the ribosomal exit tunnel geometry on the nascent chain and targeting decision. The nascent chain (red line) is born at the peptidyl transferase center (PTC) and folds in the vestibule of the lower ribosomal tunnel after passing the constriction sites build by the universally conserved ribosomal proteins uL22 and uL4 (dotted lines). Early recruitment of targeting factors involves the mRNA (blue line) and the presence and folding of the nascent chain while still residing inside the tunnel (blue and red arrow, respectively). Ribosomal surface proteins such as uL23 and uL29 (orange) which build common docking sites are mapped next to the exit port. Moreover, eL39 (dotted ellipse) that lines the interior of the ribosomal tunnel

and spans to its surface might contact the nascent chain and putative targeting factors. Similarly, NAC reaches deep into the tunnel where it scans nascent chains to coordinate between the various factors competing for the emerging nascent chain (black arrows). Accordingly, NAC may regulate the targeting priority of targeting factors including SRP, the BAG6 complex, SGTA, or the putative hSnd1 (hatched color fill) to ensure the targeting of different types of cargo. Note that NAC and SRP were mapped at overlapping sites next to the tunnel, while the location of alternative factors remains unknown. BAG6, BCL2-associated athanogene 6; GET, guided entry of TA proteins; NAC, nascent polypeptide-associated complex; SGTA, small glutamine-rich tetratricopeptide repeat co-chaperone alpha; SND, SRP-independent; SRP, signal recognition particle.

4.1. Shaping the Polypeptide Fate from within the Ribosomal Tunnel

As the protein synthesis machinery, the ribosome represents the first interaction partner that a polypeptide encounters on its way to the ER. The growing amino acid chain is born at the PTC which is buried inside the ribosome and connected to the surface and the cytosol via the ribosomal exit tunnel. Its length is typically assumed to cover 30–40 amino acids of the nascent chain [223]. The important features are the two constriction zones inside the tunnel and a vestibule near the exit port (Figure 4). The latter is evolutionary less conserved than the beginning of the tunnel close to the PTC. Moreover, due to the presence of the eucaryote-specific eL39 protein, the exit region of the tunnel is more confined in eucarya than bacteria [224]. Considering the space limitations, polypeptide folding inside the tunnel has long been assumed to be restricted to the formation of α-helices [225,226]. In procaryotes, however, the folding capacity of the exit tunnel has recently been expanded to encompass tertiary structures and small protein domains [227–229]. Regulation of these events involves the interplay between the nascent chain and the geometry and electrostatics of the exit tunnel [230–232]. Specifically, the negatively charged tunnel interior affects the translational speed and thus, the elongation rate and folding kinetics of the nascent polypeptide chain. All of these may influence the downstream interactions of the RNC and the choice of a targeting pathway. First, a decreased translational speed expands the time window for the recognition of the cargo by cytosolic factors before the release of the polypeptide chain from the ribosome. By comparison, the rate of translation elongation is slow in eucaryotes where at the same time complex chaperone systems are associated with ribosomal protein synthesis. Consequently, it has been proposed that evolution has tuned translation elongation rates to coordinate chaperone binding [233]. Speed can also be regulated by ribosome-binding proteins, such as the SRP with its Alu domain acting on the PTC. Likewise, properties associated with the nascent chain itself can modulate translation kinetics. The local slowdown of translation by non-optimal codons strategically placed at a distance of 35–40 amino acids from the N-terminus has been shown to prolong exposure of a nascent targeting signal close to the tunnel exit and improve SRP-dependent targeting [234]. In another case, translational pausing during XBP1 synthesis supports membrane targeting to the Sec61 complex by the exposure of a targeting signal [235]. Polypeptide folding may also have the capacity to slow down translation from within the tunnel, as was observed for C-terminal TMHs of TA proteins. Such a mechanism compensates for the very limited time window of TA recognition upon release from the PTC and improves recognition by the targeting factor BAG6 [236]. Furthermore, emergence from the tunnel may be delayed in case of fast and excessive folding. Varying secondary structures show distinct escape kinetics with α-helices that are usually faster than β-sheets [237,238]. Moreover, folded states of the nascent chain inside the tunnel contribute to the selective recruitment of cytosolic targeting factors to the ribosome. In particular, TMHs of strong hydrophobicity have been found to stimulate SRP binding for the subsequent targeting of respective clients to the ER [191]. Hence, translational slowdown and the early recruitment of targeting factors mediated by the nascent chain or mRNA elements (see above) ensures proper recognition by downstream interaction partners, including factors for co- and post-translational protein targeting.

4.2. Allosteric Crosstalk from within the Tunnel to the Outside

Conserved positive charges near the ribosomal exit port attract various cytosolic factors that contribute to protein folding, targeting, and quality control. However, the ribosome tunnel exit and the ribosomal surface are not just passive platforms for diverse interactions with cytosolic proteins [239]. Some ribosomal proteins that line the tunnel interior reach the outside of the ribosome and shape its surface (cf. eL39, Figure 4). Such structural organization predestines the allosteric signaling from the RNC inside the tunnel to the cytosol, thus coupling, for example, ongoing protein synthesis with targeting by selected factors or post-translational modifiers [240,241]. Moreover, it was recently uncovered that a cytosolic chaperone that occupies the exit port, the nascent-polypeptide-associated complex (NAC, see Section 4.3), reaches deep into the tunnel to sense a nascent chain before it emerges into the cytosol [242]. Similarly, the bacterial SRP has been shown to penetrate into the exit tunnel and scan for nascent chains during ribosomal protein synthesis [243]. Hence, crosstalk from the RNC to the cytosolic targeting factors may involve ribosomal tunnel proteins including NAC, SRP, and others.

4.3. Shaping the Polypeptide Fate from the Ribosomal Surface

A detailed understanding of which cytosolic factors interact with the ribosome and where on the ribosome is emerging and SRP is now understood fairly well. It shares an overlapping binding site at the tunnel exit with NAC, which implies a competition of different factors for the egressing nascent chain. Unlike the far less abundant SRP, NAC actually exists in equimolar concentration to the ribosome and seems to scan nascent polypeptides by reaching into the exit tunnel [242,244]. This priority access enables NAC to orchestrate downstream factors. For example, the occupation of ribosomes by NAC prevents the unproductive, premature interaction between ribosomes synthesizing non-secretory proteins and the Sec61 complex [245]. On the other hand, SRP promotes the binding of ribosomes that synthesize secretory proteins with proper targeting signals to the Sec61 complex. In this regard, NAC has controversially been shown to modulate the binding affinity of SRP for non-secretory RNCs, thereby influencing the specificity of SRP-mediated ER targeting [217,246]. While one model excludes the simultaneous binding of NAC and SRP to the ribosome, a second model favors the co-binding of both factors near the exit tunnel with NAC acting as a negative regulator that selectively displaces parts of SRP from the exit port [247–249]. As a consequence, NAC hampers the efficient targeting of SRP to its receptor in the ER membrane and in the case of "signal sequence-less" RNCs and other factors may gain access to the cargo. Besides the regulation by NAC, the mechanisms for nascent chain- and mRNA-dependent recruitment (Figure 4) compensate for the low abundance of SRP and prioritize the occupation of secretory RNCs with SRP [191,222,234].

SGTA and the BAG6 complex, both attributed to the GET pathway (Figure 3), are known to transiently bind the mammalian ribosome [150,166,236]. Their exact location remains unclear, but the binding sites seem to be different from SRP as simultaneous and non-competitive interactions have been observed. In accordance with the hydrophobic character of all ER targeting signals, early recruitment to the ribosome in response to the translation of respective sequences appears as a common mechanism [150,236]. Interestingly, yeast lacks a homolog of BAG6 with consequences for the hierarchical organization of the GET pathway compared to mammalian cells. Specifically, Get4/5 (GET4/5 homolog) binds the yeast ribosome overlapping with SRP via the Get5 subunit [250]. Sgt2 (SGTA homolog) is recruited to the tunnel exit only in a subsequent step. Hence, competitive scanning of the ribosome by SRP and Get4/5 seems to drive the targeting fate in yeast, in accordance with the high prevalence of the post-translational transport mode. By contrast, the ribosomal contact in mammalian cells is mediated by the BAG6 subunit and SGTA also binds to the ribosome independently of the BAG6 complex (Figure 4). Questions regarding which factor catches the cargo first and why an additional component, BAG6, is located at the mammalian ribosome are currently unanswered.

Furthermore, it has also been assumed that Snd1, the cytosolic component of the SND pathway in yeast, is recruited to the ribosome during mRNA translation [21,186]. Still, a homolog in mammalian cells has not yet been identified. How the corresponding clients with a central TMH are recognized and delivered to the mammalian ER thus remains to be clarified. In principle, several factors have the capacity to bind central targeting signals. First, SRP has been shown to be involved in all membrane protein targeting in yeast, including those with TMHs distant from the termini [125]. Second, SGTA may also bind internal TMHs, though it has been suggested that it assists SRP-mediated targeting by preventing aggregation, ubiquitination, and degradation [150].

Thus, the choice for a targeting pathway starts with the birth of the nascent chain inside the ribosome. The interplay between the elements of the cargo (mRNA motifs and codon choice, nascent chain sequence, and secondary structure), the ribosome (exit tunnel and surface), and the targeting factors influence the overall efficiency, specificity, and fidelity of the targeting process [248].

5. Different ER Protein Translocases Act as Membrane-Integrated Chaperones

Consistent with the multiplicity and complexity of targeting signals and targeting pathways described in the previous sections, recent findings in the field of ER protein import have also extended this concept to a small assortment of ER protein translocases. Different types and arrangements of protein translocases, sometimes acting in concert, manage the translocation or insertion of a dedicated subset of incoming polypeptides (Figure 5). The different multimeric protein complexes that catalyze the insertion of an unfolded polypeptide, nascent or full-length, behave in a way very similar to targeting factors acting as a temporary safe harbor. As such, a membrane-integrated protein translocase transiently shields segments of incoming cargo and thereby facilitates the partitioning of hydrophobic TMH into the ER membrane or the translocation of soluble domains into the ER lumen. Thus, incoming polypeptides are handed over from a soluble targeting factor to a membrane-integrated translocase, both of which provide a chaperone-like environment and prevent improper interactions and the aggregation of unfolded polypeptides. Here, we will summarize the central constituents of protein translocase complexes and refer the reader to excellent reviews found in this special issue on the "Mechanisms of ER Protein Import" as well as others [26,27,30,32].

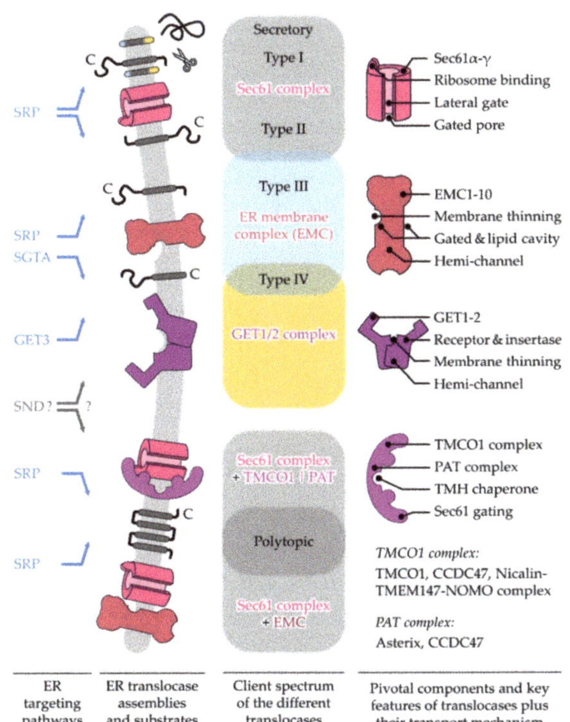

Figure 5. Features of protein translocation machines and their substrate spectrum. Depicted from the left to right are targeting pathways (blue), membrane-integrated protein translocation machines (pink, purple), their favored types of clients (boxes), and some key features for the central components of the protein translocation machines. Apart from GET1/2, the membrane-anchored receptor components for the targeting pathways are not shown. The scarcely characterized SND pathway (grey) as an alternative targeting route for a subset of GPI-anchored, tail-anchored, short secretory, and polytopic membrane proteins is indicated once. If SND delivers substrates either to a preferred ER translocase or a different one, or if its membrane-embedded component(s) might perform a dual function as receptor plus insertase similar to the GET1/2 complex remains to be seen. The auxiliary TMCO1 and PAT complexes are presented as one assembly. Of note, the PAT complex that was also described as a stand-alone intramembrane chaperone complex, is a heterodimer comprised of the indicated proteins CCDC47 and Asterix [251]. The TMCO1 containing translocon comprises the Sec61 channel and five accessory factors: TMCO1, CCDC47, and the Nicalin–TMEM147–NOMO complex [252]. For reasons of simplicity, we refer to those five accessory factors as "TMCO1 complex". The PAT and TMCO1 complex share at least one subunit, CCDC47. Further details of the TMCO1–PAT–Sec61 assembly and how they might act in concert to form an operational protein translocase can be found in the text. Signal peptides and transmembrane helices are integrated into the ER membrane and classified and colored according to Figures 1 and 2. The cleavage of signal peptides for secretory and type I membrane proteins is indicated (scissors and cleaved signal peptides are shown) as is the C-terminus (C) for each type of membrane protein. The spectra of substrates preferentially handled by assemblies that entail the Sec61 complexes are represented by grey boxes. Substrates handled by the ER membrane complex (EMC) and the GET1/2 complex are highlighted by light blue and yellow boxes, respectively. Overlapping boxes depict overlaps in the substrate range. CCDC47, coiled-coil domain containing protein 47; NOMO, nodal modulator protein; TMCO1, transmembrane and coiled-coil domain-containing protein 1; TMEM147, transmembrane protein 147.

5.1. The Sec61 Translocase—Director for SPs and the Majority of TMHs

The protein translocase embedded in the lipid bilayer of the ER described first was the heterotrimeric Sec61 complex (Figure 5). This universally conserved protein-conducting channel is composed of the pore-forming subunit Sec61α (Sec61p in yeast, SecY in bacteria) and two TA proteins, Sec61β and Sec61γ (Sbh1 and Ss1p in yeast, SecG and SecE in bacteria) [28,253–255]. Initially, a genetic screen of yeast mutants unable to translocate a SP-containing marker protein identified the yeast *SEC61* gene as an important constituent in the early stages of ER protein translocation [256,257]. Subsequently, the mammalian homolog of Sec61p was identified and its functional as well as structural characterization as the major polypeptide-conducting channel of the ER followed [79,258–261]. Complementing the functional conservation, many structural studies have also highlighted the conserved architecture of the Sec61 complex from bacteria and archaea to lower and higher eucaryotes [262–268]. In its closed conformation, the structural data depict the pore-forming Sec61α as a multi-spanning membrane protein with an hourglass-like configuration. Its funnels are oriented perpendicular to the plane of the ER membrane, thus facing the cytosol and the ER lumen [11]. The ten THMs are organized in a pseudo-symmetrical fashion with the first and last five TMHs forming an N-terminal as well as a C-terminal half of the molecule, respectively, and are connected by the short luminal loop5. The clamp-like movement of the Sec61 complex opens up the so-called lateral gate that is able to transiently harbor both types of sufficiently hydrophobic targeting signals, SPs and TMHs [80,264,269,270]. While SPs are usually cleaved off by a signal peptidase complex [271–273], TMHs are released either individually or as a small bundle into the ER membrane [112,274,275]. The lateral gate, which is mainly framed by the sterically adjacent TMH2 and TMH7 of Sec61α, also contributes to another important feature of the channel, the pore ring. The pore ring represents a constriction zone in the center of the Sec61 complex and consists of six bulky, hydrophobic residues, three of which are located in the lateral gate helices 2 and 7. Residues of the pore ring face towards the center and form a flexible gasket avoiding the excessive membrane permeability of small molecules and ions during the transport process. While certain exchanges of all six pore ring residues can be tolerated in yeast, the mutation of a single human pore ring residue such as V85D can have severe consequences causing primary antibody deficiency [276–278]. Based on the hourglass-like structural layout, the pore ring separates the two opposing funnels. In the idle state, the cytosolic funnel is water-filled, whereas the luminal funnel is occupied by another critical feature called the plug domain. The plug is formed by roughly 20 amino acids of the first luminal loop of the primary structure of Sec61α and supports the gating of the channel as a small, flexible helix. The transition from the closed to the open state of the Sec61 complex is associated with the widening of the pore ring and the displacement of the plug domain to form a conduit across the membrane [279–281]. However, some structural data also show snapshots of the transport process of TMH without major displacement of the plug [264]. Thus, the insertion of a TMH or a SP might have different requirements regarding the luminal and/or lateral opening of the Sec61 channel. Accordingly, additional membrane and soluble proteins, some of which will be described below, accompany the Sec61 complex and act as allosteric effectors that stimulate substrate-specific gating of the channel. Among different homologs of the Sec61 complex, the plug region shows low sequence conservation [31,282]. Similar to the pore ring, the plug shows a secondary role for function and cell viability in yeast, whereas a single dominant mutation in the human plug domain such as V67G can have severe consequences causing autosomal dominant tubulointerstitial kidney disease [282–284]. The polypeptides destined for transport by the Sec61 complex can be presented either co-translationally as a nascent chain in conjunction with the ribosome, or post-translationally as a full-length unfolded protein accompanied by a proper targeting factor [16,285]. Across all kingdoms of life, the Sec61 complex associates with different accessory factors to facilitate both variations of substrate transport, co- and post-translational [16,17,286–289]. Irremissible for substrate transport is the transition of the Sec61 complex from the closed to the open state. Unlike a simple lock, the opening of the Sec61 complex is a multifactorial, stepwise process mediated by the combination of a SP/TMH wedging itself into the lateral gate plus accessory factors such as the ribosome binding to a functionally conserved cytosolic docking port consisting of loops 6 and 8 [265,290,291]. In a mammalian setting, the cytosolic loops 6 and 8 of Sec61α are often referred to as the ribosome binding site. Yet, in light of existing structural and functional data from different organisms, this domain rather serves as a universal docking port for native (bacterial ATPase SecA, the translating ribosome, the ER membrane protein Sec63) as well as pseudo-native (heterologous anti-Sec61α F$_{ab}$ fragment, autologous Sec61 molecules arising from crystal packing) ligands [80,86,262,264–267,292–299]. Those ligands are auxiliary factors that act as biocatalysts and lower the activation energy to support the opening and closing of the Sec61 complex for efficient protein transport [291]. Both termini of Sec61α face the cytosol and the somewhat longer N-terminus contributes as a binding spot for other regulatory factors such as calmodulin or post-translational modifications such as N-acetylation for the efficient gating of the Sec61α protein in mammals and yeast, respectively [300,301]. Although fewer data are available for the C-terminus, it may support ER retention or, as was recently shown for the C-terminus of yeast Sec61p, it stimulates ribosome binding and co-translational protein transport [302].

5.1.1. Opening of the Sec61 Complex by Targeting Signals

The interplay of the structural key elements of the Sec61 complex together with targeting signals and accessory factors supports (i) the ER-specific entry and (ii) the proper topology of the transported substrates that are either fully translocated across, integrated via TMH into, or associated via a lipid-anchor with the ER membrane. Thus, hydrophobic targeting signals serve multiple purposes including targeting to as well as the gating of the Sec61 complex and may very well encode a post-translocation function for the topology and folding of the substrate itself. The latter is not surprising in the case of TMHs that are an integral part of the mature protein. Yet, some reports show the impact of cleavable SPs, or fragments thereof that are produced by the signal peptide peptidase that influence the folding of the downstream protein domain via specific chaperone recruitment or are used otherwise for the presentation of self-antigens via MHC class I [56,303–308]. Conclusions from research strategies such as in vitro protein import, photo-crosslinking studies, single Sec61 channel recordings from planar lipid bilayers experiments, and structural analyses of substrate-engaged Sec61 complexes have demonstrated the opening of the channel by SPs or TMHs of substrates, whereas SPs could be delivered co- or post-translationally [80,264,296,309–318]. To a certain degree, the process of opening the Sec61 complex by SPs and certain TMHs (excl. type III and type IV) resemble one another but the process has been studied best for a co-translationally directed SP, which will be outlined here. The binding of a translating ribosome to the universal docking port causes initial structural re-arrangements that prime the Sec61 complex and align the ribosomal exit tunnel with the nascent polypeptide plus SP on top of the cytosolic funnel. Moreover, the binding of the RNC causes the destabilization of a polar cluster within Sec61α which opens a crack in the cytosolic half of the lateral gate and exposes a single hydrophobic patch in the cytosolic funnel. The hydrophobic patch attracts the hydrophobic h-domain of an incoming SP and intercalates the targeting signal in the lateral gate while simultaneously supplanting the lateral gate of helix 2. Analogous to the placeholder helix that was described in client recognition by SRP above, helix 2 of Sec61α has been interpreted as a placeholder, setting the hydrophobicity threshold for productive interactions with targeting signals and productive channel gating. Some of the residues that form the polar cluster (two out of three amino acids), the hydrophobic patch (three out of four amino acids), and the pore ring (three out of six amino acids) all reside in the helices that form the lateral gate and further emphasize its importance as a critical structural element of the Sec61 complex [31]. Consequently, the intercalation of a targeting signal at the lateral gate has been extensively addressed in crosslinking reports [260,318–322] as well as computational analyses [323–325]. After engaging the Sec61 complex headfirst, biochemical assays, structural models, and coarse-grained modeling approaches have shown an inversion or a flip-turn of the SP as well as the corresponding TMH pendant of type II membrane proteins generating the N_{cyto}/C_{exo} orientation of those targeting signals housed in the lateral gate upon re-orientation [80,97,98,325]. At this point, the channel is fully open with the pore ring widened and the plug domain displaced [80,290]. While the hydrophilic sequence elements downstream of the SP can enter the ER lumen, hydrophobic TMH can partition laterally into the membrane. During the transport process, the pore ring residues surround the polypeptide in transit to preserve the permeability barrier for other small molecules and ions [276,279,292].

5.1.2. The Growing Family of Sec61 Complex-Associated Factors

This general concept of channel opening, substrate transport, and its topology determination is further influenced by many biophysical factors. For example, sequence composition as well as the length of the targeting signal and the following mature domain, the folding speed of flanking sequence elements, ribosomal translation speed, or the positive-inside rule all can affect translocation fidelity. In other words, the fruitful biogenesis of proteins that belong to the secretory pathway is not a "one molecule job". It relies on information that is encoded by the nascent chain plus additional mRNA sequence elements and is initially decoded by the ribosome, targeting factors, and the Sec61 complex plus its transiently or permanently associated proteins, as well as the lipid environment of the membrane, and is subsequently complemented by further maturation and quality control elements [27,42,95,119,270,323,326–332]. Recent structural, proteomic, and reconstituted import studies have addressed the contribution of different substrate-specific factors that support the Sec61 complex during the gating of "imperfect" SPs, a few of which are shortly highlighted. The TRAP complex supports the Sec61 complex during the co-translational transport of SPs with an above-average content of helix-breaking glycine-plus-proline residues [89,333–337]. Using classical reconstitution approaches, the translocating chain-associated membrane protein 1 (TRAM1) has been found to aid the insertion of nascent chains into the Sec61 complex when their SP has a shorter than average N-region [319,336,338,339]. Further data have implied that TRAM1 may also assist protein import by making the lipid bilayer in the vicinity of the lateral gate of the Sec61 complex conducive for accepting targeting signals [340]. Alternatively, the Sec62/63 complex assists the substrate-specific post-translational or co-translational opening of the Sec61 complex [83,85–87,293,295,296,332,341]. Proteomic abundance analysis in mammalian cells has identified SPs with longer but fewer hydrophobic h-regions plus a lower C-region polarity that is dependent on the Sec62/63 module [342]. While the TRAP and Sec62/63 complex seem to assist mainly in the transport of SP-carrying precursor proteins, other trans-acting factors have recently been shown to support the Sec61 complex during the insertion of "imperfect" TMHs, especially those of polytopic membrane proteins. The trans-acting factors in question that await a more detailed functional characterization in the near future are the TMCO1 (transmembrane and coiled-coil domain-containing protein 1) and PAT complex [112,251,252,343]. Both these modules consist of multiple ER membrane proteins. The combination of cryo-EM and crosslink-MS data upon the affinity purification of TMCO1–ribosome complexes identified proteins of the ribosome and the Sec61 complex, and also the Nicalin–TMEM147–NOMO complex as well as CCDC47 (coiled-coil domain containing protein 47). Furthermore, the use of affinity-purified TMCO1–ribosome complexes allowed McGilvray et al. to sequence ribosome-associated mRNAs that are enriched at the TMCO1 translocon. While mRNAs encoding for soluble as well as membrane-spanning secretory proteins with up to three TMHs were underrepresented, those with four and more TMHs showed a strong enrichment [252]. The beauty of the TMCO1 translocon structure (PDB: 6W6L) is amplified by its ambiguity, simultaneously answering as well as opening up questions regarding the regulation of the ER protein translocase. The TMCO1 complex occupies much of the space that is reserved during the co-translational import by the oligosaccharyltransferase (OST) complex including contacts to rRNA helices H19 and H25 in the cytosol or the volume taken by OST subunit Stt3a in the ER lumen [294,344]. Either different translocon modules rapidly exchange and rearrange during the insertion of glycosylated multi-spanning proteins or these proteins are specifically glycosylated post-translocationally by the OST paralog acting independently of the Sec61 complex [294,345,346]. Similar to OST and TRAP, the TMCO1 complex lingers on the "back side" of the Sec61 complex opposite the lateral gate as a second belt in addition to Sec61γ. This positioning makes it hard to envision a direct contact of a TMH with the lipid-filled cavity provided by the TMCO1 complex on the back side of Sec61. Yet, some of the unaccounted density of a TMH in the OST structure shown by Braunger et al. could represent the imperfect TMH1 of the bovine opsin substrate used for purification [294]. Eventually, this TMH1 meanders around the outskirts of the Sec61 complex to the back side to find the lipid-filled cavity of the TMCO1 complex. On the other hand, this positioning grants the TMEM147 subunit of the TMCO1 complex easy access to the hinge region (luminal loop 5) connecting the N- and C-terminal halves of Sec61α, thereby supporting the lateral opening and release of a TMH. The long α-helical arm of CCDC47 (Asn399-Lys474) in the cytosol moves in close proximity to the universal docking port (cytosolic loops 6 and 8) of Sec61α and the ribosomal protein uL22 at the ribosomal exit tunnel. Hence, CCDC47 might impact the opening of Sec61α, the directionality of an incoming TMH plus its flanking regions, or ribosomal translation. Last, the large luminal portion of Nicalin could aid translocated loops and domains of polytopic membrane proteins or, contrariwise, shield these domains from the luminal chaperone BiP from either ratcheting them or prematurely closing the Sec61 complex [347,348]. Overall, the players of the TMCO1 complex are arranged in a similar way to a relay system with CCDC47–TMCO1–TMEM147–Nicalin mainly contacting the ribosome-Sec61 complex ER lumen, respectively [252]. Interestingly, an independent study also identified CCDC47 with a functional link to membrane protein biogenesis [251]. The heterodimeric complex consisting of CCDC47 and Asterix (Pat-10) was termed PAT, short for proteins associated with the ER translocon. Using site-specific crosslinking and stability analyses of bicistronic reporter proteins via flow cytometry, the PAT complex was shown to facilitate the biogenesis of polytopic membrane proteins with imperfect, hydrophilic TMHs after such TMHs were inserted into the lipid bilayer [111,112,251]. Consequently, the PAT complex is considered to be an intramembrane chaperone. Via the shared subunit CCDC47, the PAT complex could act in conjunction with the TMCO1/Sec61 translocon. This "ménage à trois" creates an operational protein conducting channel whose active center (Sec61 complex) is supported by an allosteric effector (TMCO1 complex) and a folding assistant (PAT complex) for the proper biogenesis of polytopic membrane proteins with imperfect TMH (Figure 5). Interestingly, in this setup, at least one subunit of each module (Sec61α, TMCO1, CCDC47) in the Sec61–TMCO1–PAT complex has been shown to be involved in calcium homeostasis of the ER, highlighting a tight connection between protein transport, membrane permeability, and calcium signaling [349–351].

Different lines of evidence based on immunodepletion or the chemical inhibition of the Sec61 complex have revealed the limitations of the Sec61-centric translocases for the insertion of certain bi- and polytopic membrane proteins [99,352–354]. Most notably, membrane proteins carrying a type III or type IV TMH are more resistant to the inactivation of the Sec61 complex and instead can be inserted by the EMC or GET complexes. Similar to TMCO1, the EMC and GET complex harbor subunits that also belong to the Oxa1 superfamily [29,355].

5.2. The GET1/2 Complex—Post-Translational Machine for Type IV TMH of TA Proteins

A priori, the absence of an SP and the C-terminal localization of a TMH defined as type IV requires a specialized, i.e., a post-translational, targeting and insertion mechanism. Many components of this pathway are abbreviated by the acronym GET, guided entry of TA proteins [203]. TA polypeptides are fully synthesized and released from the ribosome while their type IV TMH is recognized by a pre-targeting complex and handed over to GET3 (see Section 3.2). Although some TA proteins have been reported to insert spontaneously into protein-free liposomes in vitro, the macromolecular crowding and the presence of different organellar target membranes under cellular conditions require an insertase for TA proteins [117,118,356]. In yeast and mammals, the ER proteins GET1 and GET2 form a complex (Figure 5) that first serves as a membrane-integrated receptor for incoming GET3–TA protein cargo and subsequently as an insertase for the C-terminal TMH [19,154,357–359]. Overall, the fundamentals of the GET system for the targeting and insertion of TA proteins resemble those of the Sec61 complex for secretory proteins. The four-step mechanism includes (i) the recognition of targeting signals by a targeting factor, (ii) the transfer of the cargo complex to an ER-localized receptor, (iii) the release of the cargo from the targeting factor for cargo insertion by a translocase, and (iv) the recycling of the targeting factor.

The GET1/2 Duality—Receptor and Insertase Function

Studies addressing the insertion mechanism of TA proteins by the GET1/2 complex point at a monomeric or homodimeric assembly of the GET1-GET2 heterodimer as a minimal functional unit [359,360]. Irrespective of the oligomerization, the GET1/2 complex attracts the active GET3 dimer that delivers a single TA protein. Both proteins, GET1 and GET2, have three TMHs and an extended cytoplasmic domain capable of binding to GET3. Thus, they likely cooperate in the targeting and release of a TA protein. The current model suggests that the long, unstructured N-terminal tether of GET2 captures the GET3–TA protein complex and brings it into proximity to the coiled-coil domain of GET1. In contrast to the tether of GET2, GET1's coiled-coil domain triggers a conformational change in the GET3–TA protein complex triggering the release of the cargo close to the insertase domain, which is ascribed to the TMHs of the GET1/2 complex and includes the conically shaped hydrophilic groove formed by the three TMHs of GET1 [153,154,358,359,361]. In contrast to the aqueous pore of the Sec61 complex traversing the entire ER membrane, the hydrophilic groove of the GET1/2 complex forms a discontinuous vestibule or "hemi-channel" that grants access to the cytosolic part of the membrane, but is sealed on the lumenal leaflet by an additional α-helix running parallel to the plane of the membrane [29]. Each one of the three TMHs of human GET1 carries at least one positively charged residue that rests in the cytosolic facing half of each TMH. Therefore, the hydrophilic vestibule might support both the membrane passage of the short, hydrophilic tails of the C-terminal of the type IV TMH and the "decoding" of the positive inside rule for proper TA protein topology by blocking the translocation of positive charges [93,330]. Membrane integration of a TA protein might further be stimulated by GET2-mediated membrane thinning. Unlike the archetypical 21 amino acid TMH running perpendicular through the membrane (Figure 2B), all three TMHs of human GET2 are shorter, comprising only 18–19 amino acids. The short TMHs of GET2 cause a local bilayer destabilization that promotes the integration of a TA protein. After the release and integration of the TA protein, the cargo- and nucleotide-free GET3 is recycled. The cytosolic GET components GET4/5 promote the rebinding of ATP in the empty nucleotide-binding pockets of the GET3 dimer and its dissociation from the GET1/2 complex [362]. Although both organisms, yeast and mouse, encode the TA proteins essential for viability, the deletion of the *GET* genes is tolerated by the yeast cells, whereas the constitutive genetic ablation of *GET3* or *GET2* causes embryonic lethality in mice [19,127,363]. The conditional knockout of GET components in certain tissues is viable in mice, but shows abnormalities sometimes related to improper TA protein biogenesis [129,130,364–366]. One assumption, mainly driven by the viability of GET-deficient yeast cells, was the existence of an alternative insertion machinery for TA proteins.

5.3. The EMC—Emcee for Type III, Type IV, and Charge-Containing TMHs

Similar to the central components of the GET pathway, the initial discovery of the EMC was achieved by high-throughput screening in yeast, where genetic interactions revealed their functional relationships [367]. The functional conservation and importance of EMC for the proper protein biogenesis of membrane proteins was subsequently demonstrated in many organisms including worms, zebrafish, flies, mice, and human cells [368–372]. Depending on the organism, EMC harbors 8–10 subunits, with seven of them being ER membrane proteins that are accompanied by up to three cytosolic proteins (EMC2/8/9) as is the case for the human EMC [373]. Detailed functional analyses based on reconstitution, proteomics, and proximity-specific ribosome profiling approaches narrowed down the substrate spectrum of EMC. This multimeric assembly preferentially supports the biogenesis of two classes of membrane proteins: TA proteins that are skipped by or are not ideally suited for GET3 and polytopic membrane proteins that either start with a type III TMH or carry charged and aromatic residues in the TMHs. All these types of TMH are difficult to handle for the Sec61 complex (Figure 5). In addition, mutational analyses of the TMHs of EMC substrates further supported these findings [99,157,372,374]. Considering the client spectrum, it appears that EMC can act in two different modes either as post-translational insertase for TA proteins or as co-translational insertase plus intra-membrane chaperone for individual TMHs of polytopic membrane proteins [157,251,375]. Similarly, the structure-guided mutational analyses of yeast and human EMC subunits have suggested the multifunctionality of this complex during the biogenesis of different membrane proteins [376–378]. Within the membrane-embedded core of the EMC, Miller-Vedam et al. defined two cavities, a lipid-filled and a gated one, on opposite sides of the transmembrane core and with different functionalities. The gated cavity is lined by portions of EMC3, EMC4, and EMC6 and seems to provide the actual conduit for the insertion of the terminal TMHs from either TA proteins or polytopic membrane proteins. This cavity resembles the hydrophilic vestibule of GET1/2 that promotes the access of TMHs to the lipid bilayer. Similar to TMCO1 and GET1, EMC3 also belongs to the class of Oxa1 superfamily members providing an evolutionary perspective on the conserved functional principle of these translocase subunits [29,355]. Overall, multiple cryo-EM structures of EMC and GET1/2 propose a similar mechanism for the integration of TMHs into the ER membrane. As described above for GET1/2, the EMC seems to adhere to the translocation-stimulating concept of the Oxa1 superfamily and employs local membrane thinning and a hydrophilic vestibule as a hemi-channel too [29,375–378]. With the limited size and discontinuity of the hemi-channel, one can appreciate the limited ability of EMC and GET1/2 to translocate larger flanking domains and their preference for N-terminal type III or C-terminal type IV TMHs, usually requiring the translocation of smaller flanking domains [26]. The function of the lipid-filled cavity that is built by portions of EMC1/3/5/6, is less clearly defined and more pleiotropic [376]. However, as this cavity shows a uniformly hydrophobic surface and appears to be accessible from the membrane or the ER lumen, it may provide a proper space for the intra-membrane chaperone and holdase function of the EMC.

The Intra-Membrane Handover between EMC and Sec61

A major difference between the Sec61 complex capable of opening a continuous, aqueous pore in comparison to the hemi-channels of EMC and GET1/2 is the translocation of larger domains into the ER lumen. Only the Sec61 complex seems to be efficient in transporting larger polypeptide stretches across the membrane. In the case of the Sec61 complex, this also includes soluble, secretory proteins that seem to be exclusively handled by this protein translocase. In contrast, membrane proteins with an initial type III stop-transfer or a type IV TMH that both require the translocation of shorter domains into the ER lumen are preferentially handled by EMC and GET1/2, respectively. Considering the size of EMC, ribosomes, SRP, and the Sec61 complex, a major question that might soon find an answer relates to the resulting steric constraints during the co-translational insertion of polytopic membrane proteins by EMC. This activity might entail the cooperation of EMC with the Sec61 complex and the RNC and require a uni- or bidirectional handover between the translocases. However, once the ribosome binds with high affinity to the Sec61 complex, the minimum distance between the lateral gate of Sec61α and the hemi-channel of EMC is approximately 110 Å [375]. Therefore, it appears rather unlikely that EMC captures TMHs at the Sec61 lateral gate and more likely that EMC mediates the insertion of TMHs before ribosomes bind to the Sec61 complex. This cooperation between two translocases, or intra-membrane chaperones, resembles the handover of polypeptide cargo from an upstream to a downstream targeting factor partially driven by increasing affinity, a concept that was discussed above.

The transport of polypeptides into the ER lumen or the ER membrane is usually followed by folding, modification, and eventually the assembly of proteins to achieve a native conformation. Similar to the targeting and translocation process, polypeptides that enter the ER encounter several molecular chaperones and co-chaperones that in this case support the folding and quality control process. Intriguing details about the impact of such factors on protein biogenesis have been reviewed previously and are not further discussed here [379–382].

6. Disease-Causing Mutations of Targeting and Translocation Components

As a result of the complex orchestration of the protein targeting network, the misfunctions of single players disrupt its sensitive balance and mutations of different protagonists have led to a broad spectrum of pathological phenomena. Disease-causing mutations have been identified in pivotal components in many of the protein targeting and transport pathways discussed before (Figure 6).

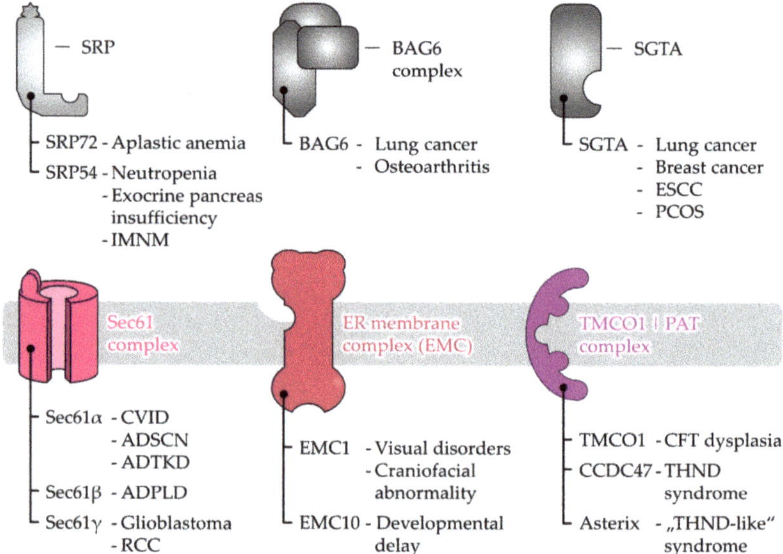

Figure 6. Disease associations of selected targeting and translocation components. The cartoon summarizes disease associations for critical subunits of targeting factors and translocation machines. Further details can be found in the text. ADPLD, autosomal dominant polycystic liver disease; ADSCN, autosomal dominant severe congenital neutropenia; ADTKD, autosomal dominant tubulo-interstitial kidney disease; CFT dysplasia, cerebro-facio-thoracic dysplasia; CVID, common variable immunodeficiency, ESCC, esophageal squamous cell carcinoma; IMNM, immune-mediated necrotizing myopathy; PCOS, polycystic ovary syndrome; RCC, renal cell carcinoma; THND syndrome, tricho-hepato-neuro-developmental syndrome.

6.1. Disease Associations of Protein Targeting Factors

Beginning with the soluble components of the targeting pathways, SRP72, one of the six proteins of the SRP, forms a complex with SRP68 and binds the 7S RNA to guide the pre-SRP complex from the nucleus to the cytosol [140,383,384]. Two mutations of the *SRP72* gene are known to cause the pathological phenotype of aplastic anemia, a developmental defect causing maturation defects of blood cells in the bone marrow that can progress into acute myeloid leukemia. The dominant SRP72 mutations result in a frame-shift causing the appearance of a premature stop codon (p.Thr355Lysfs*19) or in an amino acid exchange (p.Arg207His). While the truncated SRP72 is ineffective in binding the

7S RNA component, it has been speculated that the mutant SRP72-R207H protein binds less efficiently to ribosomal proteins, SRP68, or SRP54 [385,386]. Similarly, SRP54 appears to be disease-related. Its function has been described as an RNA binding protein and it mediates the interaction with the SRP receptor in the ER membrane [387–389]. Autosomal dominant mutations in *SRP54* are known from three different patients and result in neutropenia with similarities to the Shwachman-Diamond Syndrome. The three de novo missense variants of SRP54 all affect the conserved residues of the GTPase domain known to be critical for GTP and receptor binding. In two of the three patients, their neutropenia, due to the *SRP54* mutation, was also accompanied by an exocrine pancreatic insufficiency [390]. Of note, although not based on mutations in the *SRP54* gene, auto-antibodies directed against the SRP54 protein are considered as diagnostic biomarkers as well as pathogenic agents driving the progression of immune-mediated necrotizing myopathy, a muscle-specific autoimmune disease [391,392]. Mutations in the *SRP72* and *SRP54* genes demonstrate the importance of efficient protein secretion for the professional secretory cell types of the immune system and pancreas (Figure 6). Regarding the subunits of the GET pre-targeting complex BAG6 and co-chaperone SGTA, we wish to refer readers to a recent review on the roles of these cytosolic quality control proteins in disease [393]. Briefly, evidence from different cohorts and meta-analyses for the *BAG6* gene has linked certain single nucleotide polymorphisms to an increased risk for lung cancer and osteoarthritis [394–397]. Further experimental data have also suggested that the BAG6 protein is involved in male infertility and autoimmune disease [393]. The co-chaperone SGTA has been discussed in the context of different types of cancers including esophageal squamous cell carcinoma, breast cancer, and lung cancer. In all three cases, the SGTA protein showed an elevated abundance in cancerous tissue samples and was correlated with shorter survival rates [398–400]. Although the underlying mechanism is still unclear, these studies have reported on the impact of SGTA on cell proliferation and cell cycle progression. A single nucleotide polymorphism of the *SGTA* gene is associated with polycystic ovary syndrome, an endocrine disorder causing a hormonal imbalance in women characterized by increased levels of androgens [401]. Women suffering from the syndrome are more likely to develop endometrial cancer.

6.2. Disease Associations of Receptor and Protein Translocase Subunits

Aside from the protein targeting components, mutations have also been identified in genes encoding for subunits of the ER protein translocation machines (Figure 6). Pathologic functions and phenotypes connected with the Sec61 protein and its interaction partners Sec62 and Sec63 have recently been reviewed elsewhere [402,403]. Related diseases that arise from mutations or the overabundance of these proteins are diverse and include different types of cancer (Sec61γ, Sec62, Sec63), autosomal dominant polycystic liver disease (Sec61β, Sec63) as well as common variable immunodeficiency, neutropenia, and autosomal dominant tubulointerstitial kidney disease (Sec61α) [276,283,403–408]. Mutations in other allosteric effectors of the Sec61 channel such as the TRAP and OST complex are also the reason for severe pathologies based on disorders of N-glycosylation [409,410]. Components of the recently described TMCO1 and PAT complex are also disease associated upon mutation. As such, different autosomal recessive nonsense-mutations that entail the premature translational termination of TMCO1 (e.g., p.Ser47*, p.Arg87*, p.Ser98*, p.Arg114, etc.) have been sequenced that cause cerebro-facio-thoracic dysplasia. This multisystem developmental disorder causes intellectual disability, facial dysmorphism (a wide and short skull, highly arched eyebrows, widely spaced eyes), as well as abnormalities of the ribs and spinal bones [411–414]. The frequent malformation of bones in this context raises the question of whether the underlying pathogenic mechanism involves the functions of TMCO1 related to protein transport and/or calcium homeostasis. CCDC47, which appears to be part of the TMCO1 as well as of the PAT complex, has been associated with tricho-hepato-neuro-developmental syndrome. This multisystem disorder is characterized by woolly hair, pruritus (itching), hepatic dysfunction, general dysmorphic features, and developmental delay. Homozygous and compound heterozygous recessive alleles of the *CCDC47* gene can result in the absence of the functional CCDC47 protein. Its loss in patient fibroblasts has been shown to decrease calcium storage in the ER and insufficient calcium refilling during store-operated calcium entry [415]. While the experimental evidence demonstrates the dysregulation of calcium homeostasis upon loss of CCDC47, it is unclear if the patients with tricho-hepato-neuro-developmental syndrome also suffer from distinct abnormalities of protein transport related to the translocase and chaperone function of the TMCO1 and PAT complexes. For the other subunit of the PAT complex, Asterix (WDR83OS), a recessive variant has been reported to cause an unspecified syndrome with intractable itching, facial dysmorphia, microcephalus, hypercholanemia, short stature, and intellectual disability. Comparing these features with the ones from the loss of CCDC47, some similarities such as itching, hepatic dysfunction, dysmorphic features, and developmental delay are evident. The phenotypic overlap upon loss of either PAT complex component speaks in favor of the genetic and functional interaction of the two genes and the encoded wild-type proteins CCDC47 and Asterix.

Mutations in the subunits of the EMC are also related or causative for different diseases including neurological disorders or cancer. For instance, a defect of the EMC1 protein is related to visual disorders, craniofacial abnormalities, and epilepsy [416,417].A recessive loss of function related to the mutation of the *EMC10* gene has been reported in patients with intellectual disabilities and developmental delay [418]. Interestingly, the overexpression of either *EMC6* or *EMC10* has been reported to provide anti-tumor activity in glioblastoma cells. The tumor-suppressor activity upon the overexpression of those EMC subunits has been attributed to changes in gene expression that slow down cell proliferation, cell cycle progression, and tumor invasiveness or changes in signaling that increase autophagic flux via the inactivation of the mTOR pathway [419–421].

The majority of components of the GET pathway and its function are conserved from yeast to plants and mammals [153,422,423]. With regard to illness, the genomic locus of the *GET1* gene has been mapped to the congenital heart disease region of human chromosome 21 [424]. Yet, a clear correlation between *GET1* and its role in the development of cardiac defects and congenital heart disease has not been established in humans. However, direct evidence for this association comes from non-human studies. It has been shown that after the depletion of the GET1 protein in embryos of Xenopus frogs and Medaka fish, the development of their heart was initiated, but its morphogenesis could not be finalized [425,426]. A more targeted approach testing the tissue-specific knockout of the *GET1* gene in cardiomyocytes of mice reported hepatic damage and fibrosis, but no heart-specific phenotype was observed either [129]. Regarding the GET2 protein, studies on human cancer cell lines have presented evidence for an association of GET2 with skin and breast tumors due to the effects on either the calcium or prolactin receptor signaling pathways [427,428]. Yet, mechanistic details related to the biogenesis of TA proteins are scarce in these studies. Work on Myc-induced B-cell lymphoma cells from mice has described the relevance of GET2, in particular its C-terminal GET1 binding domain, for the survival and mitotic progression of lymphoma cells. In this case, the oncogenic potential of GET2 was shown to be independent of its TA protein insertion function [177]. Similarly, no mutation providing evidence for a gene-disease association with clinical significance of the GET3 protein is listed in the OMIM (www.omim.org, accessed 26 October 2021) or gnomAD (https://gnomad.broadinstitute.org, accessed 26 October 2021) database. Thus, recessive mutations that affect the functionality of the components of the GET pathway could eventually be compensated for and such compensation might suppress any disease phenotype. However, homozygosity (or compound heterozygosity) of mutant alleles might be detrimental as demonstrated by mouse models for *GET1-3* and the embryonic lethality of the constitutive knockouts [127,363,429].

Less is known about the SND protein targeting pathway in mammalian cells and its correlation with diseases or in vivo malfunctions. Considering that the knockout of SND components in yeast leads to defects in carboxypeptidase Y maturation (Snd2) or phosphate regulation and connectivity between nuclear envelope and vacuole (Snd3), there could be a defect in mammalian cells upon the loss of SND [430–432]. In 2019, Talbot et al.

generated a hSnd2 knockout in cultivated cells which resulted in the reduced expression of some polytopic membrane proteins including TRPC6 and KCNN4, two ion channels of the plasma membrane [185]. Maybe a connection between the human SND pathway and diseases might be based on those substrates of the pathway. As well as gain-of-function mutations, loss-of-function mutations of the *TRPC6* gene that can cause chronic kidney disease affecting the glomerulus have also been described. The corresponding disease is defined by a focal and segmental glomerulosclerosis phenotype [433–435]. The other SND substrate, KCNN4, has been associated with congenital hemolytic anemia. *KCNN4* mutations that were predicted to be deleterious perturb the cation permeability of cells and can lead to primary erythrocyte dehydration. The dysregulated volume homeostasis might affect the deformation capacity of erythrocytes and microcirculation which together could play a role as a pathogenic mechanism for hemolytic anemia [436,437]. However, direct disease associations of the SND pathway remain to be established.

All in all, the diseases caused by mutations of cytosolic or membrane-located targeting and transport factors show a broad phenotypic variance and affect different organs and tissues. This might be a result of the network organization of the different pathways and their central function in the biosynthesis of membrane or secreted proteins. Along the same line, defects of individual ER targeting, and translocation components might affect the efficient biogenesis of a limited client spectrum. The total loss or reduced abundance of tissue-specific proteins could also cause or contribute to a localized disease phenotype. As is the case of the widely present Sec61α protein, the majority of analyzed Sec61α point mutations result in distinct pathological phenotypes in different organs or tissues and eventually reflect the substrate-specific requirements found in the affected body part [276,283,404,438,439].

7. Conclusions and Perspectives

The mechanism of protein translocation through eucaryotic membranes is now more complicated than the mechanism described five decades ago. The original idea of a single targeting factor, a single receptor, and a single protein translocase has been massively expanded during the last years and now a variety of targeting routes and translocases dedicated to the transport of different polypeptide precursors has been shown. For certain proteins, it is clear how they are conducted to the ER, but some proteins can be delivered in alternative ways. The circumstance that allocates a polypeptide at a given time to use a certain transport pathway remains unclear.

The fate of the nascent polypeptide chain is majorly influenced by the type of targeting signal and hydrophobicity of the latter. It is not only the hydrophobicity that has an impact on the targeting mechanism, associated proteins and the local concentration of receptor components can all influence the transport destiny. It seems that different pathways can complement each other depending on the cellular environment. Cells may adjust to the different circumstances and fine-tune all the pathways to achieve the best-desired performance in terms of protein transport into the ER.

Clearly, more research is required to clear the possibilities and modes for protein translocation across the ER membrane. State-of-the-art structural and functional methods will unravel the enigmatic mechanisms of the different targeting routes and translocases. Even though we can find specific features for particular substrates, as always, there can be exceptions.

Author Contributions: A.T., M.S., D.H., S.H. and S.L. conceptualized the work; S.H. and S.L. acquired funding; A.T., M.S., D.H., S.H. and S.L. generated graphs and prepared the original draft; all authors reviewed and edited the manuscript. All authors have read and agreed to the published version of the manuscript.

Funding: This research was funded by the Deutsche Forschungsgemeinschaft (DFG, German Research Foundation) grants SFB 894 and IRTG 1830 (S.L.). A.T. was supported by the HOMFORexcellent program of Saarland University Medical Center. S.H. was supported by the Dean of Faculty and FGS Excellence postdoc fellowships of the Weizmann Institute of Science. We also acknowledge support by the DFG and Saarland University within the funding program Open Access Publishing.

Data Availability Statement: Sequences of human signal peptides and membrane proteins used for the analyses of Figures 1 and 2 were extracted from www.uniprot.org. The protein abundance of receptor subunits reported in Figure 3 were taken from Hein et al. [132].

Acknowledgments: We are grateful to our mentor Richard Zimmermann for his input and critical feedback over the years and for this review. We apologize to the many colleagues whose related work was inadvertently not cited here.

Conflicts of Interest: The authors declare no conflict of interest.

List of Abbreviations:

aa	amino acid
BAG6	BCL2-associated athanogene 6
cyto	cytoplasmic
CCDC47	coiled-coil domain containing protein 47
EMC	ER membrane protein complex
ER	endoplasmic reticulum
exo	exoplasmic
GET	guided entry of tail-anchored proteins
GPI	glycosylphosphatidylinositol
HSP	heat shock proteins
MP	membrane proteins
NAC	nascent-polypeptide-associated complex
NOMO	nodal modulator protein
OST	oligosaccharyltransferase
PAT	proteins associated with the ER translocon
PTC	peptidyl transferase center
rHRE	RNA hypoxia response elements
RNC	Ribosome-nascent chain complex
SGTA	small glutamine-rich tetratricopeptide repeat-containing protein alpha
SND	SRP-independent targeting pathway
SP	signal peptide
SR	SRP receptor
SRP	signal recognition particle
TA	tail-anchored
TMCO1	transmembrane and coiled-coil domain-containing protein 1
TMEM147	transmembrane protein 147
TMH	transmembrane helix
TRAM1	translocating chain-associated membrane protein 1
TRAP	translocon-associated protein

References

1. Baum, D.A.; Baum, B. An inside-out origin for the eukaryotic cell. *BMC Biol.* **2014**, *12*, 76. [CrossRef]
2. Embley, T.M.; Martin, W. Eukaryotic evolution, changes and challenges. *Nature* **2006**, *440*, 623–630. [CrossRef] [PubMed]
3. Sammels, E.; Parys, J.B.; Missiaen, L.; De Smedt, H.; Bultynck, G. Intracellular Ca2+ storage in health and disease: A dynamic equilibrium. *Cell Calcium* **2010**, *47*, 297–314. [CrossRef] [PubMed]
4. Feske, S. Calcium signalling in lymphocyte activation and disease. *Nat. Rev. Immunol.* **2007**, *7*, 690–702. [CrossRef] [PubMed]
5. Valm, A.M.; Cohen, S.; Legant, W.R.; Melunis, J.; Hershberg, U.; Wait, E.; Cohen, A.R.; Davidson, M.W.; Betzig, E.; Lippincott-Schwartz, J. Applying systems-level spectral imaging and analysis to reveal the organelle interactome. *Nature* **2017**, *546*, 162–167. [CrossRef] [PubMed]
6. Palade, G.E. The endoplasmic reticulum. *J. Biophys. Biochem. Cytol.* **1956**, *2*, 85–98. [CrossRef] [PubMed]

7. Pick, T.; Beck, A.; Gamayun, I.; Schwarz, Y.; Schirra, C.; Jung, M.; Krause, E.; Niemeyer, B.A.; Zimmermann, R.; Lang, S.; et al. Remodelling of Ca^{2+} homeostasis is linked to enlarged endoplasmic reticulum in secretory cells. *Cell Calcium* **2021**, *99*, 102473. [CrossRef] [PubMed]
8. Federovitch, C.M.; Ron, D.; Hampton, R.Y. The dynamic ER: Experimental approaches and current questions. *Curr. Opin. Cell Biol.* **2005**, *17*, 409–414. [CrossRef]
9. Aviram, N.; Schuldiner, M. Targeting and translocation of proteins to the endoplasmic reticulum at a glance. *J. Cell Sci.* **2017**, *130*, 4079–4085. [CrossRef] [PubMed]
10. Guerriero, C.J.; Brodsky, J.L. The Delicate Balance Between Secreted Protein Folding and Endoplasmic Reticulum-Associated Degradation in Human Physiology. *Physiol. Rev.* **2012**, *92*, 537–576. [CrossRef] [PubMed]
11. Zimmermann, R.; Eyrisch, S.; Ahmad, M.; Helms, V. Protein translocation across the ER membrane. *Biochim. Et Biophys. Acta* **2011**, *1808*, 912–924. [CrossRef]
12. Walter, P.; Ibrahimi, I.; Blobel, G. Translocation of proteins across the endoplasmic reticulum. I. Signal recognition protein (SRP) binds to in-vitro-assembled polysomes synthesizing secretory protein. *J. Cell Biol.* **1981**, *91*, 545–550. [CrossRef] [PubMed]
13. Gilmore, R.; Blobel, G.; Walter, P. Protein translocation across the endoplasmic reticulum. I. Detection in the microsomal membrane of a receptor for the signal recognition particle. *J. Cell Biol.* **1982**, *95*, 463–469. [CrossRef] [PubMed]
14. Panzner, S.; Dreier, L.; Hartmann, E.; Kostka, S.; Rapoport, T.A. Posttranslational protein transport in yeast reconstituted with a purified complex of Sec proteins and Kar2p. *Cell* **1995**, *81*, 561–570. [CrossRef]
15. Plath, K.; Rapoport, T.A. Spontaneous Release of Cytosolic Proteins from Posttranslational Substrates before Their Transport into the Endoplasmic Reticulum. *J. Cell Biol.* **2000**, *151*, 167–178. [CrossRef] [PubMed]
16. Johnson, N.; Powis, K.; High, S. Post-translational translocation into the endoplasmic reticulum. *Biochim. Et Biophys. Acta (BBA) Mol. Cell Res.* **2013**, *1833*, 2403–2409. [CrossRef] [PubMed]
17. Haßdenteufel, S.; Nguyen, D.; Helms, V.; Lang, S.; Zimmermann, R. ER import of small human presecretory proteins: Components and mechanisms. *FEBS Lett.* **2019**, *593*, 2506–2524. [CrossRef] [PubMed]
18. Stefanovic, S.; Hegde, R.S. Identification of a Targeting Factor for Posttranslational Membrane Protein Insertion into the ER. *Cell* **2007**, *128*, 1147–1159. [CrossRef] [PubMed]
19. Schuldiner, M.; Metz, J.; Schmid, V.; Denic, V.; Rakwalska, M.; Schmitt, H.D.; Schwappach, B.; Weissman, J.S. The GET Complex Mediates Insertion of Tail-Anchored Proteins into the ER Membrane. *Cell* **2008**, *134*, 634–645. [CrossRef]
20. Johnson, N.; Vilardi, F.; Lang, S.; Leznicki, P.; Zimmermann, R.; High, S. TRC40 can deliver short secretory proteins to the Sec61 translocon. *J. Cell Sci.* **2012**, *125*, 3612–3620. [CrossRef] [PubMed]
21. Aviram, N.; Ast, T.; Costa, E.A.; Arakel, E.C.; Chuartzman, S.G.; Jan, C.H.; Haßdenteufel, S.; Dudek, J.; Jung, M.; Schorr, S.; et al. The SND proteins constitute an alternative targeting route to the endoplasmic reticulum. *Nature* **2016**, *540*, 134–138. [CrossRef] [PubMed]
22. Haßdenteufel, S.; Sicking, M.; Schorr, S.; Aviram, N.; Fecher-Trost, C.; Schuldiner, M.; Jung, M.; Zimmermann, R.; Lang, S. hSnd2 protein represents an alternative targeting factor to the endoplasmic reticulum in human cells. *FEBS Lett.* **2017**, *591*, 3211–3224. [CrossRef] [PubMed]
23. Casson, J.; McKenna, M.; Haßdenteufel, S.; Aviram, N.; Zimmerman, R.; High, S. Multiple pathways facilitate the biogenesis of mammalian tail-anchored proteins. *J. Cell Sci.* **2017**, *130*, 3851–3861. [CrossRef] [PubMed]
24. Shao, S.; Hegde, R.S. A Calmodulin-Dependent Translocation Pathway for Small Secretory Proteins. *Cell* **2011**, *147*, 1576–1588. [CrossRef]
25. Haßdenteufel, S.; Schäuble, N.; Cassella, P.; Leznicki, P.; Müller, A.; High, S.; Jung, M.; Zimmermann, R. Ca2+-calmodulin inhibits tail-anchored protein insertion into the mammalian endoplasmic reticulum membrane. *FEBS Lett.* **2011**, *585*, 3485–3490. [CrossRef] [PubMed]
26. Hegde, R.S.; Keenan, R.J. The mechanisms of integral membrane protein biogenesis. *Nat. Rev. Mol. Cell Biol.* **2021**. [CrossRef] [PubMed]
27. O'Keefe, S.; Pool, M.R.; High, S. Membrane protein biogenesis at the ER: The highways and byways. *FEBS J.* **2021**, *n/a*. [CrossRef]
28. Park, E.; Rapoport, T.A. Mechanisms of Sec61/SecY-Mediated Protein Translocation Across Membranes. *Annu. Rev. Biophys.* **2012**, *41*, 21–40. [CrossRef] [PubMed]
29. McDowell, M.A.; Heimes, M.; Sinning, I. Structural and molecular mechanisms for membrane protein biogenesis by the Oxa1 superfamily. *Nat. Struct. Mol. Biol.* **2021**, *28*, 234–239. [CrossRef] [PubMed]
30. Gemmer, M.; Förster, F. A clearer picture of the ER translocon complex. *J. Cell Sci.* **2020**, *133*, jcs231340. [CrossRef] [PubMed]
31. Rapoport, T.A.; Li, L.; Park, E. Structural and Mechanistic Insights into Protein Translocation. *Annu. Rev. Cell Dev. Biol.* **2017**, *33*, 369–390. [CrossRef] [PubMed]
32. Chitwood, P.J.; Hegde, R.S. The Role of EMC during Membrane Protein Biogenesis. *Trends Cell Biol.* **2019**, *29*, 371–384. [CrossRef] [PubMed]
33. Mateja, A.; Keenan, R.J. A structural perspective on tail-anchored protein biogenesis by the GET pathway. *Curr. Opin. Struct. Biol.* **2018**, *51*, 195–202. [CrossRef] [PubMed]
34. Blobel, G.; Sabatini, D.D. Ribosome-Membrane Interaction in Eukaryotic Cells. In *Biomembranes: Volume 2, Manson, L.A., Ed.*; Springer: Boston, MA, USA, 1971; pp. 193–195.

35. Hannigan, M.M.; Hoffman, A.M.; Thompson, W.; Zheng, T.; Nicchitta, C.V. Quantitative proteomics links the LRRC59 interactome to mRNA translation on the ER membrane. *Mol. Cell. Proteom.* **2020**, *19*, 1826–1849. [CrossRef]
36. Bhadra, P.; Schorr, S.; Lerner, M.; Nguyen, D.; Dudek, J.; Förster, F.; Helms, V.; Lang, S.; Zimmermann, R. Quantitative Proteomics and Differential Protein Abundance Analysis after Depletion of Putative mRNA Receptors in the ER Membrane of Human Cells Identifies Novel Aspects of mRNA Targeting to the ER. *Molecules* **2021**, *26*, 3591. [CrossRef]
37. Pyhtila, B.; Zheng, T.; Lager, P.J.; Keene, J.D.; Reedy, M.C.; Nicchitta, C.V. Signal sequence- and translation-independent mRNA localization to the endoplasmic reticulum. *RNA* **2008**, *14*, 445–453. [CrossRef]
38. Béthune, J.; Jansen, R.-P.; Feldbrügge, M.; Zarnack, K. Membrane-Associated RNA-Binding Proteins Orchestrate Organelle-Coupled Translation. *Trends Cell Biol.* **2019**, *29*, 178–188. [CrossRef] [PubMed]
39. Cui, X.A.; Zhang, H.; Ilan, L.; Liu, A.X.; Kharchuk, I.; Palazzo, A.F. mRNA encoding Sec61β, a tail-anchored protein, is localized on the endoplasmic reticulum. *J. Cell Sci.* **2015**, *128*, 3398–3410. [CrossRef]
40. Cui, X.A.; Zhang, H.; Palazzo, A.F. p180 Promotes the Ribosome-Independent Localization of a Subset of mRNA to the Endoplasmic Reticulum. *PLoS Biol.* **2012**, *10*, e1001336. [CrossRef] [PubMed]
41. Hsu, J.C.C.; Reid, D.W.; Hoffman, A.M.; Sarkar, D.; Nicchitta, C.V. Oncoprotein AEG-1 is an endoplasmic reticulum RNA-binding protein whose interactome is enriched in organelle resident protein-encoding mRNAs. *RNA* **2018**, *24*, 688–703. [CrossRef] [PubMed]
42. Hegde, R.S.; Kang, S.-W. The concept of translocational regulation. *J. Cell Biol.* **2008**, *182*, 225–232. [CrossRef] [PubMed]
43. Yang, J.; Hirata, T.; Liu, Y.-S.; Guo, X.-Y.; Gao, X.-D.; Kinoshita, T.; Fujita, M. Human SND2 mediates ER targeting of GPI-anchored proteins with low hydrophobic GPI attachment signals. *FEBS Lett.* **2021**, *595*, 1542–1558. [CrossRef] [PubMed]
44. Davis, E.M.; Kim, J.; Menasche, B.L.; Sheppard, J.; Liu, X.; Tan, A.-C.; Shen, J. Comparative Haploid Genetic Screens Reveal Divergent Pathways in the Biogenesis and Trafficking of Glycophosphatidylinositol-Anchored Proteins. *Cell Rep.* **2015**, *11*, 1727–1736. [CrossRef] [PubMed]
45. Dhiman, R.; Caesar, S.; Thiam, A.R.; Schrul, B. Mechanisms of protein targeting to lipid droplets: A unified cell biological and biophysical perspective. *Semin. Cell Dev. Biol.* **2020**, *108*, 4–13. [CrossRef]
46. Allen, K.N.; Entova, S.; Ray, L.C.; Imperiali, B. Monotopic Membrane Proteins Join the Fold. *Trends Biochem. Sci.* **2019**, *44*, 7–20. [CrossRef] [PubMed]
47. Blobel, G.; Dobberstein, B. Transfer of proteins across membranes. I. Presence of proteolytically processed and unprocessed nascent immunoglobulin light chains on membrane-bound ribosomes of murine myeloma. *J. Cell Biol.* **1975**, *67*, 835–851. [CrossRef] [PubMed]
48. Blobel, G.; Dobberstein, B. Transfer of proteins across membranes. II. Reconstitution of functional rough microsomes from heterologous components. *J. Cell Biol.* **1975**, *67*, 852–862. [CrossRef] [PubMed]
49. Gilmore, R.; Blobel, G. Transient involvement of signal recognition particle and its receptor in the microsomal membrane prior to protein translocation. *Cell* **1983**, *35*, 677–685. [CrossRef]
50. Gilmore, R.; Walter, P.; Blobel, G. Protein translocation across the endoplasmic reticulum. II. Isolation and characterization of the signal recognition particle receptor. *J. Cell Biol.* **1982**, *95*, 470–477. [CrossRef] [PubMed]
51. Jackson, R.C.; Blobel, G. Post-translational cleavage of presecretory proteins with an extract of rough microsomes from dog pancreas containing signal peptidase activity. *Proc. Natl. Acad. Sci. USA* **1977**, *74*, 5598–5602. [CrossRef]
52. Walter, P.; Blobel, G. Purification of a membrane-associated protein complex required for protein translocation across the endoplasmic reticulum. *Proc. Natl. Acad. Sci. USA* **1980**, *77*, 7112–7116. [CrossRef] [PubMed]
53. Walter, P.; Blobel, G. Translocation of proteins across the endoplasmic reticulum. II. Signal recognition protein (SRP) mediates the selective binding to microsomal membranes of in-vitro-assembled polysomes synthesizing secretory protein. *J. Cell Biol.* **1981**, *91*, 551–556. [CrossRef] [PubMed]
54. Walter, P.; Blobel, G. Translocation of proteins across the endoplasmic reticulum III. Signal recognition protein (SRP) causes signal sequence-dependent and site-specific arrest of chain elongation that is released by microsomal membranes. *J. Cell Biol.* **1981**, *91*, 557–561. [CrossRef] [PubMed]
55. Walter, P.; Blobel, G. Signal recognition particle contains a 7S RNA essential for protein translocation across the endoplasmic reticulum. *Nature* **1982**, *299*, 691–698. [CrossRef] [PubMed]
56. Hegde, R.S.; Bernstein, H.D. The surprising complexity of signal sequences. *Trends Biochem. Sci.* **2006**, *31*, 563–571. [CrossRef] [PubMed]
57. von Heijne, G. Towards a comparative anatomy of N-terminal topogenic protein sequences. *J. Mol. Biol.* **1986**, *189*, 239–242. [CrossRef]
58. Nilsson, I.; Lara, P.; Hessa, T.; Johnson, A.E.; von Heijne, G.; Karamyshev, A.L. The Code for Directing Proteins for Translocation across ER Membrane: SRP Cotranslationally Recognizes Specific Features of a Signal Sequence. *J. Mol. Biol.* **2015**, *427*, 1191–1201. [CrossRef] [PubMed]
59. O'Neil, K.T.; DeGrado, W.F. A Thermodynamic Scale for the Helix-Forming Tendencies of the Commonly Occurring Amino Acids. *Science* **1990**, *250*, 646–651. [CrossRef]
60. Lyu, P.C.; Liff, M.I.; Marky, L.A.; Kallenbach, N.R. Side Chain Contributions to the Stability of Alpha-Helical Structure in Peptides. *Science* **1990**, *250*, 669–673. [CrossRef] [PubMed]

61. Lumangtad, L.A.; Bell, T.W. The signal peptide as a new target for drug design. *Bioorganic Med. Chem. Lett.* **2020**, *30*, 127115. [CrossRef] [PubMed]
62. von Heijne, G. Signal sequences: The limits of variation. *J. Mol. Biol.* **1985**, *184*, 99–105. [CrossRef]
63. von Heijne, G. The signal peptide. *J. Membr. Biol.* **1990**, *115*, 195–201. [CrossRef] [PubMed]
64. Gierasch, L.M. Signal sequences. *Biochemistry* **1989**, *28*, 923–930. [CrossRef] [PubMed]
65. Käll, L.; Krogh, A.; Sonnhammer, E.L.L. A Combined Transmembrane Topology and Signal Peptide Prediction Method. *J. Mol. Biol.* **2004**, *338*, 1027–1036. [CrossRef] [PubMed]
66. Almagro Armenteros, J.J.; Tsirigos, K.D.; Sønderby, C.K.; Petersen, T.N.; Winther, O.; Brunak, S.; von Heijne, G.; Nielsen, H. SignalP 5.0 improves signal peptide predictions using deep neural networks. *Nat. Biotechnol.* **2019**, *37*, 420–423. [CrossRef] [PubMed]
67. Kaiser, C.A.; Preuss, D.; Grisafi, P.; Botstein, D. Many Random Sequences Functionally Replace the Secretion Signal Sequence of Yeast Invertase. *Science* **1987**, *235*, 312–317. [CrossRef] [PubMed]
68. Beck, B.B.; Trachtman, H.; Gitman, M.; Miller, I.; Sayer, J.A.; Pannes, A.; Baasner, A.; Hildebrandt, F.; Wolf, M.T.F. Autosomal Dominant Mutation in the Signal Peptide of Renin in a Kindred With Anemia, Hyperuricemia, and CKD. *Am. J. Kidney Dis.* **2011**, *58*, 821–825. [CrossRef] [PubMed]
69. Živná, M.; Hůlková, H.; Matignon, M.; Hodaňová, K.; Vyleťal, P.; Kalbáčová, M.; Barešová, V.; Sikora, J.; Blažková, H.; Živný, J.; et al. Dominant Renin Gene Mutations Associated with Early-Onset Hyperuricemia, Anemia, and Chronic Kidney Failure. *Am. J. Hum. Genet.* **2009**, *85*, 204–213. [CrossRef]
70. Clissold, R.L.; Clarke, H.C.; Spasic-Boskovic, O.; Brugger, K.; Abbs, S.; Bingham, C.; Shaw-Smith, C. Discovery of a novel dominant mutation in the REN gene after forty years of renal disease: A case report. *BMC Nephrol.* **2017**, *18*, 234. [CrossRef]
71. Seppen, J.; Steenken, E.; Lindhout, D.; Bosma, P.J.; Oude Elferink, R.P.J. A mutation which disrupts the hydrophobic core of the signal peptide of bilirubin UDP-glucuronosyltransferase, an endoplasmic reticulum membrane protein, causes Crigler-Najjar type IIs. *FEBS Lett.* **1996**, *390*, 294–298. [CrossRef]
72. Sneitz, N.; Bakker, C.T.; de Knegt, R.J.; Halley, D.J.J.; Finel, M.; Bosma, P.J. Crigler-Najjar syndrome in The Netherlands: Identification of four novel UGT1A1 alleles, genotype–phenotype correlation, and functional analysis of 10 missense mutants. *Hum. Mutat.* **2010**, *31*, 52–59. [CrossRef]
73. Rothe, C.; Lehle, L. Sorting of invertase signal peptide mutants in yeast dependent and independent on the signal-recognition particle. *Eur. J. Biochem.* **1998**, *252*, 16–24. [CrossRef] [PubMed]
74. Williams, E.J.B.; Pal, C.; Hurst, L.D. The molecular evolution of signal peptides. *Gene* **2000**, *253*, 313–322. [CrossRef]
75. Rutkowski, D.T.; Lingappa, V.R.; Hegde, R.S. Substrate-specific regulation of the ribosome—Translocon junction by N-terminal signal sequences. *Proc. Natl. Acad. Sci. USA* **2001**, *98*, 7823–7828. [CrossRef] [PubMed]
76. Kang, S.-W.; Rane, N.S.; Kim, S.J.; Garrison, J.L.; Taunton, J.; Hegde, R.S. Substrate-Specific Translocational Attenuation during ER Stress Defines a Pre-Emptive Quality Control Pathway. *Cell* **2006**, *127*, 999–1013. [CrossRef] [PubMed]
77. Kim, S.J.; Mitra, D.; Salerno, J.R.; Hegde, R.S. Signal Sequences Control Gating of the Protein Translocation Channel in a Substrate-Specific Manner. *Dev. Cell* **2002**, *2*, 207–217. [CrossRef]
78. Keenan, R.J.; Freymann, D.M.; Walter, P.; Stroud, R.M. Crystal Structure of the Signal Sequence Binding Subunit of the Signal Recognition Particle. *Cell* **1998**, *94*, 181–191. [CrossRef]
79. Van den Berg, B.; Clemons, W.M., Jr.; Collinson, I.; Modis, Y.; Hartmann, E.; Harrison, S.C.; Rapoport, T.A. X-ray structure of a protein-conducting channel. *Nature* **2004**, *427*, 36–44. [CrossRef] [PubMed]
80. Voorhees, R.M.; Hegde, R.S. Structure of the Sec61 channel opened by a signal sequence. *Science* **2016**, *351*, 88–91. [CrossRef] [PubMed]
81. Ng, D.T.; Brown, J.D.; Walter, P. Signal sequences specify the targeting route to the endoplasmic reticulum membrane. *J. Cell Biol.* **1996**, *134*, 269–278. [CrossRef] [PubMed]
82. Waters, M.G.; Blobel, G. Secretory protein translocation in a yeast cell-free system can occur posttranslationally and requires ATP hydrolysis. *J. Cell Biol.* **1986**, *102*, 1543–1550. [CrossRef] [PubMed]
83. Lang, S.; Benedix, J.; Fedeles, S.V.; Schorr, S.; Schirra, C.; Schäuble, N.; Jalal, C.; Greiner, M.; Haßdenteufel, S.; Tatzelt, J.; et al. Different effects of Sec61α, Sec62 and Sec63 depletion on transport of polypeptides into the endoplasmic reticulum of mammalian cells. *J. Cell Sci.* **2012**, *125*, 1958–1969. [CrossRef] [PubMed]
84. Johnson, N.; Haßdenteufel, S.; Theis, M.; Paton, A.W.; Paton, J.C.; Zimmermann, R.; High, S. The Signal Sequence Influences Post-Translational ER Translocation at Distinct Stages. *PLoS ONE* **2013**, *8*, e75394. [CrossRef] [PubMed]
85. Lakkaraju, A.K.K.; Thankappan, R.; Mary, C.; Garrison, J.L.; Taunton, J.; Strub, K. Efficient secretion of small proteins in mammalian cells relies on Sec62-dependent posttranslational translocation. *Mol. Biol. Cell* **2012**, *23*, 2712–2722. [CrossRef]
86. Wu, X.; Cabanos, C.; Rapoport, T.A. Structure of the post-translational protein translocation machinery of the ER membrane. *Nature* **2019**, *566*, 136–139. [CrossRef] [PubMed]
87. Haßdenteufel, S.; Johnson, N.; Paton, A.W.; Paton, J.C.; High, S.; Zimmermann, R. Chaperone-Mediated Sec61 Channel Gating during ER Import of Small Precursor Proteins Overcomes Sec61 Inhibitor-Reinforced Energy Barrier. *Cell Rep.* **2018**, *23*, 1373–1386. [CrossRef] [PubMed]
88. Sicking, M.; Jung, M.; Lang, S. Lights, Camera, Interaction: Studying Protein–Protein Interactions of the ER Protein Translocase in Living Cells. *Int. J. Mol. Sci.* **2021**, *22*, 10358. [CrossRef] [PubMed]

89. Nguyen, D.; Stutz, R.; Schorr, S.; Lang, S.; Pfeffer, S.; Freeze, H.H.; Förster, F.; Helms, V.; Dudek, J.; Zimmermann, R. Proteomics reveals signal peptide features determining the client specificity in human TRAP-dependent ER protein import. *Nat. Commun.* **2018**, *9*, 3765. [CrossRef] [PubMed]

90. Krogh, A.; Larsson, B.; von Heijne, G.; Sonnhammer, E.L.L. Predicting transmembrane protein topology with a hidden markov model: Application to complete genomes11Edited by F. Cohen. *J. Mol. Biol.* **2001**, *305*, 567–580. [CrossRef] [PubMed]

91. Von Heijne, G. The membrane protein universe: What's out there and why bother? *J. Intern. Med.* **2007**, *261*, 543–557. [CrossRef]

92. Sharpe, H.J.; Stevens, T.J.; Munro, S. A Comprehensive Comparison of Transmembrane Domains Reveals Organelle-Specific Properties. *Cell* **2010**, *142*, 158–169. [CrossRef]

93. Baker, J.A.; Wong, W.-C.; Eisenhaber, B.; Warwicker, J.; Eisenhaber, F. Charged residues next to transmembrane regions revisited: "Positive-inside rule" is complemented by the "negative inside depletion/outside enrichment rule". *BMC Biol.* **2017**, *15*, 66. [CrossRef] [PubMed]

94. Höhr, A.I.C.; Lindau, C.; Wirth, C.; Qiu, J.; Stroud, D.A.; Kutik, S.; Guiard, B.; Hunte, C.; Becker, T.; Pfanner, N.; et al. Membrane protein insertion through a mitochondrial β-barrel gate. *Science* **2018**, *359*, eaah6834. [CrossRef] [PubMed]

95. Spiess, M.; Junne, T.; Janoschke, M. Membrane Protein Integration and Topogenesis at the ER. *Protein J.* **2019**, *38*, 306–316. [CrossRef] [PubMed]

96. Higy, M.; Junne, T.; Spiess, M. Topogenesis of Membrane Proteins at the Endoplasmic Reticulum. *Biochemistry* **2004**, *43*, 12716–12722. [CrossRef] [PubMed]

97. Devaraneni, P.K.; Conti, B.; Matsumura, Y.; Yang, Z.; Johnson, A.E.; Skach, W.R. Stepwise Insertion and Inversion of a Type II Signal Anchor Sequence in the Ribosome-Sec61 Translocon Complex. *Cell* **2011**, *146*, 134–147. [CrossRef] [PubMed]

98. Goder, V.; Spiess, M. Molecular mechanism of signal sequence orientation in the endoplasmic reticulum. *Embo J.* **2003**, *22*, 3645–3653. [CrossRef] [PubMed]

99. Chitwood, P.J.; Juszkiewicz, S.; Guna, A.; Shao, S.; Hegde, R.S. EMC Is Required to Initiate Accurate Membrane Protein Topogenesis. *Cell* **2018**, *175*, 1507–1519. [CrossRef]

100. O'Keefe, S.; Zong, G.; Duah, K.B.; Andrews, L.E.; Shi, W.Q.; High, S. An alternative pathway for membrane protein biogenesis at the endoplasmic reticulum. *Commun. Biol.* **2021**, *4*, 828. [CrossRef] [PubMed]

101. Junne, T.; Spiess, M. Integration of transmembrane domains is regulated by their downstream sequences. *J. Cell Sci.* **2017**, *130*, 372–381. [CrossRef]

102. von Heijne, G. Membrane-protein topology. *Nat. Rev. Mol. Cell Biol.* **2006**, *7*, 909. [CrossRef] [PubMed]

103. Nilsson, J.; Persson, B.; von Heijne, G. Comparative analysis of amino acid distributions in integral membrane proteins from 107 genomes. *Proteins Struct. Funct. Bioinform.* **2005**, *60*, 606–616. [CrossRef] [PubMed]

104. von Heijne, G. Control of topology and mode of assembly of a polytopic membrane protein by positively charged residues. *Nature* **1989**, *341*, 456–458. [CrossRef] [PubMed]

105. Kida, Y.; Morimoto, F.; Mihara, K.; Sakaguchi, M. Function of Positive Charges Following Signal-Anchor Sequences during Translocation of the N-terminal Domain *. *J. Biol. Chem.* **2006**, *281*, 1152–1158. [CrossRef] [PubMed]

106. Denzer, A.J.; Nabholz, C.E.; Spiess, M. Transmembrane orientation of signal-anchor proteins is affected by the folding state but not the size of the N-terminal domain. *EMBO J.* **1995**, *14*, 6311–6317. [CrossRef] [PubMed]

107. Arkin, I.T.; Brunger, A.T. Statistical analysis of predicted transmembrane α-helices. *Biochim. Et Biophys. Acta (BBA) Protein Struct. Mol. Enzymol.* **1998**, *1429*, 113–128. [CrossRef]

108. Dou, D.; Silva, D.V.d.; Nordholm, J.; Wang, H.; Daniels, R. Type II transmembrane domain hydrophobicity dictates the cotranslational dependence for inversion. *Mol. Biol. Cell* **2014**, *25*, 3363–3374. [CrossRef]

109. Zhang, L.; Sato, Y.; Hessa, T.; von Heijne, G.; Lee, J.-K.; Kodama, I.; Sakaguchi, M.; Uozumi, N. Contribution of hydrophobic and electrostatic interactions to the membrane integration of the Shaker K$^+$ channel voltage sensor domain. *Proc. Natl. Acad. Sci. USA* **2007**, *104*, 8263–8268. [CrossRef] [PubMed]

110. Öjemalm, K.; Halling, K.K.; Nilsson, I.; von Heijne, G. Orientational Preferences of Neighboring Helices Can Drive ER Insertion of a Marginally Hydrophobic Transmembrane Helix. *Mol. Cell* **2012**, *45*, 529–540. [CrossRef] [PubMed]

111. Ismail, N.; Crawshaw, S.G.; High, S. Active and passive displacement of transmembrane domains both occur during opsin biogenesis at the Sec61 translocon. *J. Cell Sci.* **2006**, *119*, 2826–2836. [CrossRef] [PubMed]

112. Ismail, N.; Crawshaw, S.G.; Cross, B.C.S.; Haagsma, A.C.; High, S. Specific transmembrane segments are selectively delayed at the ER translocon during opsin biogenesis. *Biochem. J.* **2008**, *411*, 495–506. [CrossRef] [PubMed]

113. Sato, M.; Hresko, R.; Mueckler, M. Testing the Charge Difference Hypothesis for the Assembly of a Eucaryotic Multispanning Membrane Protein. *J. Biol. Chem.* **1998**, *273*, 25203–25208. [CrossRef] [PubMed]

114. Sato, M.; Mueckler, M. A Conserved Amino Acid Motif (R-X-G-R-R) in the Glut1 Glucose Transporter Is an Important Determinant of Membrane Topology. *J. Biol. Chem.* **1999**, *274*, 24721–24725. [CrossRef] [PubMed]

115. Coelho, J.P.L.; Stahl, M.; Bloemeke, N.; Meighen-Berger, K.; Alvira, C.P.; Zhang, Z.-R.; Sieber, S.A.; Feige, M.J. A network of chaperones prevents and detects failures in membrane protein lipid bilayer integration. *Nat. Commun.* **2019**, *10*, 672. [CrossRef] [PubMed]

116. Doyle, D.A.; Cabral, J.M.; Pfuetzner, R.A.; Kuo, A.; Gulbis, J.M.; Cohen, S.L.; Chait, B.T.; MacKinnon, R. The Structure of the Potassium Channel: Molecular Basis of K$^+$ Conduction and Selectivity. *Science* **1998**, *280*, 69–77. [CrossRef] [PubMed]

117. Brambillasca, S.; Yabal, M.; Makarow, M.; Borgese, N. Unassisted translocation of large polypeptide domains across phospholipid bilayers. *J. Cell Biol.* **2006**, *175*, 767–777. [CrossRef]
118. Brambillasca, S.; Yabal, M.; Soffientini, P.; Stefanovic, S.; Makarow, M.; Hegde, R.S.; Borgese, N. Transmembrane topogenesis of a tail-anchored protein is modulated by membrane lipid composition. *EMBO J.* **2005**, *24*, 2533–2542. [CrossRef]
119. Dowhan, W.; Bogdanov, M. Lipid-Dependent Membrane Protein Topogenesis. *Annu. Rev. Biochem.* **2009**, *78*, 515–540. [CrossRef] [PubMed]
120. Ast, T.; Schuldiner, M. All roads lead to Rome (but some may be harder to travel): SRP-independent translocation into the endoplasmic reticulum. *Crit. Rev. Biochem. Mol. Biol.* **2013**, *48*, 273–288. [CrossRef] [PubMed]
121. Ast, T.; Cohen, G.; Schuldiner, M. A Network of Cytosolic Factors Targets SRP-Independent Proteins to the Endoplasmic Reticulum. *Cell* **2013**, *152*, 1134–1145. [CrossRef] [PubMed]
122. Schlenstedt, G.; Zimmermann, R. Import of frog prepropeptide GLa into microsomes requires ATP but does not involve docking protein or ribosomes. *EMBO J.* **1987**, *6*, 699–703. [CrossRef]
123. Wiech, H.; Sagstetter, M.; Müller, G.; Zimmermann, R. The ATP requiring step in assembly of M13 procoat protein into microsomes is related to preservation of transport competence of the precursor protein. *EMBO J.* **1987**, *6*, 1011–1016. [CrossRef] [PubMed]
124. Müller, G.; Zimmermann, R. Import of honeybee prepromelittin into the endoplasmic reticulum: Structural basis for independence of SRP and docking protein. *EMBO J.* **1987**, *6*, 2099–2107. [CrossRef] [PubMed]
125. Costa, E.A.; Subramanian, K.; Nunnari, J.; Weissman, J.S. Defining the physiological role of SRP in protein-targeting efficiency and specificity. *Science* **2018**, *359*, 689–692. [CrossRef] [PubMed]
126. Hann, B.C.; Walter, P. The signal recognition particle in S. cerevisiae. *Cell* **1991**, *67*, 131–144. [CrossRef]
127. Mukhopadhyay, R.; Ho, Y.-S.; Swiatek, P.J.; Rosen, B.P.; Bhattacharjee, H. Targeted disruption of the mouse Asna1 gene results in embryonic lethality. *FEBS Lett.* **2006**, *580*, 3889–3894. [CrossRef]
128. Daniele, L.L.; Emran, F.; Lobo, G.P.; Gaivin, R.J.; Perkins, B.D. Mutation of wrb, a Component of the Guided Entry of Tail-Anchored Protein Pathway, Disrupts Photoreceptor Synapse Structure and Function. *Investig. Ophthalmol. Vis. Sci.* **2016**, *57*, 2942–2954. [CrossRef] [PubMed]
129. Rivera-Monroy, J.; Musiol, L.; Unthan-Fechner, K.; Farkas, Á.; Clancy, A.; Coy-Vergara, J.; Weill, U.; Gockel, S.; Lin, S.-Y.; Corey, D.P.; et al. Mice lacking WRB reveal differential biogenesis requirements of tail-anchored proteins in vivo. *Sci. Rep.* **2016**, *6*, 39464. [CrossRef] [PubMed]
130. Vogl, C.; Panou, I.; Yamanbaeva, G.; Wichmann, C.; Mangosing, S.J.; Vilardi, F.; Indzhykulian, A.A.; Pangršič, T.; Santarelli, R.; Rodriguez-Ballesteros, M.; et al. Tryptophan-rich basic protein (WRB) mediates insertion of the tail-anchored protein otoferlin and is required for hair cell exocytosis and hearing. *EMBO J.* **2016**, *35*, 2536–2552. [CrossRef]
131. Coy-Vergara, J.; Rivera-Monroy, J.; Urlaub, H.; Lenz, C.; Schwappach, B. A trap mutant reveals the physiological client spectrum of TRC40. *J. Cell Sci.* **2019**, *132*, jcs230094. [CrossRef] [PubMed]
132. Hein, M.Y.; Hubner, N.C.; Poser, I.; Cox, J.; Nagaraj, N.; Toyoda, Y.; Gak, I.A.; Weisswange, I.; Mansfeld, J.; Buchholz, F.; et al. A Human Interactome in Three Quantitative Dimensions Organized by Stoichiometries and Abundances. *Cell* **2015**, *163*, 712–723. [CrossRef]
133. Carvalho, H.J.F.; Del Bondio, A.; Maltecca, F.; Colombo, S.F.; Borgese, N. The WRB Subunit of the Get3 Receptor is Required for the Correct Integration of its Partner CAML into the ER. *Sci. Rep.* **2019**, *9*, 11887. [CrossRef] [PubMed]
134. Colombo, S.F.; Cardani, S.; Maroli, A.; Vitiello, A.; Soffientini, P.; Crespi, A.; Bram, R.F.; Benfante, R.; Borgese, N. Tail-anchored protein biogenesis in mammals: Function and reciprocal interactions of the two subunits of the TRC40 receptor. *J. Biol. Chem.* **2016**, *291*, 15292–15306. [CrossRef] [PubMed]
135. Egea, P.F.; Shan, S.O.; Napetschnig, J.; Savage, D.F.; Walter, P.; Stroud, R.M. Substrate twinning activates the signal recognition particle and its receptor. *Nature* **2004**, *427*, 215–221. [CrossRef] [PubMed]
136. Lee, J.H.; Jomaa, A.; Chung, S.; Fu, Y.-H.H.; Qian, R.; Sun, X.; Hsieh, H.-H.; Chandrasekar, S.; Bi, X.; Mattei, S.; et al. Receptor compaction and GTPase rearrangement drive SRP-mediated cotranslational protein translocation into the ER. *Sci. Adv.* **2021**, *7*, eabg0942. [CrossRef] [PubMed]
137. Kobayashi, K.; Jomaa, A.; Lee, J.H.; Chandrasekar, S.; Boehringer, D.; Shan, S.-O.; Ban, N. Structure of a prehandover mammalian ribosomal SRP·SRP receptor targeting complex. *Science* **2018**, *360*, 323–327. [CrossRef] [PubMed]
138. Jomaa, A.; Fu, Y.-H.H.; Boehringer, D.; Leibundgut, M.; Shan, S.-O.; Ban, N. Structure of the quaternary complex between SRP, SR, and translocon bound to the translating ribosome. *Nat. Commun.* **2017**, *8*, 15470. [CrossRef]
139. Bohnsack, M.T.; Schleiff, E. The evolution of protein targeting and translocation systems. *Biochim. Et Biophys. Acta (BBA) Mol. Cell Res.* **2010**, *1803*, 1115–1130. [CrossRef] [PubMed]
140. Wild, K.; Becker, M.M.M.; Kempf, G.; Sinning, I. Structure, dynamics and interactions of large SRP variants. *Biol. Chem.* **2020**, *401*, 63–80. [CrossRef]
141. Halic, M.; Becker, T.; Pool, M.R.; Spahn, C.M.; Grassucci, R.A.; Frank, J.; Beckmann, R. Structure of the signal recognition particle interacting with the elongation-arrested ribosome. *Nature* **2004**, *427*, 808–814. [CrossRef] [PubMed]
142. Häsler, J.; Strub, K. Alu RNP and Alu RNA regulate translation initiation in vitro. *Nucleic Acids Res.* **2006**, *34*, 2374–2385. [CrossRef] [PubMed]
143. Lakkaraju, A.K.; Mary, C.; Scherrer, A.; Johnson, A.E.; Strub, K. SRP keeps polypeptides translocation-competent by slowing translation to match limiting ER-targeting sites. *Cell* **2008**, *133*, 440–451. [CrossRef] [PubMed]

144. Schibich, D.; Gloge, F.; Pöhner, I.; Björkholm, P.; Wade, R.C.; von Heijne, G.; Bukau, B.; Kramer, G. Global profiling of SRP interaction with nascent polypeptides. *Nature* **2016**, *536*, 219. [CrossRef] [PubMed]
145. Abell, B.M.; Pool, M.R.; Schlenker, O.; Sinning, I.; High, S. Signal recognition particle mediates post-translational targeting in eukaryotes. *EMBO J.* **2004**, *23*, 2755–2764. [CrossRef] [PubMed]
146. Reid, D.W.; Nicchitta, C.V. Diversity and selectivity in mRNA translation on the endoplasmic reticulum. *Nat. Rev. Mol. Cell Biol.* **2015**, *16*, 221–231. [CrossRef] [PubMed]
147. Jagannathan, S.; Reid, D.W.; Cox, A.H.; Nicchitta, C.V. De novo translation initiation on membrane-bound ribosomes as a mechanism for localization of cytosolic protein mRNAs to the endoplasmic reticulum. *RNA* **2014**, *20*, 1489–1498. [CrossRef] [PubMed]
148. Jan, C.H.; Williams, C.C.; Weissman, J.S. Principles of ER cotranslational translocation revealed by proximity-specific ribosome profiling. *Science* **2014**, *346*, 1257521. [CrossRef] [PubMed]
149. Chen, Q.; Jagannathan, S.; Reid, D.W.; Zheng, T.; Nicchitta, C.V. Hierarchical regulation of mRNA partitioning between the cytoplasm and the endoplasmic reticulum of mammalian cells. *Mol. Biol. Cell* **2011**, *22*, 2646–2658. [CrossRef] [PubMed]
150. Leznicki, P.; High, S. SGTA associates with nascent membrane protein precursors. *EMBO Rep.* **2020**, *21*, e48835. [CrossRef] [PubMed]
151. Kutay, U.; Hartmann, E.; Rapoport, T.A. A class of membrane proteins with a C-terminal anchor. *Trends Cell Biol.* **1993**, *3*, 72–75. [CrossRef]
152. Shan, S.-O. Guiding Tail-anchored Membrane Proteins to the ER In a Chaperone Cascade. *J. Biol. Chem.* **2019**, *294*, 16577–16586. [CrossRef] [PubMed]
153. Borgese, N.; Coy-Vergara, J.; Colombo, S.F.; Schwappach, B. The Ways of Tails: The GET Pathway and more. *Protein J.* **2019**, *38*, 289–305. [CrossRef] [PubMed]
154. Stefer, S.; Reitz, S.; Wang, F.; Wild, K.; Pang, Y.-Y.; Schwarz, D.; Bomke, J.; Hein, C.; Löhr, F.; Bernhard, F.; et al. Structural Basis for Tail-Anchored Membrane Protein Biogenesis by the Get3-Receptor Complex. *Science* **2011**, *333*, 758–762. [CrossRef]
155. Rome, M.E.; Rao, M.; Clemons, W.M.; Shan, S.-O. Precise timing of ATPase activation drives targeting of tail-anchored proteins. *Proc. Natl. Acad. Sci. USA* **2013**, *110*, 7666–7671. [CrossRef] [PubMed]
156. Rao, M.; Okreglak, V.; Chio, U.S.; Cho, H.; Walter, P.; Shan, S.-O. Multiple selection filters ensure accurate tail-anchored membrane protein targeting. *eLife* **2016**, *5*, e21301. [CrossRef]
157. Guna, A.; Volkmar, N.; Christianson, J.C.; Hegde, R.S. The ER membrane protein complex is a transmembrane domain insertase. *Science* **2018**, *359*, 470–473. [CrossRef] [PubMed]
158. Gristick, H.B.; Rome, M.E.; Chartron, J.W.; Rao, M.; Hess, S.; Shan, S.-O.; Clemons, W.M. Mechanism of Assembly of a Substrate Transfer Complex during Tail-anchored Protein Targeting. *J. Biol. Chem.* **2015**, *290*, 30006–30017. [CrossRef] [PubMed]
159. Darby, J.F.; Krysztofinska, E.M.; Simpson, P.J.; Simon, A.C.; Leznicki, P.; Sriskandarajah, N.; Bishop, D.S.; Hale, L.R.; Alfano, C.; Conte, M.R.; et al. Solution Structure of the SGTA Dimerisation Domain and Investigation of Its Interactions with the Ubiquitin-Like Domains of BAG6 and UBL4A. *PLoS ONE* **2014**, *9*, e113281. [CrossRef] [PubMed]
160. Mock, J.-Y.; Chartron, J.W.; Zaslaver, M.A.; Xu, Y.; Ye, Y.; Clemons, W.M. Bag6 complex contains a minimal tail-anchor–targeting module and a mock BAG domain. *Proc. Natl. Acad. Sci. USA* **2015**, *112*, 106–111. [CrossRef]
161. Guna, A.; Hegde, R.S. Transmembrane Domain Recognition during Membrane Protein Biogenesis and Quality Control. *Curr. Biol.* **2018**, *28*, R498–R511. [CrossRef]
162. Rodrigo-Brenni, M.C.; Gutierrez, E.; Hegde, R.S. Cytosolic Quality Control of Mislocalized Proteins Requires RNF126 Recruitment to Bag6. *Mol. Cell* **2014**, *55*, 227–237. [CrossRef] [PubMed]
163. Rodrigo-Brenni, M.C.; Hegde, R.S. Design Principles of Protein Biosynthesis-Coupled Quality Control. *Dev. Cell* **2012**, *23*, 896–907. [CrossRef] [PubMed]
164. Hessa, T.; Sharma, A.; Mariappan, M.; Eshleman, H.D.; Gutierrez, E.; Hegde, R.S. Protein targeting and degradation are coupled for elimination of mislocalized proteins. *Nature* **2011**, *475*, 394–397. [CrossRef] [PubMed]
165. Wang, F.; Brown, E.C.; Mak, G.; Zhuang, J.; Denic, V. A chaperone cascade sorts proteins for posttranslational membrane insertion into the endoplasmic reticulum. *Mol. Cell* **2010**, *40*, 159–171. [CrossRef]
166. Shao, S.; Rodrigo-Brenni, M.C.; Kivlen, M.H.; Hegde, R.S. Mechanistic basis for a molecular triage reaction. *Science* **2017**, *355*, 298–302. [CrossRef] [PubMed]
167. Wunderley, L.; Leznicki, P.; Payapilly, A.; High, S. SGTA regulates the cytosolic quality control of hydrophobic substrates. *J. Cell Sci.* **2014**, *127*, 4728–4739. [CrossRef] [PubMed]
168. Krysztofinska, E.M.; Martínez-Lumbreras, S.; Thapaliya, A.; Evans, N.J.; High, S.; Isaacson, R.L. Structural and functional insights into the E3 ligase, RNF126. *Sci. Rep.* **2016**, *6*, 26433. [CrossRef] [PubMed]
169. Leznicki, P.; Roebuck, Q.P.; Wunderley, L.; Clancy, A.; Krysztofinska, E.M.; Isaacson, R.L.; Warwicker, J.; Schwappach, B.; High, S. The Association of BAG6 with SGTA and Tail-Anchored Proteins. *PLoS ONE* **2013**, *8*, e59590. [CrossRef]
170. Figueiredo Costa, B.; Cassella, P.; Colombo, S.F.; Borgese, N. Discrimination between the endoplasmic reticulum and mitochondria by spontaneously inserting tail-anchored proteins. *Traffic* **2018**, *19*, 182–197. [CrossRef] [PubMed]
171. Cichocki, B.A.; Krumpe, K.; Vitali, D.G.; Rapaport, D. Pex19 is involved in importing dually targeted tail-anchored proteins to both mitochondria and peroxisomes. *Traffic* **2018**, *19*, 770–785. [CrossRef] [PubMed]

172. Cho, H.; Shim, W.J.; Liu, Y.; Shan, S.-O. J-domain proteins promote client relay from Hsp70 during tail-anchored membrane protein targeting. *J. Biol. Chem.* **2021**, *296*, 100546. [CrossRef]
173. Cho, H.; Shan, S.-O. Substrate relay in an Hsp70-cochaperone cascade safeguards tail-anchored membrane protein targeting. *EMBO J.* **2018**, *37*, e99264. [CrossRef]
174. Martínez-Lumbreras, S.; Krysztofinska, E.M.; Thapaliya, A.; Spilotros, A.; Matak-Vinkovic, D.; Salvadori, E.; Roboti, P.; Nyathi, Y.; Muench, J.H.; Roessler, M.M.; et al. Structural complexity of the co-chaperone SGTA: A conserved C-terminal region is implicated in dimerization and substrate quality control. *BMC Biol.* **2018**, *16*, 76. [CrossRef] [PubMed]
175. Hegde, R.S.; Zavodszky, E. Recognition and Degradation of Mislocalized Proteins in Health and Disease. *Cold Spring Harb. Perspect. Biol.* **2019**, *11*, a033902. [CrossRef] [PubMed]
176. Powis, K.; Schrul, B.; Tienson, H.; Gostimskaya, I.; Breker, M.; High, S.; Schuldiner, M.; Jakob, U.; Schwappach, B. Get3 is a holdase chaperone and moves to deposition sites for aggregated proteins when membrane targeting is blocked. *J. Cell Sci.* **2013**, *126*, 473–483. [CrossRef] [PubMed]
177. Shing, J.C.; Lindquist, L.D.; Borgese, N.; Bram, R.J. CAML mediates survival of Myc-induced lymphoma cells independent of tail-anchored protein insertion. *Cell Death Discov.* **2017**, *3*, 16098. [CrossRef] [PubMed]
178. Shing, J.C.; Bram, R.J. Yet another hump for CAML: Support of cell survival independent of tail-anchored protein insertion. *Cell Death Dis.* **2017**, *8*, e2960. [CrossRef] [PubMed]
179. Yamamoto, Y.; Sakisaka, T. The emerging role of calcium-modulating cyclophilin ligand in posttranslational insertion of tail-anchored proteins into the endoplasmic reticulum membrane. *J. Biochem.* **2015**, *157*, 419–429. [CrossRef] [PubMed]
180. Inglis, A.J.; Page, K.R.; Guna, A.; Voorhees, R.M. Differential Modes of Orphan Subunit Recognition for the WRB/CAML Complex. *Cell Rep.* **2020**, *30*, 3691–3698. [CrossRef] [PubMed]
181. Zhao, Y.; Lin, Y.; Zhang, H.; Mañas, A.; Tang, W.; Zhang, Y.; Wu, D.; Lin, A.; Xiang, J. Ubl4A is required for insulin-induced Akt plasma membrane translocation through promotion of Arp2/3-dependent actin branching. *Proc. Natl. Acad. Sci. USA* **2015**, *112*, 9644–9649. [CrossRef]
182. Wang, Y.; Ning, H.; Ren, F.; Zhang, Y.; Rong, Y.; Wang, Y.; Su, F.; Cai, C.; Jin, Z.; Li, Z.; et al. GdX/UBL4A Specifically Stabilizes the TC45/STAT3 Association and Promotes Dephosphorylation of STAT3 to Repress Tumorigenesis. *Mol. Cell* **2014**, *53*, 752–765. [CrossRef] [PubMed]
183. Liang, J.; Li, J.; Fu, Y.; Ren, F.; Xu, J.; Zhou, M.; Li, P.; Feng, H.; Wang, Y. GdX/UBL4A null mice exhibit mild kyphosis and scoliosis accompanied by dysregulation of osteoblastogenesis and chondrogenesis. *Cell Biochem. Funct.* **2018**, *36*, 129–136. [CrossRef] [PubMed]
184. Pan, X.; Ye, P.; Yuan, D.S.; Wang, X.; Bader, J.S.; Boeke, J.D. A DNA Integrity Network in the Yeast Saccharomyces cerevisiae. *Cell* **2006**, *124*, 1069–1081. [CrossRef]
185. Talbot, B.E.; Vandorpe, D.H.; Stotter, B.R.; Alper, S.L.; Schlondorff, J.S. Transmembrane insertases and N-glycosylation critically determine synthesis, trafficking, and activity of the nonselective cation channel TRPC6. *J. Biol. Chem.* **2019**, *294*, 12655–12669. [CrossRef] [PubMed]
186. Fleischer, T.C.; Weaver, C.M.; McAfee, K.J.; Jennings, J.L.; Link, A.J. Systematic identification and functional screens of uncharacterized proteins associated with eukaryotic ribosomal complexes. *Genes Dev.* **2006**, *20*, 1294–1307. [CrossRef]
187. Mehlhorn, D.G.; Asseck, L.Y.; Grefen, C. Looking for a safe haven: Tail-anchored proteins and their membrane insertion pathways. *Plant Physiol.* **2021**, *187*, 1916–1928. [CrossRef]
188. Borgese, N. Searching for remote homologs of CAML among eukaryotes. *Traffic* **2020**, *21*, 647–658. [CrossRef] [PubMed]
189. Lütcke, H.; High, S.; Römisch, K.; Ashford, A.J.; Dobberstein, B. The methionine-rich domain of the 54 kDa subunit of signal recognition particle is sufficient for the interaction with signal sequences. *EMBO J.* **1992**, *11*, 1543–1551. [CrossRef] [PubMed]
190. Bernstein, H.D.; Poritz, M.A.; Strub, K.; Hoben, P.J.; Brenner, S.; Walter, P. Model for signal sequence recognition from amino-acid sequence of 54K subunit of signal recognition particle. *Nature* **1989**, *340*, 482–486. [CrossRef]
191. Voorhees, R.M.; Hegde, R.S. Structures of the scanning and engaged states of the mammalian SRP-ribosome complex. *eLife* **2015**, *4*, e07975. [CrossRef]
192. Zopf, D.; Bernstein, H.; Johnson, A.; Walter, P. The methionine-rich domain of the 54 kd protein subunit of the signal recognition particle contains an RNA binding site and can be crosslinked to a signal sequence. *EMBO J.* **1990**, *9*, 4511–4517. [CrossRef] [PubMed]
193. Mateja, A.; Paduch, M.; Chang, H.-Y.; Szydlowska, A.; Kossiakoff, A.A.; Hegde, R.S.; Keenan, R.J. Structure of the Get3 targeting factor in complex with its membrane protein cargo. *Science* **2015**, *347*, 1152–1155. [CrossRef] [PubMed]
194. Bozkurt, G.; Stjepanovic, G.; Vilardi, F.; Amlacher, S.; Wild, K.; Bange, G.; Favaloro, V.; Rippe, K.; Hurt, E.; Dobberstein, B.; et al. Structural insights into tail-anchored protein binding and membrane insertion by Get3. *Proc. Natl. Acad. Sci. USA* **2009**, *106*, 21131–21136. [CrossRef]
195. Farkas, Á.; De Laurentiis, E.I.; Schwappach, B. The natural history of Get3-like chaperones. *Traffic* **2019**, *20*, 311–324. [CrossRef] [PubMed]
196. Chio, U.S.; Chung, S.; Weiss, S.; Shan, S.-o. A Chaperone Lid Ensures Efficient and Privileged Client Transfer during Tail-Anchored Protein Targeting. *Cell Rep.* **2019**, *26*, 37–44. [CrossRef] [PubMed]
197. Favaloro, V.; Vilardi, F.; Schlecht, R.; Mayer, M.P.; Dobberstein, B. Asna1/TRC40-mediated membrane insertion of tail-anchored proteins. *J. Cell Sci.* **2010**, *123*, 1522–1530. [CrossRef] [PubMed]

198. Suloway, C.J.; Rome, M.E.; Clemons, W.M., Jr. Tail-anchor targeting by a Get3 tetramer: The structure of an archaeal homologue. *EMBO J.* **2012**, *31*, 707–719. [CrossRef] [PubMed]
199. Chio, U.S.; Chung, S.; Weiss, S.; Shan, S.-O. A protean clamp guides membrane targeting of tail-anchored proteins. *Proc. Natl. Acad. Sci. USA* **2017**, *114*, E8585–E8594. [CrossRef] [PubMed]
200. Mateja, A.; Szlachcic, A.; Downing, M.E.; Dobosz, M.; Mariappan, M.; Hegde, R.S.; Keenan, R.J. The structural basis of tail-anchored membrane protein recognition by Get3. *Nature* **2009**, *461*, 361–366. [CrossRef] [PubMed]
201. Borgese, N.; Colombo, S.; Pedrazzini, E. The tale of tail-anchored proteins: Coming from the cytosol and looking for a membrane. *J. Cell Biol.* **2003**, *161*, 1013–1019. [CrossRef] [PubMed]
202. Borgese, N.; Brambillasca, S.; Colombo, S. How tails guide tail-anchored proteins to their destinations. *Curr. Opin. Cell Biol.* **2007**, *19*, 368–375. [CrossRef] [PubMed]
203. Borgese, N.; Fasana, E. Targeting pathways of C-tail-anchored proteins. *Biochim. Et Biophys. Acta (BBA) Biomembr.* **2011**, *1808*, 937–946. [CrossRef]
204. Vitali, D.G.; Sinzel, M.; Bulthuis, E.P.; Kolb, A.; Zabel, S.; Mehlhorn, D.G.; Figueiredo Costa, B.; Farkas, Á.; Clancy, A.; Schuldiner, M.; et al. The GET pathway can increase the risk of mitochondrial outer membrane proteins to be mistargeted to the ER. *J. Cell Sci.* **2018**, *131*, jcs211110. [CrossRef] [PubMed]
205. Lin, K.-F.; Fry, M.Y.; Saladi, S.M.; Clemons, W.M. The client-binding domain of the cochaperone Sgt2 has a helical-hand structure that binds a short hydrophobic helix. *bioRxiv* **2020**, 517573. [CrossRef]
206. Fry, M.Y.; Saladi, S.M.; Cunha, A.; Clemons, W.M., Jr. Sequence-based features that are determinant for tail-anchored membrane protein sorting in eukaryotes. *Traffic* **2021**, *22*, 306–318. [CrossRef]
207. Roberts, J.D.; Thapaliya, A.; Martínez-Lumbreras, S.; Krysztofinska, E.M.; Isaacson, R.L. Structural and Functional Insights into Small, Glutamine-Rich, Tetratricopeptide Repeat Protein Alpha. *Front. Mol. Biosci.* **2015**, *2*, 71. [CrossRef] [PubMed]
208. Lin, K.-F.; Fry, M.Y.; Saladi, S.M.; Clemons, W.M. Molecular basis of tail-anchored integral membrane protein recognition by the cochaperone Sgt2. *J. Biol. Chem.* **2021**, *296*, 100441. [CrossRef] [PubMed]
209. Tanaka, H.; Takahashi, T.; Xie, Y.; Minami, R.; Yanagi, Y.; Hayashishita, M.; Suzuki, R.; Yokota, N.; Shimada, M.; Mizushima, T.; et al. A conserved island of BAG6/Scythe is related to ubiquitin domains and participates in short hydrophobicity recognition. *FEBS J.* **2016**, *283*, 662–677. [CrossRef]
210. Jumper, J.; Evans, R.; Pritzel, A.; Green, T.; Figurnov, M.; Ronneberger, O.; Tunyasuvunakool, K.; Bates, R.; Žídek, A.; Potapenko, A.; et al. Highly accurate protein structure prediction with AlphaFold. *Nature* **2021**, *596*, 583–589. [CrossRef] [PubMed]
211. Lei, P.; Xiao, Y.; Li, P.; Xie, P.; Wang, H.; Huang, S.; Song, P.; Zhao, Y. hSnd2/TMEM208 is an HIF-1α-targeted gene and contains a WH2 motif. *Acta Biochim. Et Biophys. Sin.* **2020**, *52*, 328–331. [CrossRef] [PubMed]
212. Lee, D.W.; Jung, C.; Hwang, I. Cytosolic events involved in chloroplast protein targeting. *Biochim. Et Biophys. Acta (BBA) Mol. Cell Res.* **2013**, *1833*, 245–252. [CrossRef] [PubMed]
213. Dederer, V.; Khmelinskii, A.; Huhn, A.G.; Okreglak, V.; Knop, M.; Lemberg, M.K. Cooperation of mitochondrial and ER factors in quality control of tail-anchored proteins. *eLife* **2019**, *8*, e45506. [CrossRef] [PubMed]
214. McKenna, M.J.; Sim, S.I.; Ordureau, A.; Wei, L.; Harper, J.W.; Shao, S.; Park, E. The endoplasmic reticulum P5A-ATPase is a transmembrane helix dislocase. *Science* **2020**, *369*, eabc5809. [CrossRef] [PubMed]
215. Farkas, Á.; Bohnsack, K.E. Capture and delivery of tail-anchored proteins to the endoplasmic reticulum. *J. Cell Biol.* **2021**, *220*, e202105004. [CrossRef] [PubMed]
216. Koch, C.; Schuldiner, M.; Herrmann, J.M. ER-SURF: Riding the Endoplasmic Reticulum Surface to Mitochondria. *Int. J. Mol. Sci.* **2021**, *22*, 9655. [CrossRef] [PubMed]
217. Deuerling, E.; Gamerdinger, M.; Kreft, S.G. Chaperone Interactions at the Ribosome. *Cold Spring Harb. Perspect. Biol.* **2019**, *11*, a033977. [CrossRef]
218. Genuth, N.R.; Barna, M. The Discovery of Ribosome Heterogeneity and Its Implications for Gene Regulation and Organismal Life. *Mol. Cell* **2018**, *71*, 364–374. [CrossRef] [PubMed]
219. Melanson, G.; Timpano, S.; Uniacke, J. The eIF4E2-Directed Hypoxic Cap-Dependent Translation Machinery Reveals Novel Therapeutic Potential for Cancer Treatment. *Oxid. Med. Cell. Longev.* **2017**, *2017*, 6098107. [CrossRef]
220. Uniacke, J.; Holterman, C.E.; Lachance, G.; Franovic, A.; Jacob, M.D.; Fabian, M.R.; Payette, J.; Holcik, M.; Pause, A.; Lee, S. An oxygen-regulated switch in the protein synthesis machinery. *Nature* **2012**, *486*, 126–129. [CrossRef]
221. James, C.C.; Smyth, J.W. Alternative mechanisms of translation initiation: An emerging dynamic regulator of the proteome in health and disease. *Life Sci.* **2018**, *212*, 138–144. [CrossRef]
222. Chartron, J.W.; Hunt, K.C.L.; Frydman, J. Cotranslational signal-independent SRP preloading during membrane targeting. *Nature* **2016**, *536*, 224–228. [CrossRef]
223. Sabatini, D.D.; Blobel, G. Controlled proteolysis of nascent polypeptides in rat liver cell fractions: II. Location of the Polypeptides in Rough Microsomes. *J. Cell Biol.* **1970**, *45*, 146–157. [CrossRef]
224. Dao Duc, K.; Batra, S.S.; Bhattacharya, N.; Cate, J.H.D.; Song, Y.S. Differences in the path to exit the ribosome across the three domains of life. *Nucleic Acids Res.* **2019**, *47*, 4198–4210. [CrossRef] [PubMed]
225. Voss, N.R.; Gerstein, M.; Steitz, T.A.; Moore, P.B. The Geometry of the Ribosomal Polypeptide Exit Tunnel. *J. Mol. Biol.* **2006**, *360*, 893–906. [CrossRef] [PubMed]
226. Lu, J.; Deutsch, C. Folding zones inside the ribosomal exit tunnel. *Nat. Struct. Mol. Biol.* **2005**, *12*, 1123–1129. [CrossRef]

227. Wruck, F.; Tian, P.; Kudva, R.; Best, R.B.; von Heijne, G.; Tans, S.J.; Katranidis, A. The ribosome modulates folding inside the ribosomal exit tunnel. *Commun. Biol.* **2021**, *4*, 523. [CrossRef] [PubMed]

228. Marino, J.; von Heijne, G.; Beckmann, R. Small protein domains fold inside the ribosome exit tunnel. *FEBS Lett.* **2016**, *590*, 655–660. [CrossRef] [PubMed]

229. Liutkute, M.; Samatova, E.; Rodnina, M.V. Cotranslational Folding of Proteins on the Ribosome. *Biomolecules* **2020**, *10*, 97. [CrossRef]

230. Lu, J.; Deutsch, C. Electrostatics in the Ribosomal Tunnel Modulate Chain Elongation Rates. *J. Mol. Biol.* **2008**, *384*, 73–86. [CrossRef] [PubMed]

231. Farías-Rico, J.A.; Ruud Selin, F.; Myronidi, I.; Frühauf, M.; von Heijne, G. Effects of protein size, thermodynamic stability, and net charge on cotranslational folding on the ribosome. *Proc. Natl. Acad. Sci. USA* **2018**, *115*, E9280–E9287. [CrossRef] [PubMed]

232. Bui, P.T.; Hoang, T.X. Protein escape at the ribosomal exit tunnel: Effect of the tunnel shape. *J. Chem. Phys.* **2020**, *153*, 045105. [CrossRef] [PubMed]

233. Sharma, A.K.; O'Brien, E.P. Non-equilibrium coupling of protein structure and function to translation—Elongation kinetics. *Curr. Opin. Struct. Biol.* **2018**, *49*, 94–103. [CrossRef]

234. Pechmann, S.; Chartron, J.W.; Frydman, J. Local slowdown of translation by nonoptimal codons promotes nascent-chain recognition by SRP in vivo. *Nat. Struct. Mol. Biol.* **2014**, *21*, 1100–1105. [CrossRef]

235. Yanagitani, K.; Kimata, Y.; Kadokura, H.; Kohno, K. Translational Pausing Ensures Membrane Targeting and Cytoplasmic Splicing of XBP1u mRNA. *Science* **2011**, *331*, 586–589. [CrossRef] [PubMed]

236. Mariappan, M.; Li, X.; Stefanovic, S.; Sharma, A.; Mateja, A.; Keenan, R.J.; Hegde, R.S. A ribosome-associating factor chaperones tail-anchored membrane proteins. *Nature* **2010**, *466*, 1120–1124. [CrossRef] [PubMed]

237. Bui, P.T.; Hoang, T.X. Protein escape at the ribosomal exit tunnel: Effects of native interactions, tunnel length, and macromolecular crowding. *J. Chem. Phys.* **2018**, *149*, 045102. [CrossRef] [PubMed]

238. Bui, P.T.; Hoang, T.X. Folding and escape of nascent proteins at ribosomal exit tunnel. *J. Chem. Phys.* **2016**, *144*, 095102. [CrossRef] [PubMed]

239. Simsek, D.; Tiu, G.C.; Flynn, R.A.; Byeon, G.W.; Leppek, K.; Xu, A.F.; Chang, H.Y.; Barna, M. The Mammalian Ribo-interactome Reveals Ribosome Functional Diversity and Heterogeneity. *Cell* **2017**, *169*, 1051–1065. [CrossRef]

240. Guzel, P.; Yildirim, H.Z.; Yuce, M.; Kurkcuoglu, O. Exploring Allosteric Signaling in the Exit Tunnel of the Bacterial Ribosome by Molecular Dynamics Simulations and Residue Network Model. *Front. Mol. Biosci.* **2020**, *7*, 276. [CrossRef] [PubMed]

241. Berndt, U.; Oellerer, S.; Zhang, Y.; Johnson, A.E.; Rospert, S. A signal-anchor sequence stimulates signal recognition particle binding to ribosomes from inside the exit tunnel. *Proc. Natl. Acad. Sci. USA* **2009**, *106*, 1398–1403. [CrossRef]

242. Gamerdinger, M.; Kobayashi, K.; Wallisch, A.; Kreft, S.G.; Sailer, C.; Schlömer, R.; Sachs, N.; Jomaa, A.; Stengel, F.; Ban, N.; et al. Early Scanning of Nascent Polypeptides inside the Ribosomal Tunnel by NAC. *Mol. Cell* **2019**, *75*, 996–1006. [CrossRef] [PubMed]

243. Denks, K.; Sliwinski, N.; Erichsen, V.; Borodkina, B.; Origi, A.; Koch, H.-G. The signal recognition particle contacts uL23 and scans substrate translation inside the ribosomal tunnel. *Nat. Microbiol.* **2017**, *2*, 16265. [CrossRef] [PubMed]

244. Raue, U.; Oellerer, S.; Rospert, S. Association of Protein Biogenesis Factors at the Yeast Ribosomal Tunnel Exit Is Affected by the Translational Status and Nascent Polypeptide Sequence. *J. Biol. Chem.* **2007**, *282*, 7809–7816. [CrossRef]

245. Jungnickel, B.; Rapoport, T.A. A posttargeting signal sequence recognition event in the endoplasmic reticulum membrane. *Cell* **1995**, *82*, 261–270. [CrossRef]

246. Zhang, Y.; Berndt, U.; Gölz, H.; Tais, A.; Oellerer, S.; Wölfle, T.; Fitzke, E.; Rospert, S. NAC functions as a modulator of SRP during the early steps of protein targeting to the endoplasmic reticulum. *Mol. Biol. Cell* **2012**, *23*, 3027–3040. [CrossRef] [PubMed]

247. Wiedmann, B.; Sakai, H.; Davis, T.A.; Wiedmann, M. A protein complex required for signal-sequence-specific sorting and translocation. *Nature* **1994**, *370*, 434–440. [CrossRef] [PubMed]

248. Hsieh, H.-H.; Lee, J.H.; Chandrasekar, S.; Shan, S.-o. A ribosome-associated chaperone enables substrate triage in a cotranslational protein targeting complex. *Nat. Commun.* **2020**, *11*, 5840. [CrossRef] [PubMed]

249. del Alamo, M.; Hogan, D.J.; Pechmann, S.; Albanese, V.; Brown, P.O.; Frydman, J. Defining the specificity of cotranslationally acting chaperones by systematic analysis of mRNAs associated with ribosome-nascent chain complexes. *PLOS Biol.* **2011**, *9*, e1001100. [CrossRef]

250. Zhang, Y.; De Laurentiis, E.; Bohnsack, K.E.; Wahlig, M.; Ranjan, N.; Gruseck, S.; Hackert, P.; Wölfle, T.; Rodnina, M.V.; Schwappach, B.; et al. Ribosome-bound Get4/5 facilitates the capture of tail-anchored proteins by Sgt2 in yeast. *Nat. Commun.* **2021**, *12*, 782. [CrossRef] [PubMed]

251. Chitwood, P.J.; Hegde, R.S. An intramembrane chaperone complex facilitates membrane protein biogenesis. *Nature* **2020**, *584*, 630–634. [CrossRef] [PubMed]

252. McGilvray, P.T.; Anghel, S.A.; Sundaram, A.; Zhong, F.; Trnka, M.J.; Fuller, J.R.; Hu, H.; Burlingame, A.L.; Keenan, R.J. An ER translocon for multi-pass membrane protein biogenesis. *eLife* **2020**, *9*, e56889. [CrossRef] [PubMed]

253. Calo, D.; Eichler, J. Crossing the membrane in Archaea, the third domain of life. *Biochim. Et Biophys. Acta (BBA) Biomembr.* **2011**, *1808*, 885–891. [CrossRef] [PubMed]

254. du Plessis, D.J.F.; Nouwen, N.; Driessen, A.J.M. The Sec translocase. *Biochim. Et Biophys. Acta (BBA) Biomembr.* **2011**, *1808*, 851–865. [CrossRef] [PubMed]

255. Dalal, K.; Duong, F. The SecY complex: Conducting the orchestra of protein translocation. *Trends Cell Biol.* **2009**, *21*, 506–514. [CrossRef] [PubMed]
256. Deshaies, R.J.; Schekman, R. A yeast mutant defective at an early stage in import of secretory protein precursors into the endoplasmic reticulum. *J. Cell Biol.* **1987**, *105*, 633–645. [CrossRef] [PubMed]
257. Schekman, R. SEC mutants and the secretory apparatus. *Nat. Med.* **2002**, *8*, 1055–1058. [CrossRef] [PubMed]
258. Görlich, D.; Rapoport, T.A. Protein translocation into proteoliposomes reconstituted from purified components of the endoplasmic reticulum membrane. *Cell* **1993**, *75*, 615–630. [CrossRef]
259. Hartmann, E.; Sommer, T.; Prehn, S.; Görlich, D.; Jentsch, S.; Rapoport, T.A. Evolutionary conservation of components of the protein translocation complex. *Nature* **1994**, *367*, 654–657. [CrossRef] [PubMed]
260. Görlich, D.; Prehn, S.; Hartmann, E.; Kalies, K.-U.; Rapoport, T.A. A mammalian homolog of SEC61p and SECYp is associated with ribosomes and nascent polypeptides during translocation. *Cell* **1992**, *71*, 489–503. [CrossRef]
261. Kalies, K.U.; Görlich, D.; Rapoport, T.A. Binding of ribosomes to the rough endoplasmic reticulum mediated by the Sec61p-complex. *J. Cell Biol.* **1994**, *126*, 925–934. [CrossRef] [PubMed]
262. Egea, P.F.; Stroud, R.M. Lateral opening of a translocon upon entry of protein suggests the mechanism of insertion into membranes. *Proc. Natl. Acad. Sci. USA* **2010**, *107*, 17182–17187. [CrossRef] [PubMed]
263. Becker, T.; Bhushan, S.; Jarasch, A.; Armache, J.P.; Funes, S.; Jossinet, F.; Gumbart, J.; Mielke, T.; Berninghausen, O.; Schulten, K.; et al. Structure of monomeric yeast and mammalian Sec61 complexes interacting with the translating ribosome. *Science* **2009**, *326*, 1369–1373. [CrossRef] [PubMed]
264. Gogala, M.; Becker, T.; Beatrix, B.; Armache, J.-P.; Barrio-Garcia, C.; Berninghausen, O.; Beckmann, R. Structures of the Sec61 complex engaged in nascent peptide translocation or membrane insertion. *Nature* **2014**, *506*, 107–110. [CrossRef] [PubMed]
265. Voorhees, R.M.; Fernández, I.S.; Scheres, S.H.W.; Hegde, R.S. Structure of the Mammalian Ribosome-Sec61 Complex to 3.4 Å Resolution. *Cell* **2014**, *157*, 1632–1643. [CrossRef] [PubMed]
266. Zimmer, J.; Nam, Y.; Rapoport, T.A. Structure of a complex of the ATPase SecA and the protein-translocation channel. *Nature* **2008**, *455*, 936–943. [CrossRef] [PubMed]
267. Tanaka, Y.; Sugano, Y.; Takemoto, M.; Mori, T.; Furukawa, A.; Kusakizako, T.; Kumazaki, K.; Kashima, A.; Ishitani, R.; Sugita, Y.; et al. Crystal Structures of SecYEG in Lipidic Cubic Phase Elucidate a Precise Resting and a Peptide-Bound State. *Cell Rep.* **2015**, *13*, 1561–1568. [CrossRef] [PubMed]
268. Pfeffer, S.; Burbaum, L.; Unverdorben, P.; Pech, M.; Chen, Y.; Zimmermann, R.; Beckmann, R.; Förster, F. Structure of the native Sec61 protein-conducting channel. *Nat. Commun.* **2015**, *6*, 8403. [CrossRef] [PubMed]
269. du Plessis, D.J.F.; Berrelkamp, G.; Nouwen, N.; Driessen, A.J.M. The Lateral Gate of SecYEG Opens during Protein Translocation. *J. Biol. Chem.* **2009**, *284*, 15805–15814. [CrossRef] [PubMed]
270. Bischoff, L.; Wickles, S.; Berninghausen, O.; van der Sluis, E.O.; Beckmann, R. Visualization of a polytopic membrane protein during SecY-mediated membrane insertion. *Nat. Commun.* **2014**, *5*, 4103. [CrossRef] [PubMed]
271. Paetzel, M.; Karla, A.; Strynadka, N.C.J.; Dalbey, R.E. Signal Peptidases. *Chem. Rev.* **2002**, *102*, 4549–4580. [CrossRef]
272. Liaci, A.M.; Steigenberger, B.; Tamara, S.; de Souza, P.C.T.; Gröllers-Mulderij, M.; Ogrissek, P.; Marrink, S.J.; Scheltema, R.A.; Förster, F. Structure of the Human Signal Peptidase Complex Reveals the Determinants for Signal Peptide Cleavage. *bioRxiv* **2020**, *81*, 3934–3948. [CrossRef]
273. Kalies, K.-U.; Rapoport, T.A.; Hartmann, E. The beta-Subunit of the Sec61 Complex Facilitates Cotranslational Protein Transport and Interacts with the Signal Peptidase during Translocation. *J. Cell Biol.* **1998**, *141*, 887–894. [CrossRef] [PubMed]
274. Sadlish, H.; Pitonzo, D.; Johnson, A.E.; Skach, W.R. Sequential triage of transmembrane segments by Sec61α during biogenesis of a native multispanning membrane protein. *Nat. Struct. Mol. Biol.* **2005**, *12*, 870–878. [CrossRef] [PubMed]
275. Audigier, Y.; Friedlander, M.; Blobel, G. Multiple topogenic sequences in bovine opsin. *Proc. Natl. Acad. Sci. USA* **1987**, *84*, 5783–5787. [CrossRef] [PubMed]
276. Schubert, D.; Klein, M.-C.; Hassdenteufel, S.; Caballero-Oteyza, A.; Yang, L.; Proietti, M.; Bulashevska, A.; Kemming, J.; Kühn, J.; Winzer, S.; et al. Plasma cell deficiency in human subjects with heterozygous mutations in Sec61 translocon alpha 1 subunit (SEC61A1). *J. Allergy Clin. Immunol.* **2018**, *141*, 1427–1438. [CrossRef] [PubMed]
277. Gumbart, J.; Schulten, K. Molecular Dynamics Studies of the Archaeal Translocon. *Biophys. J.* **2006**, *90*, 2356–2367. [CrossRef] [PubMed]
278. Junne, T.; Kocik, L.; Spiess, M. The hydrophobic core of the Sec61 translocon defines the hydrophobicity threshold for membrane integration. *Mol. Biol. Cell* **2010**, *21*, 1662–1670. [CrossRef] [PubMed]
279. Park, E.; Rapoport, T.A. Preserving the membrane barrier for small molecules during bacterial protein translocation. *Nature* **2011**, *473*, 239–242. [CrossRef]
280. Saparov, S.M.; Erlandson, K.; Cannon, K.; Schaletzky, J.; Schulman, S.; Rapoport, T.A.; Pohl, P. Determining the Conductance of the SecY Protein Translocation Channel for Small Molecules. *Mol. Cell* **2007**, *26*, 501–509. [CrossRef] [PubMed]
281. Tam, P.C.; Maillard, A.P.; Chan, K.K.; Duong, F. Investigating the SecY plug movement at the SecYEG translocation channel. *EMBO J.* **2005**, *24*, 3380–3388. [CrossRef] [PubMed]
282. Li, W.; Schulman, S.; Boyd, D.; Erlandson, K.; Beckwith, J.; Rapoport, T.A. The Plug Domain of the SecY Protein Stabilizes the Closed State of the Translocation Channel and Maintains a Membrane Seal. *Mol. Cell* **2007**, *26*, 511–521. [CrossRef] [PubMed]

283. Bolar, N.A.; Golzio, C.; Živná, M.; Hayot, G.; Van Hemelrijk, C.; Schepers, D.; Vandeweyer, G.; Hoischen, A.; Huyghe, J.R.; Raes, A.; et al. Heterozygous Loss-of-Function *SEC61A1* Mutations Cause Autosomal-Dominant Tubulo-Interstitial and Glomerulocystic Kidney Disease with Anemia. *Am. J. Hum. Genet.* **2016**, *99*, 174–187. [CrossRef] [PubMed]
284. Junne, T.; Schwede, T.; Goder, V.; Spiess, M. The Plug Domain of Yeast Sec61p Is Important for Efficient Protein Translocation, but Is Not Essential for Cell Viability. *Mol. Biol. Cell* **2006**, *17*, 4063–4068. [CrossRef]
285. Haßdenteufel, S.; Klein, M.-C.; Melnyk, A.; Zimmermann, R. Protein transport into the human ER and related diseases, Sec61-channelopathies. *Biochem. Cell Biol.* **2014**, *92*, 499–509. [CrossRef] [PubMed]
286. Osborne, A.R.; Rapoport, T.A.; van den Berg, B. Protein translocation by the Sec61/SecY channel. *Annu. Rev. Cell Dev. Biol.* **2005**, *21*, 529–550. [CrossRef]
287. Aviram, N.; Schuldiner, M. Embracing the void—how much do we really know about targeting and translocation to the endoplasmic reticulum? *Curr. Opin. Cell Biol.* **2014**, *29*, 8–17. [CrossRef] [PubMed]
288. Zimmermann, R.; Sagstetter, M.; Lewis, M.J.; Pelham, H.R.B. Seventy-kilodalton heat shock proteins and an additional component from reticulocyte lysate stimulate import of M13 procoat protein into microsomes. *EMBO J.* **1988**, *7*, 2875–2880. [CrossRef] [PubMed]
289. Steinberg, R.; Koch, H.-G. The largely unexplored biology of small proteins in pro- and eukaryotes. *FEBS J.* **2021**, *7*, 7002–7004. [CrossRef] [PubMed]
290. Voorhees, R.M.; Hegde, R.S. Toward a structural understanding of co-translational protein translocation. *Curr. Opin. Cell Biol.* **2016**, *41*, 91–99. [CrossRef]
291. Lang, S.; Nguyen, D.; Pfeffer, S.; Förster, F.; Helms, V.; Zimmermann, R. Functions and Mechanisms of the Human Ribosome-Translocon Complex. In *Macromolecular Protein Complexes II: Structure and Function*; Harris, J.R., Marles-Wright, J., Eds.; Springer International Publishing: Cham, Switzerland, 2019; pp. 83–141.
292. Li, L.; Park, E.; Ling, J.; Ingram, J.; Ploegh, H.; Rapoport, T.A. Crystal structure of a substrate-engaged SecY protein-translocation channel. *Nature* **2016**, *531*, 395–399. [CrossRef] [PubMed]
293. Itskanov, S.; Park, E. Structure of the posttranslational Sec protein-translocation channel complex from yeast. *Science* **2019**, *363*, 84–87. [CrossRef] [PubMed]
294. Braunger, K.; Pfeffer, S.; Shrimal, S.; Gilmore, R.; Berninghausen, O.; Mandon, E.C.; Becker, T.; Förster, F.; Beckmann, R. Structural basis for coupling protein transport and N-glycosylation at the mammalian endoplasmic reticulum. *Science* **2018**, *360*, 215–219. [CrossRef] [PubMed]
295. Itskanov, S.; Kuo, K.M.; Gumbart, J.C.; Park, E. Stepwise gating of the Sec61 protein-conducting channel by Sec63 and Sec62. *Nat. Struct. Mol. Biol.* **2021**, *28*, 162–172. [CrossRef] [PubMed]
296. Weng, T.-H.; Steinchen, W.; Beatrix, B.; Berninghausen, O.; Becker, T.; Bange, G.; Cheng, J.; Beckmann, R. Architecture of the active post-translational Sec translocon. *EMBO J.* **2021**, *40*, e105643. [CrossRef] [PubMed]
297. Cheng, Z.; Jiang, Y.; Mandon, E.C.; Gilmore, R. Identification of cytoplasmic residues of Sec61p involved in ribosome binding and cotranslational translocation. *J. Cell Biol.* **2005**, *168*, 67–77. [CrossRef] [PubMed]
298. Tsukazaki, T.; Mori, H.; Fukai, S.; Ishitani, R.; Mori, T.; Dohmae, N.; Perederina, A.; Sugita, Y.; Vassylyev, D.G.; Ito, K.; et al. Conformational transition of Sec machinery inferred from bacterial SecYE structures. *Nature* **2008**, *455*, 988–991. [CrossRef] [PubMed]
299. Frauenfeld, J.; Gumbart, J.; Sluis, E.O.V.D.; Funes, S.; Gartmann, M.; Beatrix, B.; Mielke, T.; Berninghausen, O.; Becker, T.; Schulten, K.; et al. Cryo-EM structure of the ribosome–SecYE complex in the membrane environment. *Nat. Struct. Mol. Biol.* **2011**, *18*, 614–621. [CrossRef]
300. Elia, F.; Yadhanapudi, L.; Tretter, T.; Römisch, K. The N-terminus of Sec61p plays key roles in ER protein import and ERAD. *PLoS ONE* **2019**, *14*, e0215950. [CrossRef] [PubMed]
301. Erdmann, F.; Schäuble, N.; Lang, S.; Jung, M.; Honigmann, A.; Ahmad, M.; Dudek, J.; Benedix, J.; Harsman, A.; Kopp, A.; et al. Interaction of calmodulin with Sec61alpha limits Ca^{2+} leakage from the endoplasmic reticulum. *EMBO J.* **2011**, *30*, 17–31. [CrossRef] [PubMed]
302. Mandon, E.C.; Butova, C.; Lachapelle, A.; Gilmore, R. Conserved motifs on the cytoplasmic face of the protein translocation channel are critical for the transition between resting and active conformations. *J. Biol. Chem.* **2018**, *293*, 13662–13672. [CrossRef]
303. Kapp, K.; Schrempf, S.; Lemberg, M.; Dobberstein, B. Post-Targeting Functions of Signal Peptides. In *Protein Transport into the Endoplasmic Reticulum*; Zimmermann, R., Ed.; Landes Bioscience: Austin, TX, USA, 2009; pp. 1–16.
304. Sun, S.; Li, X.; Mariappan, M. Signal Sequences Encode Information for Protein Folding in the Endoplasmic Reticulum. *bioRxiv* **2020**. [CrossRef]
305. Martoglio, B.; Graf, R.; Dobberstein, B. Signal peptide fragments of preprolactin and HIV-1 p-gp160 interact with calmodulin. *EMBO J.* **1997**, *16*, 6636–6645. [CrossRef] [PubMed]
306. Lemberg, M.K.; Martoglio, B. On the mechanism of SPP-catalysed intramembrane proteolysis; conformational control of peptide bond hydrolysis in the plane of the membrane. *FEBS Lett.* **2004**, *564*, 213–218. [CrossRef]
307. Mentrup, T.; Cabrera-Cabrera, F.; Fluhrer, R.; Schröder, B. Physiological functions of SPP/SPPL intramembrane proteases. *Cell Mol. Life Sci.* **2020**, *77*, 2959–2979. [CrossRef] [PubMed]

308. Braud, V.; Yvonne Jones, E.; McMichael, A. The human major histocompatibility complex class Ib molecule HLA-E binds signal sequence-derived peptides with primary anchor residues at positions 2 and 9. *Eur. J. Immunol.* **1997**, *27*, 1164–1169. [CrossRef] [PubMed]
309. Wirth, A.; Jung, M.; Bies, C.; Frien, M.; Tyedmers, J.; Zimmermann, R.; Wagner, R. The Sec61p complex is a dynamic precursor activated channel. *Mol. Cell* **2003**, *12*, 261–268. [CrossRef]
310. Cross, B.C.S.; High, S. Dissecting the physiological role of selective transmembrane-segment retention at the ER translocon. *J. Cell Sci.* **2009**, *122*, 1768–1777. [CrossRef] [PubMed]
311. McCormick, P.J.; Miao, Y.; Shao, Y.; Lin, J.; Johnson, A.E. Cotranslational Protein Integration into the ER Membrane Is Mediated by the Binding of Nascent Chains to Translocon Proteins. *Mol. Cell* **2003**, *12*, 329–341. [CrossRef]
312. Heinrich, S.U.; Mothes, W.; Brunner, J.; Rapoport, T.A. The Sec61p complex mediates the integration of a membrane protein by allowing lipid partitioning of the transmembrane domain. *Cell* **2000**, *102*, 233–244. [CrossRef]
313. Mothes, W.; Heinrich, S.U.; Graf, R.; Nilsson, I.; von Heijne, G.; Brunner, J.; Rapoport, T.A. Molecular mechanism of membrane protein integration into the endoplasmic reticulum. *Cell* **1997**, *89*, 523–533. [CrossRef]
314. Simon, S.M.; Blobel, G. Signal peptides open protein-conducting channels in *E. coli*. *Cell* **1992**, *69*, 677–684. [CrossRef]
315. Simon, S.M.; Blobel, G. A protein-conducting channel in the endoplasmic reticulum. *Cell* **1991**, *65*, 371–380. [CrossRef]
316. Kida, Y.; Sakaguchi, M. Interaction mapping of the Sec61 translocon identifies two Sec61α regions interacting with hydrophobic segments in translocating chains. *J. Biol. Chem.* **2018**, *293*, 17050–17060. [CrossRef] [PubMed]
317. Reithinger, J.H.; Yim, C.; Kim, S.; Lee, H.; Kim, H. Structural and functional profiling of the lateral gate of the Sec61 translocon. *J. Biol. Chem.* **2014**, *289*, 15845–15855. [CrossRef] [PubMed]
318. Plath, K.; Mothes, W.; Wilkinson, B.M.; Stirling, C.J.; Rapoport, T.A. Signal sequence recognition in posttranslational protein transport across the yeast ER membrane. *Cell* **1998**, *94*, 795–807. [CrossRef]
319. High, S.; Martoglio, B.; Görlich, D.; Andersen, S.S.; Ashford, A.J.; Giner, A.; Hartmann, E.; Prehn, S.; Rapoport, T.A.; Dobberstein, B. Site-specific photocross-linking reveals that Sec61p and TRAM contact different regions of a membrane-inserted signal sequence. *J. Biol. Chem.* **1993**, *268*, 26745–26751. [CrossRef]
320. Mothes, W.; Jungnickel, B.; Brunner, J.; Rapoport, T.A. Signal sequence recognition in cotranslational translocation by protein components of the endoplasmic reticulum membrane. *J. Cell Biol.* **1998**, *142*, 355–364. [CrossRef]
321. Mothes, W.; Prehn, S.; Rapoport, T.A. Systematic probing of the environment of a translocating secretory protein during translocation through the ER membrane. *EMBO J.* **1994**, *13*, 3973–3982. [CrossRef] [PubMed]
322. High, S.; Andersen, S.S.L.; Görlich, D.; Hartmann, E.; Prehn, S.; Rapoport, T.A.; Dobberstein, B. Sec61p is adjecent to nascent type I and type II signal-anchor proteins during their membrane insertion. *J. Cell Biol.* **1993**, *121*, 743–750. [CrossRef] [PubMed]
323. Zhang, B.; Miller, T.F., III. Long-Timescale Dynamics and Regulation of Sec-Facilitated Protein Translocation. *Cell Rep.* **2012**, *2*, 927–937. [CrossRef] [PubMed]
324. Zhang, B.; Miller, T.F. Direct Simulation of Early-Stage Sec-Facilitated Protein Translocation. *J. Am. Chem. Soc.* **2012**, *134*, 13700–13707. [CrossRef]
325. Rychkova, A.; Warshel, A. Exploring the nature of the translocon-assisted protein insertion. *Proc. Natl. Acad. Sci. USA* **2013**, *110*, 495–500. [CrossRef] [PubMed]
326. Phillips, B.P.; Miller, E.A. Ribosome-associated quality control of membrane proteins at the endoplasmic reticulum. *J. Cell Sci.* **2020**, *133*, jcs251983. [CrossRef] [PubMed]
327. Charneski, C.A.; Hurst, L.D. Positively Charged Residues Are the Major Determinants of Ribosomal Velocity. *PLoS Biol.* **2013**, *11*, e1001508. [CrossRef]
328. Elazar, A.; Weinstein, J.J.; Prilusky, J.; Fleishman, S.J. Interplay between hydrophobicity and the positive-inside rule in determining membrane-protein topology. *Proc. Natl. Acad. Sci. USA* **2016**, *113*, 10340–10345. [CrossRef] [PubMed]
329. Seppälä, S.; Slusky, J.S.; Lloris-Garcerá, P.; Rapp, M.; von Heijne, G. Control of Membrane Protein Topology by a Single C-Terminal Residue. *Science* **2010**, *328*, 1698–1700. [CrossRef] [PubMed]
330. von Heijne, G.; Gavel, Y. Topogenic signals in integral membrane proteins. *Eur. J. Biochem.* **1988**, *174*, 671–678. [CrossRef] [PubMed]
331. Wallin, E.; Heijne, G.V. Genome-wide analysis of integral membrane proteins from eubacterial, archaean, and eukaryotic organisms. *Protein Sci.* **1998**, *7*, 1029–1038. [CrossRef] [PubMed]
332. Ziska, A.; Tatzelt, J.; Dudek, J.; Paton, A.W.; Paton, J.C.; Zimmermann, R.; Haßdenteufel, S. The signal peptide plus a cluster of positive charges in prion protein dictate chaperone-mediated Sec61 channel gating. *Biol. Open* **2019**, *8*, bio040691. [CrossRef] [PubMed]
333. Kriegler, T.; Lang, S.; Notari, L.; Hessa, T. Prion Protein Translocation Mechanism Revealed by Pulling Force Studies. *J. Mol. Biol.* **2020**, *432*, 4447–4465. [CrossRef] [PubMed]
334. Pfeffer, S.; Dudek, J.; Schaffer, M.; Ng, B.G.; Albert, S.; Plitzko, J.M.; Baumeister, W.; Zimmermann, R.; Freeze, H.H.; Engel, B.D.; et al. Dissecting the molecular organization of the translocon-associated protein complex. *Nat. Commun.* **2017**, *8*, 14516. [CrossRef]
335. Kriegler, T.; Kiburg, G.; Hessa, T. Translocon-Associated Protein Complex (TRAP) is Crucial for Co-Translational Translocation of Pre-Proinsulin. *J. Mol. Biol.* **2020**, *432*, 166694. [CrossRef]

336. Fons, R.D.; Bogert, B.A.; Hegde, R.S. Substrate-specific function of the translocon-associated protein complex during translocation across the ER membrane. *J. Cell Biol.* **2003**, *160*, 529–539. [CrossRef] [PubMed]
337. Ménétret, J.F.; Hegde, R.S.; Aguiar, M.; Gygi, S.P.; Park, E.; Rapoport, T.A.; Akey, C.W. Single copies of Sec61 and TRAP associate with a nontranslating mammalian ribosome. *Structure* **2008**, *16*, 1126–1137. [CrossRef] [PubMed]
338. Görlich, D.; Hartmann, E.; Prehn, S.; Rapoport, T.A. A protein of the endoplasmic reticulum involved early in polypeptide translocation. *Nature* **1992**, *357*, 47–52. [CrossRef] [PubMed]
339. Voigt, S.; Jungnickel, B.; Hartmann, E.; Rapoport, T.A. Signal sequence-dependent function of the TRAM protein during early phases of protein transport across the endoplasmic reticulum membrane. *J. Cell Biol.* **1996**, *134*, 25–35. [CrossRef] [PubMed]
340. Klein, M.-C.; Lerner, M.; Nguyen, D.; Pfeffer, S.; Dudek, J.; Förster, F.; Helms, V.; Lang, S.; Zimmermann, R. TRAM1 protein may support ER protein import by modulating the phospholipid bilayer near the lateral gate of the Sec61-channel. *Channels* **2020**, *14*, 28–44. [CrossRef] [PubMed]
341. Conti, B.J.; Devaraneni, P.K.; Yang, Z.; David, L.L.; Skach, W.R. Cotranslational Stabilization of Sec62/63 within the ER Sec61 Translocon Is Controlled by Distinct Substrate-Driven Translocation Events. *Mol. Cell* **2015**, *58*, 269–283. [CrossRef] [PubMed]
342. Schorr, S.; Nguyen, D.; Haßdenteufel, S.; Nagaraj, N.; Cavalié, A.; Greiner, M.; Weissgerber, P.; Loi, M.; Paton, A.W.; Paton, J.C.; et al. Identification of signal peptide features for substrate specificity in human Sec62/Sec63-dependent ER protein import. *FEBS J.* **2020**, *287*, 4612–4640. [CrossRef] [PubMed]
343. Meacock, S.L.; Lecomte, F.J.L.; Crawshaw, S.G.; High, S. Different Transmembrane Domains Associate with Distinct Endoplasmic Reticulum Components during Membrane Integration of a Polytopic Protein. *Mol. Biol. Cell* **2002**, *13*, 4114–4129. [CrossRef] [PubMed]
344. Pfeffer, S.; Dudek, J.; Gogala, M.; Schorr, S.; Linxweiler, J.; Lang, S.; Becker, T.; Beckmann, R.; Zimmermann, R.; Förster, F. Structure of the mammalian oligosaccharyl-transferase complex in the native ER protein translocon. *Nat. Commun.* **2014**, *5*, 3072. [CrossRef] [PubMed]
345. Fenech, E.J.; Ben-Dor, S.; Schuldiner, M. Double the Fun, Double the Trouble: Paralogs and Homologs Functioning in the Endoplasmic Reticulum. *Annu. Rev. Biochem.* **2020**, *89*, 637–666. [CrossRef] [PubMed]
346. Shrimal, S.; Trueman, S.F.; Gilmore, R. Extreme C-terminal sites are posttranslocationally glycosylated by the STT3B isoform of the OST. *J. Cell Biol.* **2013**, *201*, 81–95. [CrossRef] [PubMed]
347. Tyedmers, J.; Lerner, M.; Wiedmann, M.; Volkmer, J.; Zimmermann, R. Polypeptide-binding proteins mediate completion of co-translational protein translocation into the mammalian endoplasmic reticulum. *EMBO Rep.* **2003**, *4*, 505–510. [CrossRef] [PubMed]
348. Schäuble, N.; Lang, S.; Jung, M.; Cappel, S.; Schorr, S.; Ulucan, O.; Linxweiler, J.; Dudek, J.; Blum, R.; Helms, V.; et al. BiP-mediated closing of the Sec61 channel limits Ca^{2+} leakage from the ER. *EMBO J.* **2012**, *31*, 3282–3296. [CrossRef] [PubMed]
349. Lang, S.; Erdmann, F.; Jung, M.; Wagner, R.; Cavalié, A.; Zimmermann, R. Sec61 complexes form ubiquitous ER Ca($^{2+}$) leak channels. *Channels* **2011**, *5*, 228–235. [CrossRef]
350. Wang, Q.-C.; Zheng, Q.; Tan, H.; Zhang, B.; Li, X.; Yang, Y.; Yu, J.; Liu, Y.; Chai, H.; Wang, X.; et al. TMCO1 Is an ER Ca^{2+} Load-Activated Ca^{2+} Channel. *Cell* **2016**, *165*, 1454–1466. [CrossRef] [PubMed]
351. Zhang, M.; Yamazaki, T.; Yazawa, M.; Treves, S.; Nishi, M.; Murai, M.; Shibata, E.; Zorzato, F.; Takeshima, H. Calumin, a novel Ca2+-binding transmembrane protein on the endoplasmic reticulum. *Cell Calcium* **2007**, *42*, 83–90. [CrossRef] [PubMed]
352. Morel, J.-D.; Paatero, A.O.; Wei, J.; Yewdell, J.W.; Guenin-Macé, L.; Van Haver, D.; Impens, F.; Pietrosemoli, N.; Paavilainen, V.O.; Demangel, C. Proteomics Reveals Scope of Mycolactone-mediated Sec61 Blockade and Distinctive Stress Signature. *Mol Cell Proteom.* **2018**, *17*, 1750–1765. [CrossRef] [PubMed]
353. McKenna, M.; Simmonds, R.E.; High, S. Mycolactone reveals the substrate-driven complexity of Sec61-dependent transmembrane protein biogenesis. *J. Cell Sci.* **2017**, *130*, 1307–1320. [CrossRef] [PubMed]
354. Zong, G.; Hu, Z.; O'Keefe, S.; Tranter, D.; Iannotti, M.J.; Baron, L.; Hall, B.; Corfield, K.; Paatero, A.O.; Henderson, M.J.; et al. Ipomoeassin F Binds Sec61α to Inhibit Protein Translocation. *J. Am. Chem. Soc.* **2019**, *141*, 8450–8461. [CrossRef] [PubMed]
355. Anghel, S.A.; McGilvray, P.T.; Hegde, R.S.; Keenan, R.J. Identification of Oxa1 Homologs Operating in the Eukaryotic Endoplasmic Reticulum. *Cell Rep.* **2017**, *21*, 3708–3716. [CrossRef] [PubMed]
356. Yabal, M.; Brambillasca, S.; Soffientini, P.; Pedrazzini, E.; Borgese, N.; Makarow, M. Translocation of the C Terminus of a Tail-anchored Protein across the Endoplasmic Reticulum Membrane in Yeast Mutants Defective in Signal Peptide-driven Translocation. *J. Biol. Chem.* **2003**, *278*, 3489–3496. [CrossRef] [PubMed]
357. Vilardi, F.; Stephan, M.; Clancy, A.; Janshoff, A.; Schwappach, B. WRB and CAML Are Necessary and Sufficient to Mediate Tail-Anchored Protein Targeting to the ER Membrane. *PLoS ONE* **2014**, *9*, e85033. [CrossRef] [PubMed]
358. Wang, F.; Chan, C.; Weir, N.R.; Denic, V. The Get1/2 transmembrane complex is an endoplasmic-reticulum membrane protein insertase. *Nature* **2014**, *512*, 441–444. [CrossRef]
359. McDowell, M.A.; Heimes, M.; Fiorentino, F.; Mehmood, S.; Farkas, Á.; Coy-Vergara, J.; Wu, D.; Bolla, J.R.; Schmid, V.; Heinze, R.; et al. Structural Basis of Tail-Anchored Membrane Protein Biogenesis by the GET Insertase Complex. *Mol. Cell* **2020**, *80*, 72–86. [CrossRef] [PubMed]
360. Zalisko, B.E.; Chan, C.; Denic, V.; Rock, R.S.; Keenan, R.J. Tail-Anchored Protein Insertion by a Single Get1/2 Heterodimer. *Cell Rep.* **2017**, *20*, 2287–2293. [CrossRef] [PubMed]

361. Mariappan, M.; Mateja, A.; Dobosz, M.; Bove, E.; Hegde, R.S.; Keenan, R.J. The mechanism of membrane-associated steps in tail-anchored protein insertion. *Nature* **2011**, *477*, 61–66. [CrossRef]

362. Rome, M.E.; Chio, U.S.; Rao, M.; Gristick, H.; Shan, S.-O. Differential gradients of interaction affinities drive efficient targeting and recycling in the GET pathway. *Proc. Natl. Acad. Sci. USA* **2014**, *111*, E4929–E4935. [CrossRef]

363. Tran, D.D.; Russell, H.R.; Sutor, S.L.; van Deursen, J.; Bram, R.J. CAML Is Required for Efficient EGF Receptor Recycling. *Dev. Cell* **2003**, *5*, 245–256. [CrossRef]

364. Tran, D.D.; Edgar, C.E.; Heckman, K.L.; Sutor, S.L.; Huntoon, C.J.; van Deursen, J.; McKean, D.L.; Bram, R.J. CAML Is a p56Lck-Interacting Protein that Is Required for Thymocyte Development. *Immunity* **2005**, *23*, 139–152. [CrossRef] [PubMed]

365. Norlin, S.; Parekh, V.S.; Naredi, P.; Edlund, H. Asna1/TRC40 Controls β-Cell Function and Endoplasmic Reticulum Homeostasis by Ensuring Retrograde Transport. *Diabetes* **2016**, *65*, 110–119. [CrossRef] [PubMed]

366. Bryda, E.C.; Johnson, N.T.; Ohlemiller, K.K.; Besch-Williford, C.L.; Moore, E.; Bram, R.J. Conditional deletion of calcium-modulating cyclophilin ligand causes deafness in mice. *Mamm. Genome* **2012**, *23*, 270–276. [CrossRef]

367. Jonikas, M.C.; Collins, S.R.; Denic, V.; Oh, E.; Quan, E.M.; Schmid, V.; Weibezahn, J.; Schwappach, B.; Walter, P.; Weissman, J.S.; et al. Comprehensive characterization of genes required for protein folding in the endoplasmic reticulum. *Science* **2009**, *323*, 1693–1697. [CrossRef] [PubMed]

368. Satoh, T.; Ohba, A.; Liu, Z.; Inagaki, T.; Satoh, A.K. dPob/EMC is essential for biosynthesis of rhodopsin and other multi-pass membrane proteins in Drosophila photoreceptors. *eLife* **2015**, *4*, e06306. [CrossRef] [PubMed]

369. Richard, M.; Boulin, T.; Robert, V.J.P.; Richmond, J.E.; Bessereau, J.-L. Biosynthesis of ionotropic acetylcholine receptors requires the evolutionarily conserved ER membrane complex. *Proc. Natl. Acad. Sci. USA* **2013**, *110*, E1055–E1063. [CrossRef] [PubMed]

370. Tang, X.; Snowball, J.M.; Xu, Y.; Na, C.-L.; Weaver, T.E.; Clair, G.; Kyle, J.E.; Zink, E.M.; Ansong, C.; Wei, W.; et al. EMC3 coordinates surfactant protein and lipid homeostasis required for respiration. *J. Clin. Investig.* **2017**, *127*, 4314–4325. [CrossRef] [PubMed]

371. Zhou, Y.; Wu, F.; Zhang, M.; Xiong, Z.; Yin, Q.; Ru, Y.; Shi, H.; Li, J.; Mao, S.; Li, Y.; et al. EMC10 governs male fertility via maintaining sperm ion balance. *J. Mol. Cell Biol.* **2018**, *10*, 503–514. [CrossRef]

372. Shurtleff, M.J.; Itzhak, D.N.; Hussmann, J.A.; Schirle Oakdale, N.T.; Costa, E.A.; Jonikas, M.; Weibezahn, J.; Popova, K.D.; Jan, C.H.; Sinitcyn, P.; et al. The ER membrane protein complex interacts cotranslationally to enable biogenesis of multipass membrane proteins. *eLife* **2018**, *7*, e37018. [CrossRef]

373. Volkmar, N.; Christianson, J.C. Squaring the EMC—How promoting membrane protein biogenesis impacts cellular functions and organismal homeostasis. *J. Cell Sci.* **2020**, *133*, jcs243519. [CrossRef]

374. Tian, S.; Wu, Q.; Zhou, B.; Choi, M.Y.; Ding, B.; Yang, W.; Dong, M. Proteomic Analysis Identifies Membrane Proteins Dependent on the ER Membrane Protein Complex. *Cell Rep.* **2019**, *28*, 2517–2526. [CrossRef] [PubMed]

375. O'Donnell, J.P.; Phillips, B.P.; Yagita, Y.; Juszkiewicz, S.; Wagner, A.; Malinverni, D.; Keenan, R.J.; Miller, E.A.; Hegde, R.S. The architecture of EMC reveals a path for membrane protein insertion. *eLife* **2020**, *9*, e57887. [CrossRef] [PubMed]

376. Miller-Vedam, L.E.; Bräuning, B.; Popova, K.D.; Schirle Oakdale, N.T.; Bonnar, J.L.; Prabu, J.R.; Boydston, E.A.; Sevillano, N.; Shurtleff, M.J.; Stroud, R.M.; et al. Structural and mechanistic basis of the EMC-dependent biogenesis of distinct transmembrane clients. *eLife* **2020**, *9*, e62611. [CrossRef] [PubMed]

377. Pleiner, T.; Tomaleri, G.P.; Januszyk, K.; Inglis, A.J.; Hazu, M.; Voorhees, R.M. Structural basis for membrane insertion by the human ER membrane protein complex. *Science* **2020**, *369*, 433–436. [CrossRef] [PubMed]

378. Bai, L.; You, Q.; Feng, X.; Kovach, A.; Li, H. Structure of the ER membrane complex, a transmembrane-domain insertase. *Nature* **2020**, *584*, 475–478. [CrossRef] [PubMed]

379. Dudek, J.; Pfeffer, S.; Lee, P.-H.; Jung, M.; Cavalié, A.; Helms, V.; Förster, F.; Zimmermann, R. Protein Transport into the Human Endoplasmic Reticulum. *J. Mol. Biol.* **2015**, *427*, 1159–1175. [CrossRef] [PubMed]

380. Braakman, I.; Hebert, D.N. Protein folding in the endoplasmic reticulum. *Cold Spring Harb. Perspect. Biol.* **2013**, *5*, a013201. [CrossRef] [PubMed]

381. Phillips, B.P.; Gomez-Navarro, N.; Miller, E.A. Protein quality control in the endoplasmic reticulum. *Curr. Opin. Cell Biol.* **2020**, *65*, 96–102. [CrossRef] [PubMed]

382. Melnyk, A.; Rieger, H.; Zimmermann, R. Co-chaperones of the Mammalian Endoplasmic Reticulum. In *The Networking of Chaperones by Co-Chaperones: Control of Cellular Protein Homeostasis*; Blatch, G.L., Edkins, A.L., Eds.; Springer International Publishing: Cham, Switzerland, 2015; pp. 179–200.

383. Siegel, V.; Walter, P. Binding sites of the 19-kDa and 68/72-kDa signal recognition particle (SRP) proteins on SRP RNA as determined in protein-RNA "footprinting". *Proc. Natl. Acad. Sci. USA* **1988**, *85*, 1801–1805. [CrossRef] [PubMed]

384. van Nues, R.W.; Leung, E.; McDonald, J.C.; Dantuluru, I.; Brown, J.D. Roles for Srp72p in assembly, nuclear export and function of the signal recognition particle. *RNA Biol.* **2008**, *5*, 73–83. [CrossRef] [PubMed]

385. Iakhiaeva, E.; Yin, J.; Zwieb, C. Identification of an RNA-binding Domain in Human SRP72. *J. Mol. Biol.* **2005**, *345*, 659–666. [CrossRef]

386. Kirwan, M.; Walne, A.J.; Plagnol, V.; Velangi, M.; Ho, A.; Hossain, U.; Vulliamy, T.; Dokal, I. Exome Sequencing Identifies Autosomal-Dominant SRP72 Mutations Associated with Familial Aplasia and Myelodysplasia. *Am. J. Hum. Genet.* **2012**, *90*, 888–892. [CrossRef] [PubMed]

387. Pool, M.R.; Stumm, J.; Fulga, T.A.; Sinning, I.; Dobberstein, B. Distinct Modes of Signal Recognition Particle Interaction with the Ribosome. *Science* **2002**, *297*, 1345–1348. [CrossRef] [PubMed]
388. Akopian, D.; Shen, K.; Zhang, X.; Shan, S.-o. Signal Recognition Particle: An Essential Protein-Targeting Machine. *Annu. Rev. Biochem.* **2013**, *82*, 693–721. [CrossRef] [PubMed]
389. Gowda, K.; Chittenden, K.; Zwieb, C. Binding site of the M-domain of human protein SRP54 determined by systematic site-directed mutagenesis of signal recognition particle RNA. *Nucleic Acids Res.* **1997**, *25*, 388–394. [CrossRef] [PubMed]
390. Carapito, R.; Konantz, M.; Paillard, C.; Miao, Z.; Pichot, A.; Leduc, M.S.; Yang, Y.; Bergstrom, K.L.; Mahoney, D.H.; Shardy, D.L.; et al. Mutations in signal recognition particle SRP54 cause syndromic neutropenia with Shwachman-Diamond–like features. *J. Clin. Investig.* **2017**, *127*, 4090–4103. [CrossRef] [PubMed]
391. Allenbach, Y.; Benveniste, O.; Stenzel, W.; Boyer, O. Immune-mediated necrotizing myopathy: Clinical features and pathogenesis. *Nat. Rev. Rheumatol.* **2020**, *16*, 689–701. [CrossRef] [PubMed]
392. Römisch, K.; Miller, F.W.; Dobberstein, B.; High, S. Human autoantibodies against the 54 kDa protein of the signal recognition particle block function at multiple stages. *Arthritis Res.* **2006**, *8*, R39. [CrossRef]
393. Benarroch, R.; Austin, J.M.; Ahmed, F.; Isaacson, R.L. Chapter Seven—The roles of cytosolic quality control proteins, SGTA and the BAG6 complex, in disease. In *Advances in Protein Chemistry and Structural Biology*; Donev, R., Ed.; Academic Press: Cambridge, MA, USA, 2019; Volume 114, pp. 265–313.
394. Behl, C. Breaking BAG: The Co-Chaperone BAG3 in Health and Disease. *Trends Pharmacol. Sci.* **2016**, *37*, 672–688. [CrossRef] [PubMed]
395. Chen, J.; Zang, Y.-S.; Xiu, Q. BAT3 rs1052486 and rs3117582 polymorphisms are associated with lung cancer risk: A meta-analysis. *Tumor Biol.* **2014**, *35*, 9855–9858. [CrossRef] [PubMed]
396. Etokebe, G.E.; Zienolddiny, S.; Kupanovac, Z.; Enersen, M.; Balen, S.; Flego, V.; Bulat-Kardum, L.; Radojčić-Badovinac, A.; Skaug, V.; Bakke, P.; et al. Association of the FAM46A Gene VNTRs and BAG6 rs3117582 SNP with Non Small Cell Lung Cancer (NSCLC) in Croatian and Norwegian Populations. *PLoS ONE* **2015**, *10*, e0122651. [CrossRef] [PubMed]
397. Wang, T.; Liang, Y.; Li, H.; Li, H.; He, Q.; Xue, Y.; Shen, C.; Zhang, C.; Xiang, J.; Ding, J.; et al. Single Nucleotide Polymorphisms and Osteoarthritis: An Overview and a Meta-Analysis. *Medicine* **2016**, *95*, e2811. [CrossRef] [PubMed]
398. Xue, Q.; Lv, L.; Wan, C.; Chen, B.; Li, M.; Ni, T.; Liu, Y.; Liu, Y.; Cong, X.; Zhou, Y.; et al. Expression and clinical role of small glutamine-rich tetratricopeptide repeat (TPR)-containing protein alpha (SGTA) as a novel cell cycle protein in NSCLC. *J. Cancer Res. Clin. Oncol.* **2013**, *139*, 1539–1549. [CrossRef]
399. Yang, X.; Cheng, L.; Li, M.; Shi, H.; Ren, H.; Ding, Z.; Liu, F.; Wang, Y.; Cheng, C. High Expression of SGTA in Esophageal Squamous Cell Carcinoma Correlates With Proliferation and Poor Prognosis. *J. Cell. Biochem.* **2014**, *115*, 141–150. [CrossRef] [PubMed]
400. Zhu, T.; Ji, Z.; Xu, C.; Peng, Z.; Gu, L.; Zhang, R.; Liu, Y. Expression and prognostic role of SGTA in human breast carcinoma correlates with tumor cell proliferation. *J. Mol. Histol.* **2014**, *45*, 665–677. [CrossRef] [PubMed]
401. Goodarzi, M.O.; Xu, N.; Cui, J.; Guo, X.; Chen, Y.I.; Azziz, R. Small glutamine-rich tetratricopeptide repeat-containing protein alpha (SGTA), a candidate gene for polycystic ovary syndrome. *Hum. Reprod.* **2008**, *23*, 1214–1219. [CrossRef] [PubMed]
402. Sicking, M.; Lang, S.; Bochen, F.; Roos, A.; Drenth, J.P.H.; Zakaria, M.; Zimmermann, R.; Linxweiler, M. Complexity and Specificity of Sec61-Channelopathies: Human Diseases Affecting Gating of the Sec61 Complex. *Cells* **2021**, *10*, 1036. [CrossRef] [PubMed]
403. Linxweiler, M.; Schick, B.; Zimmermann, R. Let's talk about Secs: Sec61, Sec62 and Sec63 in signal transduction, oncology and personalized medicine. *Signal Transduct. Target. Ther.* **2017**, *2*, 17002. [CrossRef] [PubMed]
404. Van Nieuwenhove, E.; Barber, J.S.; Neumann, J.; Smeets, E.; Willemsen, M.; Pasciuto, E.; Prezzemolo, T.; Lagou, V.; Seldeslachts, L.; Malengier-Devlies, B.; et al. Defective Sec61α1 underlies a novel cause of autosomal dominant severe congenital neutropenia. *J. Allergy Clin. Immunol.* **2020**, *146*, 1180–1193. [CrossRef] [PubMed]
405. Davila, S.; Furu, L.; Gharavi, A.G.; Tian, X.; Onoe, T.; Qian, Q.; Li, A.; Cai, Y.; Kamath, P.S.; King, B.F.; et al. Mutations in SEC63 cause autosomal dominant polycystic liver disease. *Nat. Genet.* **2004**, *36*, 575–577. [CrossRef]
406. Greiner, M.; Kreutzer, B.; Jung, V.; Grobholz, R.; Hasenfus, A.; Stöhr, R.F.; Tornillo, L.; Dudek, J.; Stöckle, M.; Unteregger, G.; et al. Silencing of the SEC62 gene inhibits migratory and invasive potential of various tumor cells. *Int. J. Cancer* **2011**, *128*, 2284–2295. [CrossRef] [PubMed]
407. Besse, W.; Dong, K.; Choi, J.; Punia, S.; Fedeles, S.V.; Choi, M.; Gallagher, A.-R.; Huang, E.B.; Gulati, A.; Knight, J.; et al. Isolated polycystic liver disease genes define effectors of polycystin-1 function. *J. Clin. Investig.* **2017**, *127*, 1772–1785. [CrossRef] [PubMed]
408. Meng, H.; Jiang, X.; Wang, J.; Sang, Z.; Guo, L.; Yin, G.; Wang, Y. SEC61G is upregulated and required for tumor progression in human kidney cancer. *Mol. Med. Rep.* **2021**, *23*, 427. [CrossRef] [PubMed]
409. Ng, B.G.; Lourenço, C.M.; Losfeld, M.-E.; Buckingham, K.J.; Kircher, M.; Nickerson, D.A.; Shendure, J.; Bamshad, M.J.; University of Washington Center for Mendelian, G.; Freeze, H.H. Mutations in the translocon-associated protein complex subunit SSR3 cause a novel congenital disorder of glycosylation. *J. Inherit. Metab. Dis.* **2019**, *42*, 993–997. [CrossRef] [PubMed]
410. Dittner-Moormann, S.; Lourenço, C.M.; Reunert, J.; Nishinakamura, R.; Tanaka, S.S.; Werner, C.; Debus, V.; Zimmer, K.-P.; Wetzel, G.; Naim, H.Y.; et al. TRAPγ-CDG shows asymmetric glycosylation and an effect on processing of proteins required in higher organisms. *J. Med. Genet.* **2021**, *58*, 213–216. [CrossRef] [PubMed]
411. Michael Yates, T.; Ng, O.-H.; Offiah, A.C.; Willoughby, J.; Berg, J.N.; Study, D.; Johnson, D.S. Cerebrofaciothoracic dysplasia: Four new patients with a recurrent TMCO1 pathogenic variant. *Am. J. Med. Genet. Part A* **2019**, *179*, 43–49. [CrossRef] [PubMed]

412. Sharkia, R.; Zalan, A.; Jabareen-Masri, A.; Hengel, H.; Schöls, L.; Kessel, A.; Azem, A.; Mahajnah, M. A novel biallelic loss-of-function mutation in TMCO1 gene confirming and expanding the phenotype spectrum of cerebro-facio-thoracic dysplasia. *Am. J. Med. Genet. Part A* **2019**, *179*, 1338–1345. [CrossRef] [PubMed]
413. Alanay, Y.; Ergüner, B.; Utine, E.; Haçarız, O.; Kiper, P.O.S.; Taşkıran, E.Z.; Perçin, F.; Uz, E.; Sağıroğlu, M.Ş.; Yuksel, B.; et al. TMCO1 deficiency causes autosomal recessive cerebrofaciothoracic dysplasia. *Am. J. Med. Genet. Part A* **2014**, *164*, 291–304. [CrossRef] [PubMed]
414. Caglayan, A.; Per, H.; Akgumus, G.; Gumus, H.; Baranoski, J.; Canpolat, M.; Calik, M.; Yikilmaz, A.; Bilguvar, K.; Kumandas, S.; et al. Whole-exome sequencing identified a patient with TMCO1 defect syndrome and expands the phenotic spectrum. *Clin. Genet.* **2013**, *84*, 394–395. [CrossRef]
415. Morimoto, M.; Waller-Evans, H.; Ammous, Z.; Song, X.; Strauss, K.A.; Pehlivan, D.; Gonzaga-Jauregui, C.; Puffenberger, E.G.; Holst, C.R.; Karaca, E.; et al. Bi-allelic CCDC47 Variants Cause a Disorder Characterized by Woolly Hair, Liver Dysfunction, Dysmorphic Features, and Global Developmental Delay. *Am. J. Hum. Genet.* **2018**, *103*, 794–807. [CrossRef]
416. Geetha, T.S.; Lingappa, L.; Jain, A.R.; Govindan, H.; Mandloi, N.; Murugan, S.; Gupta, R.; Vedam, R. A novel splice variant in EMC1 is associated with cerebellar atrophy, visual impairment, psychomotor retardation with epilepsy. *Mol. Genet. Genom. Med.* **2018**, *6*, 282–287. [CrossRef] [PubMed]
417. Harel, T.; Yesil, G.; Bayram, Y.; Coban-Akdemir, Z.; Charng, W.-L.; Karaca, E.; Al Asmari, A.; Eldomery, M.K.; Hunter, J.V.; Jhangiani, S.N.; et al. Monoallelic and Biallelic Variants in EMC1 Identified in Individuals with Global Developmental Delay, Hypotonia, Scoliosis, and Cerebellar Atrophy. *Am. J. Hum. Genet.* **2016**, *98*, 562–570. [CrossRef]
418. Shao, D.D.; Straussberg, R.; Ahmed, H.; Khan, A.; Tian, S.; Hill, R.S.; Smith, R.S.; Majmundar, A.J.; Ameziane, N.; Neil, J.E.; et al. A recurrent, homozygous EMC10 frameshift variant is associated with a syndrome of developmental delay with variable seizures and dysmorphic features. *Genet. Med.* **2021**, *23*, 1158–1162. [CrossRef] [PubMed]
419. Junes-Gill, K.S.; Gallaher, T.K.; Gluzman-Poltorak, Z.; Miller, J.D.; Wheeler, C.J.; Fan, X.; Basile, L.A. hHSS1: A novel secreted factor and suppressor of glioma growth located at chromosome 19q13.33. *J. Neurooncol.* **2011**, *102*, 197–211. [CrossRef] [PubMed]
420. Shen, X.; Kan, S.; Hu, J.; Li, M.; Lu, G.; Zhang, M.; Zhang, S.; Hou, Y.; Chen, Y.; Bai, Y. EMC6/TMEM93 suppresses glioblastoma proliferation by modulating autophagy. *Cell Death Dis.* **2016**, *7*, e2043. [CrossRef] [PubMed]
421. Junes-Gill, K.S.; Lawrence, C.E.; Wheeler, C.J.; Cordner, R.; Gill, T.G.; Mar, V.; Shiri, L.; Basile, L.A. Human hematopoietic signal peptide-containing secreted 1 (hHSS1) modulates genes and pathways in glioma: Implications for the regulation of tumorigenicity and angiogenesis. *BMC Cancer* **2014**, *14*, 920. [CrossRef]
422. Asseck, L.Y.; Mehlhorn, D.G.; Monroy, J.R.; Ricardi, M.M.; Breuninger, H.; Wallmeroth, N.; Berendzen, K.W.; Nowrousian, M.; Xing, S.; Schwappach, B.; et al. Endoplasmic reticulum membrane receptors of the GET pathway are conserved throughout eukaryotes. *Proc. Natl. Acad. Sci. USA* **2021**, *118*, e2017636118. [CrossRef] [PubMed]
423. Yamamoto, Y.; Sakisaka, T. Molecular Machinery for Insertion of Tail-Anchored Membrane Proteins into the Endoplasmic Reticulum Membrane in Mammalian Cells. *Mol. Cell* **2012**, *48*, 387–397. [CrossRef] [PubMed]
424. Egeo, A.; Mazzocco, M.; Sotgia, F.; Arrigo, P.; Oliva, R.; Bergonòn, S.; Nizetic, D.; Rasore-Quartino, A.; Scartezzini, P. Identification and characterization of a new human cDNA from chromosome 21q22.3 encoding a basic nuclear protein. *Hum. Genet.* **1998**, *102*, 289–293. [CrossRef] [PubMed]
425. Murata, K.; Degmetich, S.; Kinoshita, M.; Shimada, E. Expression of the congenital heart disease 5/tryptophan rich basic protein homologue gene during heart development in Medaka fish, Oryzias latipes. *Dev. Growth Differ.* **2009**, *51*, 95–107. [CrossRef]
426. Sojka, S.; Amin, N.M.; Gibbs, D.; Christine, K.S.; Charpentier, M.S.; Conlon, F.L. Congenital heart disease protein 5 associates with CASZ1 to maintain myocardial tissue integrity. *Development* **2014**, *141*, 3040–3049. [CrossRef] [PubMed]
427. Lim, J.-H.; Kim, T.-Y.; Kim, W.-H.; Park, J.-W. CAML promotes prolactin-dependent proliferation of breast cancer cells by facilitating prolactin receptor signaling pathways. *Breast Cancer Res. Treat.* **2011**, *130*, 19–27. [CrossRef] [PubMed]
428. Long, T.; Su, J.; Tang, W.; Luo, Z.; Liu, S.; Liu, Z.; Zhou, H.; Qi, M.; Zeng, W.; Zhang, J.; et al. A novel interaction between calcium-modulating cyclophilin ligand and Basigin regulates calcium signaling and matrix metalloproteinase activities in human melanoma cells. *Cancer Lett.* **2013**, *339*, 93–101. [CrossRef]
429. Available online: www.mousephenotype.org/data/genes/MGI:2136882#phenotypesTab (accessed on 22 December 2021).
430. Tosal-Castano, S.; Peselj, C.; Kohler, V.; Habernig, L.; Berglund, L.L.; Ebrahimi, M.; Vögtle, F.N.; Höög, J.; Andréasson, C.; Büttner, S. Snd3 controls nucleus-vacuole junctions in response to glucose signaling. *Cell Rep.* **2021**, *34*, 108637. [CrossRef] [PubMed]
431. Ricarte, F.; Menjivar, R.; Chhun, S.; Soreta, T.; Oliveira, L.; Hsueh, T.; Serranilla, M.; Gharakhanian, E. A Genome-Wide Immunodetection Screen in S. cerevisiae Uncovers Novel Genes Involved in Lysosomal Vacuole Function and Morphology. *PLoS ONE* **2011**, *6*, e23696. [CrossRef]
432. Yompakdee, C.; Ogawa, N.; Harashima, S.; Oshima, Y. A putative membrane protein, Pho88p, involved in inorganic phosphate transport inSaccharomyces cerevisiae. *Mol. Gen. Genet. MGG* **1996**, *251*, 580–590. [CrossRef] [PubMed]
433. Mukerji, N.; Damodaran, T.V.; Winn, M.P. TRPC6 and FSGS: The latest TRP channelopathy. *Biochim. Et Biophys. Acta (BBA) Mol. Basis Dis.* **2007**, *1772*, 859–868. [CrossRef]
434. Riehle, M.; Büscher, A.K.; Gohlke, B.-O.; Kaßmann, M.; Kolatsi-Joannou, M.; Bräsen, J.H.; Nagel, M.; Becker, J.U.; Winyard, P.; Hoyer, P.F.; et al. *TRPC6 G757D Loss-of-Function Mutation Associates with FSGS. J. Am. Soc. Nephrol.* **2016**, *27*, 2771–2783. [CrossRef] [PubMed]

435. Winn, M.P.; Conlon, P.J.; Lynn, K.L.; Farrington, M.K.; Creazzo, T.; Hawkins, A.F.; Daskalakis, N.; Kwan, S.Y.; Ebersviller, S.; Burchette, J.L.; et al. A Mutation in the *TRPC6* Cation Channel Causes Familial Focal Segmental Glomerulosclerosis. *Science* **2005**, *308*, 1801–1804. [CrossRef] [PubMed]
436. Andolfo, I.; Russo, R.; Manna, F.; Shmukler, B.E.; Gambale, A.; Vitiello, G.; De Rosa, G.; Brugnara, C.; Alper, S.L.; Snyder, L.M.; et al. Novel Gardos channel mutations linked to dehydrated hereditary stomatocytosis (xerocytosis). *Am. J. Hematol.* **2015**, *90*, 921–926. [CrossRef] [PubMed]
437. Glogowska, E.; Lezon-Geyda, K.; Maksimova, Y.; Schulz, V.P.; Gallagher, P.G. Mutations in the Gardos channel (KCNN4) are associated with hereditary xerocytosis. *Blood* **2015**, *126*, 1281–1284. [CrossRef] [PubMed]
438. Lloyd, D.J.; Wheeler, M.C.; Gekakis, N. A point mutation in Sec61alpha1 leads to diabetes and hepatosteatosis in mice. *Diabetes* **2010**, *59*, 460–470. [CrossRef] [PubMed]
439. Espino-Hernández, M.; Palma Milla, C.; Vara-Martín, J.; González-Granado, L.I. De novo SEC61A1 mutation in autosomal dominant tubulo-interstitial kidney disease: Phenotype expansion and review of literature. *J. Paediatr. Child Health* **2021**, *57*, 1305–1307. [CrossRef] [PubMed]

International Journal of
Molecular Sciences

Review

Take Me Home, Protein Roads: Structural Insights into Signal Peptide Interactions during ER Translocation

A. Manuel Liaci and Friedrich Förster *

Bijvoet Centre for Biomolecular Research, Structural Biochemistry Group, Utrecht University, Universiteitsweg 99, 3584 CG Utrecht, The Netherlands; a.m.liaci@uu.nl
* Correspondence: f.g.forster@uu.nl

Abstract: Cleavable endoplasmic reticulum (ER) signal peptides (SPs) and other non-cleavable signal sequences target roughly a quarter of the human proteome to the ER. These short peptides, mostly located at the N-termini of proteins, are highly diverse. For most proteins targeted to the ER, it is the interactions between the signal sequences and the various ER targeting and translocation machineries such as the signal recognition particle (SRP), the protein-conducting channel Sec61, and the signal peptidase complex (SPC) that determine the proteins' target location and provide translocation fidelity. In this review, we follow the signal peptide into the ER and discuss the recent insights that structural biology has provided on the governing principles of those interactions.

Keywords: signal peptide; signal peptidase; ER translocon; endoplasmic reticulum; protein targeting; chaperones; protein translocation

Citation: Liaci, A.M.; Förster, F. Take Me Home, Protein Roads: Structural Insights into Signal Peptide Interactions during ER Translocation. *Int. J. Mol. Sci.* **2021**, 22, 11871. https://doi.org/10.3390/ijms222111871

Academic Editor: Masatoshi Maki

Received: 30 September 2021
Accepted: 28 October 2021
Published: 1 November 2021

Publisher's Note: MDPI stays neutral with regard to jurisdictional claims in published maps and institutional affiliations.

1. Introduction

The secretory pathway is a protein trafficking highway utilized by more than a quarter of the human proteome [1,2]. Soluble secreted proteins such as antibodies and protein hormones rely on this pathway. The pathway also delivers transmembrane proteins (TMPs) to the endoplasmic reticulum (ER), its downstream organelles such as the Golgi apparatus, and the plasma membrane.

All secretory proteins are translated by cytosolic ribosomes and must be first targeted to and then transported across (or inserted into) the ER membrane at the early stage of their life, either co- or post-translationally [3,4]. A complex network of cytosolic and ER membrane-resident macromolecules facilitate and assist the ER targeting and translocation.

Both ER targeting and translocation/insertion critically depend on so-called signal sequences (SSs), short hydrophobic peptide stretches in the amino acid sequence of the newly synthesized proteins that are recognized by the secretory machinery as trafficking signals. While SSs may appear exceedingly simple, they possess a remarkably versatile and complex physiology. Besides the choice of trafficking routes, SSs carry information about translocation efficiency, occurrence and timing of cleavage, and post-targeting functions.

There are four main classes of SSs: (i) cleavable signal peptides (SPs), found on secreted proteins such as insulin and type I membrane proteins such as HLA molecules; (ii) type II signal anchor sequences (SASs), found on single- and multi-pass transmembrane proteins (TMPs) such as the membrane-bound form of tumor necrosis factor (TNF); (iii) type III SASs found, e.g., on Sec61β; and (iv) tail anchors (TAs) found on proteins such as Sec61γ (Figure 1a). Signal peptides (SPs) are by far the most populous class of SSs. In humans alone, there are an estimated >3000 different SP-containing proteins, constituting >10% of the whole proteome. SPs are usually localized within the first 30 amino acids of the coding sequence but can in some cases also be found more internally. The defining trait of SPs is the capacity to be cleaved by the aptly named signal peptidase complex (SPC).

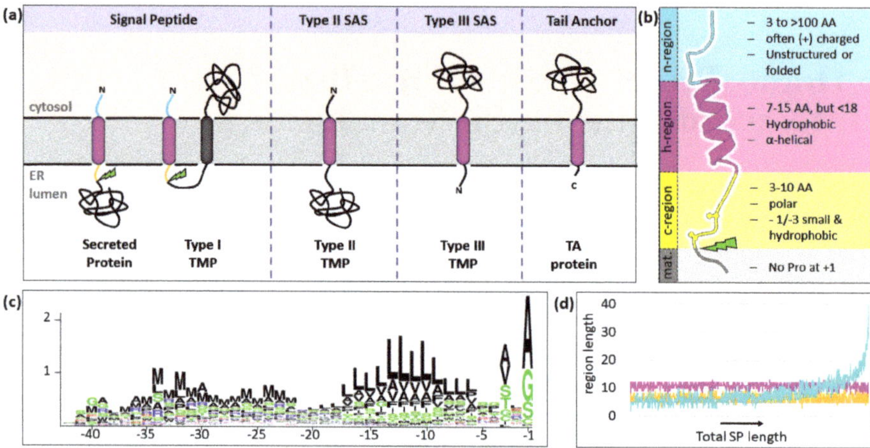

Figure 1. Types of signal sequences. (**a**) Depiction of the four types of SSs with their membrane topology indicated. Hydrophobic segments are depicted in magenta. (**b**) Signal peptides have a tripartite structure, consisting of an n- (cyan), h- (magenta), and c-region (yellow) and are cleaved in the ER lumen by the SPC (green flash). (**c**) Frequency of residue types relative to the cleavage site [5]. (**d**) Relative length of the respective regions (colored as in (**b**)) as a function of total SP length. The bulk of the length variation stems from the n-region. Panels c–d were adapted from [5].

The primary sequence of SPs is only loosely defined. In fact, approximately 20% of randomized sequences can promote protein secretion in yeast [6]. SPs are characterized by a tripartite structure (Figure 1b–d): (i) an often positively charged, N-terminal 'n-region' that faces the cytosol; (ii) a short hydrophobic core—most commonly between 7 and 15, but not longer than 18–20 amino acids called 'h-region'; and (iii) a polar luminal C-terminal 'c-region' that contains the scissile bond and must be occupied by short, hydrophobic residues at positions -1 and -3 relative to the cleavage site [7,8]. Initially, SPs are inserted into the ER membrane with the N-terminus facing towards the cytosol (N_{in}) and the mature sequence facing the organellar lumen (C_{out}). In the case of type I TMPs, the removal of the SP leads to an 'inverted' topology of the mature sequence in which the N-terminus is facing 'outside' (N_{out}), while the C-terminus is facing the cytosol (C_{in}) (Figure 1a).

In this review, we delineate the lessons learned from the structural characterization of the different secretory machineries and their interactions with SPs, starting at the ribosomal exit tunnel and ending in the ER membrane. For the targeting and translocation of non-SP SSs, we refer to several excellent recent reviews [9–11].

2. SP Recognition in the Cytosol and ER Targeting

As the first step of their life cycle, secreted proteins and TMPs must be targeted to the ER membrane (Figure 2). The timing of translation, folding, and ER transport is of particular importance: on one hand, nascent chains (NCs) can only cross the ER membrane in an unfolded state, while on the other hand, this prerequisite exposes their hydrophobic segments and makes them prone to aggregation and proteolysis. Therefore, these proteins must be shielded from the hydrophilic cytosol, which is achieved by one of two separate strategies: (i) direct recognition of the nascent protein at the ribosomal exit tunnel by the SRP, leading to co-translational translocation/insertion through the recruitment of the ribosome–nascent chain complex (RNC) to the ER; or (ii) post-translational transport to the ER, which requires the involvement of chaperones to protect the clients from the aqueous environment.

In this section, we dissect the different cellular pathways for SP delivery to the ER surface (Figure 2). The decision of which routes an NC takes largely depends on interactions between the SP and the respective components of a pathway. It should be kept in mind that these pathways are not necessarily strictly separated, and preferences for one pathway or another may reflect the physiological status of the cell [12]. Additional pathways exist for the targeting of other SSs, such as the TRC/GET pathway that mainly caters to TA proteins [13].

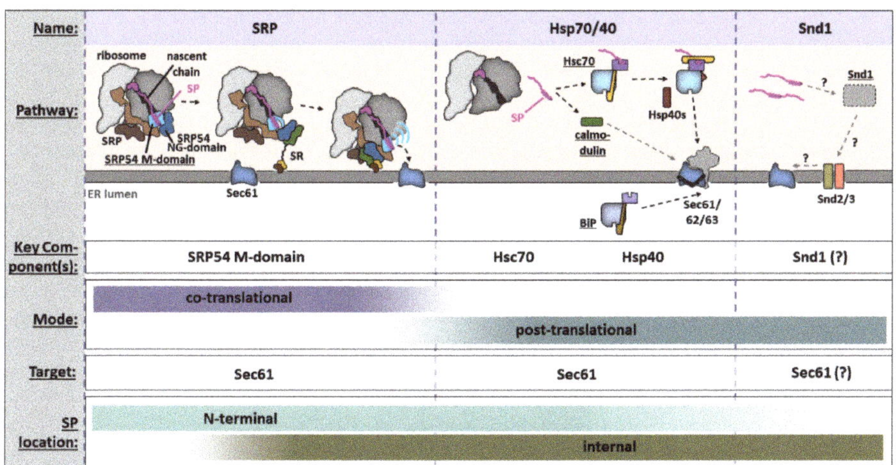

Figure 2. ER delivery pathways for SP-containing proteins. The upper panel shows a schematic of each delivery pathway. The central component of each pathway is underscored. **Left**: SRP (blue/brown subunits) recognizes SPs emerging from the ribosomal exit tunnel and shields them through the SRP54 M-domain. SP binding triggers the heterodimerization of the SRP54 NG domain (blue) with that of SRα (green), guiding the RNC to the ER. A large conformational rearrangement partially exposes the SP for handover to Sec61 [14]. **Middle**: Cytosolic calmodulin or Hsp40-assisted Hsc70 recognize SP-containing proteins and guide them to the ER. **Right**: The recently discovered Snd pathway likely consists of a cytosolic component, Snd1, which might act as a chaperone, and two ER membrane-resident components, Snd2/3, which might facilitate handover to Sec61 in some unknown way [15,16]. Lower panels show specifics of each pathway.

2.1. Life in the Fast Lane: Co-Translational Recognition and ER Targeting by the SRP

Co-translational targeting of SP-containing proteins is accomplished by the signal recognition particle (SRP) and its cognate ER-associated SRP receptor (SR). The SRP is one of the earliest factors that scan the SPs of NCs when they emerge from the ribosomal exit tunnel. The SRP competes for NC binding with the nascent chain associating complex (NAC) [17]. Thus, whether an SP can interact efficiently with SRP is one of the main determinants of their trafficking route.

The SRP pathway is by far the best-studied ER delivery pathway. It is also considered to be the most common ER delivery system in mammals, although potentially less dominant than earlier assumed [18–21]. Proteins of the SRP pathway generally have strongly hydrophobic, N-terminal SPs. As a general rule, the less hydrophobic an SP, the less it tends to depend on SRP [22–25]. However, the dependence of the SRP system on hydrophobicity seems to vary considerably between organisms [21]. In mammals, the majority of SPs are thought to rely on SRP [26], while in yeast, only ~14% of cleavable SPs (which possess shorter hydrophobic segments [5,27]) are SRP-dependent. In contrast to SPs, the overwhelming majority of SASs cannot be efficiently targeted to the ER without SRP, regardless of the organism [21,28–30]. In addition to hydrophobicity, the SRP pathway is sensitive to the localization of the SS within the NC, which is mostly relevant for other types of SSs. Because SRPs are strongly sub-stoichiometric compared to ribosomes [31], substrate recognition needs to be fast. Therefore, the optimal SRP recognition window is within the first ~95 amino acids (AA) of the substrate, where the SP is positioned in most pre-proteins [32]. Remarkably, however, internal SSs located up to ~65 AA distal from the pre-protein's C-terminus can still be recognized by the SRP [21,29]. In contrast, small proteins below 100 AA cannot be efficiently recognized. In both cases, when the SS is too close to the C-terminus or the protein is too short, the protein synthesis is terminated in mere seconds, which is insufficient for SRP targeting [18,33].

Robust targeting by the SRP pathway is estimated to require approximately 5–7 s [34]. When SRP binds to an SP at the ribosomal exit tunnel, its main functions are (i) to halt the translation to prevent cytosolic aggregation, (ii) to recruit RNCs to the ER surface via the SRP receptor (SR), and (iii) to facilitate the handover of the NC to the Sec61 translocon (Figure 2).

The SRP, a composite of one RNA and six protein molecules, binds ribosomes in a 'scanning' mode even before SPs emerge from the exit tunnel [29,35–38]. In some cases, ribosome–mRNA complexes have been found to recruit the SRP before the SP has even been translated [29], or from within the exit tunnel through an allosteric interaction if the N-terminal SS is particularly hydrophobic, which is mostly the case for SASs [39,40].

One of the SRP proteins, SRP54, contains the hydrophobic SS binding groove in its C-terminal M domain. In the scanning mode, the M domain is pre-positioned in direct proximity to the exit tunnel in an auto-inhibited state [41], in which its C-terminal helix ahM6 [14] occupies the SP binding site. As soon as an appropriate SP emerges from the ribosome exit tunnel, it is embraced by the SRP54 M domain in the 'cargo recognition' state, providing shelter from the aqueous environment. The mammalian SRP54–SS interface in the 'cargo recognition' state has been resolved in a recent structure of SRPs in complex with ribosomes translating a type II SAS with a 16 AA hydrophobic core [14] (Figure 3a–e). The ahM6 helix is displaced by the SS such that it completes a ring of α-helices around the SS helix, which remains accessible at both ends. The bulk of the interaction is mediated through non-directional, hydrophobic protein–protein contacts (mostly through flexible methionine side chains) of the SS core with the SRP54 M domain. The hydrophobic surface of the SRP54 M-domain is about 20–23 Å long, matching hydrophobic sequences of ~16 AA (Figure 3e). It appears plausible that SPs with much shorter h-regions [5,27] produce a partial hydrophobic mismatch and are less efficient in both recognition and the induction of the necessary conformational rearrangements in SRP. Indeed, deletions in the h-region of the (already short) SP of preprolactin have drastic effects in its ability to trigger both SRP binding and ER translocation [42]. Coincidentally, these shorter hydrophobic sequences also reduce the susceptibility to aggregation in the cytosol, rationalizing their reduced dependency on the pathway [21].

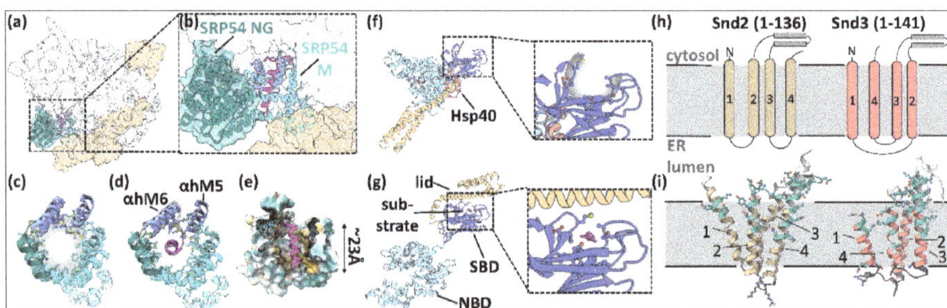

Figure 3. Details of cytosolic SP recognition and ER targeting. (**a–e**) The SRP (orange) and its central component SRP54 (teal/cyan) recognize SPs at the ribosomal exit tunnel (**a,b**) and bury the hydrophobic core in cavity formed by the SRP54 M-domain (**c,d**). αhM5/6 are colored blue. (**e**) Dimensions of the hydrophobic SRP54–SS interface. Figures based on PDB 7OBR [14]. (**f,g**) Cargo recognition by Hsp70 and Hsp40 binding, exemplified by bacterial DnaK complexed with the Hsp40 DnaJ in an open conformation [PDB 5NRO] and substrate-engaged high-affinity conformation [PDB 2KHO]. ATP hydrolysis, stimulated by substrate and Hsp40 binding, leads to a high affinity state with a closed lid domain (yellow). NBD: nucleotide binding domain; SBD: substrate binding domain. (**h,i**) AlphaFold 2 predictions of the ER-resident membrane proteins Snd2/3 from *S. cerevisiae* (entries Q99382 and P38264). TM helices are numbered Membrane topologies are based on UniProt. Both predicted structured possess hydrophilic vestibules that reach into the presumed membrane environment. Helices are numbered.

The SRP54 M-domain provides the plasticity to accommodate a broad variety of SPs. Positively charged residues in the flanking SP regions fine-tune SRP binding [42]. Indeed, a lower-resolution structure of the rabbit ribosome–SRP complex loaded with a 20 AA type II SAS [35] and crystal structures of homologous SRP/SS complexes [43–46] display differing orientations of the SS. Whether these models reflect the physiological variability of the SS binding or are affected by modeling inaccuracies or crystal packing artefacts must be addressed in more extensive sampling of SS space.

The formation of the SRP54–SP interface not only increases SRP/RNC affinity [40,47] and slows down translation [48–50], but it also enables efficient recruitment of the ER-bound SR [41]. Information of SP engagement is passed within SRP54 by the so-called fingerloop of the M domain to the N-terminal GTPase (NG) domain, which contains the SR interface as well as the GTPase site [14]. The SRP54 NG domain can now dimerize with the NG domain of the SR α-subunit (SRα) at the cytosolic surface of the ER [51], leading to a large-scale conformational change when the GTPase activity of these NG domains is activated [52–54]. This large-scale remodeling involves the withdrawal of the SRP54 helices ahM5 and ahM6 from the SP interface, resulting in the partial exposure of the SP to the aqueous environment [14,55]. This conformational change enables cargo handover to the protein conducting channel Sec61, which form the core of ER translocon complex [56]. The interaction kinetics of SRP and SR are regulated by NAC, which prevents the recruitment of SS-less ribosomes to the ER [57].

Faulty SSs, e.g., segments with deletions in their hydrophobic core, trigger a recently discovered quality control pathway, termed regulation of aberrant protein production (RAPP) [58]. As shown by photo-crosslinking, the SRP likely hands over aberrant SPs to Argonaute2 (Ago2) instead of Sec61, triggering Ago2's specific mRNA degradation. However, there is yet no evidence for a direct interaction of Ago2 with SPs.

2.2. With a Little Help from My Friends: Post-Translational Recognition and ER Targeting

Substrates that are targeted to the ER post-translationally require ATP-dependent chaperoning in the cytosol. In yeast, post-translational ER targeting of SP-containing proteins is common [21]. Some 'classic' yeast post-translational substrates are targeted co-translationally in mammals, indicating a preference of the co-translational mode in these species [26]. Nevertheless, also in mammals, SRP-independent translocation of non-TA proteins goes much beyond early discovered single cases [59], as recent studies show [18–20,33,60,61]. We are still at an early point of our understanding of post-translational import of SP-containing pre-proteins in mammals.

2.2.1. Stairway to Lumen: Hsp70/40-Mediated Chaperoning in the Cytosol and ER Delivery

The Hsp70/40 system is thought to deliver its substrates mainly to Sec61/62/63. Post-translational substrates are usually poor substrates of the SRP, and their structural motives are therefore reciprocal to those of SRP substrates—the less hydrophobic an SS, and the closer it is to the C-terminus, the more likely it is to take a chaperone-mediated post-translational delivery route to the ER [24,62]. Hence, most substrates of the pathway feature SPs, although some type II TMPs and multi-spanning TMPs are also influenced by the depletion of both Sec62 and Sec63. Prominent substrates are weakly hydrophobic, N-terminal SPs with weak statistical trends for long h-regions and low-polarity c-regions [62,63]. The post-translational pathway can process more proteins per time than the SRP-dependent route [12].

Hsp70-mediated ER targeting has so far mostly been studied in yeast. Cytosolic Hsp70 (Hsc70 in mammals, Ssa1 in yeast) has broad substrate selectivity for a degenerate, hydrophobic 4–5 AA motif [64] that occurs in SPs, but also in virtually any other protein (Figure 3f,g). Despite its broad functionality as a chaperone, the protein can carry out substrate-specific ER delivery of secretory pathway proteins. Although Ssa1 can directly interact with Sec72, a part of the fungal Sec62/63 translocon that does not exist in mammals [65], the targeting is largely regulated in a substrate-dependent way by a network of

Hsp40 (or J-domain protein) co-chaperones (Figure 2) such as the abundant, ER-tethered Ydj1 (Hdj2 in mammals) or Jjj3 [66–70]. Hsp40s (i) increase the chaperoning activity of Hsp70s [71], (ii) recognize and deliver substrates to their respective Hsp70, and (iii) target their Hsp70 to a specific location in the cell [72]. It appears that these chaperones have some redundancy, as no single one of them is essential [67]. Hsc70-mediated targeting in mammals has been demonstrated for low-hydrophobicity TA proteins, but not yet for SP-containing proteins [73].

In addition to Hsp70/40s, calmodulin mediates targeting in mammals, by recognizing short substrates through their SPs [60]. Proteins of less than 100 AA, often chemokines or hormones, are poor SRP substrates because of their short recognition window at the ribosome [18]. Simply extending their mature sequence has been shown to revert substrates to SRP-dependence [33,74]. The calmodulin-mediated pathway is better characterized for the delivery of weakly hydrophobic TA proteins to the EMC [75].

2.2.2. The Joker: The Snd Pathway

The list of ER targeting pathways was expanded when a screen in yeast revealed three proteins called Snd1-3 that can compensate for the known pathways and act as a backup ER targeting system [15]. The Snd pathway has been implicated in the biogenesis of short secretory proteins [76] as well as both single- and multi-pass transmembrane proteins [16,77]. The system appears capable of recognizing SSs at any position in the protein, from N- to C-terminus [15], allowing it to substitute SRP and other pathways, such as the TRC/GET pathway.

Of the three known yeast Snd proteins, Snd1 was found to be soluble and ribosome-interacting [78], while the other two are ER-resident membrane proteins. Snd2 can be co-immunoprecipitated with Snd1, Snd3, and Sec61, leading to the hypothesis that Snd1 may be involved in target recognition at the ribosome, while the other two subunits could act as a composite receptor and mediate substrate-handover to the Sec61 translocon for translocation, analogous to the SRP/SR system [76]. To date, only the orthologue of Snd2 (called TMEM208 or hSnd2) has been identified in mammals [16,79]. AlphaFold predictions of yeast Snd2/3 show a hydrophilic vestibule on the hydrophobic segments of both proteins (Figure 3h,i), and it is tempting to speculate that these might be key to their function.

3. Protein Translocation and Insertion at the ER Membrane

Following ER targeting, secretory pathway proteins must be either translocated through or inserted into the ER membrane. The main insertase/translocase for SPs is the protein-conducting channel Sec61, which is a part of the ER translocon and can associate with different auxiliary protein complexes (Figure 4) [80].

The ER translocon is a dynamic super-complex at the ER membrane that facilitates the translocation, folding, and post-translational modification of many NCs (Figure 4). In addition to its central component Sec61, the ER translocon features a portfolio of subcomplexes with specific auxiliary functions that can be proactively recruited in a substoichiometric manner depending on the type of SS at hand [11,80]. Some, such as the translocon-associated protein complex (TRAP), are associated near-stoichiometrically to the co-translational ER translocon (Figure 4), while others such as the SPC can act in concert but are not or only transiently recruited [81]. New configurations of the translocon are still being discovered and structurally characterized, giving a plethora of new information their interactions with SSs [5,82–90]. For example, a range of newly characterized ER-resident insertase complexes containing an Oxa1 family subunit can facilitate the insertion of other types of SSs, particularly type III SASs and TAs, but also multi-pass TMPs (reviewed in [9,10]) in concert with Sec61. The ER-resident Oxa1 family complexes comprise the ER membrane complex (EMC), WRB/CAML, which is the insertase of the TRC/GET pathway, and the TMCO1 translocon. TMCO1 proteins might be involved in the translocation of SP-containing type I membrane proteins by facilitating the insertion of downstream TMDs after initial recognition by Sec61, respectively, while the EMC is fully dispensable for the insertion of SPs and type II TMDs but may have chaperoning activities [75,91]. To date, it appears that Sec61 is the only insertase for SPs.

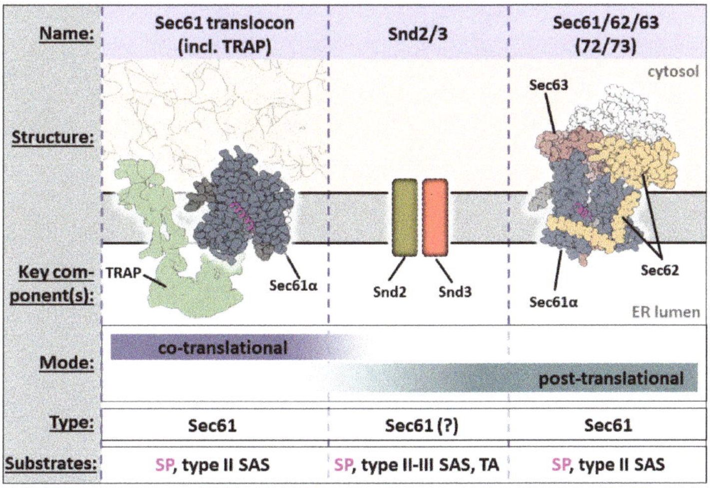

Name:	Sec61 translocon (incl. TRAP)	Snd2/3	Sec61/62/63 (72/73)
Structure:			cytosol
Key component(s):	TRAP Sec61α	Snd2 Snd3	Sec63 Sec62 Sec61α ER lumen
Mode:	co-translational	post-translational	post-translational
Type:	Sec61	Sec61 (?)	Sec61
Substrates:	SP, type II SAS	SP, type II-III SAS, TA	SP, type II SAS

Figure 4. ER translocation and insertion machineries for SP-containing proteins. Left upper panel: The Sec61 translocon comprises, among others, the ribosome (shown in the cytosolic background), Sec61 (Sec61α blue, Sec61β dark gray, Sec61γ light gray), and the yet structurally uncharacterized TRAP (green EM map). The structure is a composite of EMDB 4315 [90] and PDB 3JC2 [92]. Middle upper panel: The yet structurally uncharacterized Snd2/3 membrane complex [15]. Right upper panel: *S. cerevisiae* Sec61 opened by Sec62 (light teal), Sec63 (dark teal), and the yeast-specific Sec71/72 (white). The structure is a composite of PDBs 7AFT and 6ZZZ [93].

In the following section, we will discuss how different components of the ER translocon interact and are recruited by features in the SP, starting with its prime component, the protein conducting channel Sec61.

3.1. Through the Barricades: Co-Translational Translocation/Insertion

In the co-translational mode, the ER translocon associates with an RNC. The interaction of the RNC with the Sec61 core is conserved from prokaryotes to eukaryotes, while many accessory factors, which interact with Sec61 and often also directly with the ribosome, have evolved. Importantly, such interactions are often competitive, making the ER translocon dynamic.

3.1.1. Tunnel of Love: Protein Translocation by Sec61

The heterotrimer Sec61, consisting of Sec61α, Sec61β, and Sec61γ, is chiefly responsible for (i) co-translationally translocating or inserting nascent chains and for (ii) determining substrate topology [94–96]. Its largest subunit, Sec61α, features two pseudo-symmetrical halves (formed by TMD1-5 and 6-10, respectively) that together produce an hourglass-shaped pore across the ER membrane (Figure 5) [92,95,97,98]. The channel must be primed and activated by a partner to enable translocation or membrane insertion. Sec61 is gated in two directions: (i) opening of the two halves in a clam-like motion reveals a lateral gate to the ER membrane; (ii) the same motion also opens the vertical channel that is initially sealed by the 'plug' from the lumen, allowing the passage of the nascent chain [99] (Figure 5c,d).

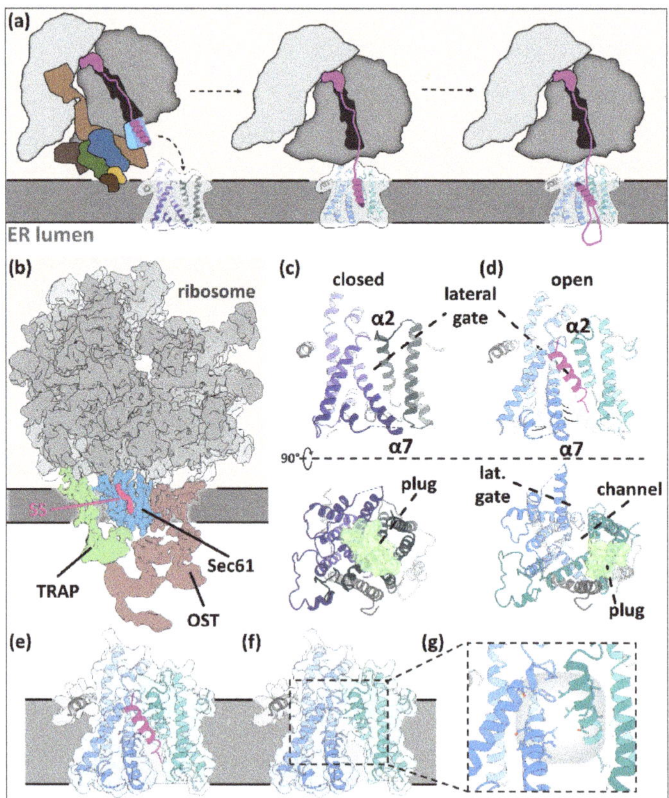

Figure 5. Co-translational ER translocation by Sec61 through the ER translocon. (**a**) Graphic description of co-translational ER insertion/translocation by the SRP-Sec61. The SP is handed over from SRP54-M and inserts head-on into primed Sec61, removing the plug, before rearranging in a hairpin with a N_{in}-C_{out} conformation. The SP is then accommodated at the lateral gate of Sec61. (**b**) Tomographic map of the translating ER translocon (EMDB-4315) [90]. TRAP is accommodated at the side of Sec61, its cytosolic portion interacting with the ribosome, its luminal portion close to the opening of the translocation channel. The oligosaccharyltransferase (OST) does not interact with SPs. (**c,d**) Open and closed conformations of Sec61, based on PDBs 3J7Q and 3JC2 [92,95]. Upon channel opening, the plug (green, seen from within the lumen facing the membrane) is removed and the vertical channel is opened. The lateral gate can now harbor the SS. (**e–g**) SP binding site at the lateral gate. The binding site is partially open to the ER lipid environment, partially formed by hydrophobic residues of helices α2 and α7.

In its co-translational mode, Sec61 can recognize type II SASs in addition to SPs. Upon RNC binding, the gap between the ribosome exit tunnel and the Sec61 translocating channel is reduced to approximately 20 Å [81,92]. The hydrophobic SPs approach the channel head-first. The docking induces conformational changes in Sec61α that, with one rigid body motion, simultaneously open the lateral gate (allowing the SP to squeeze between TMDs 2 and 7 and displace TMD2), and remove the 'plug' to vertically open the channel [92,95,97,98].

Subsequently, the NC re-organizes into a hairpin that exposes the N-terminus to the cytosol. The final orientation of the SP inside Sec61 determines the topology that the protein will ultimately assume. The outcome seems to be mainly affected by the balance of positive charges on both ends of the c-region, with positive charges at the n-region encouraging insertion with the N-terminus facing the cytosol and positive charges in the early mature region apparently impairing translocation altogether [97]. For both type I and type II membrane proteins, the N-terminus of the SS faces the cytosol according to the 'positive-inside' rule, with the distinction that the SPs of type I membrane proteins are later removed by the SPC, whereas the signal anchors of type II membrane proteins are not (see below). Nevertheless, positive charges of the n-region are not the only orientation-determining parameter as ~20% of ~900 analyzed SPs do not contain positively charged residues [5,8].

The decisive parameter for Sec61-dependent translocation is the 'gating' strength of the SP, which is the ability to autonomously induce the opening of Sec61's lateral gate. The gating strength can be measured directly as a pulling force on the nascent chain that drags the SP into the primed channel [100,101]. Weakly gating SPs may trigger the recruitment of assisting co-factors (see below). Interestingly, the gating strength is in many instances evolutionarily conserved, implying that strong or weak gating can fulfill important functions, e.g., in the timing and overall production levels of substrates, or even to maintain a cytosolic presence of the substrate, as is, e.g., the case for calreticulin [102,103]. Despite being measurable, the precise individual structural/physical features that determine the gating strength are poorly understood [104]. The hydrophobicity of the SPs core is suspected to play a key role and other potential factors might be the charge of adjacent residues in the nascent chain, the degree of N-terminal folding, and the lipid composition of the ER membrane [19,105].

When finally opened, Sec61 allows the co-translational translocation of the nascent chain's hydrophilic regions across the membrane and into the ER lumen, driven by the GTP hydrolysis of the ribosome.

During translocation, the SP of the nascent peptide remains accommodated in the lateral gate of the Sec61 protein-conducting channel [92,98,106] in a slanted orientation. The interaction is mediated by hydrophobic residues and halfway open to the surrounding lipid environment (Figure 5) [92]. Early crosslinking experiments showed that changing the amino acid sequence in the h-region can change the positioning of the SP in the gate [25], implying that the interaction is subject to considerable conformational liberty.

Interestingly, electron tomography data of pancreatic vesicles suggest that SPs seem to linger stably at the lateral gate while the ribosome is bound to the ER translocon because Sec61 is detected solely in an open state with SPs bound [92,106]. This observation suggests that SPs are not able to diffuse freely into the membrane, presumably due to their short h-regions that cause a hydrophobic mismatch with the ER bilayer. In the case of cleavable SPs, the peptides eventually need to be transferred from Sec61 to the SPC, although it is currently unclear how this handover may take place.

3.1.2. Smooth Operator: TRAP Is a Sec61 Assistant

SRP delivers SSs to Sec61 irrespective of their 'gating' efficiency [100,107]. When encountering substrates with ineffective or slow channel gating abilities, Sec61 requires assisting factors that help it to exert second pulling event that occurs during translocation. The most stoichiometrically present of several auxiliary complexes is TRAP, which specializes in co-translational enhancement of Sec61 translocation/insertion.

Some of the features of SPs that mediate important functions—such as non-standard lengths, charged residues, kinks, or reduced hydrophobicity—are also hallmarks of reduced 'gating' efficiency and therefore particularly challenging for Sec61-mediated translocation/insertion. It appears that no single parameter determines TRAP dependency—rather, it is the combination of several parameters that ultimately weaken the interaction with the primed Sec61 translocating channel [104]. Although there are subtle differences, unifying characteristics of TRAP-dependent substrates are (i) a low hydrophobicity and/or (ii) positively charged clusters in the mature portion directly following the SP that impair translocation [62,63,108], and (iii) an enriched glycine and proline content [63,109]. Additionally, TRAP may be able to assist in the translocation of charged unstructured protein domains [108].

TRAP comprises the four subunits TRAPα-δ (or SSR1-4) [104]. The helix-breaking properties of TRAP-dependent substrates may impair their ability to insert into the lateral gate of Sec61α [92], potentially resulting in SPs that 'lounge' in the ~20 Å gap between ribosome and Sec61 instead of being pulled in by primed Sec61. TRAP is associated near-stoichiometrically with the translocon and interacts with both the ribosome and Sec61 [81,90,106,110].

Mechanistic insights into TRAP function remain scarce. A recent study delineated the 'translational timeline' of TRAP-assisted translocation. Based on crosslinking experiments it postulates a weak, potentially sensory interaction of the ~13-amino-acid-long cytosolic portion of TRAPβ with a weakly gating model SS before the SS is threaded into the Sec61 channel [109]. Tomographic studies [63,81] indicate that the luminal domains of TRAPα/β might interact with loop 5 of the Sec61 hinge region to facilitate channel opening. Additionally, it has been suggested that that luminal domains of TRAP may stabilize the topology of thus far 'uncommitted' nascent TMDs by a direct interaction [111]. However, there are currently no structural data of sufficient resolution to corroborate these postulated interactions.

3.2. Insane in the Membrane: Post-Translational Translocation/Insertion

Much like the co-translational pathways, post-translational translocation/insertion of proteins with N-terminal SPs and SSs is carried out through Sec61 (aided by auxiliary factors) while Oxa1 family complexes such as EMC and Get1/2 complexes likely do not interact with SPs.

With Arms Wide Open: Post-Translational Insertion and Translocation Assisted by Sec62/63

Some secretory proteins are targeted to Sec61 post-translationally [18,24,67,74,112–114]. Together, the auxiliary proteins Sec62 and Sec63 can facilitate post-translational Sec61-dependent translocation/insertion, allowing Sec61 to translocate substrates by prying open the lateral gate [115–118]. First, Sec63 partially opens the gate through interactions on both sides of the membrane, before Sec62 can fully displace the plug and shield the channel from incoming lipid molecules [84,93,119,120].

To date, 56 Sec62/63 dependent proteins have been identified in human cell lines [62,63]. The SP features that confer Sec62/63 dependency are very similar to those for its Hsp70/40 delivery system, mainly featuring reduced hydrophobicity and weak Sec61 gating capacities, which make the pre-opening of the channel by Sec62/63 necessary [120]. Additionally, some Sec62/63-dependent substrates possess a positively charged cluster in the mature region that likely impairs the translocation across the Sec61 channel [62]. Notably, there is some overlap between Sec62/63- and TRAP-dependent SSs, as both comprise low-hydrophobicity cores. Indeed, there are specific examples that have been shown to depend on both complexes [62,63], although TRAP is thought to operate co-translationally whereas Sec62/63 is generally described to act post-translationally. It has been reported that some substrates can follow both the co- and the post-translational route, such as pre-proinsulin [121].

The Hsp70/Hsp40 system also possesses a luminal portion, which is essential for post-translational ER translocation by providing directionality and might also play a role during co-translational import. In absence of a ribosome, protein translocation is powered from within the lumen by the Hsp70 'ratchet' BiP (HSPA5, or Kar2 in yeast), which is recruited by the Sec63 (or ERj2) Hsp40 J-domain and can interact with a luminal loop of Sec61α [76,122–125]. Another example is the ER membrane-resident mammalian J-protein ERj1 (Htj1 in humans) can bind ribosomes near the ribosomal exit tunnel using its cytosolic portion, promote translational stalling, and potentially recruit ribosomes to the ER during co-translational targeting. Its J-domain is located in the ER lumen, where it aids in protein translocation by stimulating BiP, similarly to Sec63 [126]. Other ERj proteins modulate BiP functions for protein quality control or stress conditions such as the unfolded protein response or ER-associated protein degradation [127].

Some studies of specific substrates found crosslinks of mammalian Sec62/63 with the co-translational ER translocon, implying that Sec62/63 might also be involved in co-translational protein translocation [128–130]. However, structural studies of the early co-translational ER translocon suggest that there would be steric clashes between Sec61-bound Sec62/63 and both the ribosome and TRAP [81,95]. Thus, either the conformation of Sec61/62/63 must differ in the co-translational mode or the ribosome associates differently [84,120,131]. Biochemical studies corroborate the competitive binding of co- and post-translational ER translocon components by showing that both the SR and the ribosome individually induce the dissociation of Sec62 from Sec61 [131].

Together, these data suggest that Sec62/63 might only be involved in the later stages of co-translational translocation, if at all [129]. As an exemplary hypothesis for such an operating mode, TRAP might initially ensure the engagement of the SP by Sec61, before Sec62/63 eventually carry out in-channel inversion and gating in a temporally distinct step after translocon remodeling [108]. However, there is currently no mechanistic basis for these kinds of proposed mechanisms, and it is not known if and to what extent Sec62/63 can compensate for a loss of TRAP or vice versa.

4. SP Removal and Post-Targeting Functions

4.1. Time to Say Goodbye: SP Removal by the Signal Peptidase Complex

Regardless of how the specific characteristics of SSs influence the routes taken, the ultimate choice between SPs and SASs is—per definition—made after ER targeting and translocation. Completed translocation of about 80 amino acids through Sec61 marks the earliest possible time point for co-translational SP removal [132]. The SPC, a heterotetrameric serine protease, which exists in two distinct paralogs in higher eukaryotes [5], is responsible for this process.

By definition, the SPC only cleaves SPs, although TAs can also be engineered sufficiently short to be cleaved in the ER lumen [133]. SPC substrates require an h-region below 18–20 amino acids and a c-region with short, hydrophobic residues at the positions −1 and −3 relative to the cleavage site [5,133,134].

Although it is assumed that SP removal typically occurs co-translationally, post-translational SP cleavage is well within the SPCs repertoire [135]. SPs of co-translationally translocated NCs need to transfer from Sec61 to the SPC. However, the SPC is not resolved as part of native ribosome–translocon complexes by cryo-electron tomography [106,136], indicating that the SPC likely associates with the ER translocon transiently or in a structurally flexible manner [81].

The molecular workings of eukaryotic SP removal, and the distinction between SPs and SASs by the SPC, have eluded mechanistic explanation for decades. Recent structural insights reveal that the enzyme complex employs two structural motifs to make this distinction [5]: (i) a specialized transmembrane window (TM window) that locally thins the ER bilayer directly above (ii) a conserved shallow hydrophobic binding pocket that contains the cleavage site, located in the lumen about 17 Å from the luminal membrane interface (Figure 6). Using the TM window as a 'molecular ruler', the enzyme measures the length of its substrates' hydrophobic segments and excludes SSs with TM segments of more than 18–20 amino acids [133,134]. Substrates that have short enough TM segments (h-regions) to be admitted into the binding pocket then adopt a β-strand conformation in the stretch directly adjacent to the membrane (c-region) and are scanned for small, hydrophobic residues at the evolutionarily ancient −1 and −3 positions relative to the cleavage site by the SPC's catalytic subunit SEC11A/C [137]. Additionally, SPs with prolines at position +1 cannot be cleaved efficiently [138], likely because cis-peptide bonds cannot be effectively polarized in the protease's oxyanion hole. Based on this bifold recognition principle, the SPC effectively 'defines' the distinction between SPs and type II SASs by the length, but not the primary sequence of their hydrophobic segments and by the presence of a suitable set of −1/−3 residues in direct proximity to the ER membrane–lumen interface. Type III SASs cannot be cleaved by the SPC because of their reversed topology that prevent the recognition of the scissile bond.

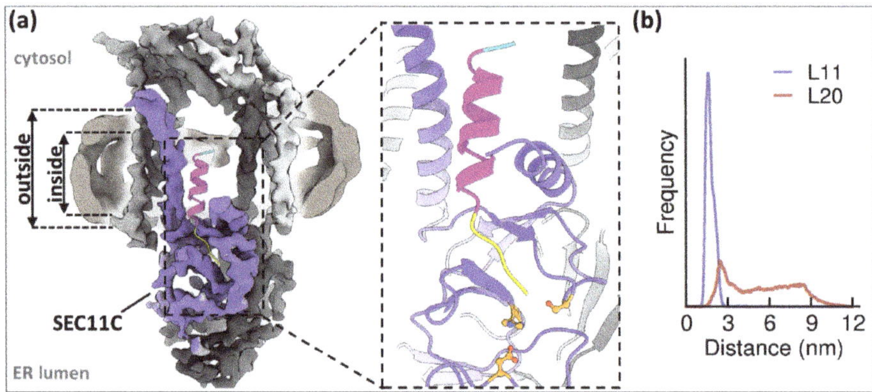

Figure 6. SP removal by the signal peptidase complex. (**a**) Atomic model of an SP fitted into the binding pocket of SPC-C (shown as EM map, catalytic subunit SEC11C purple, SEC22/23 dark gray, SPC25 gray, SPC12 light gray; EMDB-13172). The micelle (show in the background in pale yellow) is thinned inside the TM window and measures the length of the h-region. The blowup shows the binding groove for with the SP c-region and the catalytic residues of SEC11C (orange). Residues −1 and −3 point towards the bottom of the pocket and therefore need to be short and hydrophobic. (**b**) Coarse-grained molecular dynamics simulations show the mean distance of an SP with 11 residues (L11) and 20 residues (L20) from the binding pocket. Long h-regions are excluded by the thinned membrane in the TM window. Figure based on [5].

The timing of SPC-mediated SP removal can impact a protein's maturation process. Especially viral pre-proteins of flavi- and retroviruses, which heavily depend on the SPC for their maturation, utilize cleavage delay to halt the trafficking of their pre-proteins, which is crucial for maintaining the correct processing order and joint incorporation of the fragments during virion assembly [135]. Delays can be caused by a highly apolar or helical c-region, folded n-regions, or directly neighboring TM segments that may sterically obstruct access to the cleavage site [139–141]. It will be interesting to further assess the details of the SPC-SP interactions by structural biology, and to define the interactions of the SPC with the ER translocon during co-translational SP cleavage.

4.2. The Show Must Go On: Post-Targeting Functions of SPs

While SASs and TAs remain associated with their proteins, SPs can have a range of interesting and versatile functions after their cleavage that are unrelated to their original protein. While many are recycled by signal peptide peptidases, specific SPs can, for example, modulate NK-cell mediated immune responses via HLA-E, become a part of mature virions, or associate with cytosolic factors such as calmodulin as feedback loops. For further reading, we refer to [12,142].

5. Summary and Outlook

5.1. The Long and Winding Road: Principles of SP Recognition

Amongst the many interactions laid out here, the three key moments for SPs are (i) the efficiency of the interaction with SRP54, which occurs very early and effectively seems to decide whether a protein takes the co- or the post-translational route; (ii) the gating of Sec61, which determines whether auxiliary factors need to be recruited; and (iii) their removal by the SPC, which is usually necessary for the functionality and further trafficking of the protein. The loose sequence requirements for the SP targeting and translocation machinery lead to an elastic and partially overlapping system, in which subtle changes can alter pathway usage. With recent structural data, the molecular rules and responsible mechanisms that determine the SP's fate are slowly emerging.

There are a few governing principles followed by all or many of the proteins that interact with SPs. Their main interaction point is obviously the h-region. Two major types of strategy emerge for this interaction: (i) cytosolic factors such as the SRP and Hsps bury the h-region in a hydrophobic groove that can be closed by a lid. The interface is often lined by dynamic residues such as methionine and does not induce a specific shape requirement for the substrate. Therefore, it is rather the absolute hydrophobicity and the length of the h-region that determine how efficient the binding is. (ii) Intramembrane interfaces, found, e.g., on that Sec61 lateral gate and the SPC, utilize the lipid surrounding and often leave a substantial portion of the h-region surface open to the membrane, which greatly reduces the requirement for shape complementarity.

The resulting low-sequence constraint on SPs provides an optimal opportunity for an evolutionary fine-tuning of beneficial interactions. As such, the ER delivery and translocation route (and with it, e.g., the speed and absolute capacity of protein synthesis [12]), the efficiency and speed of translocation/insertion (and with it, e.g., the maintenance of a cytosolic pool [102,103]), the timing of SP cleavage (influencing, e.g., the time to add posttranslational modifications [135]), and in some cases even the post-cleavage functions can all be used to tailor to the biosynthetic needs of a specific protein. In addition, SPs can be used to generate alternatively localized versions of a protein from a single mRNA. So how are these changes accomplished, precisely? Unfortunately, too many unknowns still exist to answer this question decisively.

Lastly, the recognition principles employed by the SPC are markedly different from those of other complexes involved in targeting and translocation. As a result, the choice of co- or post-translational translocation/insertion does not affect or only moderately affects SPC cleavage. This difference in recognition for the strictly necessary step of SP removal greatly increases the evolutionary freedom for the largely optional targeting and translocation routes.

5.2. Imagine: Outlook

It is entirely possible that more ER targeting and translocation pathways await discovery. For example, it has only recently emerged that thinning of the particularly elastic ER membrane is a common principle among ER-resident Oxa1 family protein complexes, which use it to catalyze the insertion specific types of transmembrane helices, and by the SPC, which use it as a 'molecular ruler' to measure the length of the h-region [5,10,85]. These findings were rather unanticipated, but have substantially broadened our understanding of the membrane insertion of less well-studied SSs such as type III SASs, TAs, and the downstream TMDs of multi-pass membrane proteins [82,85,143].

There are, however, more gaps in our understanding [144], particularly when it comes to the handover of substrates between complexes, which is one of the most difficult problems to solve with structural biology, and to the relative usage frequency of the individual pathways, which may well differ between organisms and physiological states. As such, relatively few post-translational substrates are known to date. Additionally, the overlap between the pathways greatly complicates the experimental design, rendering a coherent understanding difficult [15].

Compared to yeast, the ER-resident protein translocation machinery has evolved substantially in mammals. It is striking that many genes have duplicated, giving rise to several paralogs at various stages of protein biogenesis in the ER [145]. For example, the mammalian ER possesses two oligosaccharyltransferase paralogs that specialize in co- or post-translational processing, respectively [146], and the SPC exists in two paralogs with overlapping function [5]. Another example would be Sec61α, which exists in two paralogs, the latter of which has been poorly characterized to date. More studies are needed that focus on the quantification of the pathway usage, particularly in mammals, the evolutionary advantages of several paralogs, whether the different paralogs could fulfill specialized roles with respect to SP interactions.

Author Contributions: A.M.L. and F.F. jointly wrote the manuscript. All authors have read and agreed to the published version of the manuscript.

Funding: This research was supported by the ERC Consolidator Grant 724425 (Biogenesis and Degradation of Endoplasmic Reticulum Proteins, to F.F.).

Conflicts of Interest: The authors declare no conflict of interests.

References

1. Palade, G.; Arch, A.; Locke, M.; Locke, E.; Palade, G. Intracellular aspects of the process of protein synthesis. *Science* **1975**, *189*, 347–358. [CrossRef]
2. Uhlén, M.; Fagerberg, L.; Hallström, B.M.; Lindskog, C.; Oksvold, P.; Mardinoglu, A.; Sivertsson, Å.; Kampf, C.; Sjöstedt, E.; Asplund, A.; et al. Tissue-based map of the human proteome. *Science* **2015**, *347*, 1260419. [CrossRef]
3. Aviram, N.; Schuldiner, M. Targeting and translocation of proteins to the endoplasmic reticulum at a glance. *J. Cell Sci.* **2017**, *130*, 4079–4085. [CrossRef] [PubMed]
4. Rapoport, T.A.; Li, L.; Park, E. Structural and mechanistic insights into protein translocation. *Annu. Rev. Cell Dev. Biol.* **2017**, *33*, 369–390. [CrossRef]
5. Liaci, A.M.; Steigenberger, B.; Tamara, S.; Telles de Souza, P.C.; Gröllers-Mulderij, M.; Ogrissek, P.; Marrink, S.J.; Scheltema, R.; Förster, F. Structure of the human signal peptidase complex reveals the determinants for signal peptide cleavage. *Mol. Cell* **2021**, *81*, 3934–3948. [CrossRef] [PubMed]
6. Kaiser, C.A.; Preuss, D.; Grisafi, P.; Botstein, D. Many random sequences functionally replace the secretion signal sequence of yeast invertase. *Science* **1987**, *235*, 312–317. [CrossRef] [PubMed]
7. Von Heijne, G. The signal peptide. *J. Membr. Biol.* **1990**, *115*, 195–201. [CrossRef] [PubMed]
8. Liaci, A.M.; Förster, F. *Empirical Analysis of Eukaryotic ER Signal Peptides*; Mendeley Data; Data Archiving and Networked Services (DANS): The Hague, The Netherlands, 2021. [CrossRef]
9. McDowell, M.A.; Heimes, M.; Sinning, I. Structural and molecular mechanisms for membrane protein biogenesis by the Oxa1 superfamily. *Nat. Struct. Mol. Biol.* **2021**, *28*, 234–239. [CrossRef]
10. Hegde, R.S.; Keenan, R.J. The mechanisms of integral membrane protein biogenesis. *Nat. Rev. Mol. Cell Biol.* **2021**, 1–18. [CrossRef]
11. O'Keefe, S.; Pool, M.R.; High, S. Membrane protein biogenesis at the ER: The highways and byways. *FEBS J.* **2021**, 1–28. [CrossRef]
12. Hegde, R.S.; Bernstein, H.D. The surprising complexity of signal sequences. *Trends Biochem. Sci.* **2006**, *31*, 563–571. [CrossRef]
13. Borgese, N.; Coy-Vergara, J.; Colombo, S.F.; Schwappach, B. The ways of tails: The GET pathway and more. *Protein J.* **2019**, *38*, 289–305. [CrossRef]
14. Jomaa, A.; Eitzinger, S.; Zhu, Z.; Chandrasekar, S.; Kobayashi, K.; Shan, S.; Ban, N. Molecular mechanism of cargo recognition and handover by the mammalian signal recognition particle. *Cell Rep.* **2021**, *36*, 109350. [CrossRef] [PubMed]
15. Aviram, N.; Ast, T.; Costa, E.A.; Arakel, E.C.; Chartzman, S.G.; Jan, C.H.; Haßdenteufel, S.; Dudek, J.; Jung, M.; Schorr, S.; et al. The SND proteins constitute an alternative targeting route to the endoplasmic reticulum. *Nature* **2016**, *540*, 134–138. [CrossRef]
16. Haßdenteufel, S.; Sicking, M.; Schorr, S.; Aviram, N.; Fecher-Trost, C.; Schuldiner, M.; Jung, M.; Zimmermann, R.; Lang, S. hSnd2 protein represents an alternative targeting factor to the endoplasmic reticulum in human cells. *FEBS Lett.* **2017**, *591*, 3211–3224. [CrossRef] [PubMed]
17. Gamerdinger, M.; Hanebuth, M.A.; Frickey, T.; Deuerling, E. The principle of antagonism ensures protein targeting specificity at the endoplasmic reticulum. *Science* **2015**, *348*, 201–207. [CrossRef] [PubMed]
18. Lakkaraju, A.K.K.; Thankappan, R.; Mary, C.; Garrison, J.L.; Taunton, J.; Strub, K. Efficient secretion of small proteins in mammalian cells relies on Sec62-dependent posttranslational translocation. *Mol. Biol. Cell* **2012**, *23*, 2712–2722. [CrossRef]
19. Shao, S.; Hegde, R.S. Membrane protein insertion at the endoplasmic reticulum. *Annu. Rev. Cell Dev. Biol.* **2011**, *27*, 25–56. [CrossRef]
20. Johnson, N.; Vilardi, F.; Lang, S.; Leznicki, P.; Zimmermann, R.; High, S. TRC40 can deliver short secretory proteins to the Sec61 translocon. *J. Cell Sci.* **2012**, *125*, 4414. [CrossRef]
21. Costa, E.A.; Subramanian, K.; Nunnari, J.; Weissman, J.S. Defining the physiological role of SRP in protein-targeting efficiency and specificity. *Science* **2018**, *359*, 689–692. [CrossRef] [PubMed]
22. Chirico, W.J.; Waters, M.G.; Blobel, G. 70K heat shock related proteins stimulate protein translocation into microsomes. *Nature* **1988**, *332*, 805–810. [CrossRef] [PubMed]
23. Deshaies, R.J.; Koch, B.D.; Werner-Washburne, M.; Craig, E.A.; Schekman, R. A subfamily of stress proteins facilitates translocation of secretory and mitochondrial precursor polypeptides. *Nature* **1988**, *332*, 800–805. [CrossRef]
24. Ng, D.T.W.; Brown, J.D.; Walter, P. Signal sequences specify the targeting route to the endoplasmic reticulum membrane. *J. Cell Biol.* **1996**, *134*, 269–278. [CrossRef] [PubMed]
25. Plath, K.; Mothes, W.; Wilkinson, B.M.; Stirling, C.J.; Rapoport, T.A. Signal sequence recognition in posttranslational protein transport across the yeast ER membrane. *Cell* **1998**, *94*, 795–807. [CrossRef]
26. Garcia, P.D.; Walter, P. Full-length prepro-α-factor can be translocated across the mammalian microsomal membrane only if translation has not terminated. *J. Cell Biol.* **1988**, *106*, 1043–1048. [CrossRef]
27. Von Heijne, G. Signal sequences. The limits of variation. *J. Mol. Biol.* **1985**, *184*, 99–105. [CrossRef]

28. Schibich, D.; Gloge, F.; Pöhner, I.; Björkholm, P.; Wade, R.C.; Von Heijne, G.; Bukau, B.; Kramer, G. Global profiling of SRP interaction with nascent polypeptides. *Nature* **2016**, *536*, 219–223. [CrossRef]

29. Chartron, J.W.; Hunt, K.C.L.; Frydman, J. Cotranslational signal-independent SRP preloading during membrane targeting. *Nature* **2016**, *536*, 224–228. [CrossRef]

30. O'Keefe, S.; Zong, G.; Duah, K.B.; Andrews, L.E.; Shi, W.Q.; High, S. An alternative pathway for membrane protein biogenesis at the endoplasmic reticulum. *Commun. Biol.* **2021**, *4*, 828. [CrossRef]

31. Raue, U.; Oellerer, S.; Rospert, S. Association of protein biogenesis factors at the yeast ribosomal tunnel exit is affected by the translational status and nascent polypeptide sequence. *J. Biol. Chem.* **2007**, *282*, 7809–7816. [CrossRef]

32. Noriega, T.R.; Tsai, A.; Elvekrog, M.M.; Petrov, A.; Neher, S.B.; Chen, J.; Bradshaw, N.; Puglisi, J.D.; Walter, P. Signal recognition particle-ribosome binding is sensitive to nascent chain length. *J. Biol. Chem.* **2014**, *289*, 19294–19305. [CrossRef] [PubMed]

33. Johnson, N.; Haßdenteufel, S.; Theis, M.; Paton, A.W.; Paton, J.C.; Zimmermann, R.; High, S. The signal sequence influences post-translational ER translocation at distinct stages. *PLoS ONE* **2013**, *8*, e75394. [CrossRef] [PubMed]

34. Goder, V.; Crottet, P.; Spiess, M. In vivo kinetics of protein targeting to the endoplasmic reticulum determined by site-specific phosphorylation. *EMBO J.* **2000**, *19*, 6704–6712. [CrossRef]

35. Voorhees, R.M.; Hegde, R.S. Structures of the scanning and engaged states of the mammalian SRP-ribosome complex. *eLife* **2015**, *4*, e07975. [CrossRef] [PubMed]

36. Bornemann, T.; Jöckel, J.; Rodnina, M.V.; Wintermeyer, W. Signal sequence-independent membrane targeting of ribosomes containing short nascent peptides within the exit tunnel. *Nat. Struct. Mol. Biol.* **2008**, *15*, 494–499. [CrossRef]

37. Denks, K.; Sliwinski, N.; Erichsen, V.; Borodkina, B.; Origi, A.; Koch, H.G. The signal recognition particle contacts uL23 and scans substrate translation inside the ribosomal tunnel. *Nat. Microbiol.* **2017**, *2*, 16265. [CrossRef]

38. Jomaa, A.; Boehringer, D.; Leibundgut, M.; Ban, N. Structures of the *E. coli* translating ribosome with SRP and its receptor and with the translocon. *Nat. Commun.* **2016**, *7*, 10471. [CrossRef] [PubMed]

39. Berndt, U.; Oellerer, S.; Zhang, Y.; Johnson, A.E.; Rospert, S. A signal-anchor sequence stimulates signal recognition particle binding to ribosomes from inside the exit tunnel. *Proc. Natl. Acad. Sci. USA* **2009**, *106*, 1398–1403. [CrossRef]

40. Flanagan, J.J.; Chen, J.C.; Miao, Y.; Shao, Y.; Lin, J.; Bock, P.E.; Johnson, A.E. Signal recognition particle binds to ribosome-bound signal sequences with fluorescence-detected subnanomolar affinity that does not diminish as the nascent chain lengthens. *J. Biol. Chem.* **2003**, *278*, 18628–18637. [CrossRef]

41. Lee, J.H.; Chandrasekar, S.; Chung, S.Y.; Fu, Y.H.H.; Liu, D.; Weiss, S.; Shan, S.O. Sequential activation of human signal recognition particle by the ribosome and signal sequence drives efficient protein targeting. *Proc. Natl. Acad. Sci. USA* **2018**, *115*, E5487–E5496. [CrossRef]

42. Nilsson, I.; Lara, P.; Hessa, T.; Johnson, A.E.; von Heijne, G.; Karamyshev, A.L. The code for directing proteins for translocation across ER membrane: SRP cotranslationally recognizes specific features of a signal sequence. *J. Mol.* **2015**, *427*, 1191–1201. [CrossRef] [PubMed]

43. Janda, C.Y.; Li, J.; Oubridge, C.; Hernández, H.; Robinson, C.V.; Nagai, K. Recognition of a signal peptide by the signal recognition particle. *Nature* **2010**, *465*, 507–510. [CrossRef]

44. Hainzl, T.; Huang, S.; Meriläinen, G.; Brännström, K.; Sauer-Eriksson, A.E. Structural basis of signal-sequence recognition by the signal recognition particle. *Nat. Struct. Mol. Biol.* **2011**, *18*, 389–391. [CrossRef] [PubMed]

45. Jomaa, A.; Fu, Y.H.H.; Boehringer, D.; Leibundgut, M.; Shan, S.O.; Ban, N. Structure of the quaternary complex between SRP, SR, and translocon bound to the translating ribosome. *Nat. Commun.* **2017**, *8*, 15470. [CrossRef]

46. Wang, S.; Jomaa, A.; Jaskolowski, M.; Yang, C.I.; Ban, N.; Shan, S. The molecular mechanism of cotranslational membrane protein recognition and targeting by SecA. *Nat. Struct. Mol. Biol.* **2019**, *26*, 919–929. [CrossRef]

47. Powers, T.; Walter, P. The nascent polypeptide-associated complex modulates interactions between the signal recognition particle and the ribosome. *Curr. Biol.* **1996**, *6*, 331–338. [CrossRef]

48. Walter, P.; Blobel, G. Translocation of proteins across the endoplasmic reticulum. III. Signal recognition protein (SRP) causes signal sequence-dependent and site-specific arrest of chain elongation that is released by microsomal membranes. *J. Cell Biol.* **1981**, *91*, 557–561. [CrossRef] [PubMed]

49. Mason, N.; Ciufo, L.F.; Brown, J.D. Elongation arrest is a physiologically important function of signal recognition particle. *EMBO J.* **2000**, *19*, 4164–4174. [CrossRef]

50. Lakkaraju, A.K.K.; Mary, C.; Scherrer, A.; Johnson, A.E.; Strub, K. SRP keeps polypeptides translocation-competent by slowing translation to match limiting ER-targeting sites. *Cell* **2008**, *133*, 440–451. [CrossRef] [PubMed]

51. Ariosa, A.R.; Duncan, S.S.; Saraogi, I.; Lu, X.; Brown, A.; Phillips, G.J.; Shan, S.O. Fingerloop activates cargo delivery and unloading during cotranslational protein targeting. *Mol. Biol. Cell* **2013**, *24*, 63–73. [CrossRef]

52. Egea, P.F.; Shan, S.O.; Napetschnig, J.; Savage, D.F.; Walter, P.; Stroud, R.M. Substrate twinning activates the signal recognition particle and its receptor. *Nature* **2004**, *427*, 215–221. [CrossRef]

53. Freymann, D.M.; Keenan, R.J.; Stroud, R.M.; Walter, P. Structure of the conserved GTPase domain of the signal recognition particle. *Nature* **1997**, *385*, 361–364. [CrossRef] [PubMed]

54. Padmanabhan, S.; Freymann, D.M. The conformation of bound GMPPNP suggests a mechanism for gating the active site of the SRP GTPase. *Structure* **2001**, *9*, 859–867. [CrossRef]

55. Kobayashi, K.; Jomaa, A.; Lee, J.H.; Chandrasekar, S.; Boehringer, D.; Shan, S.; Ban, N. Structure of a prehandover mammalian ribosomal SRP · SRP receptor targeting complex. *Science* **2018**, *327*, 323–327. [CrossRef] [PubMed]

56. Shen, K.; Arslan, S.; Akopian, D.; Ha, T.; Shan, S. Activated GTPase movement on an RNA scaffold drives co-translational protein targeting. *Nature* **2012**, *492*, 271–275. [CrossRef]

57. Hsieh, H.-H.; Lee, J.H.; Chandrasekar, S.; Shan, S. A ribosome-associated chaperone enables substrate triage in a cotranslational protein targeting complex. *Nat. Commun.* **2020**, *11*, 5840. [CrossRef] [PubMed]

58. Karamyshev, A.L.; Patrick, A.E.; Karamysheva, Z.N.; Griesemer, D.S.; Hudson, H.; Tjon-Kon-Sang, S.; Nilsson, I.; Otto, H.; Liu, Q.; Rospert, S.; et al. Inefficient SRP interaction with a nascent chain triggers a MRNA quality control pathway. *Cell* **2014**, *156*, 146–157. [CrossRef]

59. Zimmermann, R.; Sagstetter, M.; Lewis, M.J.; Pelham, H.R. Seventy-kilodalton heat shock proteins and an additional component from reticulocyte lysate stimulate import of M13 procoat protein into microsomes. *EMBO J.* **1988**, *7*, 2875–2880. [CrossRef] [PubMed]

60. Shao, S.; Hegde, R.S. A calmodulin-dependent translocation pathway for small secretory proteins. *Cell* **2011**, *147*, 1576–1588. [CrossRef]

61. Johnson, N.; Powis, K.; High, S. Post-translational translocation into the endoplasmic reticulum. *Biochim. Biophys. Acta—Mol. Cell Res.* **2013**, *1833*, 2403–2409. [CrossRef] [PubMed]

62. Schorr, S.; Nguyen, D.; Haßdenteufel, S.; Nagaraj, N.; Cavalié, A.; Greiner, M.; Weissgerber, P.; Loi, M.; Paton, A.W.; Paton, J.C.; et al. Identification of signal peptide features for substrate specificity in human Sec62/Sec63-dependent ER protein import. *FEBS J.* **2020**, *287*, 4612–4640. [CrossRef] [PubMed]

63. Nguyen, D.; Stutz, R.; Schorr, S.; Lang, S.; Pfeffer, S.; Freeze, H.H.; Förster, F.; Helms, V.; Dudek, J.; Zimmermann, R. Proteomics reveals signal peptide features determining the client specificity in human TRAP-dependent ER protein import. *Nat. Commun.* **2018**, *9*, 3765. [CrossRef]

64. Rüdiger, S.; Germeroth, L.; Schneider-Mergener, J.; Bukau, B. Substrate specificity of the DnaK chaperone determined by screening cellulose-bound peptide libraries. *EMBO J.* **1997**, *16*, 1501–1507. [CrossRef]

65. Tripathi, A.; Mandon, E.C.; Gilmore, R.; Rapoport, T.A. Two alternative binding mechanisms connect the protein translocation Sec71-Sec72 complex with heat shock proteins. *J. Biol. Chem.* **2017**, *292*, 8007–8018. [CrossRef] [PubMed]

66. Ngosuwan, J.; Wang, N.M.; Fung, K.L.; Chirico, W.J. Roles of cytosolic Hsp70 and Hsp40 molecular chaperones in post-translational translocation of presecretory proteins into the endoplasmic reticulum. *J. Biol. Chem.* **2003**, *278*, 7034–7042. [CrossRef]

67. Ast, T.; Cohen, G.; Schuldiner, M. A network of cytosolic factors targets SRP-independent proteins to the endoplasmic reticulum. *Cell* **2013**, *152*, 1134–1145. [CrossRef]

68. Becker, T.; Song, J.; Pfanner, N. Versatility of preprotein transfer from the cytosol to mitochondria. *Trends Cell Biol.* **2019**, *29*, 534–548. [CrossRef]

69. Caplan, A.J.; Cyr, D.M.; Douglas, M.G. YDJ1p facilitates polypeptide translocation across different intracellular membranes by a conserved mechanism. *Cell* **1992**, *71*, 1143–1155. [CrossRef]

70. Craig, E.A. Hsp70 at the membrane: Driving protein translocation. *BMC Biol.* **2018**, *16*, 11. [CrossRef] [PubMed]

71. Kityk, R.; Kopp, J.; Mayer, M.P. Molecular mechanism of J-domain-triggered ATP hydrolysis by Hsp70 chaperones. *Mol. Cell* **2018**, *69*, 227–237. [CrossRef]

72. Craig, E.A.; Marszalek, J. How do J-proteins get Hsp70 to do so many different things? *Trends Biochem. Sci.* **2017**, *42*, 355–368. [CrossRef] [PubMed]

73. Rabu, C.; Wipf, P.; Brodsky, J.L.; High, S. A precursor-specific role for Hsp40/Hsc70 during tail-anchored protein integration at the endoplasmic reticulum. *J. Biol. Chem.* **2008**, *283*, 27504–27513. [CrossRef]

74. Müller, G.; Zimmermann, R. Import of honeybee prepromelittin into the endoplasmic reticulum: Structural basis for independence of SRP and docking protein. *EMBO J.* **1987**, *6*, 2099–2107. [CrossRef] [PubMed]

75. Guna, A.; Volkmar, N.; Christianson, J.C.; Hegde, R.S. The ER membrane protein complex is a transmembrane domain insertase. *Science* **2018**, *359*, 470–473. [CrossRef]

76. Haßdenteufel, S.; Johnson, N.; Paton, A.W.; Paton, J.C.; High, S.; Zimmermann, R. Chaperone-mediated Sec61 channel gating during ER import of small precursor proteins overcomes Sec61 inhibitor-reinforced energy barrier. *Cell Rep.* **2018**, *23*, 1373–1386. [CrossRef]

77. Casson, J.; McKenna, M.; Haßdenteufel, S.; Aviram, N.; Zimmerman, R.; High, S. Multiple pathways facilitate the biogenesis of mammalian tail-anchored proteins. *J. Cell Sci.* **2017**, *130*, 3851–3861. [CrossRef]

78. Fleischer, T.C.; Weaver, C.M.; McAfee, K.J.; Jennings, J.L.; Link, A.J. Systematic identification and functional screens of uncharacterized proteins associated with eukaryotic ribosomal complexes. *Genes Dev.* **2006**, *20*, 1294–1307. [CrossRef] [PubMed]

79. Lei, P.; Xiao, Y.; Li, P.; Xie, P.; Wang, H.; Huang, S.; Song, P.; Zhao, Y. hSnd2/TMEM208 is an HIF-1α-targeted gene and contains a WH2 motif. *Acta Biochim. Biophys. Sin.* **2020**, *52*, 328–331. [CrossRef]

80. Gemmer, M.; Förster, F. A clearer picture of the ER translocon complex. *J. Cell Sci.* **2020**, *133*, jcs231340. [CrossRef] [PubMed]

81. Pfeffer, S.; Dudek, J.; Schaffer, M.; Ng, B.G.; Albert, S.; Plitzko, J.M.; Baumeister, W.; Zimmermann, R.; Freeze, H.H.; Engel, B.D.; et al. Dissecting the molecular organization of the translocon-associated protein complex. *Nat. Commun.* **2017**, *8*, 14516. [CrossRef]

82. McGilvray, P.T.; Anghel, S.A.; Sundaram, A.; Zhong, F.; Trnka, M.J.; Fuller, J.R.; Hu, H.; Burlingame, A.L.; Keenan, R.J. An ER translocon for multi-pass membrane protein biogenesis. *eLife* **2020**, *9*, e56889. [CrossRef] [PubMed]

83. Wu, X.; Siggel, M.; Ovchinnikov, S.; Mi, W.; Svetlov, V.; Nudler, E.; Liao, M.; Hummer, G.; Rapoport, T.A. Structural basis of ER-associated protein degradation mediated by the Hrd1 ubiquitin ligase complex. *Science* **2020**, *368*, eaaz2449. [CrossRef]

84. Itskanov, S.; Park, E. Structure of the posttranslational Sec protein-translocation channel complex from yeast. *Science* **2019**, *363*, 84–87. [CrossRef] [PubMed]
85. Pleiner, T.; Tomaleri, G.P.; Januszyk, K.; Inglis, A.J.; Hazu, M.; Voorhees, R.M. Structural basis for membrane insertion by the human ER membrane protein complex. *Science* **2020**, *369*, 433–436. [CrossRef]
86. Bai, L.; You, Q.; Feng, X.; Kovach, A.; Li, H. Structure of the ER membrane complex, a transmembrane-domain insertase. *Nature* **2020**, *584*, 475–478. [CrossRef]
87. Miller-Vedam, L.E.; Bräuning, B.; Popova, K.D.; Oakdale, N.T.S.; Bonnar, J.L.; Prabu, J.R.; Boydston, E.A.; Sevillano, N.; Shurtleff, M.J.; Stroud, R.M.; et al. Structural and mechanistic basis of the EMC-dependent biogenesis of distinct transmembrane clients. *eLife* **2020**, *9*, e62611. [CrossRef] [PubMed]
88. O'Donnell, J.P.; Phillips, B.P.; Yagita, Y.; Juszkiewicz, S.; Wagner, A.; Malinverni, D.; Keenan, R.J.; Miller, E.A.; Hegde, R.S. The architecture of EMC reveals a path for membrane protein insertion. *eLife* **2020**, *9*, e57887. [CrossRef]
89. Wild, R.; Kowal, J.; Eyring, J.; Ngwa, E.M.; Aebi, M.; Locher, K.P. Structure of the yeast oligosaccharyltransferase complex gives insight into eukaryotic N-glycosylation. *Science* **2018**, *5140*, 545–550. [CrossRef]
90. Braunger, K.; Pfeffer, S.; Shrimal, S.; Gilmore, R.; Berninghausen, O.; Mandon, E.C.; Becker, T.; Förster, F.; Beckmann, R. Structural basis for coupling protein transport and N-glycosylation at the mammalian endoplasmic reticulum. *Science* **2018**, *360*, 215–219. [CrossRef]
91. Chitwood, P.J.; Juszkiewicz, S.; Guna, A.; Shao, S.; Hegde, R.S. EMC is required to initiate accurate membrane protein topogenesis. *Cell* **2018**, *175*, 1507–1519. [CrossRef]
92. Voorhees, R.M.; Hegde, R.S. Structure of the Sec61 channel opened by a signal sequence. *Science* **2016**, *351*, 88–91. [CrossRef] [PubMed]
93. Weng, T.; Steinchen, W.; Beatrix, B.; Berninghausen, O.; Becker, T.; Bange, G.; Cheng, J.; Beckmann, R. Architecture of the active post-translational Sec translocon. *EMBO J.* **2021**, *40*, e105643. [CrossRef]
94. Van Den Berg, B.; Clemons, W.M.; Collinson, I.; Modis, Y.; Hartmann, E.; Harrison, S.C.; Rapoport, T.A. X-ray structure of a protein-conducting channel. *Nature* **2004**, *427*, 36–44. [CrossRef]
95. Voorhees, R.M.; Fernández, I.S.; Scheres, S.H.W.; Hegde, R.S. Structure of the mammalian ribosome-Sec61 complex to 3.4 Å resolution. *Cell* **2014**, *157*, 1632–1643. [CrossRef] [PubMed]
96. Gogala, M.; Becker, T.; Beatrix, B.; Armache, J.P.; Barrio-Garcia, C.; Berninghausen, O.; Beckmann, R. Structures of the Sec61 complex engaged in nascent peptide translocation or membrane insertion. *Nature* **2014**, *506*, 107–110. [CrossRef]
97. Park, E.; Rapoport, T.A. Mechanisms of Sec61SecY-mediated protein translocation across membranes. *Annu. Rev. Biophys.* **2012**, *41*, 21–40. [CrossRef] [PubMed]
98. Li, L.; Park, E.; Ling, J.J.; Ingram, J.; Ploegh, H.; Rapoport, T.A. Crystal structure of a substrate-engaged SecY protein-translocation channel. *Nature* **2016**, *531*, 395–399. [CrossRef] [PubMed]
99. Gamayun, I.; O'Keefe, S.; Pick, T.; Klein, M.C.; Nguyen, D.; McKibbin, C.; Piacenti, M.; Williams, H.M.; Flitsch, S.L.; Whitehead, R.C.; et al. Eeyarestatin compounds selectively enhance Sec61-mediated Ca2+ leakage from the endoplasmic reticulum. *Cell Chem. Biol.* **2019**, *26*, 571–583. [CrossRef] [PubMed]
100. Kriegler, T.; Magoulopoulou, A.; Amate Marchal, R.; Hessa, T. Measuring endoplasmic reticulum signal sequences translocation efficiency using the Xbp1 arrest peptide. *Cell Chem. Biol.* **2018**, *25*, 880–890. [CrossRef] [PubMed]
101. Cymer, F.; Von Heijne, G. Cotranslational folding of membrane proteins probed by arrest-peptide- mediated force measurements. *Proc. Natl. Acad. Sci. USA* **2013**, *110*, 14640–14645. [CrossRef]
102. Kim, S.J.; Mitra, D.; Salerno, J.R.; Hegde, R.S. Signal sequences control gating of the protein translocation channel in a substrate-specific manner. *Dev. Cell* **2002**, *2*, 207–217. [CrossRef]
103. Shaffer, K.L.; Sharma, A.; Snapp, E.L.; Hegde, R.S. Regulation of protein compartmentalization expands the diversity of protein function. *Dev. Cell* **2005**, *9*, 545–554. [CrossRef]
104. Fons, R.D.; Bogert, B.A.; Hegde, R.S. Substrate-specific function of the translocon-associated protein complex during translocation across the ER membrane. *J. Cell Biol.* **2003**, *160*, 529–539. [CrossRef]
105. Spiess, M.; Junne, T.; Janoschke, M. Membrane Protein Integration and Topogenesis at the ER. *Protein J.* **2019**, *38*, 306–316. [CrossRef]
106. Pfeffer, S.; Burbaum, L.; Unverdorben, P.; Pech, M.; Chen, Y.; Zimmermann, R.; Beckmann, R.; Forster, F. Structure of the native Sec61 protein-conducting channel. *Nat. Commun.* **2015**, *6*, 8403. [CrossRef] [PubMed]
107. Niesen, M.J.M.; Müller-Lucks, A.; Hedman, R.; von Heijne, G.; Miller, T.F. Forces on nascent polypeptides during membrane insertion and translocation via the Sec translocon. *Biophys. J.* **2018**, *115*, 1885–1894. [CrossRef] [PubMed]
108. Kriegler, T.; Lang, S.; Notari, L.; Hessa, T. Prion protein translocation mechanism revealed by pulling force studies. *J. Mol. Biol.* **2020**, *432*, 4447–4465. [CrossRef]
109. Kriegler, T.; Kiburg, G.; Hessa, T. Translocon-associated protein complex (TRAP) is crucial for co-translational translocation of pre-proinsulin. *J. Mol. Biol.* **2020**, *432*, 166694. [CrossRef]
110. Pfeffer, S.; Dudek, J.; Zimmermann, R.; Forster, F. Organization of the native ribosome-translocon complex at the mammalian endoplasmic reticulum membrane. *Biochim. Biophys. Acta* **2016**, *1860*, 2122–2129. [CrossRef]
111. Sommer, N.; Junne, T.; Kalies, K.U.; Spiess, M.; Hartmann, E. TRAP assists membrane protein topogenesis at the mammalian ER membrane. *Biochim. Biophys. Acta—Mol. Cell Res.* **2013**, *1833*, 3104–3111. [CrossRef]
112. Cabelli, R.J.; Chen, L.; Tai, P.C.; Oliver, D.B. SecA protein is required for secretory protein translocation into *E. coli* membrane vesicles. *Cell* **1988**, *55*, 683–692. [CrossRef]

113. Brundage, L.; Hendrick, J.P.; Schiebel, E.; Driessen, A.J.M.; Wickner, W. The purified *E. coli* integral membrane protein SecY E is sufficient for reconstitution of SecA-dependent precursor protein translocation. *Cell* **1990**, *62*, 649–657. [CrossRef]

114. Chen, L.; Rhoads, D.; Tai, P.C. Alkaline phosphatase and OmpA protein can be translocated posttranslationally into membrane vesicles of Escherichia coli. *J. Bacteriol.* **1985**, *161*, 973–980. [CrossRef]

115. Rothblatt, J.A.; Deshaies, R.J.; Sanders, S.L.; Daum, G.; Schekman, R. Multiple genes are required for proper insertion of secretory proteins into the endoplasmic reticulum in yeast. *J. Cell Biol.* **1989**, *109*, 2641–2652. [CrossRef]

116. Meyer, H.A.; Grau, H.; Kraft, R.; Kostka, S.; Prehn, S.; Kalies, K.U.; Hartmann, E. Mammalian Sec61 is associated with Sec62 and Sec63. *J. Biol. Chem.* **2000**, *275*, 14550–14557. [CrossRef]

117. Tyedmers, J.; Lerner, M.; Bies, C.; Dudek, J.; Skowronek, M.H.; Haas, I.G.; Heim, N.; Nastainczyk, W.; Volkmer, J.; Zimmermann, R. Homologs of the yeast Sec complex subunits Sec62p and Sec63p are abundant proteins in dog pancreas microsomes. *Proc. Natl. Acad. Sci. USA* **2000**, *97*, 7214–7219. [CrossRef]

118. Deshaies, R.J.; Sanders, S.L.; Feldheim, D.A.; Schekman, R. Assembly of yeast Sec proteins involved in translocation into the endoplasmic reticulum into a membrane-bound multisubunit complex. *Nature* **1991**, *349*, 806–808. [CrossRef] [PubMed]

119. Wu, X.; Cabanos, C.; Rapoport, T.A. Structure of the post-translational protein translocation machinery of the ER membrane. *Nature* **2019**, *566*, 136–139. [CrossRef]

120. Itskanov, S.; Kuo, K.M.; Gumbart, J.C.; Park, E. Stepwise gating of the Sec61 protein-conducting channel by Sec63 and Sec62. *Nat. Struct. Mol. Biol.* **2021**, *28*, 162–172. [CrossRef]

121. Okun, M.M.; Eskridge, E.M.; Shields, D. Truncations of a secretory protein define minimum lengths required for binding to signal recognition particle and translation across the endoplasmic reticulum membrane. *J. Biol. Chem.* **1990**, *265*, 7478–7484. [CrossRef]

122. Feldheim, D.; Rothblatt, J.; Schekman, R. Topology and functional domains of Sec63p, an endoplasmic reticulum membrane protein required for secretory protein translocation. *Mol. Cell. Biol.* **1992**, *12*, 3288–3296. [CrossRef]

123. Matlack, K.E.S.; Misselwitz, B.; Plath, K.; Rapoport, T.A. BIP acts as a molecular ratchet during posttranslational transport of prepro-α factor across the ER membrane. *Cell* **1999**, *97*, 553–564. [CrossRef]

124. Sanders, S.L.; Whitfield, K.M.; Vogel, J.P.; Rose, M.D.; Schekman, R.W. Sec61p and BiP directly facilitate polypeptide translocation into the ER. *Cell* **1992**, *69*, 353–365. [CrossRef]

125. Schäuble, N.; Lang, S.; Jung, M.; Cappel, S.; Schorr, S.; Ulucan, Ö.; Linxweiler, J.; Dudek, J.; Blum, R.; Helms, V.; et al. BiP-mediated closing of the Sec61 channel limits Ca2+ leakage from the ER. *EMBO J.* **2012**, *31*, 3282–3296. [CrossRef] [PubMed]

126. Blau, M.; Mullapudi, S.; Becker, T.; Dudek, J.; Zimmermann, R.; Penczek, P.A.; Beckmann, R. ERj1p uses a universal ribosomal adaptor site to coordinate the 80S ribosome at the membrane. *Nat. Struct. Mol. Biol.* **2005**, *12*, 1015–1016. [CrossRef]

127. Melnyk, A.; Rieger, H.; Zimmermann, R. *The Networking of Chaperones by Co-Chaperones: Control of Cellular Protein Homeostasis*; Blatch, G.L., Edkins, A.L., Eds.; Springer International Publishing: Cham, Switzerland, 2015; pp. 179–200.

128. Collins, P.G.; Gilmore, R. Ribosome binding to the endoplasmic reticulum: A 180-kD protein identified by crosslinking to membrane-bound ribosomes is not required for ribosome binding activity. *J. Cell Biol.* **1991**, *114*, 639–649. [CrossRef]

129. Conti, B.J.; Devaraneni, P.K.; Yang, Z.; David, L.L.; Skach, W.R. Cotranslational stabilization of Sec62/63 within the ER Sec61 translocon is controlled by distinct substrate-driven translocation events. *Mol. Cell* **2015**, *58*, 269–283. [CrossRef] [PubMed]

130. Lang, S.; Benedix, J.; Fedeles, S.V.; Schorr, S.; Schirra, C.; Schäuble, N.; Jalal, C.; Greiner, M.; Haßdenteufel, S.; Tatzelt, J.; et al. Different effects of Sec61α, Sec62 and Sec63 depletion on transport of polypeptides into the endoplasmic reticulum of mammalian cells. *J. Cell Sci.* **2012**, *125*, 1958–1969.

131. Jadhav, B.; McKenna, M.; Johnson, N.; High, S.; Sinning, I.; Pool, M.R. Mammalian SRP receptor switches the Sec61 translocase from Sec62 to SRP-dependent translocation. *Nat. Commun.* **2015**, *6*, 10133. [CrossRef]

132. Mothes, W.; Prehn, S.; Rapoport, T.A. Systematic probing of the environment of a translocating secretory protein during translocation through the ER membrane. *EMBO J.* **1994**, *13*, 3973–3982. [CrossRef]

133. Nilsson, I.M.; Johnson, A.E.; Von Heijne, G. Cleavage of a tail-anchored protein by signal peptidase. *FEBS Lett.* **2002**, *516*, 106–108. [CrossRef]

134. Nilsson, I.M.; Whitley, P.; Von Heijne, G. The COOH-terminal ends of internal signal and signal-anchor sequences are positioned differently in the ER translocase. *J. Cell Biol.* **1994**, *126*, 1127–1132. [CrossRef] [PubMed]

135. Alzahrani, N.; Wu, M.J.; Shanmugam, S.; Yi, M.K. Delayed by design: Role of suboptimal signal peptidase processing of viral structural protein precursors in flaviviridae virus assembly. *Viruses* **2020**, *12*, 1090. [CrossRef]

136. Pfeffer, S.; Dudek, J.; Gogala, M.; Schorr, S.; Linxweiler, J.; Lang, S.; Becker, T.; Beckmann, R.; Zimmermann, R.; Förster, F. Structure of the mammalian oligosaccharyl-transferase complex in the native ER protein translocon. *Nat. Commun.* **2014**, *5*, 3072. [CrossRef]

137. Paetzel, M.; Karla, A.; Strynadka, N.C.J.; Dalbey, R.E. Signal peptidases. *Chem. Rev.* **2002**, *102*, 4549–4579. [CrossRef] [PubMed]

138. Cui, J.; Chen, W.; Sun, J.; Guo, H.; Madley, R.; Xiong, Y.; Pan, X.; Wang, H.; Tai, A.W.; Weiss, M.A.; et al. Competitive inhibition of the endoplasmic reticulum signal peptidase by non-cleavable mutant preprotein cargos. *J. Biol. Chem.* **2015**, *290*, 28131–28140. [CrossRef] [PubMed]

139. Lobigs, M.; Lee, E.; Ng, M.L.; Pavy, M.; Lobigs, P. A flavivirus signal peptide balances the catalytic activity of two proteases and thereby facilitates virus morphogenesis. *Virology* **2010**, *401*, 80–89. [CrossRef]

140. Snapp, E.L.; McCaul, N.; Quandte, M.; Cabartova, Z.; Bontjer, I.; Källgren, C.; Nilsson, I.; Land, A.; Von Heijne, G.; Sanders, R.W.; et al. Structure and topology around the cleavage site regulate post-translational cleavage of the HIV-1 gp160 signal peptide. *eLife* **2017**, *6*, e26067. [CrossRef]

141. Carrère-Kremer, S.; Montpellier, C.; Lorenzo, L.; Brulin, B.; Cocquerel, L.; Belouzard, S.; Penin, F.; Dubuisson, J. Regulation of hepatitis C virus polyprotein processing by signal peptidase involves structural determinants at the p7 sequence junctions. *J. Biol. Chem.* **2004**, *279*, 41384–41392. [CrossRef]

142. Kapp, K.; Schrempf, S.; Lemberg, M.K.; Dobberstein, B. Protein Transport into the Endoplasmic Reticulum. In *Protein Transport into the Endoplasmic Reticulum*; Landes Bioscience: Austin, TX, USA, 2009; Volume 1, pp. 1–16.

143. McDowell, M.A.; Heimes, M.; Fiorentino, F.; Mehmood, S.; Farkas, Á.; Coy-Vergara, J.; Wu, D.; Bolla, J.R.; Schmid, V.; Heinze, R.; et al. Structural basis of tail-anchored membrane protein biogenesis by the GET insertase complex. *Mol. Cell* **2020**, *80*, 72–86. [CrossRef]

144. Aviram, N.; Schuldiner, M. Embracing the void-how much do we really know about targeting and translocation to the endoplasmic reticulum? *Curr. Opin. Cell Biol.* **2014**, *29*, 8–17. [CrossRef] [PubMed]

145. Fenech, E.J.; Ben-Dor, S.; Schuldiner, M. Double the Fun, Double the Trouble: Paralogs and Homologs Functioning in the Endoplasmic Reticulum. *Annu. Rev. Biochem.* **2020**, *89*, 637–666. [CrossRef] [PubMed]

146. Ruiz-Canada, C.; Kelleher, D.J.; Gilmore, R. Cotranslational and posttranslational N-glycosylation of polypeptides by distinct mammalian OST isoforms. *Cell* **2009**, *136*, 272–283. [CrossRef] [PubMed]

International Journal of
Molecular Sciences

MDPI

Review

Molecular Modeling of Signal Peptide Recognition by Eukaryotic Sec Complexes

Pratiti Bhadra and Volkhard Helms *

Center for Bioinformatics, Saarland Informatics Campus, Saarland University, Postfach 15 11 50,
66041 Saarbruecken, Germany; pratiti.bhadra@bioinformatik.uni-saarland.de
* Correspondence: volkhard.helms@bioinformatik.uni-saarland.de

Abstract: Here, we review recent molecular modelling and simulation studies of the Sec translocon, the primary component/channel of protein translocation into the endoplasmic reticulum (ER) and bacterial periplasm, respectively. Our focus is placed on the eukaryotic Sec61, but we also mention modelling studies on prokaryotic SecY since both systems operate in related ways. Cryo-EM structures are now available for different conformational states of the Sec61 complex, ranging from the idle or closed state over an inhibited state with the inhibitor mycolactone bound near the lateral gate, up to a translocating state with bound substrate peptide in the translocation pore. For all these states, computational studies have addressed the conformational dynamics of the translocon with respect to the pore ring, the plug region, and the lateral gate. Also, molecular simulations are addressing mechanistic issues of insertion into the ER membrane vs. translocation into the ER, how signal-peptides are recognised at all in the translocation pore, and how accessory proteins affect the Sec61 conformation in the co- and post-translational pathways.

Keywords: signal peptide; Sec61 complex; protein translocation; nascent peptide chain; membrane insertion; molecular modelling; molecular dynamics simulations; molecular docking

Citation: Bhadra, P.; Helms, V. Molecular Modeling of Signal Peptide Recognition by Eukaryotic Sec Complexes. *Int. J. Mol. Sci.* **2021**, *22*, 10705. https://doi.org/10.3390/ijms221910705

Academic Editor: Veronica Esposito

Received: 31 August 2021
Accepted: 27 September 2021
Published: 2 October 2021

Publisher's Note: MDPI stays neutral with regard to jurisdictional claims in published maps and institutional affiliations.

1. Introduction

In eukaryotes, the majority of protein biosynthesis is carried out either by cytosolic ribosomes or by ribosomes that are attached to the ER membrane [1,2]. In the latter case, the newly synthesised "nascent polypeptide chain" (NC) is typically either translocated into the ER or laterally inserted into the ER membrane via an integral membrane protein complex termed translocon (with the exception of tail-anchored membrane proteins). In eukaryotes, the central component of this integral membrane protein complex is termed Sec61 complex. Prokaryotes also possess Sec channels named SecYEG which are responsible for protein secretion into the periplasm of the cell and for inserting membrane proteins into the cell membrane.

Proteins that should be translocated across the ER membrane typically carry a "signal peptide" (SP) at their N-terminus. In the ER, this sequence is then cleaved off by the enzyme complex signal peptidase. Alternatively, membrane proteins devoid of a SP are recognised by their first transmembrane (TM) helix. In both cases, a lateral gate opens in the translocation pore and either the TM portions of these proteins or the cleaved SP exits laterally into the lipid bilayer. In the latter case, the cleaved peptide chain is released into the ER interior. Here, we will concentrate on the translocation through Sec pores.

When a newly synthesised NC exits from the ribosome, often a protein termed signal recognition particle (SRP) binds to its SP, see Figure 1. Subsequently, upon interaction of SRP with the SRP receptor at the ER membrane, SRP dissociates from SP which is then free to insert into the Sec61 channel. This pathway is named the co-translational, SRP-dependent pathway. In another route, the so-called post-translational pathway, the fully synthesised precursor is first freely released from the ribosome and then, supported by chaperones and/or SRP-independent machineries, enters the Sec channel. In *S. cerevisiae*, the Sec61 complex mediates both co- and posttranslational translocation whereas the mammalian Sec61 complex functions primarily on the co-translational pathway [3].

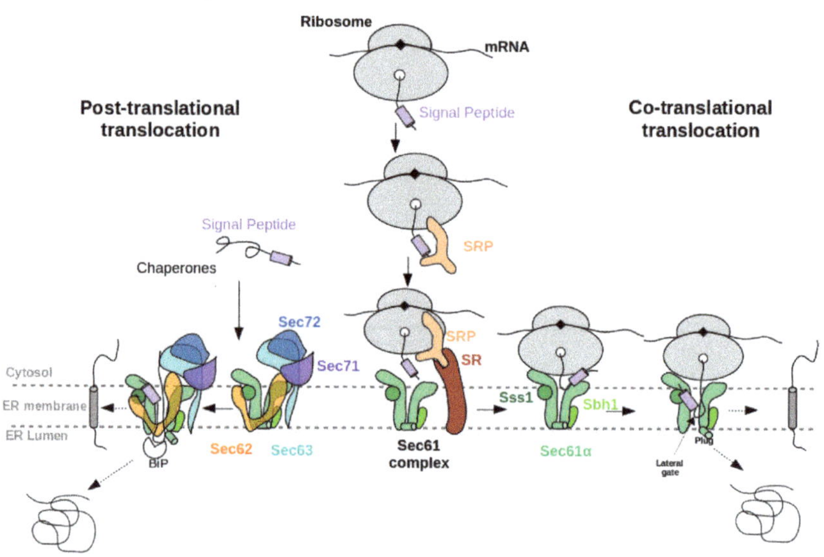

Figure 1. Schematic illustration of (left) the post-translational and (right) co-translational translocation of a eukaryotic secretory protein through the Sec61 complex. In both co- and post-translational pathways, accessory proteins such as Sec62 and Sec63 help in the translocation of precursor proteins with "weak" signal peptides (see section "properties of signal peptides"). The accessory proteins Sec71 and Sec72 exist only in yeast. Sbh1 and Sss1 are the name of the β and γ sub-units of yeast Sec61 complex.

The eukaryotic Sec61 complex consists of three subunits, Sec61α, Sec61β, and Sec61γ. Sec61α, the central channel pore, is a multiple-membrane-spanning protein. Sec61β and Sec61γ are single-membrane-spanning proteins (Sbh1 and Sss1 in *S. cerevisiae*, respectively) (see Figure 1). Sec61α and its prokaryotic homologue SecY consist of two halves that form an hourglass-shaped pore with a constriction in the middle of the membrane, a plug domain in the luminal or extracellular cavity, and a lateral gate, which opens to the surrounding lipid phase [4,5]. The Sec61 complex in the ER membrane represents the major entry point for precursor polypeptides. Besides, the open Sec61 complex forms a Ca^{2+}-permeable channel. Furthermore, it was found that different human hereditary and tumour diseases are associated with Sec61 channel gating [6]. Interestingly, various human genetic diseases result from single point mutations in the SPs of certain precursor polypeptides [6–8].

2. Properties of Signal Peptides

Comparative sequence analysis showed that SPs have a typical length of ca. 20–30 residues, but may be up to 60 residues long [9]. They have a recognisable three-domain structure involving a net basic 'N region', a 7–13 residues long hydrophobic 'H region' and a slightly polar 'C region'. Other than this, SPs of different precursor proteins have no noticeable sequence similarity to each other. In fact, SPs are often readily interchangeable, may tolerate a wide range of mutations and are capable of directing secretion in evolutionarily distant organisms. Figure 2 shows a multiple sequence alignment of the signal peptides of prion proteins from different organisms. Only two out of 21/27 positions are fully conserved, but eleven other positions are almost fully conserved.

In mammals, many signal sequences that promote successful targeting to the Sec61 complex cannot initiate translocation without further accessory membrane proteins such as TRAP or Sec62-63 being present at the site of translocation [5,10]. For example, the mechanistic role of the Sec62/63 complex is starting to be unveiled thanks to recent high-resolution cryo-EM structures of the heptameric Sec complex from *S. cerevisiae* [5,11–13], proteomic identification of putative Sec62- and Sec63-dependent precursor proteins [14], and molecular dynamics simulations [15]. The combination of such complementary techniques aids in solving the central puzzle of SP recognition by the Sec61 complex – how selectivity is achieved yet sequence diversity accommodated.

In water, functional SPs do not show substantial significant secondary structure [16]. Yet, in interfacial environments or in the specific environment of ribosome/SRP/translocon pore, e.g. when bound to SRP54-M from *M. jannaschii* [17], they have a high propensity for an α-helical conformation [16]. Their helical content is actually largest in the hydrophobic core of the signal sequence. For example, Yi and co-workers studied the conformation of a 25-residue-long functional SP of *E. coli* ribose binding protein using Circular Dichroism (CD) and Nuclear Magnetic Resonance (NMR) in solvents mimicking the amphiphilic environments around and in the Sec61 pore and compared this SP to a nonfunctional mutant-SP [18]. The functional peptide formed an 18-residue-long α-helix in a mixed 10% DMSO, 40% water, and 50% TFE solvent. The nonfunctional L9P mutant peptide did not have any secondary structure in that solvent but formed a 12-residue-long α-helix in 50% water: 50% TFE solvent. To correlate the α-helix-forming propensity of SPs with their translocation efficiency, Chi et al. determined the conformations of the SPs of two revertant ribose binding proteins in TFE/water solvent with CD and NMR spectroscopy [19]. In their CD experiments, both revertant SPs showed an intermediate helicity between those of wild-type and mutant SPs suggesting that the overall α-helical content is important for the functioning of SPs. Nguyen et al. reported that Gly/Pro amino acids are enriched in TRAP-dependent SPs [20]. Considering that Gly and Pro are known "helix breakers", this observation suggests lower α-helical propensities of such SPs. In this sense, being a "weak" SP could be associated with having a weak α-helical propensity. Related to this, Schorr and co-workers found that Sec62/63-dependent SPs have an SP with a comparatively longer but less hydrophobic H-region and a less polar C-region [14].

Entry name in UniProt	Signal Peptide
PRIO_HUMAN	M--ANL--GCWMLVLFVATWSDLGLC------
PRIO_BOVIN	MVKSHI--GSWILVLFVAMWSDVGLC------
PRIO_RAT	M--ANL--GYWLLALFVTTCTDVGLCKKRPKP
PRIO_RABIT	M--AHL--GYWMLLLFVATWSDVGLCKKRPKP
PRIO_CHICK	M--ARLLTTCCLLALLLAACTDVALS------
PRIO_PIG	MVKSHI--GGWILVLFVAAWSDIGLC------
PRIO_MOUSE	M--ANL--GYWLLALFVTMWTDVGLC------
PRIO_CAMDR	MVKSHM--GSWILVLFVVTWSDVGLC------
PRIO_CRIGR	M--ANL--SYWLLALFVATWTDVGLC------
PRIO_NEOVI	MVKSHI--GSWLLVLFVATWSDIGFC------
PRIO_VULLA	MVKSHI--GGWILLLFVATWSDVGLC------
PRIO_TRIVU	MGKIQL--GYWILVLFIVTWSDLGLC------
PRIO_SIGHI	M--ANL--GYWLLALFVATWTDVGLC------
PRIO_PONPY	M--ANL--GCWMLVLFVATWSNLGLC------
PRIO_MESAU	M--ANL--SYWLLALFVAMWTDVGLC------
PRIO_CHLAE	M--ANL--GCWMLVVFVATWSDLGLC------
PRIO_CALJA	M--ANL--GCWMLFLFVATWSDLGLC------
PRIO_BUBBU	MVKRHI--GSWILVLFVVMWSDVGLC------
PRIO_ATEPA	M--ANL--GYWMLVLFVATWSDLGLC------

Figure 2. Multiple sequence alignment, produced by Kalign [21], of the signal peptides of the prion protein (PRNP) [22] from different organisms. For this, the signal peptide sequences of the prion protein from 68 organisms were collected from UniProt. CD-Hit [23] was used to remove sequences.

with 100% sequence similarity. The 19 unique sequences (<100% sequence similarity) were used for the multiple sequence alignment. The alighnment was color-coded by conservation using the Sequence Manipulation Suite [24]. Residues that are identical among the sequences are given a black background, and those that are similar among the sequences are given a grey background. The N-, H- and C-regions of the human signal peptide, identified using Phobius [25], are highlighted in yellow, blue and red, respectively. A consensus sequence from the protein multiple sequence alignment is MANLxxWMLxLFVxxWSDVGLCxxxxxx.

3. Structural Studies of Eukaryotic Sec61:SP Complexes

The Sec61 complex consists of three subunits, namely α, β and γ. The α-subunit, Sec61α in eukaryotes, is the central component of the protein-conducting channel. Cryo-electron microscopy structures of the Sec61 complex have established its conformation in several functional states [4,5,11–13,26–31]. It contains ten transmembrane (TM) helices arranged in the central pore. The small β and γ subunits peripherally associate with the α-subunit and both of them have a single TM helix. In cryo-EM structures of Sec61 for the idle or inactive state, the helical plug domain, a helical structure formed in a sequence segment located immediately after TM1 of the α-subunit, is clearly visible inside the pore [27,29]. In translocating states, the plug moves away from its idle state position to allow transit of the SP [4,5,28]. Although its precise position could not be located, it is clear that the plug density was not present anymore in the pore. Recently, cryo-EM structures [4,5,28] demonstrated that the channel releases SPs to the lipid phase through the lateral-gate, a region between TM2 and TM7 of the α subunit (see Figure 3B).

Gogala et al. presented a cryo-EM structure of a mammalian ribosome-bound Sec61 complex from *Canis lupus familiaris* engaged in membrane insertion of nascent peptides (PDB entry 4CG6) [28]. The authors tested how hydrophobic and hydrophilic peptides translocate through the Sec61 channel on the example of hydrophilic (LepT) and hydrophobic (LepM) variants of the leader peptidase (Lep) protein. In the Sec61α complex engaged with the hydrophilic LepT peptide, the plug did not show a detectable shift compared to the idle state, the lateral-gate was partially opened (Figure 3 of [28]), and the lumenal part of the TM10 helix was shifted outward by $\sim$6 Å. The authors speculated that this shift of TM10 would be sufficient to provide the required opening for the accommodation of an extended translocating peptide segment between the plug and TM10. This would match previous crosslinking data showing that the translocating peptide is positioned near the plug helices, TM10 and TM5 of SecY (the α-subunit of prokaryotic SecYEG) [32]. However, Gogala et al. [28] were not able to model the location of the hydrophilic peptide within Sec61α in the electron density map. In contrast, the Sec61α complex engaged with the hydrophobic LepM peptide adopted a different conformation where the plug shifted compared to the idle state and the lateral-gate adopted an open conformation with displaced TM2 and TM7 helices. In that case, a rod-like extra electron density corresponding to three to four turns of an α-helix could be identified at the lateral-gate (4CG6) (see Figure 3A).

Later on, Voorhees et al. [4] presented a 3.6 Å resolution structure of the canine ribosome-Sec61 complex engaged with the secretory protein pre-prolactin (3JC2) (see Figure 3B). The signal sequence was again identified at the lateral-gate of Sec61α, similar to the previous hydrophobic peptide engaged structure (4CG6). Also here, the lateral-gate adopted an open conformation with the plug displaced from its idle position. The TM7 helix of Sec61α rotated as a rigid body relative to the plane of the membrane, thereby creating space between TM2 and TM7 for insertion of the SP (see Figure 3B, 3JC2). This cryo-EM structure demonstrated that the lateral-gate opening is asymmetric: the lumenal end of the gate shifts by a few Angstroms, whereas the cytosolic side remains closed. The authors compared the conformations of open/translocating state and idle state of Sec61α trying to explain how the SP reached this position. They suggested that the altered conformation of the lateral gate may result from a destabilized hydrogen bond network of the pore ring residues located in TM2, TM5, TM7 and TM10. Furthermore, they suggested that the interactions between the external hydrophilic surface of Sec61α and the hydrophobic lipid bilayer may also play an important role in the conformational dynamics of the lateral-gate. They did not notice any electron density of the plug domain inside the pore. Additionally, Voorhees et al. [4] reported that the density visible inside the ribosomal exit tunnel and in parts of the Sec61 channel suggests a looped configuration of the nascent chain, consistent with earlier crosslinking studies [33]. In both cryo-EM structures of mammalian ribosome-bound Sec61 complexes (co-translational mode), it was observed that hydrophobic SPs like to occupy the space between TM2 and TM7 and eventually insert into the lipid layer having a helical structure.

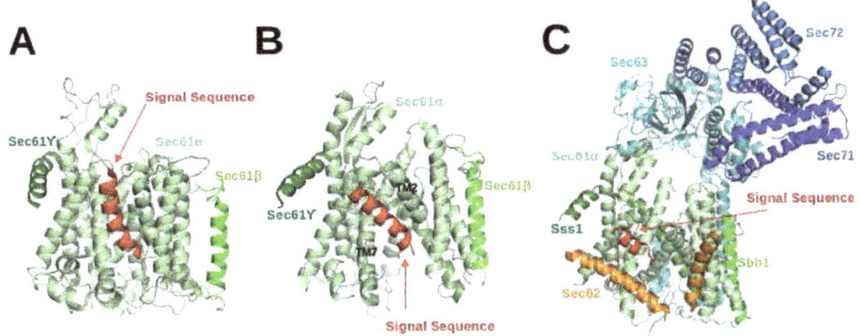

Figure 3. Shown are cryo-EM structures of the signal sequence-bound Sec61 complex (**A**) in the co-translational mode (4CG6; *Canis lupus*) [28], (**B**) in the co-translational mode (3JC2; *Canis lupus*) [4] and (**C**) in the post-translational mode (7AFT; *Saccharomyces cerevisiae*) [5].

The abovementioned cryo-EM structures characterised the position of SPs in co-translational translocation. Recently, Weng et al. [5] presented a 4.4 Å resolution cryo-EM structure of the heptameric Sec complex from yeast (the Sec complex of yeast is consists of seven subunits, namely Sec61α, Sbh1, Sss1, Sec62, Sec63, Sec71, Sec72) bound to a substrate carrying the SP of prepro-α factor (see Figure 3C, 7AFT). As shown in Figure 1, the heptameric complex is formed only in the post-translational pathway. Three recent cryo-EM studies characterised the overall architecture of this heptameric post-translational translocon in other functional states without bound SPs [11–13]. Compared to the structures of the idle state [11], Sec61α adopted an even more open conformation with a relocated plug domain when a SP was located in the groove of the lateral gate of Sec61α [5]. The SP bound conformation of Sec61α is referred to as the translocating state. In that case, the SP adopts a α-helical conformation and appears at a similar position to those previously seen in co-translationally operating Sec61α [4] where SP is located close to TM7 and parallel to the TM2 helix of Sec61α. Its position is also fully consistent with previous cross-linking data that suggested that the prepro-α factor signal sequence localises near TM2 and TM7 of the Sec complex [34,35]. Furthermore, the SP is also nearby and oriented parallel to the TM2 helix of Sec62 which is consistent with previous chemical cross-linking data [35].

Despite this spectacular progress from structural biology, our molecular understanding of how substrates open the channel for translocation and how SPs interact specifically with the translocon is still incomplete.

4. Conformational Dynamics of the Sec61 Translocon and Bound SPs in Molecular Dynamics Simulations

Over the last few years, molecular dynamics (MD) studies provided detailed insight into co-translational and post-translational translocation. Since the structural changes of the translocon are essential for translocon-assisted protein insertion, many MD simulation studies focused on the conformational changes of the prokaryotic and eukaryotic Sec channels [15,36–40]. Furthermore, MD simulation studies explored how the Sec61 channel recognizes substrate peptides and directs them either to membrane insertion or translocation [41–46]. The contribution of thermodynamics and kinetics and the detailed role of the translocon in the exit mechanism were also explored by 2D and 3D coarse-grained simulations [47,48]. Initially, the structure of prokaryotic SecY (1RHZ, 3.5 Å) from 2004 [49] was used in many early simulation studies due to the lack of high-resolution structures for eukaryotic Sec61. When Voorhees et al. presented the first high-resolution structure of mammalian Sec61 in 2014 (3J7Q, 3.5 Å resolution) [29], it turned out that the structures of SecY and Sec61 channels are closely related (TM-align score 0.21 [50]) (see Figure 4A). This matches the facts that they share 53.6% sequence similarity, they are universally conserved protein-conducting channels and that they both either translocate proteins across or integrate them into the eukaryotic ER membrane and the prokaryotic plasma membrane [1].

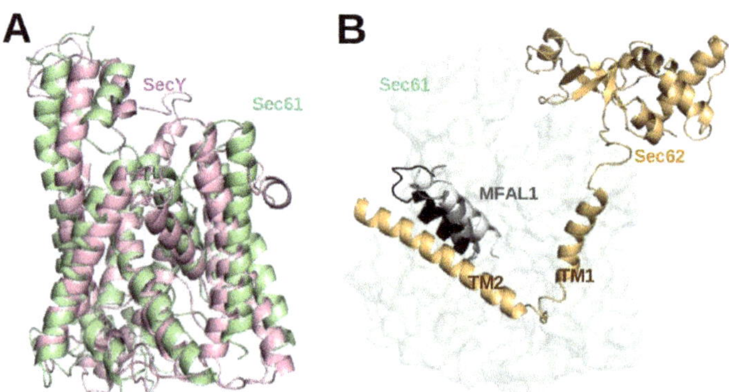

Figure 4. (**A**) Structural superposition of atomistic structures of SecY (light violet) and Sec61 (green). (**B**) Representative structures from a molecular dynamics simulation of SP bound Sec61 from yeast (unpublished). The conformations of the SPs at 0 ns, 500 ns, and 1 μs are coloured white, grey and black, respectively. For this, PDB entry 7AFT [5] detailing the cryo-EM structure of the signal sequence of Mating factor alpha-1 (MFAL1; UniProt ID: P01149) was used to model the SP (M1-A19) at the lateral-gate. The MD simulation was conducted in the presence of a part of Sec62 (orange) in a similar manner to [15].

Based on conformational analysis of the SecY channel using MD simulations, Gumbart et al. were able to explain the mechanism of lateral-gate opening, the roles of pore-ring and plug, and the effect of ribosome binding on the prokaryotic SecY channel [36–38]. Using steered MD simulations, they also pulled a deca-alanine peptide through the translocation pore of the archeal translocon and observed its consequences [46]. Due to the force exerted by the pulled deca-alanine, the plug was pushed out of the pore over a distance of 10 to 25 Å, allowing the deca-alanine to pass. Interestingly, this occurred without severe deformations of the helical structure of the peptide or the remainder of SecY. The pore ring expanded from its original size of ∼3.5–5.5 Å to a diameter of 7–12 Å. In contrast, a 19-residue long alanine/leucine helix (AL19) unfolded rapidly while being pulled through the pore in the same manner. Both deca-alanine and AL19 interacted strongly with the TM2 and TM7 helices of the translocon, suggesting this as another signal sequence recognition centre beside the lateral gate. Also, Tian and Andricioaei [51] mimicked the push-through of polypeptides across the translocon by pulling a virtual soft ball which demonstrated that the diameter of the SecY pore can be expanded to ∼16 Å without significant loss of its resilience.

Recently, Sun and coworkers [40] performed coarse-grained MD simulations of the mammalian structure of Sec61 to analyse its channel conformation. Their results suggest that the lateral gate is able to rapidly recover its partially-closed state after the nascent chain segment enters the bilayer. They also showed that the conformational dynamics of the lateral gate, pore ring and plug are interlinked. Moreover, Bhadra and Helms used all atomistic MD simulations to characterise conformational shifts in the Sec61 translocon from yeast due to the presence of the accessory protein Sec63 in the post-translational mode [15]. To collect representative statistics, they performed five independent simulations of 1 μs in length started either from the cryo-EM conformation of Sec61 bound to Sec63 or from uncomplexed Sec61. Their simulation results revealed that the wide pore opening is due to a reorientation of TM4 of Sec61 when it is attached to Sec63. This orientation shift of TM4 was governed by interactions between TM3 of Sec63 and TM1 of Sec61. Also, the simulations revealed that Sec63 affects the conformation of the lateral gate, and the plug moiety adopts different conformations within the channel in the presence and the absence of Sec63, respectively. Related to this finding, Mori et al. [39] proposed that the conformational transition from closed to pre-open states of SecY is associated with binding of SecA (an accessory protein in prokaryotes).

Apart from conformational changes of the translocon, several MD simulation studies also investigated the partitioning of SPs into the ER membrane. Zhang and Miller [45] explored the conformational landscape of the Sec translocon in the presence of hydrophobic (polyleucine, Leu30) versus hydrophilic (polyglutamine, Gln30) peptide substrates using coarse-grained enhanced sampling MD simulations of SecY. Based on the free energy cost of the structural fluctuations of the lateral gate, the authors concluded that a hydrophobic peptide substrate present in the translocon pore stabilises an open conformation of the lateral-gate, whereas a hydrophilic peptide substrate induces the closed conformation of the lateral gate. They reported that, for the hydrophilic substrate, the plug was preferentially positioned between the peptide substrate and the lateral gate, whereas for the hydrophobic substrate, the orientation was reversed such that the plug is behind the substrate with respect to the lateral gate. Thus, their results indicated that the hydrophobic substrate prefers an orientation where it is more exposed to the hydrophobic lipids of the membrane interior, whereas the hydrophilic substrate favours an orientation in which it remains fully inside the channel and shielded from the membrane by the plug.

To investigate how the hydrophobicity of SPs affects their insertion propensity into the membrane, Gumbart, Schulten and coworkers carried out all-atom MD simulations and umbrella sampling of different nascent transmembrane segments (a native Signal anchor, polyLeu, polySer, and polyGln) embedded in a ribosome-bound bacterial SecY translocon [52,53]. They tested the relationship between hydrophobicity of substrate and opening of the lateral-gate by performing all-atom classical MD simulations starting from different SecY conformations (closed-state, open-state and intermediate state). They observed that the degree of opening of the SecY lateral gate apparently did not depend on the hydrophobicity of the nascent chain inside the channel, at least not on the 1–2 μs time scale. Yet, the SP was able to spontaneously move into the membrane or back into the channel, depending on its hydrophobicity. Hence, the interactions between SP and hydrophobic lipid likely play an important role for the insertion of SPs into the membrane. Zhang and Miller [47] also investigated the role of the hydrophobicity of single anchor peptides on their insertion mechanism by coarse-graining the ribosome–nascent chain–membrane system and representing it in two spatial dimensions. They found that particular salt-bridge contacts between the nascent-protein N-terminus, cytosolic translocon residues, and phospholipid head groups favoured conformations of the nascent protein chain consistent with the type II topology. A type II topology refers to single-pass transmembrane proteins having an extracellular (or luminal) C-terminus and cytoplasmic N-terminus. In contrast, increasing the SP hydrophobicity stabilised nascent-protein configurations consistent with the type III topology [48]. A Type III topology membrane proteins have their N-terminal domains targeted to the ER lumen. Their detailed simulation studies provided a mechanistic basis for understanding experimentally observed correlations between the topology of integral membrane proteins and their amino-acid sequence. Recently, Niesen and coworkers [43] presented a novel coarse-grained model for co-translational membrane protein integration via the Sec translocon that enables simulation of long time-scales relevant for protein biosynthesis. The model was parameterised to reproduce sequence-specific

NC-translocon interactions and enabled simulating practically any nascent peptide chain by providing only its amino-acid sequence information as input. They showed that a more pronounced hydrophobicity of the signal anchor increases the probability of membrane integration. Their simulation results also demonstrated that increasing the length of the C-terminal loop of the TM segment raises the probability that the TM segment adopts a type II topology (N_{cyt}/C_{ER}). Their studies helped to elucidate the effects of sequence, translation rate, and external forces on the probability of nascent-chain integration into the membrane and to rationalise the resulting orientation of TMDs with respect to the membrane [44]. Besides, by combining site-directed mutagenesis in the yeast Sec61 translocon (in vivo) and MD simulations of SecY, it was shown that the membrane insertion propensity of signal-anchor sequences strongly depends on the positions of hydrophobic residues in the sequence segment and on its interaction with the six pore ring residues [41]. Furthermore, Rychkova and Warshel presented quantitative insertion free-energy profiles where they investigated the effect of SP and translocon mutations on the insertion free-energy using a coarse-grained model of the SecY translocon [42]. Their systematic analysis of the free-energy profile revealed that increasing the positive charge in the N-terminus increases the fraction of C-translocated peptides (N_{cyt}/C_{ER}, type II topology), whereas increasing the length of the helix reduced this fraction.

The role of water molecules in protein translocation via SecY was also investigated using all-atomistic MD simulations [54]. This study showed that the water molecules in the translocon pore do not behave as in bulk phase since they exhibited anomalous diffusion, had highly retarded rotational dynamics, and aligned their dipoles along the SecY The authors suggested that the water molecules facilitate the interaction between lipids and a peptide located inside the SecY and the dipole alignment of water molecules along the SecY axis may crucially affect the interaction of the positively charged N-terminus of a signal sequence with the translocon and may hence support membrane integration.

Recently, a structure of the active post-translational Sec complex from *Saccharomyces cerevisiae* bound to a SP was determined by cryo-electron microscopy [5]. We performed a 1 μs long atomistic MD simulation of this structure whereby the 13 amino-acid long SP of Mating factor alpha-1 (MFAL1) was modelled at the lateral gate. Interestingly, the SP shifted toward the TM2 helix of Sec62 (see Figure 4B) during the 1 μs simulation (unpublished). This hints at a direct interaction between SP and Sec62 which is in agreement with previous chemical cross-linking data [35]. Recently, Itskanov et al. suggested based on atomistic MD simulations that the presence of Sec62 also prevents lipids from invading the channel pore through the open lateral gate [13].

5. Molecular Docking Helps in Understanding the Interaction between Signal Peptide and Sec61 Translocon

Molecular docking is a well-known computational approach to investigate or predict the potential interactions between two molecules [55]. Our extensive MD simulations mentioned before [15] revealed that TM2, TM4, TM7 and the plug of Sec61 adopt different conformations in both Sec63-bound and free states. After observing these conformational transitions, we wanted to check whether those Sec61 structural elements affected by Sec63 are indeed related to substrate translocation across the membrane. To this aim, we performed molecular docking of flexible SPs in the translocon pore that was treated as rigid protein using the ADCP docking tool [56] from the Autodock developers that was optimised for docking of flexible peptides. Docking indeed showed that the hydrophobic core of the signal anchors of SRP-dependent substrates has a higher propensity for the volume between the C-terminus of TM2 and the N-terminus of TM7 than the hydrophobic core of SPs of SRP-independent substrates [15]. This suggests that the translocation process depends on the interaction of targeting sequences with the lateral gate which is in good agreement with experimental findings [57].

This study was a first attempt to use molecular docking to characterize the binding modes of SPs. Since the developers of the docking tool ADCP recommended its use only for peptides with a maximal length of 20 residues [56], only relatively short SPs can be tested in this way. The SP of *Carboxypeptidase Y* (CPY) is such a short SP with a length of only 21 amino acids. Hence, we performed several docking [15] runs for this SP using the Sec61 conformation when it is bound to the Sec62-Sec63 complex, which is believed to reflect its conformation during post-translational protein translocation. Figure 5A shows the top-ten ranked docking positions of the CPY-SP in the channel pore. However, the most favourable conformation of the docked CPY signal peptide differs somehow from the cryo-EM structure; 7AFT (Figure 5B). As a test we performed a single 1 microsecond-long MD simulation starting from the lowest-energy docking pose. The technical details of the MD protocol match those in [15]. In the MD simulation, the N-terminus of CPY-SP shifted from TM2 to TM7 (see Figure 5C), and the SP gradually shifted to occupy a very similar position as in the cryo-EM structure, 7AFT (see Figure 5D).Interestingly, over time, the SP conformation adopts an increasingly larger portion of alpha-helical conformation during the MD simulation (Figure 5E) which is in agreement with the cryo-EM structure. Figure 5E suggests that the region 13–21 a.a. of SP remains alpha-helical throughout the MD simulation, whereas, the region 3–12 a.a. of SP gradually adopts α-helical conformation. Although the MD simulation results have not been published so far, we like to add this study here as an example where docking was used in combination with MD simulations.

The MD simulation results suggest that the combination of docking and MD simulation is a suitable approach to characterise signal peptide recognition and conformational dynamics of the signal peptide bound to Sec61. Recently, the combined approach was used, for example, to provide valuable insight into drug discovery [58]. Docking protocols generally do not consider conformational flexibility of the receptor (here, Sec61 translocon), presence of water and lipids what may affect the reliability of the resulting docking poses. In contrast, MD simulations can treat both ligand and protein as flexible and the effects of explicit water molecules and lipid are directly captured. However, MD simulations are very time-consuming and can get trapped in local minima. Therefore, combining the two techniques in one protocol, where docking generates the most favorable conformation of the complex (lowest binding energy and largest cluster size) and MD simulation is then applied to explore the effect of flexibility of the molecules, water and lipid, could be an efficient approach to understand the conformation of SPs within the channel pore. Moreover, it would be appropriate to collect representative statistics from several independent MD simulations started with different starting velocities or from different docking positions.

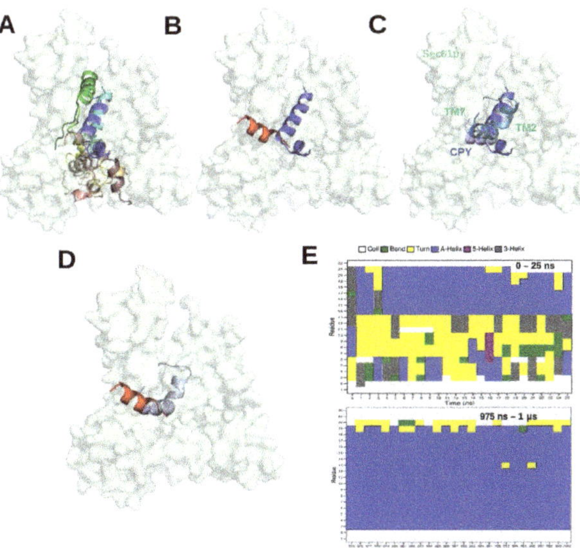

Figure 5. (**A**) The top ten ranked docking positions of the signal peptide (SP) of *Carboxypeptidase Y*

(CPY) inside the Sec61 pore from yeast in its conformation when bound to Sec62 and Sec63 in the Sec complex. The ADCP software [56] was used for molecular docking [15] (**B**) Structural superimposition of the most favourable docking pose with lowest binding energy (3 conformations out of 10 (blue variant in (**A**)) and of MFAL1-SP in the cryo-EM structure (red representing 7AFT) (**C**) The most favourable docking position (blue) was used as starting structure for a subsequent MD simulation. The conformations of the SP at 0 ns, 500 ns, and 1 µs from the MD simulation trajectory are represented in blue, skyblue and lightblue, respectively. (**D**) Structural superimposition of CPY-SP in the final snapshot (1µs) of the MD simulation (lightclue; CPY) and of MFAL1-SP in the cryo-EM structure (red, 7AFT). (**E**) DSSP [59] profiles of the signal peptide of Carboxypeptidase Y obtained from the first 25 ns and last 25 ns (975 ns–1µs) of the simulation.

6. Conclusions

The computational studies reviewed here illustrate the important contributions made by molecular dynamics simulations to the aim of better understanding Sec-SP dynamics and energetics during translocation. Application of these methods also enables us to discover mechanistic details how SP characteristics and allosteric effectors such as the accessory proteins (Sec62 and Sec63) influence protein translocation mediated by the Sec61 complex. Molecular simulations also help in connecting aspects of translation kinetics to the conformational sampling of peptide chains during membrane insertion or translocation to the ER. Thus, simulations of such systems help explain a number of experimental observations and can also serve as a catalyst to motivate new experimental studies. However, several open issues remain unanswered to date. How does the interaction of SPs and the Sec61 complex (except the lateral gate region which has been widely studied) help in protein translocation? What are the conformations of those parts of the Sec complex that are missing in the cryo-EM structures? Experimentally it has been noticed that some of these parts are essential for protein translocation (as an example, C-terminus of Sec63, N-terminus of Sec62 and N-terminus of Sec61 etc.) [3,60]. How do those parts interact with SPs? Thus, it can be expected that molecular simulations in combination with complementary experimental techniques will continue to explore new avenues in order to offer new insights into the complex behaviour of the nascent proteome and how this is influenced by translation dynamics.

Author Contributions: Conceptualization, writing—original draft preparation, P.B. and V.H.; supervision, project administration, funding acquisition, V.H. All authors have read and agreed to the published version of the manuscript.

Funding: This study was supported by Deutsche Forschungsgemeinschaft through grant He3875/15-1 to VH.

Acknowledgments: Furthermore, the authors acknowledge support by the Deutsche Forschungsgemeinschaft (DFG, German Research Foundation) and Saarland University within the funding programme Open Access Publishing.

Conflicts of Interest: The authors declare no conflict of interest.

References

1. Park, E.; Rapoport, T.A. Mechanisms of Sec61/SecY-mediated protein translocation across membranes. *Annu. Rev. Biophys.* **2012**, *41*, 21–40. [CrossRef]
2. Voorhees, R.M.; Hegde, R.S. Toward a structural understanding of co-translational protein translocation. *Curr. Opin. Cell Biol.* **2016**, *41*, 91–99. [CrossRef] [PubMed]
3. Ng, D.; Brown, J.D.; Walter, P. Signal sequences specify the targeting route to the endoplasmic reticulum membrane. *J. Cell Biol.* **1996**, *134*, 269–278. [CrossRef] [PubMed]
4. Voorhees, R.M.; Hegde, R.S. Structure of the Sec61 channel opened by a signal sequence. *Science* **2016**, *351*, 88–91. [CrossRef] [PubMed]
5. Weng, T.H.; Steinchen, W.; Beatrix, B.; Berninghausen, O.; Becker, T.; Bange, G.; Cheng, J.; Beckmann, R. Architecture of the active post-translational Sec translocon. *Embo. J.* **2021**, *40*, e105643. [CrossRef]

6. Lang, S.; Pfeffer, S.; Lee, P.H.; Cavalié, A.; Helms, V.; Förster, F.; Zimmermann, R. An update on Sec61 channel functions, mechanisms, and related diseases. *Front. Physiol.* **2017**, *8*, 887. [CrossRef] [PubMed]

7. Linxweiler, M.; Schick, B.; Zimmermann, R. Let's talk about Secs: Sec61, Sec62 and Sec63 in signal transduction, oncology and personalized medicine. *Signal Transduct. Target. Ther.* **2017**, *2*, 1–10. [CrossRef] [PubMed]

8. Sicking, M.; Lang, S.; Bochen, F.; Roos, A.; Drenth, J.P.; Zakaria, M.; Zimmermann, R.; Linxweiler, M. Complexity and Specificity of Sec61-Channelopathies: Human Diseases Affecting Gating of the Sec61 Complex. *Cells* **2021**, *10*, 1036. [CrossRef]

9. Hiss, J.A.; Schneider, G. Architecture, function and prediction of long signal peptides. *Briefings Bioinform.* **2009**, *10*, 569–578. [CrossRef] [PubMed]

10. Pfeffer, S.; Dudek, J.; Schaffer, M.; Ng, B.G.; Albert, S.; Plitzko, J.M.; Baumeister, W.; Zimmermann, R.; Freeze, H.H.; Engel, B.D.; et al. Dissecting the molecular organization of the translocon-associated protein complex. *Nat. Commun.* **2017**, *8*, 1–9. [CrossRef]

11. Wu, X.; Cabanos, C.; Rapoport, T.A. Structure of the post-translational protein translocation machinery of the ER membrane. *Nature* **2019**, *566*, 136–139. [CrossRef]

12. Itskanov, S.; Park, E. Structure of the posttranslational Sec protein-translocation channel complex from yeast. *Science* **2019**, *363*, 84–87. [CrossRef]

13. Itskanov, S.; Kuo, K.M.; Gumbart, J.C.; Park, E. Stepwise gating of the Sec61 protein-conducting channel by Sec63 and Sec62. *Nat. Struct. Mol. Biol.* **2021**, *28*, 162–172. [CrossRef]

14. Schorr, S.; Nguyen, D.; Haßdenteufel, S.; Nagaraj, N.; Cavalié, A.; Greiner, M.; Weissgerber, P.; Loi, M.; Paton, A.W.; Paton, J.C.; et al. Identification of signal peptide features for substrate specificity in human Sec62/Sec63-dependent ER protein import. *FEBS J.* **2020**, *287*, 4612–4640. [CrossRef] [PubMed]

15. Bhadra, P.; Yadhanapudi, L.; Römisch, K.; Helms, V. How does Sec63 affect the conformation of Sec61 in yeast? *PLoS Comput. Biol.* **2021**, *17*, e1008855. [CrossRef] [PubMed]

16. Clérico, E.M.; Maki, J.L.; Gierasch, L.M. Use of synthetic signal sequences to explore the protein export machinery. *Pept. Sci.* **2008**, *90*, 307–319. [CrossRef] [PubMed]

17. Hainzl, T.; Sauer-Eriksson, A.E. Signal-sequence induced conformational changes in the signal recognition particle. *Nat. Commun.* **2015**, *6*, 1–7. [CrossRef] [PubMed]

18. Yi, G.S.; Choi, B.S.; Kim, H. Structures of wild-type and mutant signal sequences of Escherichia coli ribose binding protein. *Biophys. J.* **1994**, *66*, 1604–1611. [CrossRef]

19. Chi, S.W.; Yi, G.S.; Suh, J.Y.; Choi, B.S.; Kim, H. Structures of revertant signal sequences of Escherichia coli ribose binding protein. *Biophys. J.* **1995**, *69*, 2703–2709. [CrossRef]

20. Nguyen, D.; Stutz, R.; Schorr, S.; Lang, S.; Pfeffer, S.; Freeze, H.H.; Förster, F.; Helms, V.; Dudek, J.; Zimmermann, R. Proteomics reveals signal peptide features determining the client specificity in human TRAP-dependent ER protein import. *Nat. Commun.* **2018**, *9*, 1–15. [CrossRef]

21. Lassmann, T.; Sonnhammer, E.L. Kalign–an accurate and fast multiple sequence alignment algorithm. *BMC Bioinform.* **2005**, *6*, 1–9. [CrossRef] [PubMed]

22. van Rheede, T.; Smolenaars, M.M.; Madsen, O.; de Jong, W.W. Molecular evolution of the mammalian prion protein. *Mol. Biol. Evol.* **2003**, *20*, 111–121. [CrossRef] [PubMed]

23. Fu, L.; Niu, B.; Zhu, Z.; Wu, S.; Li, W. CD-HIT: Accelerated for clustering the next-generation sequencing data. *Bioinformatics* **2012**, *28*, 3150–3152. [CrossRef] [PubMed]

24. Stothard, P. The sequence manipulation suite: JavaScript programs for analyzing and formatting protein and DNA sequences. *Biotechniques* **2000**, *28*, 1102–1104. [CrossRef]

25. Käll, L.; Krogh, A.; Sonnhammer, E.L. Advantages of combined transmembrane topology and signal peptide prediction—the Phobius web server. *Nucleic Acids Res.* **2007**, *35*, W429–W432. [CrossRef]

26. Ménétret, J.F.; Hegde, R.S.; Aguiar, M.; Gygi, S.P.; Park, E.; Rapoport, T.A.; Akey, C.W. Single copies of Sec61 and TRAP associate with a nontranslating mammalian ribosome. *Structure* **2008**, *16*, 1126–1137. [CrossRef]

27. Becker, T.; Bhushan, S.; Jarasch, A.; Armache, J.P.; Funes, S.; Jossinet, F.; Gumbart, J.; Mielke, T.; Berninghausen, O.; Schulten, K.; et al. Structure of monomeric yeast and mammalian Sec61 complexes interacting with the translating ribosome. *Science* **2009**, *326*, 1369–1373. [CrossRef]

28. Gogala, M.; Becker, T.; Beatrix, B.; Armache, J.P.; Barrio-Garcia, C.; Berninghausen, O.; Beckmann, R. Structures of the Sec61 complex engaged in nascent peptide translocation or membrane insertion. *Nature* **2014**, *506*, 107–110. [CrossRef]

29. Voorhees, R.M.; Fernández, I.S.; Scheres, S.H.; Hegde, R.S. Structure of the mammalian ribosome-Sec61 complex to 3.4 Å resolution. *Cell* **2014**, *157*, 1632–1643. [CrossRef]

30. Pfeffer, S.; Burbaum, L.; Unverdorben, P.; Pech, M.; Chen, Y.; Zimmermann, R.; Beckmann, R.; Förster, F. Structure of the native Sec61 protein-conducting channel. *Nat. Commun.* **2015**, *6*, 1–7. [CrossRef]

31. Gérard, S.F.; Hall, B.S.; Zaki, A.M.; Corfield, K.A.; Mayerhofer, P.U.; Costa, C.; Whelligan, D.K.; Biggin, P.C.; Simmonds, R.E.; Higgins, M.K. Structure of the inhibited state of the Sec translocon. *Mol. Cell* **2020**, *79*, 406–415. [CrossRef]

32. Cannon, K.S.; Or, E.; Clemons Jr, W.M.; Shibata, Y.; Rapoport, T.A. Disulfide bridge formation between SecY and a translocating polypeptide localizes the translocation pore to the center of SecY. *J. Cell Biol.* **2005**, *169*, 219–225. [CrossRef]

33. Mothes, W.; Prehn, S.; Rapoport, T.A. Systematic probing of the environment of a translocating secretory protein during translocation through the ER membrane. *EMBO J.* **1994**, *13*, 3973–3982. [CrossRef]

34. Plath, K.; Mothes, W.; Wilkinson, B.M.; Stirling, C.J.; Rapoport, T.A. Signal sequence recognition in posttranslational protein transport across the yeast ER membrane. *Cell* **1998**, *94*, 795–807. [CrossRef]
35. Plath, K.; Wilkinson, B.M.; Stirling, C.J.; Rapoport, T.A. Interactions between Sec complex and prepro-α-factor during posttranslational protein transport into the endoplasmic reticulum. *Mol. Biol. Cell* **2004**, *15*, 1–10. [CrossRef] [PubMed]
36. Gumbart, J.; Schulten, K. Structural determinants of lateral gate opening in the protein translocon. *Biochemistry* **2007**, *46*, 11147–11157. [CrossRef] [PubMed]
37. Gumbart, J.; Schulten, K. The roles of pore ring and plug in the SecY protein-conducting channel. *J. Gen. Physiol.* **2008**, *132*, 709–719. [CrossRef] [PubMed]
38. Gumbart, J.; Trabuco, L.G.; Schreiner, E.; Villa, E.; Schulten, K. Regulation of the protein-conducting channel by a bound ribosome. *Structure* **2009**, *17*, 1453–1464. [CrossRef]
39. Mori, T.; Ishitani, R.; Tsukazaki, T.; Nureki, O.; Sugita, Y. Molecular mechanisms underlying the early stage of protein translocation through the Sec translocon. *Biochemistry* **2010**, *49*, 945–950. [CrossRef]
40. Sun, S.; Wang, S.; Tong, Z.; Yao, X.; Gao, J. A molecular dynamics study on the resilience of Sec61 channel from open to closed state. *RSC Adv.* **2019**, *9*, 14876–14883. [CrossRef]
41. Demirci, E.; Junne, T.; Baday, S.; Bernèche, S.; Spiess, M. Functional asymmetry within the Sec61p translocon. *Proc. Natl. Acad. Sci. USA* **2013**, *110*, 18856–18861. [CrossRef]
42. Rychkova, A.; Warshel, A. Exploring the nature of the translocon-assisted protein insertion. *Proc. Natl. Acad. Sci. USA* **2013**, *110*, 495–500. [CrossRef] [PubMed]
43. Niesen, M.J.; Wang, C.Y.; Van Lehn, R.C.; Miller III, T.F. Structurally detailed coarse-grained model for Sec-facilitated co-translational protein translocation and membrane integration. *PLoS Comput. Biol.* **2017**, *13*, e1005427. [CrossRef] [PubMed]
44. Niesen, M.J.; Zimmer, M.H.; Miller III, T.F. Dynamics of co-translational membrane protein integration and translocation via the Sec translocon. *J. Am. Chem. Soc.* **2020**, *142*, 5449–5460. [CrossRef] [PubMed]
45. Zhang, B.; Miller, T.F. Hydrophobically stabilized open state for the lateral gate of the Sec translocon. *Proc. Natl. Acad. Sci. USA* **2010**, *107*, 5399–5404. [CrossRef]
46. Gumbart, J.; Schulten, K. Molecular dynamics studies of the archaeal translocon. *Biophys. J.* **2006**, *90*, 2356–2367. [CrossRef] [PubMed]
47. Zhang, B.; Miller III, T.F. Long-timescale dynamics and regulation of Sec-facilitated protein translocation. *Cell Rep.* **2012**, *2*, 927–937. [CrossRef]
48. Zhang, B.; Miller III, T.F. Direct simulation of early-stage Sec-facilitated protein translocation. *J. Am. Chem. Soc.* **2012**, *134*, 13700–13707. [CrossRef]
49. Van den Berg, B.; Clemons, W.M.; Collinson, I.; Modis, Y.; Hartmann, E.; Harrison, S.C.; Rapoport, T.A. X-ray structure of a protein-conducting channel. *Nature* **2004**, *427*, 36–44. [CrossRef]
50. Zhang, Y.; Skolnick, J. TM-align: A protein structure alignment algorithm based on the TM-score. *Nucleic Acids Res.* **2005**, *33*, 2302–2309. [CrossRef]
51. Tian, P.; Andricioaei, I. Size, motion, and function of the SecY translocon revealed by molecular dynamics simulations with virtual probes. *Biophys. J.* **2006**, *90*, 2718–2730. [CrossRef]
52. Gumbart, J.C.; Teo, I.; Roux, B.; Schulten, K. Reconciling the roles of kinetic and thermodynamic factors in membrane–protein insertion. *J. Am. Chem. Soc.* **2013**, *135*, 2291–2297. [CrossRef] [PubMed]
53. Gumbart, J.C.; Chipot, C. Decrypting protein insertion through the translocon with free-energy calculations. *Biochim. Biophys. Acta-(Bba)-Biomembr.* **2016**, *1858*, 1663–1671. [CrossRef] [PubMed]
54. Capponi, S.; Heyden, M.; Bondar, A.N.; Tobias, D.J.; White, S.H. Anomalous behavior of water inside the SecY translocon. *Proc. Natl. Acad. Sci. USA* **2015**, *112*, 9016–9021. [CrossRef]
55. Salmaso, V.; Moro, S. Bridging molecular docking to molecular dynamics in exploring ligand-protein recognition process: An overview. *Front. Pharmacol.* **2018**, *9*, 923. [CrossRef] [PubMed]
56. Zhang, Y.; Sanner, M.F. AutoDock CrankPep: Combining folding and docking to predict protein–peptide complexes. *Bioinformatics* **2019**, *35*, 5121–5127. [CrossRef] [PubMed]
57. Reithinger, J.H.; Yim, C.; Kim, S.; Lee, H.; Kim, H. Structural and functional profiling of the lateral gate of the Sec61 translocon. *J. Biol. Chem.* **2014**, *289*, 15845–15855. [CrossRef]
58. Santos, L.H.; Ferreira, R.S.; Caffarena, E.R. Integrating molecular docking and molecular dynamics simulations. In *Docking Screens for Drug Discovery*; Springer: Berlin/Heidelberg, Germany, 2019; pp. 13–34.
59. Kabsch, W.; Sander, C. Dictionary of protein secondary structure: Pattern recognition of hydrogen-bonded and geometrical features. *Biopolym. Orig. Res. Biomol.* **1983**, *22*, 2577–2637. [CrossRef]
60. Elia, F.; Yadhanapudi, L.; Tretter, T.; Römisch, K. The N-terminus of Sec61p plays key roles in ER protein import and ERAD. *PLoS ONE* **2019**, *14*, e0215950.

International Journal of
Molecular Sciences

Review

Emerging View on the Molecular Functions of Sec62 and Sec63 in Protein Translocation

Sung-jun Jung and Hyun Kim *

School of Biological Sciences and Institute of Microbiology, Seoul National University, Seoul 08826, Korea; lifecard@snu.ac.kr
* Correspondence: joy@snu.ac.kr; Tel.: +82-2-880-4440; Fax: +82-2-872-1993

Abstract: Most secreted and membrane proteins are targeted to and translocated across the endoplasmic reticulum (ER) membrane through the Sec61 protein-conducting channel. Evolutionarily conserved Sec62 and Sec63 associate with the Sec61 channel, forming the Sec complex and mediating translocation of a subset of proteins. For the last three decades, it has been thought that ER protein targeting and translocation occur via two distinct pathways: signal recognition particle (SRP)-dependent co-translational or SRP-independent, Sec62/Sec63 dependent post-translational translocation pathway. However, recent studies have suggested that ER protein targeting and translocation through the Sec translocon are more intricate than previously thought. This review summarizes the current understanding of the molecular functions of Sec62/Sec63 in ER protein translocation.

Keywords: Sec61; Sec62; Sec63; protein translocation; endoplasmic reticulum

Citation: Jung, S.-j.; Kim, H. Emerging View on the Molecular Functions of Sec62 and Sec63 in Protein Translocation. *Int. J. Mol. Sci.* **2021**, *22*, 12757. https://doi.org/10.3390/ijms222312757

Academic Editor: Frank Thévenod

Received: 9 November 2021
Accepted: 23 November 2021
Published: 25 November 2021

Publisher's Note: MDPI stays neutral with regard to jurisdictional claims in published maps and institutional affiliations.

1. Introduction

Approximately one-third of the eukaryotic proteome is directed to the ER for localization in the organelles of the secretory pathway or secreted out of the cells. Most translocated proteins have a signal sequence (SS) or a transmembrane domain (TMD) in their N-terminus, which targets them to the ER. These proteins are thought to be translocated across or inserted into the ER membrane via the Sec61 protein-conducting channel co-translationally or through the Sec complex post-translationally that additionally contains Sec62/Sec63 [1].

However, recent ER-proximity ribosome profiling studies have shown that most ER protein targeting occurs in a co-translational manner [2,3]. In contrast to the ribosome profiling results, cryo-electron microscopy (EM) structures of the yeast Sec complex show that the Sec complex is not compatible for co-translational translocation because the ribosome and Sec63 share the same binding site on the Sec61 channel [4–7]. In the subsequent sections, we summarize the classical and emerging views on protein targeting and translocation via the Sec complex, and discuss the underlying mechanisms and molecular functions of the Sec62/Sec63 complex in translocation of nascent chains.

2. Classical View on Protein Targeting and Translocation in the ER

The signal recognition particle (SRP) recognizes the N-terminal SS or TMD of secretory and membrane proteins when it emerges from the ribosome, arrests translation, and escorts the ribosome-nascent chain-SRP complex to the SRP receptor (SR) in the ER membrane. After delivering the complex to the ER membrane, SRP is released from the ribosome-nascent chain complex, which is then docked on the Sec61 channel. Synthesis of a nascent chain continues from the ribosome; the N-terminal SS or TMD binds to TM helices 2 and 7 of Sec61, the lateral gate, which triggers the opening of the channel and translocation of a nascent chain occurs across the ER membrane. This co-translational translocation pathway has been synonymously referred to as the SRP-dependent targeting and translocation pathway (Figure 1A) [1].

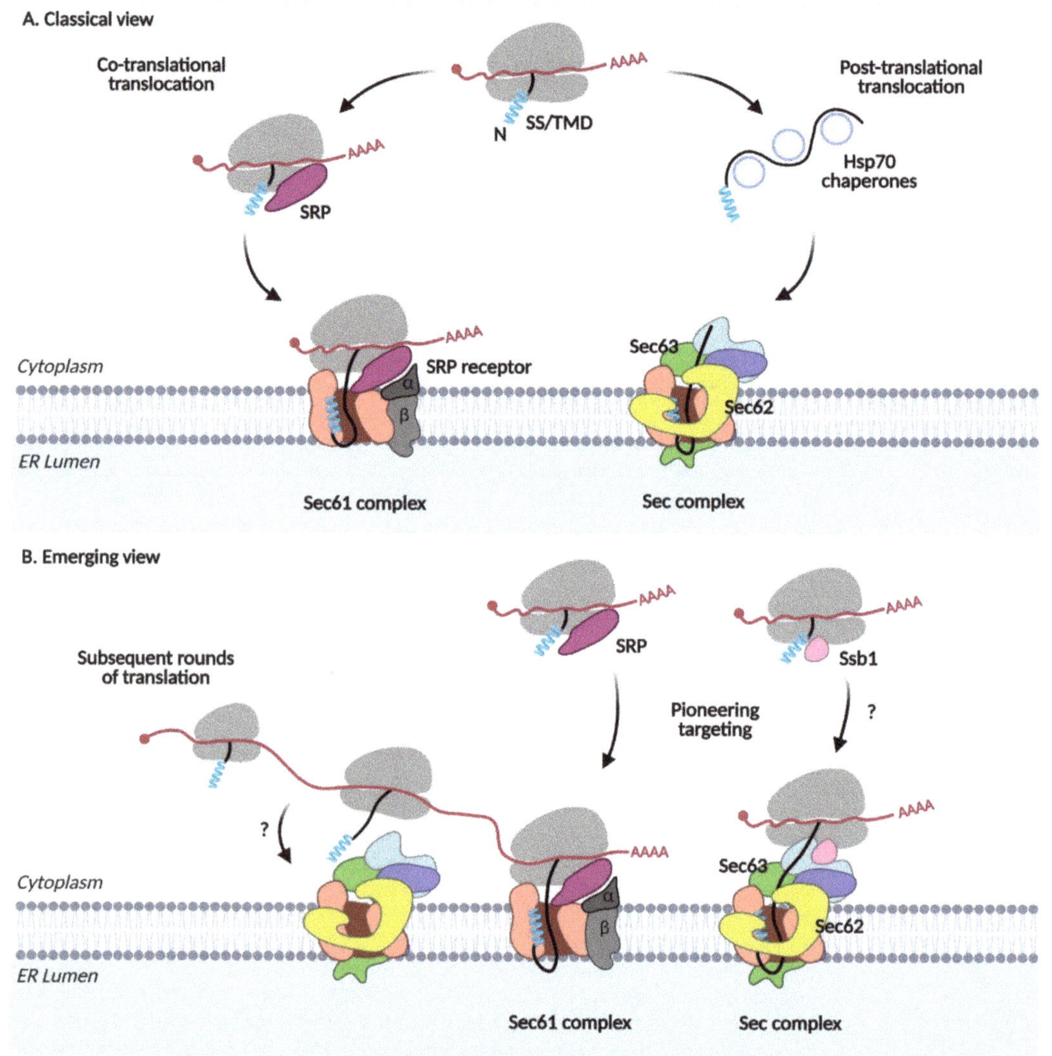

Figure 1. Protein targeting and translocation in the ER. (**A**) Classical view. SRP, signal recognition particle; SS, signal sequence; TMD, transmembrane domain. (**B**) Emerging view.

In the post-translational translocation pathway, nascent chains with less hydrophobic SSs are not recognized by SRP and completely synthesized in the cytosol (Figure 1A). Cytosolic Hsp70 chaperones bind to these proteins, preventing premature aggregation of the fully translated proteins and escort them to the Sec complex containing Sec62/Sec63 in the ER membrane. The yeast Sec complex contains additional Sec71(66) and Sec72 subunits, which are not found in mammals [8,9]. Sec63 interacts with an ER lumenal chaperone, Kar2 (Bip in mammals), which facilitates the unidirectional movement of a nascent chain into the lumen [10,11]. Since SRP is not involved here, this pathway has been referred to as the SRP-independent targeting and translocation pathway [1].

3. Emerging View on Protein Targeting and Translocation in the ER

A deep-sequence-based proximity-specific ribosome profiling technique has made it possible to investigate the local translation of the proteome. An ER-enriched ribosome profiling study revealed that not only SRP-dependent but also SRP-independent proteins are translated at the ER membrane [3]. Secretome mRNAs are associated with SRP even before the target signal is exposed and targeted to the ER membrane [2]. Thus, these studies suggest that the majority of the secretome is co-translationally targeted to the ER. Subsequent rounds of translation can occur on the mRNAs that are already present after pioneering targeting of the ribosome-nascent chain complex to the ER membrane (Figure 1B). These studies reveal that the SRP-dependent and independent targeting are not synonyms to the co- and post-translational targeting, respectively, implicating that the ER protein targeting is more dynamic and intricate than previously thought.

Hsp70 cytosolic chaperones guide the post-translational translocation of proteins. Among them, Ssb1 (Stress-Seventy subfamily B 1) in yeast is associated with the ribosome-nascent chain complex, mediating the co-translational folding of proteins [12–15]. Ribosome profiling studies on Ssb substrates revealed that ~40% of all ER-targeted proteins interact with Ssb [16]. Ssb1 also interacts with Sec72 in the Sec complex [17]. These observations suggest an interplay between co-translational targeting and translocation via the Sec complex (Figure 1B).

4. The Sec62/Sec63 Complex in Yeast

4.1. Discovery

Sec62 and Sec63 in the baker's yeast, *Saccharomyces cerevisiae*, were identified by genetic screening to isolate yeast mutants that accumulate reporter secretory proteins in the cytosol [18,19]. Translocation of a subset of secretory precursors was defective in Sec62 and Sec63 mutant strains [18,19]. Association of Sec62 and Sec63 with Sec61 channel was determined by chemical cross-linking and immunoprecipitation (IP) [20]. When DSP crosslinking of the yeast's crude membrane fraction solubilized with 1% Triton-X 100, it was followed by immunoprecipitation with Sec62 antibodies, Sec61 and Sec63 in addition to a glycosylated 31.5 kDa and an unglycosylated 23 kDa proteins were found to be associated. However, Sec61 was no longer associated with Sec62 when the cross-linking reaction was omitted prior to IP. It suggests that Sec62 interacts with Sec61 rather weakly whereas it forms a relatively stable tetramer complex with Sec63, 31.5, and 23 kDa proteins. The 31.5 and 23 kDa proteins were later identified as Sec71(66) and Sec72 in a genetic screening [21].

For reconstitution of the translocation process in vitro, the Sec translocon was purified by solubilization of yeast microsomes with 3% digitonin and subjected to incubation with Concanavalin A beads that specifically bind to glycoproteins, such as Sec71 [22]. This study identified a heptameric Sec complex, consisting of the Sec61/Sbh1/Sss1 trimer and Sec62/Sec63/Sec71/Sec72 tetramer. To determine whether the Sec61 and the Sec62/Sec63 complexes have independent binding activity with a substrate, an in vitro binding assay was carried out by reconstituting the Sec61 and Sec62/Sec63 complexes together or separately with prepro alpha factor (ppαF, yeast mating factor α), a Sec62/Sec63 dependent substrate [23]. Each complex showed weak binding whereas together they showed efficient binding to ppαF as in intact Sec complex, suggesting that the Sec heptamer is a functional complex required for binding of Sec62/Sec63 substrates. Relative stability of the sub-complexes of the Sec translocon was assessed by co-immunoprecipitation of Sec62-HA using the 3% digitonin solubilized microsomes, incubation of the purified Sec complex in buffer containing increasing concentration of Triton-X 100, and separation of the complexes on BN-PAGE [8]. The heptameric complex started to dissociate to the Sec61 trimer and the Sec62/Sec63 tetramer from the Triton-X 100 concentration above 0.4%. The Sec63 trimer containing Sec63/Sec71/Sec72 also appeared from the Triton-X 100 concentration 0.4~0.8%, indicating that Sec62 dissociated from the Sec62/Sec63 tetramer. The Sec61 trimer, the Sec62/Sec63/Sec71/Sec72 tetramer, and the Sec63/Sec71/Sec72 trimer were found in the presence of 1% Triton-X 100. A relatively week interaction of Sec62 with the remaining

Sec translocon was also observed in the BN-PAGE analysis of the 2% digitonin solubilized yeast microsomes [24]. Here, the Sec heptamer (SEC complex) and the hexamer lacking Sec62 (SEC' complex) were resolved on BN-PAGE. These experiments indicate dynamic nature of Sec62 and the possibility of the existence of the Sec complex without Sec62 in the ER membrane.

Meanwhile, Ssh1 (Sec sixty-one homolog 1) was discovered from the sequence homology search. It shares ~50% sequence identity with Sec61 [25]. Ssh1 forms a trimer complex with Sbh2 (Sbh1 homolog) and Sss1 as similar to the Sec61 trimer, but is not associated with the Sec62/Sec63 complex [26]. Hence, the Ssh1 complex has been proposed to function exclusively on the co-translational translocation.

These studies have identified that the Sec62/Sec63 complex function on protein translocation in the ER membrane and characterized the Sec heptameric complex consisting of Sec61 trimer and the Sec63 trimer complexes in addition to a loosely associated Sec62.

4.2. Structure

Cryo-EM structures of the Sec complex have been obtained recently [4–7]. The structures revealed that Sec63 tightly interacts with Sec61 through the cytosolic, membrane, and lumenal domains (Figure 2A–C). On the membrane side, three TMDs of Sec63 were found at the back-side of Sec61, opposite to the lateral gate (Figure 2C). Sec63-Sec61 interaction causes the lateral gate helices to separate; the pore is wider than that observed in any other Sec61/Y structures [27–32]. At the cytosolic side, soluble domains of Sec63, Sec71, and Sec72 interact with each other and are located above the Sec61 pore. Sec62 was poorly resolved in the first two structures [6,7]; however, two recent structures mapped Sec62 TMDs at the lateral gate of Sec61 [4,5]. An SS-bound Sec complex shows that the SS is sandwiched between the lateral gate helix 7 of Sec61 and TMD2 of Sec62 on the membrane side [4]. Sec62-bound Sec translocon shows further opening of the pore and displacement of the luminal plug domain in the lumen (Figure 2C).

These structures suggest two important functions of the Sec62/Sec63 complex: (1) its binding to the Sec61 complex opens the protein-conducting channel, and (2) Sec62 is likely the subunit that recognizes the SS and TMD of nascent chains at the lateral gate of Sec61.

4.3. Substrate Specificity

A study by Ng et al. showed that a few secretory precursors with low hydrophobic SSs that are not recognized by SRP require Sec62 and Sec63 [33]. Since then, various mutants of the Sec62/Sec63 complex have been characterized, wherein the translocation of certain secretory and membrane precursors is impaired [24,34–36]. These studies have helped to deduce the substrate specificities of the Sec62/Sec63 complex. The findings are summarized in the subsequent sections.

4.3.1. Sec62

Sec62 has two TMDs with N- and C-termini facing the cytosol (Figure 2A) [18]. The N-terminus of Sec62 contains a cluster of basic amino acids. When these residues were substituted with acidic residues (Sec62_35DDD), the interaction of Sec62 with Sec63 was disrupted. The cytosolic, C-terminal flanking region of the second TMD has been proposed to constitute a potential SS binding site [36,37]. When three residues in this region were replaced with alanines, yeast cells bearing this mutation no longer survived, and those bearing single residue replacement to alanine showed a growth defect (Sec62_P219A) [34]. However, the interaction between Sec62_P219A mutant and Sec63 was intact, unlike the Sec62_35DDD mutant [38]. Recent cryo-EM structure shows that this region is important in anchoring the Sec62 TMD2 in the cytosolic side of the ER membrane [5]. In these Sec62-mutant strains, translocation of secretory precursors and membrane proteins with moderately hydrophobic TMDs was defective [34].

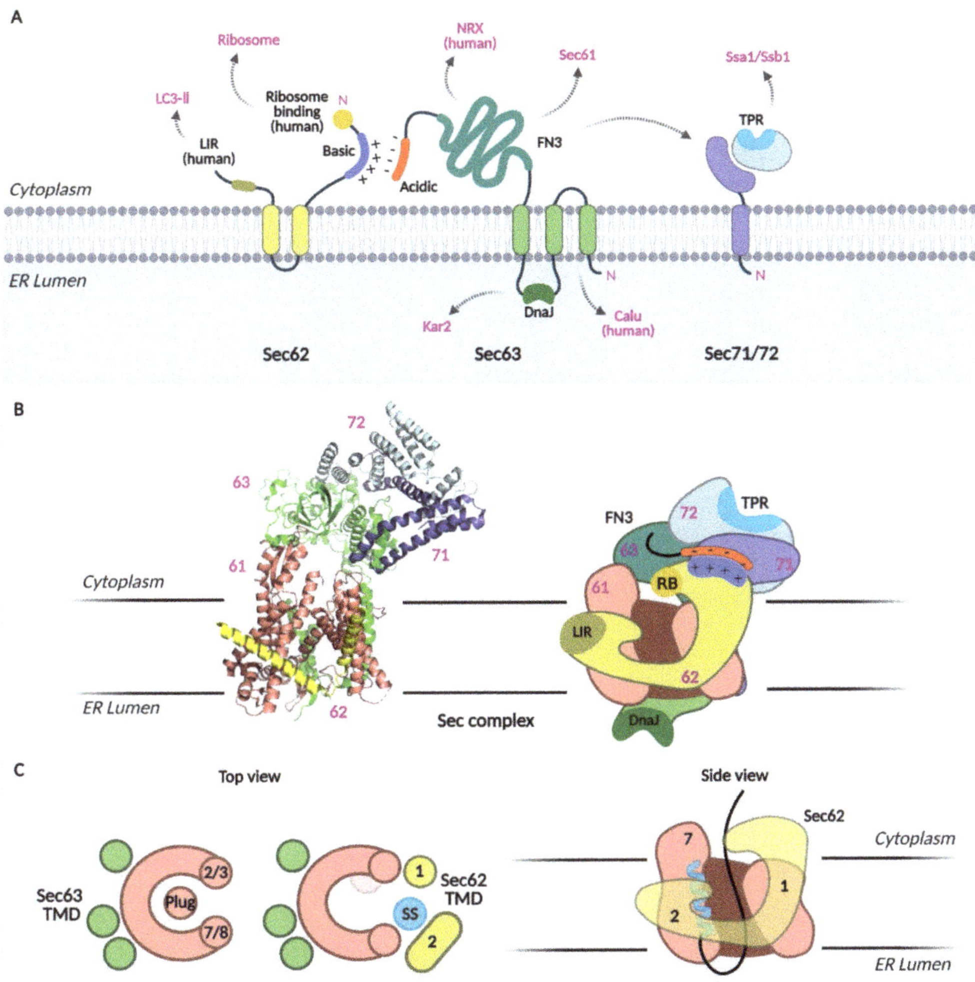

Figure 2. The Sec complex. (**A**) Schematics of Sec62, 63, 71, and 72. Domains and interacting proteins are indicated. LIR, LC3-interacting region; TPR, tetratricopeptide repeat; FN3, fibronectic type III; NRX, nucleoredoxin; calumenin, calu. (**B**) Cryo-EM structure of the Sec complex from *S. cerevisiae*. Itskanov, S., Park, E. (2020) Cryo-EM structure of the Sec complex from *S. cerevisiae*, wild-type, class with Sec62, conformation 1 (C1) doi: 10.2210/pdb7KAI/pdb (PDB code 7KAI) created with PyMOL [5]. Schematic of the Sec complex is based on the cryo-EM structure in [5]. (**C**) Channel opening by binding of Sec63, 62 and signal sequence to Sec61 [4].

For signal-anchored proteins, two modes of topogenesis have been shown: (1) the TMD inserts as the N-terminus facing the lumen (head-on) and then reorients to form the final topology, and (2) the TMD inserts as the N-terminus facing the cytosol, forming a loop conformation with the downstream region [39,40]. Our recent study showed that topogenesis of moderately hydrophobic N-terminal signal-anchored proteins was defective in the Sec62_35DDD mutant [41]. While the head-on insertion of the test N-terminal signal-anchored protein occurred, the subsequent reorientation step was defective in the Sec62_35DDD mutant but not in the Sec62_P219A mutant, indicating that association of Sec62 with Sec63 is needed for this step [41]. When positively charged residues were introduced in the C-terminal flanking region of the signal anchor of the test protein, making its inversion slower, topogenesis of the test protein became defective in both Sec62 mutant strains, suggesting that Sec62 is especially needed in inversion of the signal anchor with unfavorable topogenic signal [41]. The head-on inserted form of the test signal-anchored protein was co-immunoprecipitated with the Sec62 mutant; therefore, we proposed a model in which Sec62 recognizes the head-on inserted signal-anchored protein and mediates its reorientation as N_{in}-C_{out} membrane topology.

4.3.2. Sec63

Sec63 has three TMDs with the N- and C-termini facing the ER lumen and cytosol, respectively (Figure 2A). The lumenal loop between the second and third TMDs contains the DnaJ domain, through which an ER lumenal chaperone, Kar2 (Bip in mammals) interacts with. The DnaJ-domain of Sec63 is indispensable for the translocation of both SRP-dependent and -independent precursors (a test membrane protein with a hydrophobic TMD and secretory proteins having moderately hydrophobic SSs) [11,42,43].

Truncation of the N-terminal 40 residues including the first TMD of Sec63 destabilizes the Sec complex, judging by BN-PAGE and impaired insertion of membrane proteins [38]. Sec63 has a large cytosolic C-terminal region that associates with Sec62, Sec71 and Sec72. Deletion of the FN3 (or the Brl) domain in the cytosolic region impairs the assembly the Sec complex and translocation of both SRP-dependent and -independent substrates (test precursors with hydrophobic TMD and those with less hydrophobic cleavable SS, respectively) [6,7,24]. The C-terminal end is enriched with acidic amino acids that interact with the N-terminal basic residues of Sec62 [24,44]. Threonine at positions 652 and 654 at the C-terminus can be phosphorylated, strengthening the interaction with Sec62 [45]. Deletion of the acidic region of Sec63 impairs the translocation of Sec62-dependent substrates. Sec62 is required for the translocation of precursors with moderately hydrophobic SS or TMDs, whereas Sec63 is required for translocation and membrane insertion of most test proteins regardless of their SS hydrophobicity, implicating its general role in translocation of all types of proteins in yeast.

4.3.3. Sec71 and Sec72

Sec71 is a single-pass membrane protein with N_{out}-C_{in} orientation, and its C-terminus interacts with Sec72 (Figure 2A) [4–7,16]. Sec72 has a tetratricopeptide repeat (TPR) domain that binds to cytosolic Hsp70 chaperones, Ssa1 (Stress-Seventy subfamily A 1) and Ssb1 [15,17]. While Ssa1 binds to fully translated proteins, Ssb1 associates with translating ribosomes [13]. Mutations in the TPR domain of Sec72 lead to defects in its interaction with Ssa1 and Ssb1 and cause a translocation defect in vacuolar carboxypeptidase Y (CPY) [17].

In the systematic assessment of the Sec62/Sec63 dependent SS characteristics, the SS of CPY, a representative secretory protein, varied in its hydrophobicity and the length of the N-terminus preceding the SS hydrophobic core. Translocation efficiencies of CPY variants at an early stage were assessed using 5-min metabolic labeling of Sec62-, Sec63-defective and Sec71-, Sec72-deletion cells [46]. Deletion of Sec72 affected the translocation of a subset of CPY variants with less hydrophobic SSs, as observed in the Sec62 mutant strain. In comparison, translocation of the CPY variants with hydrophobic internal SSs, which are not dependent on Sec62, was severely impaired in the Sec71 deletion strain. A ribosome profiling study showed that targeting and translocation of precursors with internal SSs were defective in the Sec71(66) deletion strain [3]. These data suggest that Sec71(66) is involved in mediating translocation of precursors having internal SSs that insert as a loop conformation (Figure 3).

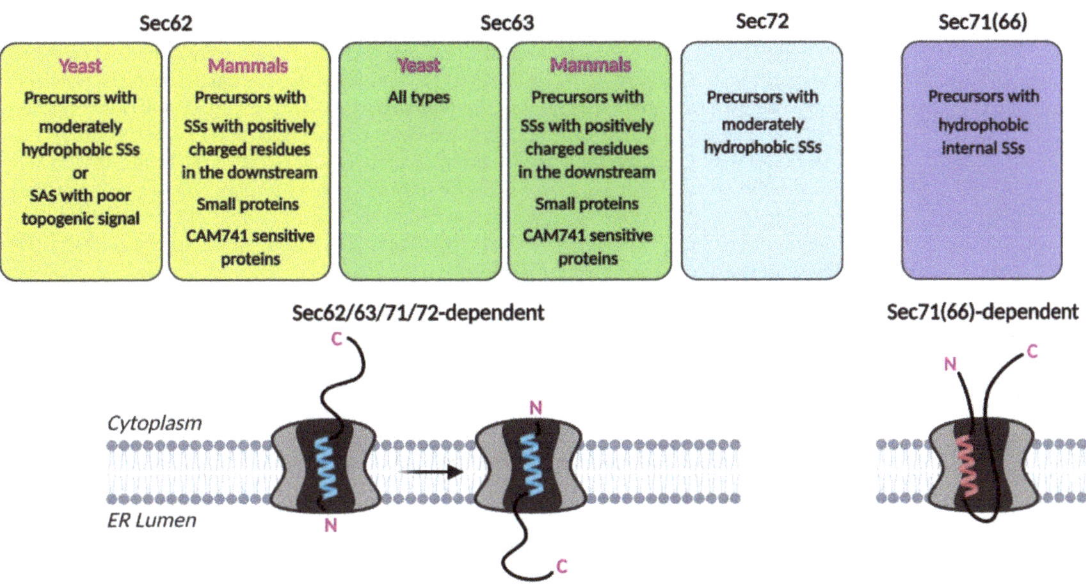

Figure 3. Sec62/Sec63 substrate specificity. Precursors having indicated characteristics may preferably insert the translocon as head-on and then invert (Sec62/63/71/72-dependent) or as a loop conformation (Sec71(66)-dependent). SS, signal sequence; SAS, signal-anchored sequence.

Observations that substrate specificities differ among the four components of the Sec62/Sec63 complex suggest the possibility that individual components of the Sec62/Sec63 complex may have distinct functions in aiding translocation, membrane insertion, and/or folding of different types of incoming nascent chains.

5. The Sec62/Sec63 Complex in Higher Eukaryotes

5.1. Discovery

Sec62 of *Drosophila melanogaster* was discovered as Dtrp1 (Drosophila translocation protein 1) [47]. Dtrp1 rescues defects in cell growth and protein translocation due to Sec62 deletion in yeast. Thereafter, human Sec62 was identified by sequence homology to *Drosophila* Sec62 (HTP, human translocation protein 1) [48]. Human Sec63 was identified by sequence homology of the human cDNA to the yeast homolog [49].

Two groups have reported that Sec62 and Sec63 are associated with the Sec61 complex in bovine and dog pancreas rough microsomes [50,51]. Meyer et al. showed that Sec62 and Sec63 are ubiquitously expressed in all rat and bovine tissues. The C-terminal acidic residues of Sec63 were found to interact with the N-terminal basic residues of Sec62 in the cytosolic side of the ER membrane as in yeast, and expression of human Sec62 rescued the growth defect of yeast cells carrying a defective Sec62, demonstrating that human and yeast Sec62s are structural and functional homologs [52].

5.2. Substrate Specificity

5.2.1. Small Proteins

Translocation of secretory precursors shorter than ~160 amino acids was impaired in mammalian cells depleted of Sec62 [53]. Interestingly, longer proteins were defective in translocation in the cells depleted of SRP receptor α (SRα) but not in the Sec62 depleted cells whereas shorter proteins were defective in translocation in the Sec62 depleted cells but not in the SRα depleted cells. Small proteins were partially translocated post-translationally in vitro. This study demonstrated functional conservation of Sec62 in post-translational translocation of secretory proteins in mammals as in yeast. However, unlike in yeast where precursors having less hydrophobic SSs are Sec62 dependent, small size precursors regardless of the characteristics of SSs are found to be dependent on Sec62 in mammals.

When different test proteins of varying hydrophobicity and the C-terminal length were assessed for their translocation efficiency in microsomes isolated from or semi-permeabilized human cells depleted of Sec62, post-translational translocation of pre-proapelin, a small secretory protein, was reduced, whereas both co- and post-translational translocation of preproapelin was impaired in the cells depleted of Sec63 [54,55]. The dependence of Sec62 and Sec63 was lost when the C-terminus of preproapelin was lengthened with dihydrofolate reductase (DHFR).

5.2.2. Signal Sequence Characteristics

Potential Sec62 and Sec63 substrates were searched using quantitative mass spectrometry analysis of proteomes from Sec62 or Sec63 knocked-down and knocked-out human cells [56,57]. Although not many were found, negatively affected proteins contain less hydrophobic SSs [56]. Post-translational translocation of preproapelin, which has relatively less hydrophobic SSs, was impaired in human cells depleted of Sec62 and Sec63 [54].

5.2.3. Mature Domain Region

For preproapelin, positively charged residues downstream of the SS are important for Sec62- or Sec63-dependent translocation [54]. When these residues were substituted to eliminate positive charges, translocation efficiency improved in the Sec62/Sec63-depleted cells. ERdj3, another Sec62 and Sec63 substrate, and prion protein have positively charged residues in their mature domains adjacent to the SS that affect their dependency on Sec62/Sec63 and Sec63, respectively [56,58]. In the Sec62 or Sec63 depleted cells, a pre-ERdj3 form was found in the membrane pellets upon carbonate extraction, suggesting that the head-on inserted precursor was unable to reorient in the absence of Sec62/Sec63.

5.2.4. Secretory Precursors That Are Inhibited by CAM741 in the ER Translocation

The cyclic heptadepsipeptide CAM741 (CPD A) is a selective translocation inhibitor for a subset of secretory precursors [55,59]. Secretory precursors with lower hydrophobicity and positively charged residues downstream of the SS, the types that depend on Sec62/Sec63, were especially sensitive to CAM741.

6. Additional Functions of Sec62 and Sec63 in Mammals

6.1. Association of Sec62 with Ribosome

Human Sec62 has longer N- and C-termini than the yeast homolog. Residues 1–15 and 156–170 of human Sec62 contain clusters of positively charged residues that resemble the ribosome binding domain found in other ribosome-interacting proteins [52]. An in vitro-binding study showed that the N-terminal cytosolic fragment of Sec62 binds to ribosomes [52]. Yeast Sec62 does not bind to the ribosome, but when the N-terminal 12 residues of human Sec62 were fused to the yeast homolog, it bound to the ribosome.

However, Sec62/Sec63 was co-immunoprecipitated with Sec61β in the ribosome-free fraction of bovine microsomes [51]. Muller et al. suggested that the failed interaction of Sec62 with ribosomes in the earlier study [51] may be due to the high salt concentration (400 nM) in the microsome solubilization buffer as the binding of the N-terminal fragment of human Sec62 and the ribosome weakened when the salt concentration was higher than 300 nM [52].

To capture the interacting partners of the translating ribosome-nascent chain, Conti et al. designed an experiment using a translation-arrested nascent chain at increasing length on the ribosome and detected using BN-PAGE [60]. The ribosome-bound prion precursor, which has an inefficient SS, was found to be associated with Sec62/Sec63 in addition to Sec61.

6.2. Competitive Binding of Sec62 and SR to Sec61

Chemical crosslinking of rough microsomes and co-immunoprecipitation with Sec61β pulled down the SRα subunit, Sec61α, Sec61β, and SPC25, a subunit of the signal peptidase complex [61]. When purified SRα was added to the microsomes, crosslinking between Sec61β and Sec62 was reduced, whereas crosslinking between Sec61β and SRα was enhanced, suggesting competitive binding between Sec62 and SRα to Sec61β. The Sec61 structures show that Sec61β is positioned near the lateral exit site where the TM segment is inserted into the membrane, indicating that multiple components (SRα, Sec62, SPC25) dynamically associate with the Sec61 complex near the lateral exit via Sec61β. The functional relationship between Sec62 and SR has also been observed in the quantitative mass spectrometry analysis of Sec62-depleted human cells. The abundance of SR subunits was upregulated upon depletion of Sec62 [56].

6.3. Role of Sec62 in Autophagy

Autophagy receptors are characterized by an LC3-interacting region (LIR). Binding of the ubiquitin-like protein LC3-II to LIR triggers selective autophagy [62]. The conserved LIR motif was found in the C-terminus of human Sec62 (NDFEMIT, residues 461–367) and is important for Sec62-mediated selective autophagy of the ER components after an unfolded protein response. This region is dispensable for protein translocation function and is not present in yeast Sec62.

6.4. Sec63 Interacting Proteins

In addition to Sec61 and Sec62, co-immunoprecipitation of Sec63 using the detergent solubilized dog pancreas microsomes yielded two additional proteins, calumenin and reticulocalbin [50]. They are calcium binding proteins residing in the ER lumen and have EF hand motifs [63,64]. A number of studies has shown that calcium leaks from the ER lumen to the cytosol via the Sec61 translocon [65–69], hinting the interplay between calcium and the translocon. However, the functional significance of the Sec63 interaction with these calcium binding proteins awaits to be revealed.

Further, a yeast two-hybrid screening of a human placenta cDNA library using the C-terminal domain of human Sec63 as a bait identified a cytosolic nucleoredoxin (NRX) [70]. GST pull-down and peptide-binding assays between the C-terminal region of Sec63 (residues 509 to 559 within the Brl domain) and the C-terminal part of NRX (residues 411 to 430) confirmed their interactions. Since NRX is involved in the Wnt signaling pathway, the authors suggested a possible link between Sec63 and Wnt signaling.

7. Human Diseases Associated with Sec62/Sec63

Considering that the Sec61 channel and its associated protein complexes mediate early protein biogenesis of about 30% of the proteome, it is not surprising that defects in the components of these machineries are linked to various diseases [71–73].

Elevated expression of Sec62 has been observed in some cancer tissues [74–80]. Hence, Sec62 has been suggested as a potential cancer marker, cancer-causing, or anti-cancer drug resistant factor, although whether the expression levels of Sec62 influence protein translocation and how the changes of its abundance are related to cell physiology and the development of diseases remain elusive. Since human Sec62 has dual functions as a translocation component and a receptor for ER-specific autophagy, its role in cancer requires further investigation.

Sec63 is found to be linked to diabetes, cancers, and autosomal dominant polycystic liver and kidney diseases [71–73]. Autosomal dominant polycystic liver and kidney diseases result from defective biogenesis of polycystin 1/2 that are cilia membrane proteins and function as a calcium-permeable receptor-channel complex. Sec63 is involved in the biogenesis of polycystin 1/2, and cyst formation in the liver is increased in Sec63 defective mice [81–83]. Sec63 interacts with calcium binding proteins in the lumen and nucleoredoxin in the cytosol, thus it is possible that signaling pathways through these interacting proteins (impaired interactions with defective Sec63) may contribute to the development of diseases. Human diseases associated with Sec translocon are summarized in [71–73].

8. Conclusions

Studies to date have shown that evolutionarily conserved Sec62 mediates the translocation of proteins with specific characteristics. These characteristics are SSs and TMDs with low hydrophobicity and poor topogenic signals in yeast, and small size, moderately hydrophobic SSs, or the presence of positively charged residues in the downstream of the SS in mammals (Figure 3). Precursors having these characteristics insert as head-on (N_{out}-C_{in} orientation), and Sec62/Sec63 mediate inversion of the head-on inserted SS and TMD as in N_{in}-C_{out} orientation. The cryo-EM structures of the yeast Sec complex show that Sec62 is located at the lateral gate of Sec61 where the SS and TMD bind to, suggesting that it recognizes the SS or TMD features of Sec62 clients [4,5]. Sec63 binds at the back-side and the cytosolic side of the Sec61 translocon, widening the pore of the Sec61 channel (Figure 2C) [4–7]. These studies collectively suggest that Sec62 and Sec63 function in Sec61-channel gating for those proteins that are insufficient to open the channel on their own.

9. Perspectives

In spite of the progress in elucidating the functions of Sec62/Sec63 in protein translocation, unresolved questions remain to be addressed in future studies. Biochemical and structural studies indicate that Sec62 is flexible, thus interacts with the Sec complex less tightly compared to the other subunits. Sec63 mediates translocation of broader types of precursors than Sec62. Although Sec62/Sec63 forms a complex, these observations raise the question of whether the Sec complex lacking Sec62 exits and functions in the translocation of Sec62-independent precursors. Further, it is unclear whether certain features of the Sec62/Sec63 clients are recognized by distinct components/domains of the Sec62/Sec63 complex or whether these features cause the nascent chain to be in particular intermediate forms (e.g., head-on inserted form), which are then recognized by Sec62/Sec63.

The CAM741 selectively inhibits translocation of a subset of secretory precursors possessing similar features as the Sec62/Sec63 clients [54]. It is elusive how the CAM741 achieves such substrate selectivity for the general translocon that handles a bulk of proteins that pass through the ER membrane. It raises the question of whether CAM741 inhibits proper association of Sec62/Sec63 with Sec61 and the incoming nascent chain, thus selectively impairing ER translocation of Sec62/Sec63 clients.

Sec71 and Sec72 were first discovered from a genetic screen searching the components involved in membrane protein biogenesis [21]; however, their functions in membrane protein biogenesis as well as the existence of their functional homologs in higher eukaryotes remain to be revealed.

Studies have shown that human Sec62 can bind to ribosomes and yeast Sec72 interacts with the ribosome-nascent chain associated Ssb1 [17,52], implying the function of the Sec complex in co-translational translocation. The underlying mechanisms of how the Sec complex mediates co-translational translocation in the ER membrane await further investigation.

Lastly, expression levels of Sec62 and Sec63 are found to be altered in various cancer cells [74–80]. Future studies are needed to clarify how the expression levels of Sec62 and Sec63 influence protein translocation, and how the changes of their abundance are related to cell physiology and the development of diseases.

Author Contributions: S.-j.J. and H.K. wrote the manuscript. S.-j.J. and H.K. have read and approved the submitted version. All authors have read and agreed to the published version of the manuscript.

Funding: This research was funded by the National Research Foundation of Korea grant number NRF-2019R1A2C2087701 and 2021R1H1A2009879. S.-j.J. is funded by the BK21 Postdoctoral Fellowship Program from the National Research Foundation of Korea.

Acknowledgments: All the figures are Created with biorender.com (accessed on 23 November 2021).

Conflicts of Interest: The authors declare no conflict of interest.

References

1. Rapoport, T.A.; Li, L.; Park, E. Structural and Mechanistic Insights into Protein Translocation. *Annu. Rev. Cell Dev. Biol.* **2017**, *33*, 369–390. [CrossRef] [PubMed]
2. Chartron, J.W.; Hunt, K.C.; Frydman, J. Cotranslational signal-independent SRP preloading during membrane targeting. *Nature* **2016**, *536*, 224–228. [CrossRef] [PubMed]
3. Jan, C.H.; Williams, C.C.; Weissman, J.S. Principles of ER cotranslational translocation revealed by proximity-specific ribosome profiling. *Science* **2014**, *346*, 1257521. [CrossRef] [PubMed]
4. Weng, T.H.; Steinchen, W.; Beatrix, B.; Berninghausen, O.; Becker, T.; Bange, G.; Cheng, J.; Beckmann, R. Architecture of the active post-translational Sec translocon. *EMBO J.* **2021**, *40*, e105643. [CrossRef]
5. Itskanov, S.; Kuo, K.M.; Gumbart, J.C.; Park, E. Stepwise gating of the Sec61 protein-conducting channel by Sec63 and Sec62. *Nat. Struct. Mol. Biol.* **2021**, *28*, 162–172. [CrossRef]
6. Wu, X.; Cabanos, C.; Rapoport, T.A. Structure of the post-translational protein translocation machinery of the ER membrane. *Nature* **2019**, *566*, 136–139. [CrossRef]
7. Itskanov, S.; Park, E. Structure of the posttranslational Sec protein-translocation channel complex from yeast. *Science* **2019**, *363*, 84–87. [CrossRef]
8. Harada, Y.; Li, H.; Wall, J.S.; Li, H.; Lennarz, W.J. Structural studies and the assembly of the heptameric post-translational translocon complex. *J. Biol. Chem.* **2011**, *286*, 2956–2965. [CrossRef]
9. Fang, H.; Green, N. Nonlethal sec71-1 and sec72-1 mutations eliminate proteins associated with the Sec63p-BiP complex from *S. cerevisiae*. *Mol. Biol. Cell* **1994**, *5*, 933–942. [CrossRef]
10. Lyman, S.K.; Schekman, R. Interaction between BiP and Sec63p is required for the completion of protein translocation into the ER of *Saccharomyces cerevisiae*. *J. Cell Biol.* **1995**, *131*, 1163–1171. [CrossRef] [PubMed]
11. Brodsky, J.L.; Goeckeler, J.; Schekman, R. BiP and Sec63p are required for both co- and posttranslational protein translocation into the yeast endoplasmic reticulum. *Proc. Natl. Acad. Sci. USA* **1995**, *92*, 9643–9646. [CrossRef] [PubMed]
12. Willmund, F.; del Alamo, M.; Pechmann, S.; Chen, T.; Albanese, V.; Dammer, E.B.; Peng, J.; Frydman, J. The cotranslational function of ribosome-associated Hsp70 in eukaryotic protein homeostasis. *Cell* **2013**, *152*, 196–209. [CrossRef] [PubMed]
13. Pfund, C.; Lopez-Hoyo, N.; Ziegelhoffer, T.; Schilke, B.A.; Lopez-Buesa, P.; Walter, W.A.; Wiedmann, M.; Craig, E.A. The molecular chaperone Ssb from *Saccharomyces cerevisiae* is a component of the ribosome-nascent chain complex. *EMBO J.* **1998**, *17*, 3981–3989. [CrossRef] [PubMed]
14. Nelson, R.J.; Ziegelhoffer, T.; Nicolet, C.; Werner-Washburne, M.; Craig, E.A. The translation machinery and 70 kd heat shock protein cooperate in protein synthesis. *Cell* **1992**, *71*, 97–105. [CrossRef]
15. Werner-Washburne, M.; Stone, D.E.; Craig, E.A. Complex interactions among members of an essential subfamily of hsp70 genes in *Saccharomyces cerevisiae*. *Mol. Cell. Biol.* **1987**, *7*, 2568–2577. [CrossRef] [PubMed]
16. Doring, K.; Ahmed, N.; Riemer, T.; Suresh, H.G.; Vainshtein, Y.; Habich, M.; Riemer, J.; Mayer, M.P.; O'Brien, E.P.; Kramer, G.; et al. Profiling Ssb-Nascent Chain Interactions Reveals Principles of Hsp70-Assisted Folding. *Cell* **2017**, *170*, 298–311.e220. [CrossRef]
17. Tripathi, A.; Mandon, E.C.; Gilmore, R.; Rapoport, T.A. Two alternative binding mechanisms connect the protein translocation Sec71-Sec72 complex with heat shock proteins. *J. Biol. Chem.* **2017**, *292*, 8007–8018. [CrossRef] [PubMed]
18. Deshaies, R.J.; Schekman, R. SEC62 encodes a putative membrane protein required for protein translocation into the yeast endoplasmic reticulum. *J. Cell Biol.* **1989**, *109*, 2653–2664. [CrossRef] [PubMed]
19. Deshaies, R.J.; Schekman, R. A yeast mutant defective at an early stage in import of secretory protein precursors into the endoplasmic reticulum. *J. Cell Biol.* **1987**, *105*, 633–645. [CrossRef] [PubMed]
20. Deshaies, R.J.; Sanders, S.L.; Feldheim, D.A.; Schekman, R. Assembly of yeast Sec proteins involved in translocation into the endoplasmic reticulum into a membrane-bound multisubunit complex. *Nature* **1991**, *349*, 806–808. [CrossRef]
21. Green, N.; Fang, H.; Walter, P. Mutants in three novel complementation groups inhibit membrane protein insertion into and soluble protein translocation across the endoplasmic reticulum membrane of *Saccharomyces cerevisiae*. *J. Cell Biol.* **1992**, *116*, 597–604. [CrossRef]
22. Panzner, S.; Dreier, L.; Hartmann, E.; Kostka, S.; Rapoport, T.A. Posttranslational protein transport in yeast reconstituted with a purified complex of Sec proteins and Kar2p. *Cell* **1995**, *81*, 561–570. [CrossRef]
23. Matlack, K.E.; Plath, K.; Misselwitz, B.; Rapoport, T.A. Protein transport by purified yeast Sec complex and Kar2p without membranes. *Science* **1997**, *277*, 938–941. [CrossRef] [PubMed]
24. Jermy, A.J.; Willer, M.; Davis, E.; Wilkinson, B.M.; Stirling, C.J. The Brl domain in Sec63p is required for assembly of functional endoplasmic reticulum translocons. *J. Biol. Chem.* **2006**, *281*, 7899–7906. [CrossRef] [PubMed]
25. Feldmann, H.; Aigle, M.; Aljinovic, G.; Andre, B.; Baclet, M.C.; Barthe, C.; Baur, A.; Becam, A.M.; Biteau, N.; Boles, E.; et al. Complete DNA sequence of yeast chromosome II. *EMBO J.* **1994**, *13*, 5795–5809. [CrossRef] [PubMed]
26. Finke, K.; Plath, K.; Panzner, S.; Prehn, S.; Rapoport, T.A.; Hartmann, E.; Sommer, T. A second trimeric complex containing homologs of the Sec61p complex functions in protein transport across the ER membrane of *S. cerevisiae*. *EMBO J.* **1996**, *15*, 1482–1494. [CrossRef] [PubMed]

27. Voorhees, R.M.; Hegde, R.S. Structure of the Sec61 channel opened by a signal sequence. *Science* **2016**, *351*, 88–91. [CrossRef]

28. Li, L.; Park, E.; Ling, J.; Ingram, J.; Ploegh, H.; Rapoport, T.A. Crystal structure of a substrate-engaged SecY protein-translocation channel. *Nature* **2016**, *531*, 395–399. [CrossRef] [PubMed]

29. Voorhees, R.M.; Fernandez, I.S.; Scheres, S.H.; Hegde, R.S. Structure of the mammalian ribosome-Sec61 complex to 3.4 A resolution. *Cell* **2014**, *157*, 1632–1643. [CrossRef]

30. Egea, P.F.; Stroud, R.M. Lateral opening of a translocon upon entry of protein suggests the mechanism of insertion into membranes. *Proc. Natl. Acad. Sci. USA* **2010**, *107*, 17182–17187. [CrossRef]

31. Zimmer, J.; Nam, Y.; Rapoport, T.A. Structure of a complex of the ATPase SecA and the protein-translocation channel. *Nature* **2008**, *455*, 936–943. [CrossRef]

32. Van den Berg, B.; Clemons, W.M., Jr.; Collinson, I.; Modis, Y.; Hartmann, E.; Harrison, S.C.; Rapoport, T.A. X-ray structure of a protein-conducting channel. *Nature* **2004**, *427*, 36–44. [CrossRef] [PubMed]

33. Ng, D.T.; Brown, J.D.; Walter, P. Signal sequences specify the targeting route to the endoplasmic reticulum membrane. *J. Cell Biol.* **1996**, *134*, 269–278. [CrossRef] [PubMed]

34. Jung, S.J.; Kim, J.E.; Reithinger, J.H.; Kim, H. The Sec62-Sec63 translocon facilitates translocation of the C-terminus of membrane proteins. *J. Cell Sci.* **2014**, *127*, 4270–4278. [CrossRef]

35. Reithinger, J.H.; Kim, J.E.; Kim, H. Sec62 protein mediates membrane insertion and orientation of moderately hydrophobic signal anchor proteins in the endoplasmic reticulum (ER). *J. Biol. Chem.* **2013**, *288*, 18058–18067. [CrossRef] [PubMed]

36. Lyman, S.K.; Schekman, R. Binding of secretory precursor polypeptides to a translocon subcomplex is regulated by BiP. *Cell* **1997**, *88*, 85–96. [CrossRef]

37. Dunnwald, M.; Varshavsky, A.; Johnsson, N. Detection of transient in vivo interactions between substrate and transporter during protein translocation into the endoplasmic reticulum. *Mol. Biol. Cell* **1999**, *10*, 329–344. [CrossRef] [PubMed]

38. Jung, S.J.; Jung, Y.; Kim, H. Proper insertion and topogenesis of membrane proteins in the ER depend on Sec63. *Biochim. Biophys. Acta Gen. Subj.* **2019**, *1863*, 1371–1380. [CrossRef]

39. Spiess, M.; Junne, T.; Janoschke, M. Membrane Protein Integration and Topogenesis at the ER. *Protein J.* **2019**, *38*, 306–316. [CrossRef]

40. Kocik, L.; Junne, T.; Spiess, M. Orientation of internal signal-anchor sequences at the Sec61 translocon. *J. Mol. Biol.* **2012**, *424*, 368–378. [CrossRef] [PubMed]

41. Jung, S.J.; Kim, J.E.H.; Junne, T.; Spiess, M.; Kim, H. Cotranslational Targeting and Posttranslational Translocation can Cooperate in Spc3 Topogenesis. *J. Mol. Biol.* **2021**, *433*, 167109. [CrossRef] [PubMed]

42. Willer, M.; Jermy, A.J.; Steel, G.J.; Garside, H.J.; Carter, S.; Stirling, C.J. An in vitro assay using overexpressed yeast SRP demonstrates that cotranslational translocation is dependent upon the J-domain of Sec63p. *Biochemistry* **2003**, *42*, 7171–7177. [CrossRef] [PubMed]

43. Young, B.P.; Craven, R.A.; Reid, P.J.; Willer, M.; Stirling, C.J. Sec63p and Kar2p are required for the translocation of SRP-dependent precursors into the yeast endoplasmic reticulum in vivo. *EMBO J.* **2001**, *20*, 262–271. [CrossRef]

44. Willer, M.; Jermy, A.J.; Young, B.P.; Stirling, C.J. Identification of novel protein-protein interactions at the cytosolic surface of the Sec63 complex in the yeast ER membrane. *Yeast* **2003**, *20*, 133–148. [CrossRef] [PubMed]

45. Wang, X.; Johnsson, N. Protein kinase CK2 phosphorylates Sec63p to stimulate the assembly of the endoplasmic reticulum protein translocation apparatus. *J. Cell Sci.* **2005**, *118*, 723–732. [CrossRef] [PubMed]

46. Yim, C.; Jung, S.J.; Kim, J.E.H.; Jung, Y.; Jeong, S.D.; Kim, H. Profiling of signal sequence characteristics and requirement of different translocation components. *Biochim. Biophys. Acta Mol. Cell Res.* **2018**, *1865*, 1640–1648. [CrossRef]

47. Noel, P.; Cartwright, I.L. A Sec62p-related component of the secretory protein translocon from *Drosophila* displays developmentally complex behavior. *EMBO J.* **1994**, *13*, 5253–5261. [CrossRef] [PubMed]

48. Daimon, M.; Susa, S.; Suzuki, K.; Kato, T.; Yamatani, K.; Sasaki, H. Identification of a human cDNA homologue to the Drosophila translocation protein 1 (Dtrp1). *Biochem. Biophys. Res. Commun.* **1997**, *230*, 100–104. [CrossRef] [PubMed]

49. Skowronek, M.H.; Rotter, M.; Haas, I.G. Molecular characterization of a novel mammalian DnaJ-like Sec63p homolog. *Biol. Chem.* **1999**, *380*, 1133–1138. [CrossRef] [PubMed]

50. Tyedmers, J.; Lerner, M.; Bies, C.; Dudek, J.; Skowronek, M.H.; Haas, I.G.; Heim, N.; Nastainczyk, W.; Volkmer, J.; Zimmermann, R. Homologs of the yeast Sec complex subunits Sec62p and Sec63p are abundant proteins in dog pancreas microsomes. *Proc. Natl. Acad. Sci. USA* **2000**, *97*, 7214–7219. [CrossRef]

51. Meyer, H.A.; Grau, H.; Kraft, R.; Kostka, S.; Prehn, S.; Kalies, K.U.; Hartmann, E. Mammalian Sec61 is associated with Sec62 and Sec63. *J. Biol. Chem.* **2000**, *275*, 14550–14557. [CrossRef] [PubMed]

52. Muller, L.; de Escauriaza, M.D.; Lajoie, P.; Theis, M.; Jung, M.; Muller, A.; Burgard, C.; Greiner, M.; Snapp, E.L.; Dudek, J.; et al. Evolutionary gain of function for the ER membrane protein Sec62 from yeast to humans. *Mol. Biol. Cell* **2010**, *21*, 691–703. [CrossRef]

53. Lakkaraju, A.K.; Thankappan, R.; Mary, C.; Garrison, J.L.; Taunton, J.; Strub, K. Efficient secretion of small proteins in mammalian cells relies on Sec62-dependent posttranslational translocation. *Mol. Biol. Cell* **2012**, *23*, 2712–2722. [CrossRef] [PubMed]

54. Hassdenteufel, S.; Johnson, N.; Paton, A.W.; Paton, J.C.; High, S.; Zimmermann, R. Chaperone-Mediated Sec61 Channel Gating during ER Import of Small Precursor Proteins Overcomes Sec61 Inhibitor-Reinforced Energy Barrier. *Cell Rep.* **2018**, *23*, 1373–1386. [CrossRef] [PubMed]

55. Johnson, N.; Hassdenteufel, S.; Theis, M.; Paton, A.W.; Paton, J.C.; Zimmermann, R.; High, S. The signal sequence influences post-translational ER translocation at distinct stages. *PLoS ONE* **2013**, *8*, e75394. [CrossRef]
56. Schorr, S.; Nguyen, D.; Hassdenteufel, S.; Nagaraj, N.; Cavalie, A.; Greiner, M.; Weissgerber, P.; Loi, M.; Paton, A.W.; Paton, J.C.; et al. Identification of signal peptide features for substrate specificity in human Sec62/Sec63-dependent ER protein import. *FEBS J.* **2020**, *287*, 4612–4640. [CrossRef] [PubMed]
57. Lang, S.; Benedix, J.; Fedeles, S.V.; Schorr, S.; Schirra, C.; Schauble, N.; Jalal, C.; Greiner, M.; Hassdenteufel, S.; Tatzelt, J.; et al. Different effects of Sec61alpha, Sec62 and Sec63 depletion on transport of polypeptides into the endoplasmic reticulum of mammalian cells. *J. Cell Sci.* **2012**, *125*, 1958–1969. [CrossRef] [PubMed]
58. Ziska, A.; Tatzelt, J.; Dudek, J.; Paton, A.W.; Paton, J.C.; Zimmermann, R.; Hassdenteufel, S. The signal peptide plus a cluster of positive charges in prion protein dictate chaperone-mediated Sec61 channel gating. *Biol. Open* **2019**, *8*, bio040691. [CrossRef]
59. Mackinnon, A.L.; Paavilainen, V.O.; Sharma, A.; Hegde, R.S.; Taunton, J. An allosteric Sec61 inhibitor traps nascent transmembrane helices at the lateral gate. *Elife* **2014**, *3*, e01483. [CrossRef] [PubMed]
60. Conti, B.J.; Devaraneni, P.K.; Yang, Z.; David, L.L.; Skach, W.R. Cotranslational stabilization of Sec62/63 within the ER Sec61 translocon is controlled by distinct substrate-driven translocation events. *Mol. Cell* **2015**, *58*, 269–283. [CrossRef]
61. Jadhav, B.; McKenna, M.; Johnson, N.; High, S.; Sinning, I.; Pool, M.R. Mammalian SRP receptor switches the Sec61 translocase from Sec62 to SRP-dependent translocation. *Nat. Commun.* **2015**, *6*, 10133. [CrossRef] [PubMed]
62. Fumagalli, F.; Noack, J.; Bergmann, T.J.; Cebollero, E.; Pisoni, G.B.; Fasana, E.; Fregno, I.; Galli, C.; Loi, M.; Solda, T.; et al. Corrigendum: Translocon component Sec62 acts in endoplasmic reticulum turnover during stress recovery. *Nat. Cell Biol.* **2016**, *19*, 76. [CrossRef] [PubMed]
63. Yabe, D.; Nakamura, T.; Kanazawa, N.; Tashiro, K.; Honjo, T. Calumenin, a Ca2+-binding protein retained in the endoplasmic reticulum with a novel carboxyl-terminal sequence, HDEF. *J. Biol. Chem.* **1997**, *272*, 18232–18239. [CrossRef]
64. Ozawa, M.; Muramatsu, T. Reticulocalbin, a novel endoplasmic reticulum resident Ca(2+)-binding protein with multiple EF-hand motifs and a carboxyl-terminal HDEL sequence. *J. Biol. Chem.* **1993**, *268*, 699–705. [CrossRef]
65. Schauble, N.; Lang, S.; Jung, M.; Cappel, S.; Schorr, S.; Ulucan, O.; Linxweiler, J.; Dudek, J.; Blum, R.; Helms, V.; et al. BiP-mediated closing of the Sec61 channel limits Ca2+ leakage from the ER. *EMBO J.* **2012**, *31*, 3282–3296. [CrossRef]
66. Lang, S.; Erdmann, F.; Jung, M.; Wagner, R.; Cavalie, A.; Zimmermann, R. Sec61 complexes form ubiquitous ER Ca2+ leak channels. *Channels* **2011**, *5*, 228–235. [CrossRef] [PubMed]
67. Erdmann, F.; Schauble, N.; Lang, S.; Jung, M.; Honigmann, A.; Ahmad, M.; Dudek, J.; Benedix, J.; Harsman, A.; Kopp, A.; et al. Interaction of calmodulin with Sec61alpha limits Ca2+ leakage from the endoplasmic reticulum. *EMBO J.* **2011**, *30*, 17–31. [CrossRef]
68. Ong, H.L.; Liu, X.; Sharma, A.; Hegde, R.S.; Ambudkar, I.S. Intracellular Ca(2+) release via the ER translocon activates store-operated calcium entry. *Pflugers Arch.* **2007**, *453*, 797–808. [CrossRef]
69. Van Coppenolle, F.; Vanden Abeele, F.; Slomianny, C.; Flourakis, M.; Hesketh, J.; Dewailly, E.; Prevarskaya, N. Ribosome-translocon complex mediates calcium leakage from endoplasmic reticulum stores. *J. Cell Sci.* **2004**, *117*, 4135–4142. [CrossRef]
70. Muller, L.; Funato, Y.; Miki, H.; Zimmermann, R. An interaction between human Sec63 and nucleoredoxin may provide the missing link between the SEC63 gene and polycystic liver disease. *FEBS Lett.* **2011**, *585*, 596–600. [CrossRef]
71. Lang, S.; Pfeffer, S.; Lee, P.H.; Cavalie, A.; Helms, V.; Forster, F.; Zimmermann, R. An Update on Sec61 Channel Functions, Mechanisms, and Related Diseases. *Front. Physiol.* **2017**, *8*, 887. [CrossRef]
72. Linxweiler, M.; Schick, B.; Zimmermann, R. Let's talk about Secs: Sec61, Sec62 and Sec63 in signal transduction, oncology and personalized medicine. *Signal Transduct. Target. Ther.* **2017**, *2*, 17002. [CrossRef]
73. Sicking, M.; Lang, S.; Bochen, F.; Roos, A.; Drenth, J.P.H.; Zakaria, M.; Zimmermann, R.; Linxweiler, M. Complexity and Specificity of Sec61-Channelopathies: Human Diseases Affecting Gating of the Sec61 Complex. *Cells* **2021**, *10*, 1036. [CrossRef] [PubMed]
74. Casper, M.; Linxweiler, M.; Linxweiler, J.; Zimmermann, R.; Glanemann, M.; Lammert, F.; Weber, S.N. SEC62 and SEC63 Expression in Hepatocellular Carcinoma and Tumor-Surrounding Liver Tissue. *Visc. Med.* **2021**, *37*, 110–115. [CrossRef] [PubMed]
75. Takacs, F.Z.; Radosa, J.C.; Linxweiler, M.; Kasoha, M.; Bohle, R.M.; Bochen, F.; Unger, C.; Solomayer, E.F.; Schick, B.; Juhasz-Boss, I. Identification of 3q oncogene SEC62 as a marker for distant metastasis and poor clinical outcome in invasive ductal breast cancer. *Arch. Gynecol. Obstet.* **2019**, *299*, 1405–1413. [CrossRef]
76. Wemmert, S.; Lindner, Y.; Linxweiler, J.; Wagenpfeil, S.; Bohle, R.; Niewald, M.; Schick, B. Initial evidence for Sec62 as a prognostic marker in advanced head and neck squamous cell carcinoma. *Oncol. Lett.* **2016**, *11*, 1661–1670. [CrossRef] [PubMed]
77. Linxweiler, M.; Bochen, F.; Schick, B.; Wemmert, S.; Al Kadah, B.; Greiner, M.; Hasenfus, A.; Bohle, R.M.; Juhasz-Boss, I.; Solomayer, E.F.; et al. Identification of SEC62 as a potential marker for 3q amplification and cellular migration in dysplastic cervical lesions. *BMC Cancer* **2016**, *16*, 676. [CrossRef] [PubMed]
78. Linxweiler, M.; Linxweiler, J.; Barth, M.; Benedix, J.; Jung, V.; Kim, Y.J.; Bohle, R.M.; Zimmermann, R.; Greiner, M. Sec62 bridges the gap from 3q amplification to molecular cell biology in non-small cell lung cancer. *Am. J. Pathol.* **2012**, *180*, 473–483. [CrossRef]
79. Greiner, M.; Kreutzer, B.; Lang, S.; Jung, V.; Cavalie, A.; Unteregger, G.; Zimmermann, R.; Wullich, B. Sec62 protein level is crucial for the ER stress tolerance of prostate cancer. *Prostate* **2011**, *71*, 1074–1083. [CrossRef]
80. Jung, V.; Kindich, R.; Kamradt, J.; Jung, M.; Muller, M.; Schulz, W.A.; Engers, R.; Unteregger, G.; Stockle, M.; Zimmermann, R.; et al. Genomic and expression analysis of the 3q25-q26 amplification unit reveals TLOC1/SEC62 as a probable target gene in prostate cancer. *Mol. Cancer Res.* **2006**, *4*, 169–176. [CrossRef]

81. Janssen, M.J.; Salomon, J.; Te Morsche, R.H.; Drenth, J.P. Loss of heterozygosity is present in SEC63 germline carriers with polycystic liver disease. *PLoS ONE* **2012**, *7*, e50324. [CrossRef]
82. Waanders, E.; te Morsche, R.H.; de Man, R.A.; Jansen, J.B.; Drenth, J.P. Extensive mutational analysis of PRKCSH and SEC63 broadens the spectrum of polycystic liver disease. *Hum. Mutat.* **2006**, *27*, 830. [CrossRef] [PubMed]
83. Davila, S.; Furu, L.; Gharavi, A.G.; Tian, X.; Onoe, T.; Qian, Q.; Li, A.; Cai, Y.; Kamath, P.S.; King, B.F.; et al. Mutations in SEC63 cause autosomal dominant polycystic liver disease. *Nat. Genet.* **2004**, *36*, 575–577. [CrossRef] [PubMed]

International Journal of
Molecular Sciences

Review

Folding and Insertion of Transmembrane Helices at the ER

Paul Whitley [1], Brayan Grau [2], James C. Gumbart [3], Luis Martínez-Gil [2] and Ismael Mingarro [2,*]

[1] Department of Biology and Biochemistry, Centre for Regenerative Medicine, University of Bath,
    Bath BA2 7AY, UK; bssprw@bath.ac.uk
[2] Department of Biochemistry and Molecular Biology, Institute of Biotechnology and
    Biomedicine (BIOTECMED), Universitat de València, E-46100 Burjassot, Spain; Brayan.Grau@uv.es (B.G.);
    Luis.Martinez-gil@uv.es (L.M.-G.)
[3] School of Physics, School of Chemistry and Biochemistry, Parker H. Petit Institute for Bioengineering and
    Bioscience, Georgia Institute of Technology, Atlanta, GA 30332, USA; gumbart@physics.gatech.edu
*   Correspondence: mingarro@uv.es; Tel.: +34-963543796

Abstract: In eukaryotic cells, the endoplasmic reticulum (ER) is the entry point for newly synthesized proteins that are subsequently distributed to organelles of the endomembrane system. Some of these proteins are completely translocated into the lumen of the ER while others integrate stretches of amino acids into the greasy 30 Å wide interior of the ER membrane bilayer. It is generally accepted that to exist in this non-aqueous environment the majority of membrane integrated amino acids are primarily non-polar/hydrophobic and adopt an α-helical conformation. These stretches are typically around 20 amino acids long and are known as transmembrane (TM) helices. In this review, we will consider how transmembrane helices achieve membrane integration. We will address questions such as: Where do the stretches of amino acids fold into a helical conformation? What is/are the route/routes that these stretches take from synthesis at the ribosome to integration through the ER translocon? How do these stretches 'know' to integrate and in which orientation? How do marginally hydrophobic stretches of amino acids integrate and survive as transmembrane helices?

Keywords: folding; insertion; membrane protein; translocon; ribosome; transmembrane segment

Citation: Whitley, P.; Grau, B.;
Gumbart, J.C.; Martínez-Gil, L.;
Mingarro, I. Folding and Insertion of
Transmembrane Helices at the ER. *Int.
J. Mol. Sci.* **2021**, *22*, 12778. https://
doi.org/10.3390/ijms222312778

Academic Editor:
Richard Zimmermann

Received: 14 October 2021
Accepted: 23 November 2021
Published: 26 November 2021

Publisher's Note: MDPI stays neutral
with regard to jurisdictional claims in
published maps and institutional affiliations.

1. Introduction

The majority of, if not all, integral membrane proteins distributed throughout the endomembrane network in eukaryotic cells first assemble into the endoplasmic reticulum (ER) membrane. Following this assembly, these proteins are distributed to their intended destinations via specific trafficking pathways. The signals and machineries that direct complex endomembrane trafficking pathways are beyond the scope of this article. Thus, we will concentrate on the initial folding and assembly of proteins into the ER membrane and not their subsequent cellular distribution.

Most integral membrane proteins require proteinaceous machinery known as an integrase for their insertion into the ER [1]. The Sec61 translocon/integrase is the primary integration machinery of the ER although there are others, notably the Get1/Get2 integrase that is responsible for tail-anchored (TA) protein integration [2,3]. A further integrase called the ER membrane protein complex (EMC) is also implicated in post-translational membrane integration of a subset of TA proteins but also seems to play additional roles in co-translational membrane protein assembly [4]. These integrases have been evolutionary conserved in eukaryotes [5] with some of them, such as Sec61 and Get1/2, having homologs in prokaryotes. In this article, we will exclusively consider membrane protein assembly mediated by the Sec61 translocon.

2. Structure-Function of the Translocon

The mammalian Sec61 translocon consists of a core heterotrimeric Sec61α, β, and γ complex which have 10, 1, and 1 transmembrane (TM) domains respectively [6]. In addition, numerous accessory proteins are associated with this core complex which may modulate translocon activity or provide functionality that complements its translocation and integrase function. Accessory protein complexes such as TRAP seems to be involved in membrane protein topogenesis [7–9] and oligosaccharyltransferase, which adds oligosaccharides to asparagine residues in the lumen of the ER (at NXS/T sequons), are commonly found associated with the translocon core and/or are present in approximately stoichiometric amounts [10,11]. Others such as signal peptidase complex, TRAM, or Sec62/63 are present in sub-stoichiometric amounts and may only associate transiently or under certain circumstances when particular substrates are present in the translocon [11]. Recently, it was discovered that the EMC is also present as a cooperative partner of the Sec61 translocon machinery during co-translational membrane-protein insertion [12]. Thus, the translocon should be considered as a dynamic rather than a well-defined complex.

The structure of the core Sec61 translocon and homologous complexes such as SecYEG of *Escherichia coli* and SecYEβ of the archaeal *Methanococcus janaschii* have been extensively studied [6]. The structures are all fundamentally similar, suggesting a common mode of action in facilitating the translocation of polypeptide chains across membranes and the integration of appropriate stretches of amino acids into the lipid bilayer. The structure/function relationship of all translocons that is most widely accepted is as follows: Looking from the top, onto the membrane the 10 TM subunit (Sec61α/SecY) is pseudosymmetrical with TM domains 1–5 and 6–10 forming the two halves in a clamshell arrangement (Figure 1A–C); A hinge region at one interface between the two halves consisting of TM5 and TM6 allows for the opening and closing (breathing) of a lateral gate between TM2b and TM7 at the opposite side of the clamshell; this breathing is proposed to allow lateral movement of polypeptides out of the translocon and into the membrane providing the integrase function. From a different view, looking into the membrane from the side (Figure 1D) there appears to be a continuous channel through the membrane that is constricted at the center of the membrane and wider at both extremities, resembling an hourglass. The constriction is lined with six hydrophobic amino acid residues [13]. Furthermore, in structures without a translocating polypeptide, there is a small 'plug' helix (on the lumenal/non-cytoplasmic side) blocking the continuous channel through the membrane. It is proposed that this small helix prevents the leakage of ions when the translocon is not occupied by a translocating polypeptide chain [14]. When there is a translocating or inserting polypeptide chain it is proposed that the plug helix is displaced so that it no longer blocks the continuous channel [15] (Figure 1C). Ion leakage is still prevented by the presence of the translocating polypeptide chain and the hydrophobic amino acids in the central constriction, which has been proposed to act as a greasy gasket forming a tight seal around the translocating chain [16,17]. However, the gasket role of the central constriction is seemingly dispensable as yeast expressing mutant Sec61 with the hydrophobic residues replaced with even charged residues can grow and perform translocation efficiently [18]. Regarding the β- and γ-subunits of the Sec61 translocon, the former is not essential for TM insertion, and the latter acts as a clamp that brings both halves of Sec61α together [19].

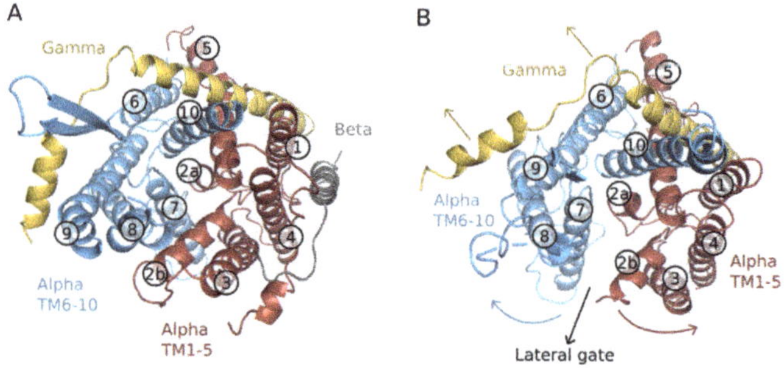

Figure 1. *Cont.*

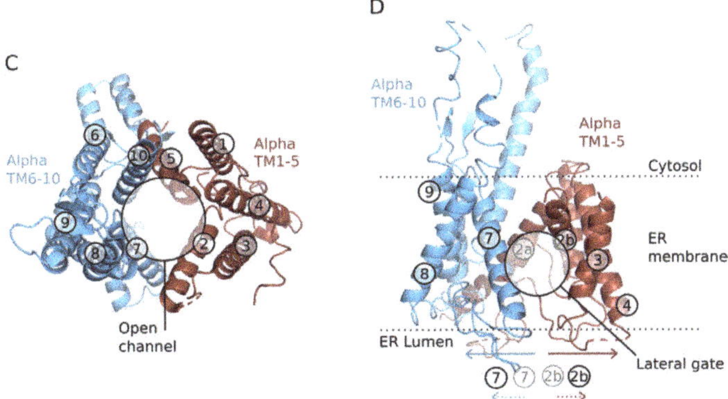

Figure 1. Structure of the translocon. Structure-based cartoon representations of the translocon. All TM segments of Sec61α are colored (red and blue for each half) except for the β-subunits and γ-subunits, which are shown in gray and yellow respectively. All TM segments are numbered for easy comparison between the open and closed structures. (**A**) Top view of the traslocon at a resting/close position from *M. jannaschii* (PDB: 1RHZ). (**B**) Top view of the partially open structure of the translocon from *P. furiosus* (PDB: 3MP7). The colored arrows in red and blue indicate the helix displacements required for the widening of the channel and opening of the lateral gate. The black arrow shows the lateral gate exit pathway of a TM segment from the interior of the channel into the membrane. (**C**) Top view of the translocon with the plug (TM 2a) open from *Saccharomyces cerevisiae* (PDB: 7KAL). The open channel facilitates translocation into the ER lumen of secreted and TM proteins segments. (**D**) Lateral view of the partially open structure of the translocon from *P. furiosus* (PDB: 3MP7). The colored arrows (red and blue) indicate the helix displacements required for the widening of the channel and opening of the lateral gate. The dotted colored arrows at the base of the panel indicate the movement of helices 2b and 7 (compared to a close state, panel A). The limits of the membrane are shown with a dotted line, data from the OPM database (https://opm.phar.umich.edu/ (accessed on 18 November 2021)).

3. Targeting to the Translocon

In higher eukaryotes such as mammals, the majority of proteins utilizing the Sec61 translocon for entry into the ER do so in a co-translational mode [20–22]. It should be noted, however, that there is also a post-translational mode of targeting to the ER that is more prevalent in lower eukaryotes such as fungi that will not be considered here [22]. In co-translational mode, proteins that are to be targeted to the ER are synthesized with a hydrophobic N-terminal sequence. This sequence can either be a signal peptide (signal sequence) that is eventually cleaved from the mature protein once it reaches the ER or an uncleaved TM segment (signal anchor) which comes in two different flavors, conventional (N-terminal cytoplasmic) (Figure 2) or reverse (N-terminal ER lumen). Upon emerging from the ribosome exit tunnel, the N-terminal hydrophobic sequence on the newly synthesizing (nascent) protein is recognized by a ribonucleoprotein chaperone complex called signal recognition particle (SRP). The binding of SRP arrests translation by the ribosome and a ribosome/nascent chain/SRP complex is formed. Targeting to the ER membrane is facilitated by a specific interaction between SRP and an SRP receptor, which is integral in the ER membrane. SRP releases from the complex upon hydrolysis of GTP, reversing the arrest in translation with the ribosome/nascent chain situated on top of a translocon (Figure 2).

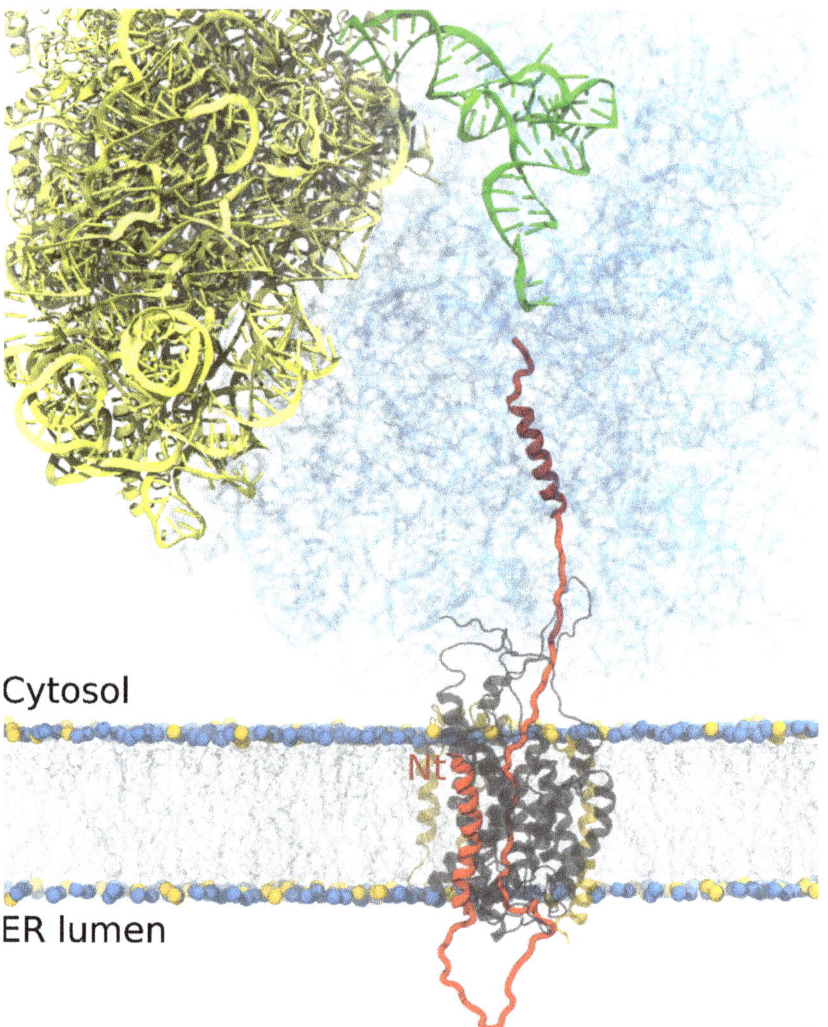

Figure 2. Pathway of a nascent protein. The ribosome is shown in yellow (small subunit) and light blue (large subunit, transparent). A P-site tRNA is shown in green. The nascent chain (red) traverses the ribosome's exit tunnel before encountering the membrane-bound translocon complex (transparent orange and grey). The nascent chain has been shown to form a secondary structure in the translocon as well as at the ribosome's exit tunnel. Most of the structure shown is taken from PDB 4V6M [23], although the nascent-chain helix inside the exit tunnel is modeled.

4. Getting across the Membrane

When translation resumes in the vicinity of the translocon, signal sequences/anchors have to orient themselves appropriately within the translocon. Positively charged residues positioned at one end of the hydrophobic stretch of amino acids play an important role in establishing the orientation of this initial interaction with the translocon together with other ER components, i.e., lipids [24] or potentially accessory proteins. Positive charges at the N-terminus of signal sequences and signal anchors position these types of topogenic sequences with their N-termini in the cytoplasm and direct translocation of C-terminal sequences into the lumen of the ER (N_{in}-C_{out}) (Figure 2). When situated at the C-terminus of a hydrophobic sequence, a C-terminus cytoplasmic orientation is favored and the N-terminus is translocated (N_{out}-C_{in}) [25]. It should be noted that other factors such as the length of hydrophobic stretch of amino acids or folding of previous sequence domains can affect this initial topogenic decision [26,27].

A cryo-EM structure of a ribosome/Sec61 complex in the process of translating/translocating pre-prolactin stalled at 86 residues shows the signal sequence positioned next to TM2 and TM7 in the open lateral gate of Sec61α and density in the translocon channel and ribosome tunnel are consistent with a looped configuration of the nascent chain [28]. If translation were to proceed, the loop would get longer on the lumenal side of the membrane as the polypeptide chain extends (Figure 2), at least until the signal sequence is cleaved and the translocating sequence is no longer a loop.

It has been shown using glycosylation mapping that translocating polypeptides can span the Sec61 translocon in an extended conformation but that α-helices can also be accommodated in the translocon [29,30]. It is broadly accepted that more hydrophobic polypeptides in an α-helical conformation move from the translocon channel into the lipid bilayer more easily than less hydrophobic segments in an extended conformation that will be translocated to the ER lumen. The potential routes that topogenic signals in nascent chains take into the translocon are the topic of later sections in this article.

5. Integrating into the Membrane

Membrane-spanning domains are typically made up of around 20 hydrophobic amino acids in an α-helical conformation. This is sufficient to span the 30 Å hydrophobic core of a model lipid bilayer as each amino acid contributes 1.5 Å to the length of a helix. Not only does this feature of TM domains make them stable in a hydrophobic rather than aqueous environment, but it is also apparent that high hydrophobicity is a major driving force for integration [31]. There seems to be a hydrophobicity threshold above which stretches of amino acids partition into the membrane from the translocon, and below which they continue their translocation through the translocon and into the lumen of the ER. The majority of single-span membrane proteins have TM domains that have a hydrophobicity above this threshold [32] and therefore are theoretically capable of integration in an autonomous (unassisted) way.

This all seems very straightforward. A sufficiently hydrophobic sequence will integrate while others will not. However, things are not that simple. Depending on the topology of the nascent polypeptide, inversions can occur via two sequential energetic transitions of the TM segment: first, the insertion, driven by the hydrophobic effect, and second, the inversion that has been proposed to be driven by electrostatic interactions between the nascent chain and the translocon (or associated proteins) and/or membrane lipids [33]. Furthermore, in multi-spanning proteins, at least 25% of TM helices from proteins of known structure do not reach the threshold that would allow them to integrate autonomously by a simple hydrophobic partitioning [32,34]. Low-hydrophobic segments could be helped by interacting with upstream or downstream hydrophobic segments in the nascent chain to adopt a TM disposition [35] since photocrosslinking experiments have shown that the translocon can simultaneously accommodate more than one TM helix [36–38] (Figure 3).

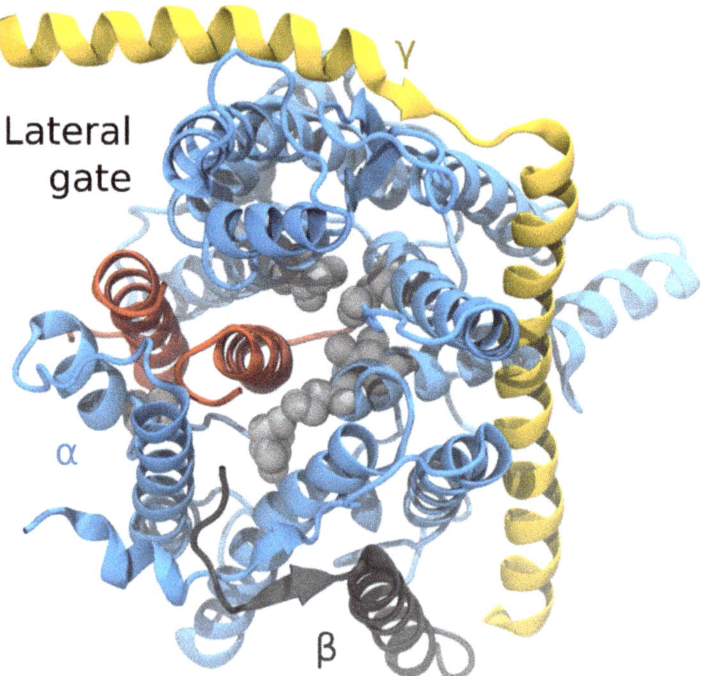

Figure 3. The translocon is shown from the cytoplasmic side in light blue (Sec61α), orange (Sec61γ), and grey (Sec61β). Two helices of a nascent chain are shown in red, one residing in the translocon channel and one at the open lateral gate. Pore ring residues are shown in a white space-filling representation. The structure was modeled based on MD simulations of PDB 1RHZ [13].

6. Where Does Folding of TM Helices Occur?

First, we want the reader to reflect on why TM domains of the type of integral membrane protein we are considering in this review are exclusively α-helical. The answer is that a regular pattern of hydrogen bonding between carbonyl oxygens and amide hydrogens in peptide bonds in an α-helix secondary structure ensures maximal hydrophobicity by effectively neutralizing polar groups in the peptide bonds. It has been calculated that a stretch of hydrophobic amino acids in an extended conformation could not reach the hydrophobic threshold to partition through the lateral gate into the lipid environment [39–42]. Thus, we assume that hydrophobic stretches of amino acids must be in an α-helical conformation to partition into the lipid environment in a thermodynamically favorable way. Hydrophobic stretches of amino acids may fold into α-helices in the translocon tunnel as they enter. However, it is also feasible that the secondary structure is formed before entering the translocon (Figure 2).

During co-translational membrane insertion secondary structures can be acquired in the ribosome specifically in the ribosome exit tunnel. This tunnel accommodates nascent peptides from the ribosomal peptidyl transferase site (P-site) to the ribosomal exit site, providing a protective shield. The first evidence for folding in the ribosome before insertion came more than two decades ago, indicating that the constrained environment of the ribosome-translocon complex has an active role in the propensity of some nascent chains to acquire an extended or a more compact conformation [29]. Furthermore, the authors determined that this compaction was acquired co-translationally in an amino acid-dependent manner. A few years later, it was discovered that the ribosome exit tunnel itself provides enough space to allow the folding of secondary structures as α-helices as Johnson and colleagues demonstrated attaching fluorescent FRET partners at different positions in nascent peptides. The results indicated that folding of TM sequences in an α-helix-like structure was induced and stabilized far inside the ribosome exit tunnel and close to the ribosomal P-site, whereas a nascent secretory protein remained in an extended conformation within the ribosome exit tunnel [43]. These results were further supported by biochemical assays based on pegylation of cysteine reporters [44]. These studies also established that there are seemingly different folding zones within the ribosome exit tunnel where the secondary structure formation can occur, as alanine-replaced peptide segments used for the pegylation experiments folded more compactly when located near the P-site than when located more distally in the tunnel [45].

Following these biochemical studies where compaction could be inferred, atomic structures of α-helical nascent polypeptide chains visualized within the ribosome exit tunnel were quick to follow. Utilizing single-particle cryo-electron microscopy reconstitutions of eukaryotic 80S ribosomes containing nascent chains with an α-helical propensity, Beckmann and collaborators visualized helix density inside the tunnel as well as interaction sites with the tunnel wall components [46].

It is thought that α-helix formation occurs first wherever it is possible and that the ribosome may be selective in the types of sequences that it allows to form secondary structures. Using glycosylation mapping and molecular dynamics (MD) simulations [47] it was shown that in combination, hydrophobicity, helical propensity, and length of hydrophobic stretches of amino acids are major determinants for α-helix adoption within the ribosome exit tunnel (Figure 2). In these experiments, nascent chains harboring α-helix sequences from TM segments were able to fold inside the ribosome exit tunnel whereas those with α-helical sequences of soluble proteins were not.

It is, however, not only secondary structures that can be adopted inside the ribosome tunnel. Small domains with a tertiary structure such as zinc-finger domains have been observed to fold close to the ribosome exit site, also called the vestibule, where the ribosome exit tunnel widens substantially [48], with this folding accelerated and stabilized by the tunnel [49]. Furthermore, helical TM hairpin formation is also possible in the vestibule of the ribosome exit tunnel [50]. Currently, we think of the ribosome/translocon complex as specialized chaperone-like machinery implicated in the formation of secondary or tertiary structure as a platform to overcome the huge energy barrier required to insert unfolded hydrophobic sequences and to prevent their exposure on the cytosolic side of the ER membrane [36,47].

The implication of the ribosome in promoting the secondary structure of TM domains raises a question, 'could this be an evolutionarily conserved function of ribosomes? Presumably, the very first TM proteins had to insert into membranes without the assistance of translocon machinery. There are numerous examples of α-helical membrane proteins that can insert into simplified biomimetic systems without translocon assistance [51–53]. Recently proteorhodopsin has been shown to spontaneously integrate into a simple lipid membrane in the absence of chaperones or a translocon [54]. Other evidence of the pre-translocon era is that yeast mitochondrial ribosomes permanently attach to the mitochondrial inner membrane which lacks the translocon [55–57]. Nevertheless, other components tethering translating ribosomes to the mitochondrial inner membrane have been identified [55]. The permanent docking of these ribosomes facilitates the insertion of membrane proteins encoded by mitochondrial DNA efficiently. Therefore, not only the ribosome but also the recruitment of translating ribosomes to membranes seem to be crucial to a successful insertion and folding process. All these examples highlight the importance of the ribosome in the process of membrane insertion, however, we should be mindful that it is not only TM domains and ribosomes that have co-evolved. There is evidence that extant biological membranes have arisen as a consequence of a lipid bilayer and membrane protein co-evolution process [58]. Altogether, it is highly probable that the ribosome's ability to assist in α-helix folding is not a coincidence exploited by membrane proteins for rapid structure acquisition before membrane insertion; rather it seems likely it is the result of a fine-tuned co-evolution process that continues to this day.

7. Route Into, Through, and Out of the Translocon

In addition to its insertase activity, the translocon is responsible for the secretion of proteins to the ER lumen. The insertion or secretion of proteins can occur both co-translationally or post-translationally depending on the presence of an N-terminal signal sequence (SS). The SS is characterized by a hydrophobic core, often the first TM segment for membrane proteins, flanked by positively charged amino acid residues and polar but uncharged residues at the N- and C-termini, respectively [59]. If the SS is present, once the ribosome is brought to the membrane, it engages with the Sec61 translocon while the nascent chain is still being translated (Figure 2). As mentioned above, the presence of positively charged residues at the N-terminus of the SS causes inversion of its topology, aided through its interactions with lipids and possibly nearby ribosomal RNA [23,33,60].

Within the ribosome/translocon complex the cytoplasmic entrance of the Sec61α channel has a diameter of 20–25 Å [13] (Figure 1D), which nicely matches with the diameter of the ribosome exit tunnel vestibule [61], allowing ribosome-acquired helices to enter into the aqueous translocon channel [62]. When the nascent chain arrives at the laterally closed translocon, it has been demonstrated that the presence of the SS triggers channel opening of the translocon at the lateral gate, which is also linked to the partitioning of TM α-helical structures of membrane proteins into the lipid phase [13,63]. Subsequent hydrophobic regions of the nascent chains will be exposed to the lipid phase and can partition into it as experimental data, structures, and simulations have revealed [64–68] (Figure 3). Recent single-molecule FRET experiments indicate that the lateral gate is highly dynamic even in the absence of a membrane-inserting SS or TM segment, although binding of the ribosome and insertion of TM segments increases the probability of the open state [69]. Indeed, earlier cryo-tomography and fluorescence studies also revealed that binding of the ribosome induces channel opening [70,71].

As noted earlier, for multi-pass membrane proteins, the topology is typically established by the first TM, which adopts either an N_{in}-C_{out} (N-terminus in the cytoplasm and C-terminus in the ER lumen) or an N_{out}-C_{in} topology in the translocon. Each subsequent TM segment adopts the opposite topology of the previous one as it enters the membrane. Recent evidence has implicated the EMC in helping to establish the topology of the first TM of many GPCRs and likely other proteins [12,72]. The translocon's lateral gate allows direct sampling of the membrane environment by the TM [73], possibly from the moment of its encounter with the translocon [39,74]. The code for insertion, i.e., what sequences partition to the membrane vs. remain in the channel, has been determined in exquisite detail with extensive, elegant experiments from von Heijne, White, and colleagues in the mid-2000s [31,32]. Nonetheless, in addition to thermodynamics, a role for kinetics has also been found from both simulations [65,75,76] and experiments [18,77,78]. In particular, slowing down the rate of protein synthesis can increase the probability of membrane insertion.

While most membrane proteins have a fixed topology that alternates from one TM to the next, some defy this straightforward expectation and are known as "dual-topology proteins". These proteins can be found in roughly equal numbers in an N_{in}-C_{out} or an N_{out}-C_{in} topology [79]. In particular, the topology of EmrE, a canonical example of a dual-topology protein, has been found to be very sensitive to the presence of charges, not just at its N-terminus but in any of its loops and even at its C-terminus [80]. This surprisingly suggests that even after synthesis is completed, dual-topology proteins can exist in a dynamic equilibrium where TM helices flip in and out of the membrane, probably in the vicinity of the translocon [81]. Coarse-grained simulations indicated this unusual ability arises from a lack of full integration of some TMs until well after the completion of synthesis [82].

8. Exploring the Limits of TM Domain Insertion

We have previously described how TM helices are proposed to partition from the translocon into the non-polar core of the lipid bilayer driven by hydrophobicity with, the limits for the insertion of TM segments being explored using computationally designed segments with naturally occurring amino acid distributions [83]. However, features found in naturally occurring TM domains such as limited length, charged residues, or a low hydrophobicity profile challenge this rather simplistic view of what a TM segment is.

Given that exposure of hydrophobic groups in proteins and lipids to water is highly unfavorable, membrane proteins tend to minimize their free energy by maximizing the match between the hydrophobic width of the bilayer and the length of a TM segment, a phenomenon called hydrophobic matching [84,85]. Indeed, the average length of a TM segment is 24 ± 5.6 residues long (36 Å in a 3,6 α-helix) [86], while for instance, the width of the hydrophobic core of an ER hepatocyte membrane is close to 38 Å [87–89]. A mismatch in length between the hydrophobic section of a membrane-spanning protein and the bilayer in which it is located results in lipid and peptide rearrangements to compensate [90]. Ultimately a major hydrophobic mismatch might prevent insertion into the membrane, but how much is too much?

The minimum hydrophobic length necessary to form a TM segment in lipid bilayers has been investigated using short hydrophobic peptides in dioleoylphosphatidylcholine (DOPC) and 1,2-dierucoyl-*sn*-glycero-3-phosphocholine (DeuPC, a shorter lipid) vesicles [91]. Peptides composed of Leu residues were compared to sequences of the same length containing alternating Leu and Ala residues (which have a hydrophobicity typical of natural TM helices) [91]. The authors observed that peptides composed exclusively of Leu residues were able to adopt a TM disposition with just 11–12 residues. In this case, the bilayer width exceeded the hydrophobic length of the peptide by ~11–12 Å. For the alternating Leu/Ala sequence 13 residues representing a negative mismatch of ~9 Å were required. The minor differences indicate that the minimum length necessary to form a TM segment is only modestly hydrophobicity-dependent, at least for the sequences tested in this study.

In vitro expression and MD simulations have also been used to systematically examine the insertion efficiency of TM segments consisting primarily of leucine residues [92,93]. Depending on the flanking residues the minimum length to achieve a ~100% insertion efficiency varied from 12, with Gly, Asn, or Asp rich flanking sequences, to 10 residues when Lys was used as a flanking residue. The MD simulations suggest that the insertion efficiency of these sequences is determined primarily by the energetic cost of distorting the bilayer in the vicinity of a short TM segment. The presence of Lys residues flanking the hydrophobic core can reduce the energetic cost by extensive hydrogen bonding with water and lipid phosphate groups (snorkeling) and by partial backbone unfolding. The unfolding is stabilized in the simulation by water molecules entering the bilayer along the peptide backbone.

The studies cited above utilized model hydrophobic sequences composed of only a few different kinds of amino acids. However, TM segments in native membrane proteins vary significantly not only in length but also in composition, revealing more complex scenarios [86]. A computational analysis of the composition and location of amino acids in TM helices found, in membrane proteins of known structure, a strong reverse correlation between the composition/overall hydrophobicity and the required length for their insertion in the lipid bilayer [83]. These results were in accordance with in vitro studies, in which the length dependence of a TM segment (varying from 10 to 25) strongly depends on its amino acid composition [32]. Furthermore, the analysis of naturally occurring residues in TM segments put the focus on the importance of residue positioning, particularly Pro and charged residues.

At the other extreme, the longest TM segment found in naturally occurring membrane proteins is close to 40 residues long [86]. Long TM segments usually adapt their hydrophobic length to the lipid membrane by tilting [86,94]. Accordingly, tilting of long hydrophobic sequences should occur before their insertion in the lipid bilayer, that is, before or during its partition from the translocon into the membrane. An extensively long TM segment might not be easily accommodated within the translocon in its tilted disposition which could prevent its insertion as a single TM domain. Interestingly, a 40-residue long hydrophobic sequence could potentially span a membrane twice in a helical hairpin conformation. The transformation from a long TM domain into a helical hairpin depends primarily on the presence of turn propensity residue(s) in the middle of the sequence both in natural and polyLeu sequences [95,96]. The probability of possessing one of these residues increases tremendously as sequence length extends beyond the 40-residue mark. Increasing the number of residues beyond 40 will facilitate the introduction of a turn propensity residue while maintaining the minimum distance to cross the membrane twice.

Another challenge to the idealized view of a TM segment is the presence of charged residues. The prevalence of these residues within the buried sections of membrane proteins is very low, as expected based on their polarity [86]. However, sequence analysis of membrane protein databases shows that ionizable amino acid residues are indeed present in TM domains, often with functional and/or structural roles [86,97]. How then, are TM domains containing charged residues inserted into the hydrophobic core of the lipid bilayer? The strong hydrophobic contribution of the abundant non-polar residues in a TM segment might be greater than the energy penalty of introducing a charge into the membrane. This negative $\Delta\Delta G$ should be enough to promote the insertion. Interestingly, computer simulation studies suggest that the transfer of four leucines from water to the bilayer interior could be sufficient to compensate for the transfer of a cationic residue [98]. Experiments have also corroborated these predictions [99]. The position within the membrane and the polarity of the residue is also a factor that needs to be taken into consideration when determining the energy necessary for their insertion. Residues close to the water lipid interface can be hydrated due to the presence of "water defects" that reduce the insertion penalty. The closer the polar residue is to the center of the bilayer the larger (and increasingly unfavorable) the water defects become.

The presence of water molecules within the lipid bilayer negatively affects membrane integrity. Interestingly, destabilization by the presence of water molecules has been exploited by some proteins to permeabilize the membrane. Water defect-inducing residues, that is, charged and polar residues surrounded by hydrophobic amino acids, are indeed found in pore-forming proteins [100–102]. Importantly, once in the membrane, polar residues strongly influence the folding or association of integral membrane proteins [97,103–105] and their activity [106,107].

The presence of polar residues within TM segments cannot always be explained by the large hydrophobicity of the surrounding amino acids compensating for the energy penalty of their membrane integration. As mentioned previously, around 25% of TM segments do not reach the hydrophobicity threshold required for autonomous partitioning into the membrane. In these cases, other forces should facilitate the localization of polar amino acids in the hydrophobic environment of the lipid bilayer. Interaction between TM segments within the translocon (Figure 3) provides an opportunity for polarity masking (see Section 5). Although rare, it has been demonstrated that polar interactions between neighboring helices can facilitate insertion [95,108]. Since these interactions are necessary for the partitioning into the lipid bilayer, a cooperative insertion of the TM segments is predicted which requires the presence at the same time of, at least, two helices in the translocon [109] (Figure 3). Similarly, it has been suggested that salt-bridge formation between residues located on the same face of a single TM domain may reduce the free energy of membrane partitioning.

Of note, in a low dielectric constant environment such as the membrane core, the force of an electrostatic interaction increases tremendously, thus creating very stable associations [103,104]. It has been shown that hydrogen bonding between TM segments gives stronger associations than the packing of surfaces in glycophorin A helices driven by the GxxG interaction domain [104,110].

In channel-forming proteins, such as aquaporins (Figure 4), the presence of polar/charged residues close to the membrane hydrophobic core is not explained either by the strong hydrophobicity of the accompanying lipid-facing residues or by electrostatic interactions with other residues. Membrane channels create an amphipathic environment.

Some of the residues in them (the apolar ones), regardless of the depth they are found within the membrane, will be facing the hydrophobic core of the membrane while others (the polar/charged residues) will be exposed to the water-filled tunnel of the channel (Figure 4). The nature of the residues in those two different situations varies accordingly, so the overall amino acid composition of the helices that constitute these channels resemble that of interfacial (amphipathic) helices where one side of the helix is polar while the other is filled with hydrophobic residues. In this case, there is no energy penalty for the inclusion of a charged residue in a TM domain as long as it is lining the aqueous pore of the channel. However, since the amphipathic environment is created by the channel itself, the assembly (or partial formation) of its tertiary structure must presumably involve some co-operativity between TM domains or require chaperones to avoid exposure of the hydrophilic residues to the lipids' hydrocarbon chains. Insertion of multiple helices at once represents a challenge for integrases such as the translocon and associated components that will probably need to expand capacity to accommodate helical bundles. In the case of aquaporin 1 (AQP1) the second TM domain of six is fully translocated into the ER lumen and only adopts a transmembrane orientation after TM4 has been synthesized [111]. Membrane insertion of TM2 also requires a 180 degree flip of TM3 in the membrane presumably, but not necessarily, facilitated by the Sec61 translocon [112]. Such TM domain gymnastics during membrane protein assembly further highlight the potential flexibility of the Sec61 translocon in facilitating different insertion modalities.

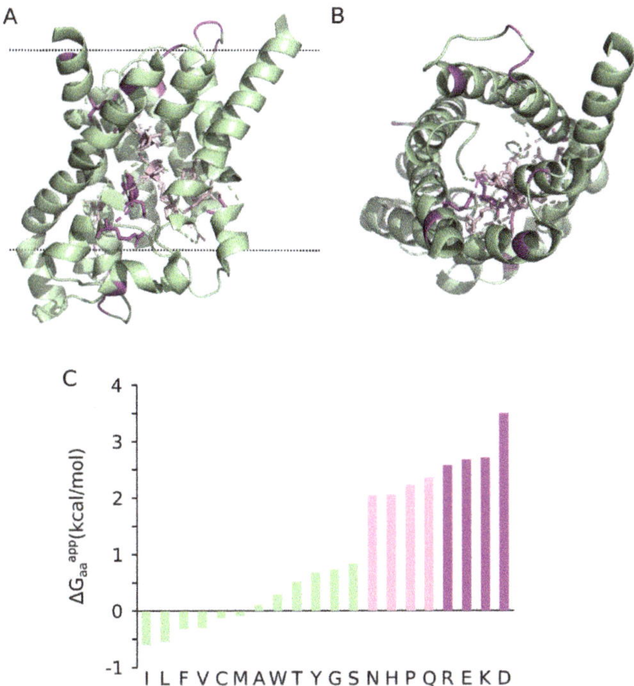

Figure 4. Crystal structure of human aquaporin 10. Cartoon representation of the human aquaporin 10 (PDB:6F7H), (**A**) lateral view, (**B**) top view. Residues are colored based on their membrane insertion propensity: Residues with a low insertion propensity are shown in deep purple and pink. Residues with a mild and high insertion propensity are depicted in green. The side chain of polar and charged residues within the protein's pore are shown in a stick representation. In panel A the limits of the membrane are shown with a dotted line, data from the OPM database. (**C**) Apparent free energy for the insertion in biological membranes of the 20 natural amino acids. Data from [31].

9. Concluding Remarks

Tremendous progress has been made over the last two decades in the field of membrane protein insertion and folding, ranging from biochemical and structural data on the acquisition of secondary structure in nascent chains to the quantitative understanding of the energetic forces for sequence-dependent membrane insertion through the ER translocon and the structures of translocons engaged in nascent chain integration. This does not mean that we have a complete understanding of all aspects of membrane protein assembly and there are still challenges for the field. For example, it will be challenging to explore the dynamics and mechanisms of recruitment of translocon-associated proteins. Thus, it is difficult to envisage how proteins such as TRAM or other accessory components in the ER, which are present in sub-stoichiometric amounts can be present only at those translocons where its suggested chaperone function is needed [113]. Deducing detailed pathways of insertion of marginally hydrophobic TM domains such as TM2 of AQP1 remains another challenge. We hope that the development of new and existing technologies that can answer these remaining questions regarding membrane protein insertion and assembly is not too distant on the horizon. In recent years, new developments in cryo-EM sample preparation and data acquisition have been very fruitful in determining structures of integrase protein complexes [114] and, hopefully, they will be able to shed light on complete dynamics of the insertion process by determining intermediate states of membrane protein assembly through the aforementioned complexes. The increase in experimental data and computational power, in addition to the design of user-friendly interfaces, make MD simulations well positioned to play a very important role in the coming years in the understanding of membrane protein insertion. Simulations of ribosome–translocon complexes on the ns-μs time scale have been possible for a decade [65,115], yet full insertion and maturation take place over seconds. Coarse-grained simulations make physiologically realistic time scales achievable, albeit at the cost of detail [67]. Balancing the need for atomic resolution with that for long time scales will be an ongoing effort, made more challenging by the increasing number of players characterized structurally in integrase complexes.

Author Contributions: P.W., B.G., J.C.G., L.M.-G. and I.M. jointly wrote the manuscript. All authors have read and agreed to the published version of the manuscript.

Funding: Work in the authors' laboratories was supported by grants PID2020-119111GB-I00 from the Spanish Ministry of Science and Innovation (to I.M.), PROMETEU/2019/065 from Generalitat Valenciana (to I.M.) and R01-GM123169 from the US National Institutes of Health (to J.C.G.).

Institutional Review Board Statement: Not applicable.

Informed Consent Statement: Not applicable.

Conflicts of Interest: The authors declare no conflict of interest.

Abbreviations

EMC: ER membrane–protein complex; ER: Endoplasmic reticulum; MD: molecular dynamic; P-site: Peptidyl transferase site; SRP: Signal recognition particle; SS: Signal sequence; TA: Tail anchored; TM: Transmembrane.

References

1. Ito, K.; Shimokawa-Chiba, N.; Chiba, S. Sec translocon has an insertase-like function in addition to polypeptide conduction through the channel. *F1000Research* **2019**, *8*, 2126. [CrossRef]
2. Farkas, A.; Bohnsack, K.E. Capture and delivery of tail-anchored proteins to the endoplasmic reticulum. *J. Cell Biol.* **2021**, *220*, e202105004. [CrossRef] [PubMed]
3. Mateja, A.; Keenan, R.J. A structural perspective on tail-anchored protein biogenesis by the GET pathway. *Curr. Opin. Struct. Biol.* **2018**, *51*, 195–202. [CrossRef] [PubMed]
4. Chitwood, P.J.; Hegde, R.S. The role of EMC during membrane protein biogenesis. *Trends Cell Biol.* **2019**, *29*, 371–384. [CrossRef] [PubMed]

5.   Wideman, J.G. The ubiquitous and ancient ER membrane protein complex (EMC): Tether or not? *F1000Research* **2015**, *4*, 624. [CrossRef]
6.   Rapoport, T.A.; Li, L.; Park, E. Structural and mechanistic insights into protein translocation. *Annu. Rev. Cell Dev. Biol.* **2017**, *33*, 369–390. [CrossRef]
7.   Bano-Polo, M.; Martinez-Garay, C.A.; Grau, B.; Martinez-Gil, L.; Mingarro, I. Membrane insertion and topology of the translocon-associated protein (TRAP) gamma subunit. *Biochim. Biophys. Acta Biomembr.* **2017**, *1859*, 903–909. [CrossRef]
8.   Pfeffer, S.; Dudek, J.; Schaffer, M.; Ng, B.G.; Albert, S.; Plitzko, J.M.; Baumeister, W.; Zimmermann, R.; Freeze, H.H.; Engel, B.D.; et al. Dissecting the molecular organization of the translocon-associated protein complex. *Nat. Commun.* **2017**, *8*, 14516. [CrossRef]
9.   Sommer, N.; Junne, T.; Kalies, K.U.; Spiess, M.; Hartmann, E. TRAP assists membrane protein topogenesis at the mammalian ER membrane. *Biochim. Biophys. Acta* **2013**, *1833*, 3104–3111. [CrossRef]
10. Braunger, K.; Pfeffer, S.; Shrimal, S.; Gilmore, R.; Berninghausen, O.; Mandon, E.C.; Becker, T.; Forster, F.; Beckmann, R. Structural basis for coupling protein transport and N-glycosylation at the mammalian endoplasmic reticulum. *Science* **2018**, *360*, 215–219. [CrossRef]
11. Gemmer, M.; Forster, F. A clearer picture of the ER translocon complex. *J. Cell Sci.* **2020**, *133*, jcs231340. [CrossRef] [PubMed]
12. Chitwood, P.J.; Juszkiewicz, S.; Guna, A.; Shao, S.; Hegde, R.S. EMC is required to initiate accurate membrane protein topogenesis. *Cell* **2018**, *175*, 1507–1519.e16. [CrossRef] [PubMed]
13. Van den Berg, B.; Clemons, W.M., Jr.; Collinson, I.; Modis, Y.; Hartmann, E.; Harrison, S.C.; Rapoport, T.A. X-ray structure of a protein-conducting channel. *Nature* **2004**, *427*, 36–44. [CrossRef] [PubMed]
14. Osborne, A.R.; Rapoport, T.A.; van den Berg, B. Protein translocation by the Sec61/SecY channel. *Annu. Rev. Cell Dev. Biol.* **2005**, *21*, 529–550. [CrossRef]
15. Voorhees, R.M.; Hegde, R.S. Toward a structural understanding of co-translational protein translocation. *Curr. Opin. Cell Biol.* **2016**, *41*, 91–99. [CrossRef]
16. Gumbart, J.; Schulten, K. The roles of pore ring and plug in the SecY protein-conducting channel. *J. Gen. Physiol.* **2008**, *132*, 709–719. [CrossRef]
17. Park, E.; Rapoport, T.A. Preserving the membrane barrier for small molecules during bacterial protein translocation. *Nature* **2011**, *473*, 239–242. [CrossRef] [PubMed]
18. Junne, T.; Kocik, L.; Spiess, M. The hydrophobic core of the Sec61 translocon defines the hydrophobicity threshold for membrane integration. *Mol. Biol. Cell* **2010**, *21*, 1662–1670. [CrossRef]
19. Martinez-Gil, L.; Sauri, A.; Marti-Renom, M.A.; Mingarro, I. Membrane protein integration into the endoplasmic reticulum. *FEBS J.* **2011**, *278*, 3846–3858. [CrossRef]
20. Johnson, A.E.; van Waes, M.A. The translocon: A dynamic gateway at the ER membrane. *Annu. Rev. Cell Dev. Biol.* **1999**, *15*, 799–842. [CrossRef]
21. Keenan, R.J.; Freymann, D.M.; Stroud, R.M.; Walter, P. The signal recognition particle. *Annu. Rev. Biochem.* **2001**, *70*, 755–775. [CrossRef] [PubMed]
22. Shao, S.; Hegde, R.S. Membrane protein insertion at the endoplasmic reticulum. *Annu. Rev. Cell Dev. Biol.* **2011**, *27*, 25–56. [CrossRef] [PubMed]
23. Frauenfeld, J.; Gumbart, J.; Sluis, E.O.; Funes, S.; Gartmann, M.; Beatrix, B.; Mielke, T.; Berninghausen, O.; Becker, T.; Schulten, K.; et al. Cryo-EM structure of the ribosome-SecYE complex in the membrane environment. *Nat. Struct. Mol. Biol.* **2011**, *18*, 614–621. [CrossRef] [PubMed]
24. Dowhan, W.; Vitrac, H.; Bogdanov, M. Lipid-assisted membrane protein folding and topogenesis. *Protein J.* **2019**, *38*, 274–288. [CrossRef]
25. Whitley, P.; Mingarro, I. Stitching proteins into membranes, not sew simple. *Biol. Chem.* **2014**, *395*, 1417–1424. [CrossRef] [PubMed]
26. Higy, M.; Junne, T.; Spiess, M. Topogenesis of membrane proteins at the endoplasmic reticulum. *Biochemistry* **2004**, *43*, 12716–12722. [CrossRef]
27. Sauri, A.; Tamborero, S.; Martinez-Gil, L.; Johnson, A.E.; Mingarro, I. Viral membrane protein topology is dictated by multiple determinants in its sequence. *J. Mol. Biol.* **2009**, *387*, 113–128. [CrossRef] [PubMed]
28. Voorhees, R.M.; Hegde, R.S. Structure of the Sec61 channel opened by a signal sequence. *Science* **2016**, *351*, 88–91. [CrossRef]
29. Mingarro, I.; Nilsson, I.; Whitley, P.; von Heijne, G. Different conformations of nascent polypeptides during translocation across the ER membrane. *BMC Cell Biol.* **2000**, *1*, 3. [CrossRef]
30. Whitley, P.; Nilsson, I.M.; von Heijne, G. A nascent secretory protein may traverse the ribosome/endoplasmic reticulum translocase complex as an extended chain. *J. Biol. Chem.* **1996**, *271*, 6241–6244. [CrossRef]
31. Hessa, T.; Kim, H.; Bihlmaier, K.; Lundin, C.; Boekel, J.; Andersson, H.; Nilsson, I.; White, S.H.; von Heijne, G. Recognition of transmembrane helices by the endoplasmic reticulum translocon. *Nature* **2005**, *433*, 377–381. [CrossRef] [PubMed]
32. Hessa, T.; Meindl-Beinker, N.M.; Bernsel, A.; Kim, H.; Sato, Y.; Lerch-Bader, M.; Nilsson, I.; White, S.H.; von Heijne, G. Molecular code for transmembrane-helix recognition by the Sec61 translocon. *Nature* **2007**, *450*, 1026–1030. [CrossRef] [PubMed]
33. Devaraneni, P.K.; Conti, B.; Matsumura, Y.; Yang, Z.; Johnson, A.E.; Skach, W.R. Stepwise insertion and inversion of a type II signal anchor sequence in the ribosome-Sec61 translocon complex. *Cell* **2011**, *146*, 134–147. [CrossRef] [PubMed]

34. Ojemalm, K.; Watson, H.R.; Roboti, P.; Cross, B.C.; Warwicker, J.; von Heijne, G.; High, S. Positional editing of transmembrane domains during ion channel assembly. *J. Cell Sci.* **2013**, *126*, 464–472. [CrossRef]
35. Ota, K.; Sakaguchi, M.; von Heijne, G.; Hamasaki, N.; Mihara, K. Forced transmembrane orientation of hydrophilic polypeptide segments in multispanning membrane proteins. *Mol. Cell* **1998**, *2*, 495–503. [CrossRef]
36. Sadlish, H.; Pitonzo, D.; Johnson, A.E.; Skach, W.R. Sequential triage of transmembrane segments by Sec61alpha during biogenesis of a native multispanning membrane protein. *Nat. Struct. Mol. Biol.* **2005**, *12*, 870–878. [CrossRef]
37. Sauri, A.; McCormick, P.J.; Johnson, A.E.; Mingarro, I. Sec61alpha and TRAM are sequentially adjacent to a nascent viral membrane protein during its ER integration. *J. Mol. Biol.* **2007**, *366*, 366–374. [CrossRef]
38. Sauri, A.; Saksena, S.; Salgado, J.; Johnson, A.E.; Mingarro, I. Double-spanning plant viral movement protein integration into the endoplasmic reticulum membrane is signal recognition particle-dependent, translocon-mediated, and concerted. *J. Biol. Chem.* **2005**, *280*, 25907–25912. [CrossRef]
39. Cymer, F.; von Heijne, G.; White, S.H. Mechanisms of integral membrane protein insertion and folding. *J. Mol. Biol.* **2015**, *427*, 999–1022. [CrossRef]
40. White, S.H.; Ladokhin, A.S.; Jayasinghe, S.; Hristova, K. How membranes shape protein structure. *J. Biol. Chem.* **2001**, *276*, 32395–32398. [CrossRef]
41. Wimley, W.C.; Creamer, T.P.; White, S.H. Solvation energies of amino acid side chains and backbone in a family of host-guest pentapeptides. *Biochemistry* **1996**, *35*, 5109–5124. [CrossRef] [PubMed]
42. Wimley, W.C.; White, S.H. Designing transmembrane alpha-helices that insert spontaneously. *Biochemistry* **2000**, *39*, 4432–4442. [CrossRef] [PubMed]
43. Woolhead, C.A.; McCormick, P.J.; Johnson, A.E. Nascent membrane and secretory proteins differ in FRET-detected folding far inside the ribosome and in their exposure to ribosomal proteins. *Cell* **2004**, *116*, 725–736. [CrossRef]
44. Lu, J.; Deutsch, C. Secondary structure formation of a transmembrane segment in Kv channels. *Biochemistry* **2005**, *44*, 8230–8243. [CrossRef] [PubMed]
45. Lu, J.; Deutsch, C. Folding zones inside the ribosomal exit tunnel. *Nat. Struct. Mol. Biol.* **2005**, *12*, 1123–1129. [CrossRef]
46. Bhushan, S.; Gartmann, M.; Halic, M.; Armache, J.P.; Jarasch, A.; Mielke, T.; Berninghausen, O.; Wilson, D.N.; Beckmann, R. alpha-Helical nascent polypeptide chains visualized within distinct regions of the ribosomal exit tunnel. *Nat. Struct. Mol. Biol.* **2010**, *17*, 313–317. [CrossRef]
47. Bano-Polo, M.; Baeza-Delgado, C.; Tamborero, S.; Hazel, A.; Grau, B.; Nilsson, I.; Whitley, P.; Gumbart, J.C.; von Heijne, G.; Mingarro, I. Transmembrane but not soluble helices fold inside the ribosome tunnel. *Nat. Commun.* **2018**, *9*, 5246. [CrossRef]
48. Nilsson, O.B.; Hedman, R.; Marino, J.; Wickles, S.; Bischoff, L.; Johansson, M.; Muller-Lucks, A.; Trovato, F.; Puglisi, J.D.; O'Brien, E.P.; et al. Cotranslational protein folding inside the ribosome exit tunnel. *Cell Rep.* **2015**, *12*, 1533–1540. [CrossRef]
49. Wruck, F.; Tian, P.; Kudva, R.; Best, R.B.; von Heijne, G.; Tans, S.J.; Katranidis, A. The ribosome modulates folding inside the ribosomal exit tunnel. *Commun. Biol.* **2021**, *4*, 523. [CrossRef]
50. Tu, L.; Khanna, P.; Deutsch, C. Transmembrane segments form tertiary hairpins in the folding vestibule of the ribosome. *J. Mol. Biol.* **2014**, *426*, 185–198. [CrossRef]
51. Baars, L.; Wagner, S.; Wickstrom, D.; Klepsch, M.; Ytterberg, A.J.; van Wijk, K.J.; de Gier, J.W. Effects of SecE depletion on the inner and outer membrane proteomes of *Escherichia coli*. *J. Bacteriol.* **2008**, *190*, 3505–3525. [CrossRef]
52. Berhanu, S.; Ueda, T.; Kuruma, Y. Artificial photosynthetic cell producing energy for protein synthesis. *Nat. Commun.* **2019**, *10*, 1325. [CrossRef] [PubMed]
53. Matsubayashi, H.; Kuruma, Y.; Ueda, T. In vitro synthesis of the *E. coli* Sec translocon from DNA. *Angew Chem.* **2014**, *53*, 7535–7538. [CrossRef]
54. Eaglesfield, R.; Madsen, M.A.; Sanyal, S.; Reboud, J.; Amtmann, A. Cotranslational recruitment of ribosomes in protocells recreates a translocon-independent mechanism of proteorhodopsin biogenesis. *iScience* **2021**, *24*, 102429. [CrossRef] [PubMed]
55. Doan, K.N.; Grevel, A.; Martensson, C.U.; Ellenrieder, L.; Thornton, N.; Wenz, L.S.; Opalinski, L.; Guiard, B.; Pfanner, N.; Becker, T. The mitochondrial import complex MIM functions as main translocase for alpha-helical outer membrane proteins. *Cell Rep.* **2020**, *31*, 107567. [CrossRef] [PubMed]
56. Glick, B.S.; Von Heijne, G. Saccharomyces cerevisiae mitochondria lack a bacterial-type sec machinery. *Protein Sci.* **1996**, *5*, 2651–2652. [CrossRef] [PubMed]
57. Jia, L.; Dienhart, M.; Schramp, M.; McCauley, M.; Hell, K.; Stuart, R.A. Yeast Oxa1 interacts with mitochondrial ribosomes: The importance of the C-terminal region of Oxa1. *EMBO J.* **2003**, *22*, 6438–6447. [CrossRef] [PubMed]
58. Mulkidjanian, A.Y.; Galperin, M.Y.; Koonin, E.V. Co-evolution of primordial membranes and membrane proteins. *Trends Biochem. Sci.* **2009**, *34*, 206–215. [CrossRef]
59. Cross, B.C.; High, S. Dissecting the physiological role of selective transmembrane-segment retention at the ER translocon. *J. Cell Sci.* **2009**, *122*, 1768–1777. [CrossRef]
60. Park, E.; Rapoport, T.A. Mechanisms of Sec61/SecY-mediated protein translocation across membranes. *Annu. Rev. Biophys.* **2012**, *41*, 21–40. [CrossRef]
61. Voss, N.R.; Gerstein, M.; Steitz, T.A.; Moore, P.B. The geometry of the ribosomal polypeptide exit tunnel. *J. Mol. Biol.* **2006**, *360*, 893–906. [CrossRef] [PubMed]

62. Gumbart, J.; Chipot, C.; Schulten, K. Free energy of nascent-chain folding in the translocon. *J. Am. Chem. Soc.* **2011**, *133*, 7602–7607. [CrossRef] [PubMed]
63. Egea, P.F.; Stroud, R.M. Lateral opening of a translocon upon entry of protein suggests the mechanism of insertion into membranes. *Proc. Natl. Acad. Sci. USA* **2010**, *107*, 17182–17187. [CrossRef] [PubMed]
64. Gogala, M.; Becker, T.; Beatrix, B.; Armache, J.P.; Barrio-Garcia, C.; Berninghausen, O.; Beckmann, R. Structures of the Sec61 complex engaged in nascent peptide translocation or membrane insertion. *Nature* **2014**, *506*, 107–110. [CrossRef] [PubMed]
65. Gumbart, J.C.; Teo, I.; Roux, B.; Schulten, K. Reconciling the roles of kinetic and thermodynamic factors in membrane-protein insertion. *J. Am. Chem. Soc.* **2013**, *135*, 2291–2297. [CrossRef] [PubMed]
66. Mackinnon, A.L.; Paavilainen, V.O.; Sharma, A.; Hegde, R.S.; Taunton, J. An allosteric Sec61 inhibitor traps nascent transmembrane helices at the lateral gate. *eLife* **2014**, *3*, e01483. [CrossRef]
67. Niesen, M.J.M.; Zimmer, M.H.; Miller, T.F., 3rd. Dynamics of co-translational membrane protein integration and translocation via the sec translocon. *J. Am. Chem. Soc.* **2020**, *142*, 5449–5460. [CrossRef]
68. Park, E.; Menetret, J.F.; Gumbart, J.C.; Ludtke, S.J.; Li, W.; Whynot, A.; Rapoport, T.A.; Akey, C.W. Structure of the SecY channel during initiation of protein translocation. *Nature* **2014**, *506*, 102–106. [CrossRef]
69. Mercier, E.; Wang, X.; Maiti, M.; Wintermeyer, W.; Rodnina, M.V. Lateral gate dynamics of the bacterial translocon during cotranslational membrane protein insertion. *Proc. Natl. Acad. Sci. USA* **2021**, *118*, e2100474118. [CrossRef]
70. Ge, Y.; Draycheva, A.; Bornemann, T.; Rodnina, M.V.; Wintermeyer, W. Lateral opening of the bacterial translocon on ribosome binding and signal peptide insertion. *Nat. Commun.* **2014**, *5*, 5263. [CrossRef]
71. Pfeffer, S.; Burbaum, L.; Unverdorben, P.; Pech, M.; Chen, Y.; Zimmermann, R.; Beckmann, R.; Forster, F. Structure of the native Sec61 protein-conducting channel. *Nat. Commun.* **2015**, *6*, 8403. [CrossRef] [PubMed]
72. Shurtleff, M.J.; Itzhak, D.N.; Hussmann, J.A.; Schirle Oakdale, N.T.; Costa, E.A.; Jonikas, M.; Weibezahn, J.; Popova, K.D.; Jan, C.H.; Sinitcyn, P.; et al. The ER membrane protein complex interacts cotranslationally to enable biogenesis of multipass membrane proteins. *eLife* **2018**, *7*, e37018. [CrossRef] [PubMed]
73. Bowie, J.U. Solving the membrane protein folding problem. *Nature* **2005**, *438*, 581–589. [CrossRef] [PubMed]
74. Nicolaus, F.; Metola, A.; Mermans, D.; Liljenstrom, A.; Krc, A.; Abdullahi, S.M.; Zimmer, M.; Miller Iii, T.F.; von Heijne, G. Residue-by-residue analysis of cotranslational membrane protein integration in vivo. *eLife* **2021**, *10*, e64302. [CrossRef] [PubMed]
75. Gumbart, J.C.; Chipot, C. Decrypting protein insertion through the translocon with free-energy calculations. *Biochim. Biophys. Acta* **2016**, *1858*, 1663–1671. [CrossRef]
76. Zhang, B.; Miller, T.F., 3rd. Hydrophobically stabilized open state for the lateral gate of the Sec translocon. *Proc. Natl. Acad. Sci. USA* **2010**, *107*, 5399–5404. [CrossRef]
77. Duong, F.; Wickner, W. Sec-dependent membrane protein biogenesis: SecYEG, preprotein hydrophobicity and translocation kinetics control the stop-transfer function. *EMBO J.* **1998**, *17*, 696–705. [CrossRef]
78. Trueman, S.F.; Mandon, E.C.; Gilmore, R. Translocation channel gating kinetics balances protein translocation efficiency with signal sequence recognition fidelity. *Mol. Biol. Cell* **2011**, *22*, 2983–2993. [CrossRef]
79. Rapp, M.; Granseth, E.; Seppala, S.; von Heijne, G. Identification and evolution of dual-topology membrane proteins. *Nat. Struct. Mol. Biol.* **2006**, *13*, 112–116. [CrossRef]
80. Seppala, S.; Slusky, J.S.; Lloris-Garcera, P.; Rapp, M.; von Heijne, G. Control of membrane protein topology by a single C-terminal residue. *Science* **2010**, *328*, 1698–1700. [CrossRef]
81. Seurig, M.; Ek, M.; von Heijne, G.; Fluman, N. Dynamic membrane topology in an unassembled membrane protein. *Nat. Chem. Biol.* **2019**, *15*, 945–948. [CrossRef] [PubMed]
82. Van Lehn, R.C.; Zhang, B.; Miller, T.F., 3rd. Regulation of multispanning membrane protein topology via post-translational annealing. *eLife* **2015**, *4*, e08697. [CrossRef]
83. Baeza-Delgado, C.; von Heijne, G.; Marti-Renom, M.A.; Mingarro, I. Biological insertion of computationally designed short transmembrane segments. *Sci. Rep.* **2016**, *6*, 23397. [CrossRef] [PubMed]
84. Andersen, O.S.; Koeppe, R.E., 2nd. Bilayer thickness and membrane protein function: An energetic perspective. *Annu. Rev. Biophys. Biomol. Struct.* **2007**, *36*, 107–130. [CrossRef] [PubMed]
85. Mouritsen, O.G.; Bloom, M. Mattress model of lipid-protein interactions in membranes. *Biophys. J.* **1984**, *46*, 141–153. [CrossRef]
86. Baeza-Delgado, C.; Marti-Renom, M.A.; Mingarro, I. Structure-based statistical analysis of transmembrane helices. *Eur. Biophys. J.* **2013**, *42*, 199–207. [CrossRef] [PubMed]
87. Chen, A.; Moy, V.T. Cross-linking of cell surface receptors enhances cooperativity of molecular adhesion. *Biophys. J.* **2000**, *78*, 2814–2820. [CrossRef]
88. Mitra, K.; Ubarretxena-Belandia, I.; Taguchi, T.; Warren, G.; Engelman, D.M. Modulation of the bilayer thickness of exocytic pathway membranes by membrane proteins rather than cholesterol. *Proc. Natl. Acad. Sci. USA* **2004**, *101*, 4083–4088. [CrossRef]
89. Rawicz, W.; Olbrich, K.C.; McIntosh, T.; Needham, D.; Evans, E. Effect of chain length and unsaturation on elasticity of lipid bilayers. *Biophys. J.* **2000**, *79*, 328–339. [CrossRef]
90. Killian, J.A. Hydrophobic mismatch between proteins and lipids in membranes. *Biochim. Biophys. Acta* **1998**, *1376*, 401–415. [CrossRef]
91. Krishnakumar, S.S.; London, E. Effect of sequence hydrophobicity and bilayer width upon the minimum length required for the formation of transmembrane helices in membranes. *J. Mol. Biol.* **2007**, *374*, 671–687. [CrossRef]

92. Jaud, S.; Fernandez-Vidal, M.; Nilsson, I.; Meindl-Beinker, N.M.; Hubner, N.C.; Tobias, D.J.; von Heijne, G.; White, S.H. Insertion of short transmembrane helices by the Sec61 translocon. *Proc. Natl. Acad. Sci. USA* **2009**, *106*, 11588–11593. [CrossRef]
93. Ulmschneider, M.B.; Ulmschneider, J.P.; Schiller, N.; Wallace, B.A.; von Heijne, G.; White, S.H. Spontaneous transmembrane helix insertion thermodynamically mimics translocon-guided insertion. *Nat. Commun.* **2014**, *5*, 4863. [CrossRef] [PubMed]
94. Grau, B.; Javanainen, M.; Garcia-Murria, M.J.; Kulig, W.; Vattulainen, I.; Mingarro, I.; Martinez-Gil, L. The role of hydrophobic matching on transmembrane helix packing in cells. *Cell Stress* **2017**, *1*, 90–106. [CrossRef] [PubMed]
95. Bano-Polo, M.; Martinez-Gil, L.; Wallner, B.; Nieva, J.L.; Elofsson, A.; Mingarro, I. Charge pair interactions in transmembrane helices and turn propensity of the connecting sequence promote helical hairpin insertion. *J. Mol. Biol.* **2013**, *425*, 830–840. [CrossRef] [PubMed]
96. Monne, M.; Hermansson, M.; von Heijne, G. A turn propensity scale for transmembrane helices. *J. Mol. Biol.* **1999**, *288*, 141–145. [CrossRef] [PubMed]
97. Bano-Polo, M.; Baeza-Delgado, C.; Orzaez, M.; Marti-Renom, M.A.; Abad, C.; Mingarro, I. Polar/Ionizable residues in transmembrane segments: Effects on helix-helix packing. *PLoS ONE* **2012**, *7*, e44263. [CrossRef]
98. MacCallum, J.L.; Bennett, W.F.; Tieleman, D.P. Distribution of amino acids in a lipid bilayer from computer simulations. *Biophys. J.* **2008**, *94*, 3393–3404. [CrossRef] [PubMed]
99. Martinez-Gil, L.; Perez-Gil, J.; Mingarro, I. The surfactant peptide KL4 sequence is inserted with a transmembrane orientation into the endoplasmic reticulum membrane. *Biophys. J.* **2008**, *95*, L36–L38. [CrossRef]
100. Garcia-Saez, A.J.; Coraiola, M.; Dalla Serra, M.; Mingarro, I.; Menestrina, G.; Salgado, J. Peptides derived from apoptotic Bax and Bid reproduce the poration activity of the parent full-length proteins. *Biophys. J.* **2005**, *88*, 3976–3990. [CrossRef]
101. Garcia-Saez, A.J.; Coraiola, M.; Serra, M.D.; Mingarro, I.; Muller, P.; Salgado, J. Peptides corresponding to helices 5 and 6 of Bax can independently form large lipid pores. *FEBS J.* **2006**, *273*, 971–981. [CrossRef]
102. Koneru, J.K.; Prakashchand, D.D.; Dube, N.; Ghosh, P.; Mondal, J. Spontaneous transmembrane pore formation by short-chain synthetic peptide. *Biophys. J.* **2021**, *120*, 4557–4574. [CrossRef]
103. Choma, C.; Gratkowski, H.; Lear, J.D.; DeGrado, W.F. Asparagine-mediated self-association of a model transmembrane helix. *Nat. Struct. Biol.* **2000**, *7*, 161–166. [CrossRef]
104. Zhou, F.X.; Cocco, M.J.; Russ, W.P.; Brunger, A.T.; Engelman, D.M. Interhelical hydrogen bonding drives strong interactions in membrane proteins. *Nat. Struct. Biol.* **2000**, *7*, 154–160. [CrossRef] [PubMed]
105. Zhou, F.X.; Merianos, H.J.; Brunger, A.T.; Engelman, D.M. Polar residues drive association of polyleucine transmembrane helices. *Proc. Natl. Acad. Sci. USA* **2001**, *98*, 2250–2255. [CrossRef]
106. Almada, J.C.; Bortolotti, A.; Ruysschaert, J.M.; de Mendoza, D.; Inda, M.E.; Cybulski, L.E. Interhelical H-bonds modulate the activity of a polytopic transmembrane kinase. *Biomolecules* **2021**, *11*, 938. [CrossRef] [PubMed]
107. Li, R.; Mitra, N.; Gratkowski, H.; Vilaire, G.; Litvinov, R.; Nagasami, C.; Weisel, J.W.; Lear, J.D.; DeGrado, W.F.; Bennett, J.S. Activation of integrin alphaIIbbeta3 by modulation of transmembrane helix associations. *Science* **2003**, *300*, 795–798. [CrossRef] [PubMed]
108. Zhang, L.; Sato, Y.; Hessa, T.; von Heijne, G.; Lee, J.K.; Kodama, I.; Sakaguchi, M.; Uozumi, N. Contribution of hydrophobic and electrostatic interactions to the membrane integration of the Shaker K+ channel voltage sensor domain. *Proc. Natl. Acad. Sci. USA* **2007**, *104*, 8263–8268. [CrossRef]
109. Hou, B.; Lin, P.J.; Johnson, A.E. Membrane protein TM segments are retained at the translocon during integration until the nascent chain cues FRET-detected release into bulk lipid. *Mol. Cell* **2012**, *48*, 398–408. [CrossRef]
110. Russ, W.P.; Engelman, D.M. The GxxxG motif: A framework for transmembrane helix-helix association. *J. Mol. Biol.* **2000**, *296*, 911–919. [CrossRef]
111. Lu, Y.; Turnbull, I.R.; Bragin, A.; Carveth, K.; Verkman, A.S.; Skach, W.R. Reorientation of aquaporin-1 topology during maturation in the endoplasmic reticulum. *Mol. Biol. Cell* **2000**, *11*, 2973–2985. [CrossRef] [PubMed]
112. Skach, W.R. Cellular mechanisms of membrane protein folding. *Nat. Struct. Mol. Biol.* **2009**, *16*, 606–612. [CrossRef] [PubMed]
113. Tamborero, S.; Vilar, M.; Martinez-Gil, L.; Johnson, A.E.; Mingarro, I. Membrane insertion and topology of the translocating chain-associating membrane protein (TRAM). *J. Mol. Biol.* **2011**, *406*, 571–582. [CrossRef] [PubMed]
114. Pleiner, T.; Tomaleri, G.P.; Januszyk, K.; Inglis, A.J.; Hazu, M.; Voorhees, R.M. Structural basis for membrane insertion by the human ER membrane protein complex. *Science* **2020**, *369*, 433–436. [CrossRef] [PubMed]
115. Gumbart, J.; Trabuco, L.G.; Schreiner, E.; Villa, E.; Schulten, K. Regulation of the protein-conducting channel by a bound ribosome. *Structure* **2009**, *17*, 1453–1464. [CrossRef] [PubMed]

International Journal of
Molecular Sciences

MDPI

Review

Inhibitors of the Sec61 Complex and Novel High Throughput Screening Strategies to Target the Protein Translocation Pathway

Eva Pauwels [1], Ralf Schülein [2] and Kurt Vermeire [1,*]

[1] KU Leuven Department of Microbiology, Immunology and Transplantation, Rega Institute for Medical Research, Laboratory of Virology and Chemotherapy, B-3000 Leuven, Belgium; eva.pauwels@kuleuven.be

[2] Leibniz-Forschungsinstitut für Molekulare Pharmakologie, Robert-Rössle-Str. 10, 13125 Berlin, Germany; Schuelein@fmp-berlin.de

* Correspondence: kurt.vermeire@kuleuven.be

Abstract: Proteins targeted to the secretory pathway start their intracellular journey by being transported across biological membranes such as the endoplasmic reticulum (ER). A central component in this protein translocation process across the ER is the Sec61 translocon complex, which is only intracellularly expressed and does not have any enzymatic activity. In addition, Sec61 translocon complexes are difficult to purify and to reconstitute. Screening for small molecule inhibitors impairing its function has thus been notoriously difficult. However, such translocation inhibitors may not only be valuable tools for cell biology, but may also represent novel anticancer drugs, given that cancer cells heavily depend on efficient protein translocation into the ER to support their fast growth. In this review, different inhibitors of protein translocation will be discussed, and their specific mode of action will be compared. In addition, recently published screening strategies for small molecule inhibitors targeting the whole SRP-Sec61 targeting/translocation pathway will be summarized. Of note, slightly modified assays may be used in the future to screen for substances affecting SecYEG, the bacterial ortholog of the Sec61 complex, in order to identify novel antibiotic drugs.

Keywords: signal recognition particle dependent protein targeting; Sec61 dependent translocation; co-translational translocation; endoplasmic reticulum; inhibitor; high throughput screening

Citation: Pauwels, E.; Schülein, R.; Vermeire, K. Inhibitors of the Sec61 Complex and Novel High Throughput Screening Strategies to Target the Protein Translocation Pathway. *Int. J. Mol. Sci.* **2021**, *22*, 12007. https://doi.org/10.3390/ijms222112007

Academic Editors: Richard Zimmermann and Sven Lang

Received: 30 September 2021
Accepted: 29 October 2021
Published: 5 November 2021

Publisher's Note: MDPI stays neutral with regard to jurisdictional claims in published maps and institutional affiliations.

1. Introduction

With the evolution of simple cellular structures to multi organelle compartmentalized cells, the transport of proteins across biological membranes has become an unavoidable challenge. Extracellular and integral membrane proteins—synthesized in the cytosol—need to be translocated either across or integrated into bilipid membranes, in order to reach their final destination. Since the discovery of the secretory pathway [1–5], numerous studies have shed light on the different targeting signals, translocation modes, and pathways used by proteins to cross the endoplasmic reticulum (ER) membrane, which is the first and decisive step in the secretory pathway for protein biogenesis (see Figure 1) [6–22]. After maturation in the ER lumen, the proteins are embedded in vesicles and travel through the Golgi apparatus to the cell membrane. Here, the vesicles fuse with the cell membrane, resulting in the expression of the membrane protein at the cell surface or in the secretion of the soluble protein into the extracellular environment.

The targeting signals that drive proteins toward the secretory pathway include N-terminal (cleavable) signal peptides (SPs) as well as transmembrane domains (TMDs). As SPs and TMDs are intrinsic targeting signals for the Sec61 dependent pathway for protein co- and post-translational translocation [8,14,23–27], C-terminally located targeting signals route the respective protein to a different translocation pathway. In tail-anchored (TA) proteins, for instance, the TMD serves as the ER membrane targeting signal. The specific C-terminal location of the targeting TMD, however, restrict TA proteins to the transmembrane recognition complex subunit of the 40 kDa (TRC40) pathway for post-translational translocation as the targeting TMD emerges from the ribosome only when translation is completed [22,28–34]. In fact, single pass membrane proteins are often classified based on their targeting signal and topology after ER translocation (see Figure 1).

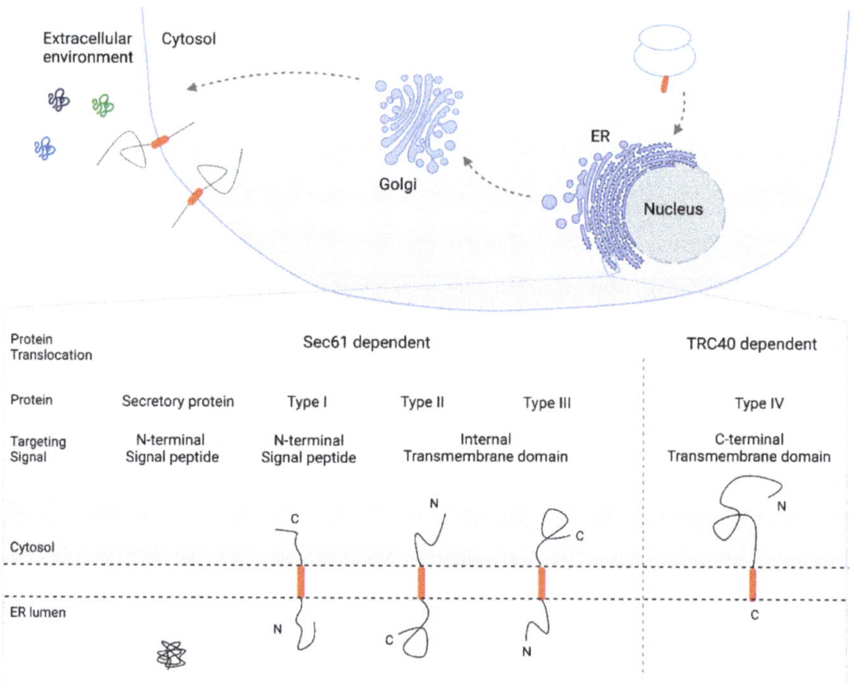

Figure 1. Overview of the secretory pathway for protein biogenesis and topology-based classification of secretory and single pass membrane proteins. Cleavable N-terminal SPs target secretory and type I membrane proteins to the ER membrane, resulting in an N-terminally translocated topology. In the case of type II and type III membrane proteins, the TMD functions as a targeting sequence. Depending on the overall hydrophobicity and charge of the topological sequences adjacent to the TMD, the C- or N-terminal end of the protein is translocated into the ER lumen (type II and type III single pass membrane protein, respectively). Type IV, or TA proteins, are targeted to the ER membrane via the C-terminal TMD. As a result, TA proteins are post-translationally translocated via the TRC-40 pathway. SP: signal peptide, ER: endoplasmic reticulum, TMD: transmembrane domain, TA: tail-anchored, TRC-40: transmembrane recognition complex subunit of 40 kDa.

Evidenced by the evolutionary conservation of the translocation pathways over the different domains of life, correct translocation of proteins across the ER membrane is essential for the proper functioning of cells [21,35–37].

Small molecule inhibitors have therefore become an attractive tool to gain insights into the complex multistep nature of the different translocation routes known today [38–42]. However, new insights in this domain have quickly arisen. In this review, we update the previous knowledge of Sec61 dependent protein translocation in higher eukaryotes, and discuss the newest insights and mode of action of specific translocation inhibitors. Additional focus is placed on the interaction sites of each inhibitor within the translocation complex and screening methods to identify novel signal recognition particle (SRP)-Sec61 specific pathway inhibitors.

2. The Sec61 Dependent Pathway for Co- and Post-Translational Protein Translocation

The Sec61 dependent translocation can occur in two modes (i.e., co- or post-translational translocation). Intrinsic to the terms, they refer to the translocation process occurring simultaneous with or after completion of protein translation, respectively. Post-translational translocation is most commonly used by small secretory proteins (SSP) (less than approximately 100 amino acid residues) and is best understood in fungi and bacteria. Co-translational translocation of proteins over the ER membrane, on the other hand, is a complex multistep process, as shown in Figure 2, which is mostly employed by higher eukaryotes [16,21,25,43–45].

In short, when the targeting signal emerges from the ribosome, it is recognized and bound to by SRP. Upon binding of SRP, protein translation is stalled and the ribosome-nascent chain (RNC) complex is targeted to the ER membrane through binding of SRP to its receptor [46–50]. Next, the targeting sequence interacts with the Sec61 translocon (i.e., the protein conducting channel). Binding of the ribosome to the translocon reinitiates translation of the nascent chain and induces translocation of the preprotein into the ER lumen [27,49–54]. In the lumen, the signal peptidase and oligosaccharyl transferase (OST) complex allow for further maturation of the translocated preprotein by cleaving the protein's signal peptide and by glycosylation of the mature protein part, respectively [55–59].

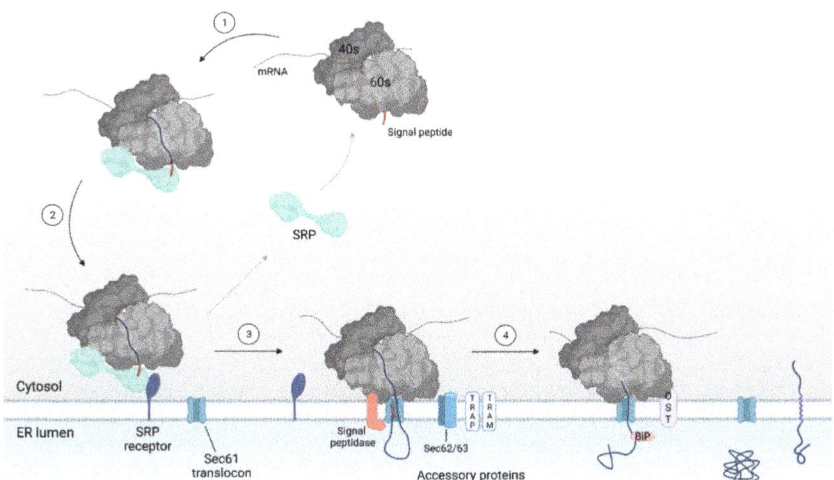

Figure 2. Overview of the SRP dependent pathway for co-translational translocation via the Sec61 translocon. A secretory or integral membrane protein is targeted toward the ER membrane by means of SRP binding to the signal sequence (i.e., the SP or TMD (steps 1–2)). SRP binding stalls protein translation to keep the nascent chain in a translocation competent state. At the ER membrane, SRP interacts with the SRP receptor. The RNC complex is then transferred to the Sec61 translocon (step 3). Interaction of the ribosome with the translocon reinitiates translation and induces conformational changes within Sec61α, eventually leading to protein translocation. In the case of a weak hydrophobic SP or TMD, the protein requires help from accessory proteins such as TRAP, TRAM, Sec62, and/or Sec63 for protein translocation. In the ER lumen, the SP is cleaved by the signal peptidase complex and the protein is glycosylated by the OST complex (step 4). SRP: signal recognition particle, ER: endoplasmic reticulum, SP: signal peptide, TMD: transmembrane domain, RNC: ribosomal nascent chain, TRAP: translocon-associated protein, TRAM: translocating chain-associating membrane protein, OST: oligosaccharyl transferase.

SRP mediated protein targeting to the ER membrane is the most common in eukaryotes and therefore forms the focus of this review. However, proteins can also be SRP independently targeted to the ER membrane, in which case specific chaperone activity is required. For an overview of SRP-independent pathways for protein targeting to the ER, the reader is referred to other publications [11,12,16,43,60–63].

2.1. SRP Dependent Protein Targeting to the ER Membrane Keeps the Protein in a Translocation Competent State

When a secretory or integral membrane protein is translated in the cytosol and the targeting signal (i.e., SP or TMD) emerges from the ribosomal exit tunnel, it is recognized and bound to by SRP [46,47,64–68] (see Figure 2). SRP is a ribonucleoprotein complex consisting of six subunits (SRP9, 14, 19, 54, 68 and 72 m) and a 7S RNA molecule, which assemble into two SRP domains [22,46,48]. SRP 19, SRP54, SRP68, and SRP72 as well as the majority of the SRP RNA make up the S domain of SRP, which holds the recognition and binding site for the emerging SP. The remaining two proteins SRP9 and SRP14 as well as the 5′ and 3′ end of the RNA molecule form the Alu domain of SRP [46]. The Alu domain interacts with the ribosome elongation site, resulting in the transient retardation of protein translation [46–48]. Hence, SRP binding to the RNC complex locks the nascent chain in a translocation competent (i.e., unfolded) state by inducing a translational arrest.

Next, the RNC complex is targeted toward the ER membrane. Here, SRP interacts with the SRP receptor (SR) [46,47,64–68]. The SR then mediates the transfer of the RNC complex to the Sec61 translocon, the central component, and protein-conducting channel of the Sec61 dependent pathway for protein translocation [22,46–48].

2.2. Binding of the RNC Complex Induces Dynamic Conformational Changes in the Translocon

The Sec61 translocon is a heterotrimeric complex that consists of Sec61α, β, and γ monomers (see Figure 3). The Sec61α subunit, composed of ten transmembrane helices (TMH), forms the central pore of the translocon [27,51–54,69,70]. In the quiescent, or native state, the translocon is axially closed by a lumenal plug domain in the central pore of the complex (see Figure 3, depicted as a single helix in red). In addition, the translocon is also laterally sealed by the lateral gate formed by the interhelical interactions between TMH2 and TMH3 (blue helices in Figure 3) and TMH7 and TMH8 (green helices in Figure 3) [51–53]. The interface between TMH2 and TMH7 near the cytosolic side of the translocon also serves as the recognition site for the targeting sequence of the protein nascent chain [27].

Structural studies have shown that binding of the RNC complex to the translocon triggers dynamic conformational changes within Sec61α, resulting in the interrupted interhelical contact between the lateral gate TMH3 and TMH8 (see Figure 3) 'primed Sec61' [35,55–59]. Interestingly, the position of the plug domain, which seals the translocon on the lumenal side of the ER membrane, is almost unaltered upon ribosome binding [51,70]. Hence, ribosome binding to the Sec61 translocon reinitiates protein translation by the release of SRP, and primes the translocon to accept an incoming nascent chain.

The inserting nascent chain can then interact with the recognition site in the lateral gate, which further opens the lateral gate, and displaces the plug domain so that the translocon is opened toward the lipid bilayer for TMD insertion, and toward the lumen for protein translocation [14,27,51,53,54,71].

2.3. Assisted Opening of the Sec61 Translocon

With the rise in structural models explaining the dynamic interactions of the Sec61 translocon upon protein insertion, it has become clear that the hydrophobic strength of the targeting signal is crucial for protein translocation. After all, the SP and/or TMD needs to be sufficiently hydrophobic to disrupt the interhelical hydrophobic interaction between the TMHs of the lateral gate to open the translocon for lateral escape into the ER membrane [27,51,52,69]. In addition, the SP and/or TMD need to displace the plug domain in order for the protein to translocate over the ER membrane.

Hence, proteins with a—so-called—weak hydrophobic SP and/or TMD require additional accessory components such as the translocon-associated protein (TRAP), translocating chain-associated membrane protein (TRAM), Sec62, and/or Sec63 for the translocation into the ER lumen. The specific accessory translocation machinery that is required, is thought to be protein, and thus SP/TMD specific [14,54,71–79].

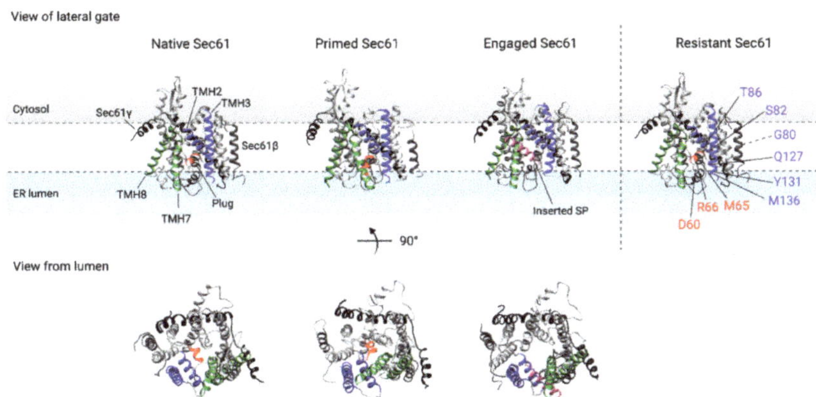

Figure 3. Dynamics of the TMHs of the Sec61 translocon (PDB 5A6U [69]) upon binding of the ribosome (primed state, PDB 3J7Q [52]) and insertion of the SP (engaged state, PDB 3JC2 [27]). Sec61α is shown in grey, Sec61β is shown in dark grey, and Sec61γ is shown in black. The interhelical interaction between TMH2 and TMH3 (shown in blue) on one half of the translocon, and TMH7 and TMH8 (shown in green) on the other half of the translocon form the lateral gate of Sec61α. Additionally, the translocon is closed axially by the lumenal plug domain of TMH2 (shown in red). Binding of the ribosome disrupts the interaction of TMH3 and TMH8 of the lateral gate and primes the translocon for insertion of the nascent protein chain, while the plug domain remains in place. The SP of the nascent chain (shown in pink) interacts with the lateral gate of the translocon, resulting in lateral escape from the translocon and insertion of the TMD into the ER membrane. In addition, the plug domain is displaced to allow for protein translocation into the ER lumen. Resistance conferring mutations located in the lateral gate or plug domain of Sec61α are shown in 'Resistant Sec61'. TMH: transmembrane helix, SP: signal peptide, TMD: transmembrane domain, ER: endoplasmic reticulum.

2.4. Chaperone Mediated Completion of Protein Translocation and Post-Translational Modifications in the ER Lumen

For the translocation of the last amino acid residues that remain in the ribosomal exit tunnel when translation is completed, proteins rely on the binding immunoglobin protein (BiP), a lumenal translocation chaperone. BiP acts as a molecular ratchet by binding to the preprotein and pulling it toward the ER lumen to complete translocation in an ATP dependent manner [80–83]. Once translocated, the proteins are post-translationally modified in the ER lumen. For instance, the SP is cleaved from the preprotein by the signal peptidase complex and the preprotein is glycosylated by the oligosaccharyl-transferase (OST) complex [55,57–59].

3. Translocation Inhibitors of the Sec61 Dependent Protein Translocation Pathway

Being a multistep process, ER protein transport provides many pitfalls for protein mis-translocation that are mostly corrected by cellular control systems and specialized clean-up pathways such as ER associated protein degradation (ERAD) [84–89]. The correct translocation of proteins is crucial for the proper functioning of cells. In fact, inefficient protein translocation has been linked to many liver, kidney, and metabolic diseases [36,90]. Cancer cells, on the other hand, depend heavily on efficient protein translocation into the

ER to support their fast growth. As such, correct protein translocation is key for many fast-growing cancers [37,91–97]. In addition, viruses exploit the host ER protein translocation machinery for the synthesis of viral proteins and host related entry receptors [98–103]. It is therefore no surprise that different inhibitors have been identified that interact with the Sec61 dependent protein translocation process.

The inhibitors known today are natural products and synthetic small molecules that inhibit Sec61 dependent protein translocation with differential substrate selectivity. Evidenced by the fact that many inhibitors originate from therapeutic screening programs, the Sec61 translocon forms a promising target for therapeutic intervention (e.g., for anticancer, immunosuppressive, and/or antiviral treatment).

In the following section, we present an overview of the Sec61 inhibitors of protein translocation known today, with a focus on the discovery, structure–activity relationship (SAR), therapeutic activity, and (putative) interaction sites within the Sec61 translocon.

3.1. Sec61 Inhibitors of Natural Origin

3.1.1. HUN7293, CAM741, and Cotransin

Cell adhesion molecules play a critical role in the immune response by regulating leucocyte migration and cell-to-cell interaction at the site of inflammation. Therefore, the expression of cell adhesion molecules has become an interesting therapeutic target in a variety of inflammatory and autoimmune diseases that are characterized by the overexpression of cell adhesion molecules [104]. With this rationale, a screening program for the inhibition of cell adhesion molecule expression was set up and led to the identification of HUN-7293 [105]. HUN-7293 is a fungal cyclic heptadepsipeptide that selectively inhibits the expression of three cell adhesion molecules (i.e., vascular cell adhesion molecule 1 (VCAM-1), intracellular adhesion molecule 1 (ICAM-1), and E-selectin) [104,105]. With the HUN-7293 compound as the lead molecule of this new class of therapeutic agents, a complete library of analogs was synthesized to study the SAR [105–107]. This led to the identification of new inhibitors of cell adhesion molecule expression that have eventually also paved the way to study Sec61 dependent protein translocation [105,108–110].

A first HUN7293 analog is CAM741, a cyclopeptolide that selectively inhibits the expression of VCAM-1 by inhibition of the VCAM-1 co-translational translocation in the ER lumen [111,112]. By means of chemical cross-linking experiments, the authors showed that in the presence of CAM741, the VCAM-1 SP adopts an altered positioning relative to the Sec61α subunit of the translocon [108]. Later, vascular endothelial growth factor (VEGF) was identified as a second substrate for CAM741 [113]. As seen for VCAM-1, the VEGF SP is also diverted to a different position within Sec61α [113]. Hence, it was suggested that CAM741 interferes with the interaction between the SP and the SP recognition site in the lateral gate of the translocon, resulting in the incorrect insertion of the nascent chain and subsequent inhibition of protein translocation.

Around the same time as the identification of CAM741, another HUN-7293 analog, cotransin, was identified [109]. Cotransin selectively inhibits the expression of VCAM-1 and p-selectin via the inhibition of the co-translational translocation of these proteins across the ER membrane. In parallel to the CAM741 study, Garrison et al. showed that the orientation of the VCAM-1 nascent chain, with regard to the different translocon subunits, was altered in the presence of cotransin [109]. Photoaffinity labelling of cotransins confirmed earlier experiments that were performed on minimal liposomes (containing only the components that are crucial for translocation, i.e., Sec61 and SR), namely that the Sec61α subunit serves as target site for cotransin activity [109]. Later, a small subset of secretory and membrane proteins were identified as additional substrates for cotransin. Among these were angiotensinogen, β-lactamase, corticotropin releasing factor 1, endothelin B receptor, and aquaporin 2 (see Table 1) [105,109,110,114–116]. Initially, researchers believed that cotransin acts in a SP discriminatory manner, as so far, only secretory and type I membrane proteins with a SP targeting signal have been identified as cotransin substrates. CT08 and CT09, two cotransin analogs, however, showed activity against tumor necrosis factor α (TNFα), a single pass type II membrane protein with a non-cleavable TMD as a targeting signal [115]. From a proteomics study on cotrasin, Klein et al. concluded that the biosynthesis of almost all secreted proteins was cotransin-sensitive at a saturating concentration, whereas only a small subset of integral membrane proteins was affected at this concentration. Interestingly, for the integral membrane protein fraction, a conformational TMD consensus motif mediating cotransin sensitivity could be identified [110]. Hence, a cleavable SP is not a strict requirement for cotransin activity, leading to an unanticipated breadth of additional cotransin substrates.

Resistance studies on cotransin and analogs have shown that the lumenal region between the plug domain and lateral gate of the translocon serves as the active site of cotransin (see Table 1 and Figure 3) [115,116]. MacKinnon et al. showed that cotransin binding nearby the plug domain stabilizes the partially opened gate of Sec61α. In this model, the SP is prevented from entering the translocon, and TMD integration is hampered by blocking displacement of the plug domain [112,116,117].

Given the prominent role of VCAM-1, ICAM-1, and TNFα in the cellular immune response, HUN7293 as well as the related molecules CAM741 and cotransins might also be interesting as immunosuppressive agents [115,118]. A more recently identified cotransin substrate is the oncoprotein human epidermal growth factor receptor 3 (HER3), suggesting a potential anticancer activity for cotransin [119]. In addition, by blocking the Sec61 translocon with cotransin, researchers were able to show the importance of the translocon to support viral replication of the influenza A virus (IAV), the human immunodeficiency virus (HIV), and Dengue virus, implicating ER protein transport as a potential antiviral strategy [100].

3.1.2. Decatransin

In contrast to the earlier described inhibitors, fungal cyclic decadepsipeptide decatransin inhibits protein translocation independent of the targeting sequence, and translocation mode, suggesting a broad-spectrum activity. Resistance profiling studies indicate that decatransin binds to Sec61α in a similar, yet distinct manner than cotransin (see Table 1) [120]. Interestingly, cotransin and decatransin also showed cross inhibitory activity with the prokaryotic SecYEG translocon [120].

3.1.3. Apratoxin A and Coibamide A

Apratoxin A and Coibamide A are small molecules isolated from marine cyanobacteria that were originally investigated for their anticancer activity [121–128]. Natural products from marine organisms have a track record of antiproliferative activity in a variety of cancer cells that has led to the development of several clinical candidates [129]. Examples of such candidates from marine cyanobacteria are anti-tubulin agents, the cryptophycins, dolastatins 10 and 15, and curacin A [129,130]. Marine cyanobacteria have been shown to be an inexhaustible source of cytotoxic depsipeptides applicable to cancer research and potential pharmaceutical development [131,132].

Of the five naturally occurring apratoxins, apratoxin A exhibits the highest potency in various cancer cell lines, as the antiproliferative activity was found to be in the low nanomolar range. The antiproliferative activity was later assigned to the apratoxin A induced G1-phase cell cycle arrest and apoptosis [124]. Proteomics revealed that apratoxin A has a broad-spectrum activity as it reversibly downmodulates the expression of numerous ER resident proteins and cancer associated receptors via the inhibition of the co-translational translocation process [125]. Substrates of apratoxin A include gp130, c-MET, HER-2, PDGFR-β, insulin-like growth factor 1β, FGFR, and VEGFR2 [125]. The biological activity and structure have prompted researchers to study the total synthesis of apratoxins [125,133]. Hence, SAR studies have further investigated the selectivity profile of apratoxins, giving rise to apratoxin S4, a synthetic analog, with a more favorable cytotoxicity profile in vivo [121].

Based on the knowledge of other translocation inhibitors (i.e., CAM741 and cotransin), Sec61α was the suggested target candidate of apratoxins. In fact, via a radioactively labelled analog, the Sec61 complex was indeed identified as the molecular target of apratoxins [121]. A competitive binding assay with HUN7293 showed that apratoxins and HUN7293 likely have different binding sites within the translocon [121]. These results were confirmed by an additional study: mutagenesis and competitive photocrosslinking indicate that apratoxin A binds to the Sec61α lateral gate in a distinct manner as was seen for cotransins [122]. In fact, a mutagenesis study revealed that T86 and Y131, two residues located near the lumenal end of TMH2 and TMH3, respectively, are important for apratoxin A activity (see Table 1 and Figure 3).

A recent study suggests an antiviral potential of apratoxins, namely against the SARS-CoV-2 virus [134]. Since many of the apratoxin substrates are receptors that are validated targets for anticancer therapy [125], apratoxin A was thought to be the first anticancer agent to act through the mechanism of co-translational translocation inhibition. Around the same time, however, coibamide A has prompted scientists to investigate it for its unprecedented anticancer activity in vitro [127].

Coibamide A inhibits the migration, invasion, and cell cycle progression of glioblastoma cells [123] and has a broad-spectrum activity that shows substrate overlap with apratoxin A [128,135]. The anticancer activity of coibamide A was also shown in in vivo murine models, however, medicinal chemistry approaches are required to limit the observed dose induced toxicity. SAR analysis showed that the cyclization of the coibamide peptide is crucial for the biological activity, as two linear analogs no longer showed antiproliferative activity against glio- and neuroblastoma cancer cells [123].

By means of a photoaffinity labelled coibamide analog, researchers were able to identify the Sec61 translocon as the main target for coibamide A [135]. Later, resistance profiling suggested a distinct binding mode of coibamide A to Sec61α compared to the other known inhibitors [135]. In fact, the S71 residue that conferred coibamide A resistance upon mutation is located near the plug domain, and is shared only with decatransin, in contrast to the binding site of other inhibitors that are located in the area of the lateral gate (see Table 1 and Figure 3).

Interestingly, a recent study showed impaired autophagy to underly the anticancer activity of coibamide A [136].

3.1.4. Mycolactone

Mycolactone is a virulence factor produced by the Mycobacterium ulcerans and is responsible for the pathogenesis of Buruli ulcers, predominantly seen in West Africa, Australia, Asia, and South America. The immunosuppressive effect caused by mycolactone upon infection of Mycobacterium ulcerans was later assigned to a broad-spectrum inhibition of Sec61 dependent co-translational translocation of secretory proteins that are important in the innate and adaptive immune response such as cytokines, chemokines, and homing receptors into the ER [137–144]. Mycolactone has a complex chemical structure consisting of a 12-membered lactone ring and two polyketide-derived chains that branch from the core in a north and south position [144]. In fact, SAR studies on mycolactone show that the northern chain of the structure is crucial for the biological activity of mycolactone [144].

Competitive binding assays with cotransin showed that mycolactone dose dependently competes with cotransin for binding to the Sec61 translocon. Resistance studies later confirmed the Sec61 translocon, specifically residues near the plug domain of the translocon as the binding partner of mycolactone [145,146]. As summarized in Table 1, these binding sites overlap with other Sec61 inhibitors, suggesting a shared mechanism of action.

A proteomics study conducted on T-cells later confirmed the broad-spectrum activity of mycolactone, as 52 proteins were significantly downregulated in the presence of mycolactone. In fact, mycolactone substrates predominantly consist of single pass type I and type II membrane proteins, containing either a SP or TMD to target the proteins to the ER membrane [138,145,147]. Hence, mycolactone is indiscriminatory for the targeting signal. Later, a more selective effect of mycolactone was seen on the expression of SSPs, which translocate in a post-translational manner. At this point, mycolactone was hypothesized to stabilize the closed conformation of the Sec61 translocon. SSPs with a sufficiently hydrophobic SP were hypothesized to overcome mycolactone activity because of the short nature of their mature protein. The large mature protein part of substrates that are co-translationally translocated, however, retains them in the translocon, independent of the strength of their SP.

This hypothesis, however, was questioned with the determination of the 3D structure of a mycolactone inhibited Sec61 translocon. To date, mycolactone is the only known inhibitor for which a high-resolution structure of the inhibited Sec61 translocon exists [148]. Here, Gérard et al. showed that the conformation of Sec61 in the presence of mycolactone is actually favorable for SP engagement as mycolactone binding induces conformational changes that open the cytoplasmic end of TMH2 and TMH3 of the translocon [148]. The broad-spectrum activity of mycolactone, however, implies that the compound occupies a site in the translocon that is important for SP or TMD binding. In fact, structural analysis confirmed the mycolactone binding site in the cytosolic entrance of the translocon, normally occupied by the SP. Mycolactone is therefore thought to prevent the SP mediated opening of the translocon and subsequent dislocation of the plug domain [148]. Strikingly, Sec61α mutations that confer resistance to mycolactone activity are not located within the mycolactone binding pocket of Sec61α (see Table 1 and Figure 3). These resistance mutations, in fact, induce conformational changes in Sec61α that reduce the formation of the mycolactone binding site near the cytoplasmic end of the translocon [148]. As a consequence, mycolactone binds less efficiently to the mutant Sec61 translocons and no longer inhibits protein translocation into the ER lumen.

3.1.5. Ipomoeassin F

Ipomoeassin F (IpoF) is a natural plant derived resin glycoside cytotoxin that showed a high anticancer potency in different cell lines [149–151]. Later, IpoF was shown to be a non-selective inhibitor of protein secretion via the co-translational translocation process [151–153]. In vitro translocation assays showed that IpoF is specific for the inhibition of Sec61 dependent protein translocation as tail-anchored proteins, but also type III single pass membrane proteins were resistant to IpoF activity [151]. Furthermore, SAR studies showed that the ring size of the IpoF structure is correlated to the biological activity, as ring expansion enhanced IpoF cytotoxicity in cells and its potency to reduce in vitro protein translocation [154].

Resistance profiling shows that IpoF competes with other known inhibitors such as cotransin, apratoxin A, and mycolactone for Sec61α binding, suggesting that these inhibitors have at least partially overlapping interaction sites within the translocon [151].

In addition to the anticancer potency of IpoF, an in vitro SARS-CoV-2 antiviral activity was recently reported for IpoF through the inhibition of the co-translational translocation process of the SARS-CoV-2 spike proteins and the host cell membrane receptor ACE2 [152].

Table 1. Overview of Sec61 translocon inhibitors, substrate specificity, active concentration, and resistance conferring mutations.

Compound	Substrate Selectivity [1]	Active Concentration [2]	Sec61α Resistance Conferring Mutations	Reference
HUN-7293	VCAM-1, ICAM-1, E-selectin	IC_{50} 1–24 nM	Undefined	[104,107]
CAM741	VCAM-1, VEGF	IC_{50} 5–200 nM	Undefined	[105,108,111,113,118]
Cotransin	VCAM-1, P-selectin, Angiotensinogen, β-lactamase, CRF1, ET_BR, AQP2, HER-3, TNF-α	IC_{50} 0.5–5 μM	R66, G80, S82, M136	[105,109,110,114–116,119]
Decatransin	Broad-spectrum	CC_{50} 30–40 nM	I41, D60, M65, R66, S71, G80, S82, M136	[120]
Apratoxin A	Broad-spectrum	CC_{50} 13 nM	T86, Y131	[122,155]
Coibamide A	Broad-spectrum	CC_{50} 10–100 nM	S71	[135,155,156]
Mycolactone	Broad-spectrum	IC_{50} 3–12 nM	R66, S71, G80, S82, T86, Q127, M136	[137–142,145]
Ipomoeassin F	Broad-spectrum	IC_{50} 50–120 μM [3]	R66, S82	[151–153]
CADA	huCD4, SORT, CD137, DNAJC3, PTK7, ERLEC1	IC_{50} 0.2–2 μM	Undefined	[103,157–159]
Eeyarestatin	Broad-spectrum	IC_{50} 70–200 μM [3]	Undefined	[160]
KZR-261/834	Broad-spectrum	IC_{50} nanomolar range	Undefined	[161]

[1] VCAM-1: Vascular cell adhesion molecule 1, ICAM-1 Intercellular adhesion molecule 1, VEGF: Vascular endothelial growth factor, CRF1: Corticotropin releasing factor 1, ET_BR: Endothelin B receptor, AQP2: Aquaporin 2, HER-3: Human epidermal growth factor receptor 3, TNF-α: Tumor necrosis factor α, huCD4: human cluster of differentiation 4, SORT: Sortilin, PTK7: Protein tyrosin kinase 7, ERLEC1: Endoplasmic reticulum lectin 1. [2] IC_{50}: inhibitory concentration producing 50% reduction in biological activity. CC_{50}: cytotoxic concentration causing 50% cell death. [3] IC_{50} in cell-free in vitro translocation assay.

3.2. Synthetic Sec61 Inhibitors

3.2.1. Cyclotriazadisulfonamide

In contrast to the previously discussed Sec61 inhibitors, cyclotriazadisulfonamide (CADA) is a synthetic small molecule translocation inhibitor that was first discovered during a human immunodeficiency virus (HIV)-screening program [103]. It was shown that CADA downmodulates huCD4 expression on a wide range of cells [99,102,103]. Since huCD4 is the main entry receptor for HIV, the reduced expression of huCD4 in the presence of CADA explains the observed antiviral effect of the compound [102,103,162]. In addition to the reported CD4-mediated antiviral effect for CADA, recently, a CD8+ T-cell mediated immunosuppressive effect was described that is related to the CADA-induced suppression of CD137 upregulation [158]. Furthermore, partial downmodulation of the sortilin protein by CADA [159] has recently been linked to a reduction in progranulin-induced breast cancer stem cell propagation [163], thus, suggesting an additional anticancer effect for CADA.

The relatively small size of CADA stimulated the synthesis of numerous analogs that could be implemented in SAR studies [102,163–169]. These SAR studies were all based on the biological effect of CADA analogs on the cellular expression of the huCD4 receptor, and structure optimization resulted in improved activity going from μM to the nM range [165]. An important condition for the preservation of activity is the closed 12-membered ring structure of the compound, given that open ring analogs did not exert any activity on huCD4 [166]. A first quantitative SAR study pointed to the importance of a relatively large, hydrophobic tail group for high impact on huCD4 [164]. In contrast to the symmetrical nature of the lead compound CADA, a subsequent SAR study revealed that unsymmetrical CADA analogs with two different side arms exerted the highest activity [168].

Mechanistic studies showed that CADA directly interacts with huCD4 SP and its reorientation within the Sec61 translocon during the co-translational translocation process of the human CD4 preprotein [170]. In fact, CADA was the first translocation inhibitor for which a direct binding to a SP was shown [170], which distinguishes it from the group of Sec61 translocon binding inhibitors described above. Specific residues in the vicinity of the hydrophobic h-region of the huCD4 SP were identified as being critical for the sensitivity to CADA [171]. Furthermore, a proteomics study on T-cells was performed and identified only five substrates for CADA (see Table 1), suggesting a selective nature of the compound [103,157–159]. Importantly, all substrates carried a cleavable SP as a targeting sequence, implicating that these proteins are Sec61 selective proteins for co-translational translocation. One can thus speculate that the common factor, Sec61α, is a target for CADA binding, however, the importance of direct interaction of CADA with the protein SP cannot be ruled out [170]. Evidence to confirm these hypotheses is awaited as well as the analysis of more potent CADA analogs on substrate selectivity and translocation inhibition.

3.2.2. Eeyarestatin

The ER to cytosol degradation pathway for the disposal of misfolded proteins is an attractive target of intervention for diseases characterized by impaired protein degradation such as Alzheimer's, Parkinson's, prion, and Huntington's disease [172–174]. It was in this regard that Eeyarestatin (ES) I and II, two structurally related chemical molecules, were identified from a library to screen for ERAD inhibitors [172,173]. ESI and ESII were shown to bind with the ER membrane bound p97 complex of ERAD, finally resulting in hampered deubiquitination of misfolded proteins, an essential step for proper proteasomal degradation [175]. As a result, misfolded proteins accumulate and rapidly induce ER stress [175,176]. However, it became clear that ESI and ESII also interfere at a step prior to proteasomal degradation. In fact, studies on ESR35, an ESI analog, showed a broad-spectrum inhibition of protein translocation [160]. Further analysis of the ES compounds suggests that ES targets a component in the Sec61 translocon and thereby sterically prevents the transfer of the RNC complex from the SRP targeting machinery to the Sec61 translocation machinery [160].

Since ES, and other inhibitors for that matter, interacts with the Sec61 translocon to prevent protein translocation into the ER lumen, they may indirectly induce Ca^{2+} leakage from the ER lumen, the major intracellular Ca^{2+} storage [177]. In fact, it was shown that ES, via its 5-NF moiety, induces Ca^{2+} leakage from cells [178,179]. ESI, ESII, and ES24, a minimal analog that closely resembles the 5-NF moiety of ES [179], have been shown to prevent protein translocation, however, while keeping Sec61 in a Ca^{2+} permeable state [178]. This apparent contradiction was reconciled in a mechanistic model for the action of ES compounds on Sec61 complexes [178]. Docking analysis of the Sec61 translocon structure revealed that ES, and particularly the 5-NF group, putatively interacts with the cytosolic end of the Sec61α lateral gate. The binding of ES in the space between TMH2 and TMH7 hampers conformational changes of Sec61α that are required for protein translocation [178]. Hence, this model suggests that ES binding stabilizes the primed, Ca^{2+} permeable, state of the Sec61 translocon by preventing the lateral gate to close [178].

Since the Sec61 translocon is tightly linked to the ERAD pathway for protein degradation, it is suggested that the inhibition of the ER translocation machinery might simultaneously also block both the retrotranslocation of misfolded proteins to the cytosol [160]. The resulting accumulation of cytosolic and misfolded proteins subsequently induces ER stress, which explains the cytotoxic effect of ESI in cellula, and even suggests an anticancer activity of ES [160,180].

In fact, induced tumor cell death upon ES treatment was reported in vitro, and appeared to be enhanced upon co-treatment with proteasomal inhibitors such as bortezomib [160,173,175,176,181–183]. Recently, an antibacterial activity of ES24, a smaller analog of ESI, was described. ES24 impairs protein translocation in E. coli via interaction with the SecYEG translocon, the prokaryotic orthologue of the Sec61 translocon [184].

3.2.3. KZR-261 and KZR-834

The most recent Sec61 dependent protein translocation inhibitors are KZR-261 and KZR-834, two structural analogs that were identified in an anticancer medicinal chemistry screening program [161]. A proteomics study that assayed KZR-261 and KZR-834 activity on different tumor cell lines showed that both compounds had a broad-spectrum activity in vitro in the nanomolar range, with a preference for secreted and type I membrane proteins [161]. In fact, in vivo studies have been performed and have even led to the selection of KZR-261 for clinical development to profile the safety and early efficacy of this novel compound [161]. Among the other substrates are more therapeutic targets such as VEGF, VEGFR, and EGFR, suggesting, besides the anticancer activity, also an immunosuppressive potency for KZR-261 and KZR-834 [161].

4. High Throughput Screening Assays to Define Novel Inhibitors of the Sec61 Complex

Novel small molecule inhibitors of protein translocation at the Sec61 complex may be important innovations in pharmacology and may have several medicinal indications in the future. From the previous section, it is clear that Sec61 inhibitors act by a selective or non-selective mechanism of action [39,90].

Broad-spectrum or non-selective inhibitors impair translocation of many or even all proteins targeted to the Sec61 complex [39,90]. They may represent novel cancer drugs slowing down tumor cell growth. The advantage of Sec61 complex inhibitors over the still widely used cancer drugs affecting nucleic acid biosynthesis would be that they should not be mutagenic by themselves. Non-selective compounds may also be used to decelerate the production of viruses in infected cells, in particular, that of the enveloped viruses that possess integral membrane proteins at the surface using the Sec pathway. It is conceivable that such non-selective substances should target the Sec61 complex alone, rather than the signal sequences of the substrate proteins.

Selective inhibitors, instead, target translocation of a small subset of proteins at the Sec61 complex in a signal sequence discriminatory manner [39,90]. Selective inhibitors may be used in the future to downregulate the biosynthesis of proteins of interest. While a number of selective compounds have been described in the past decade, a specific substance inhibiting translocation of only one protein is not known thus far. Of note, the term selective inhibitor should only be used when proteomic experiments are performed, supporting this classification.

The inhibitors of translocation outlined above were described for the eukaryotic Sec61 complex present in the ER membrane. The orthologous bacterial SecYEG complex translocates proteins across the prokaryotic plasma membrane, therefore, inhibitors of the SecYEG complex may represent novel antibiotics that are urgently needed. A proof of principle for SecYEG inhibition was published, for example, decatransin [99] and ee-yarestatin [81], although these compounds inhibit both the eukaryotic Sec61 complex and prokaryotic SecYEG.

The setup of high throughput screening assays for small molecule inhibitors of the Sec61 complex is notoriously difficult. This is essentially due to three properties of the Sec61 complex. (i) Most importantly, the heterotrimeric Sec61 core complex (Sec61$\alpha\beta\gamma$) has no enzymatic activity by itself. Sec61α gating (i.e., the switch from the closed to the open state) is facilitated by ribosomes, signal sequences, and auxiliary factors such as TRAP and/or Sec62/Sec63 (recent reviews: [131,133]). Translocation itself is facilitated by the BiP chaperone ratchet mechanism at the ER lumenal side [83]. (ii) The accessibility of the complex for compounds in live cells is limited because it is only expressed intracellularly in the ER membrane. (iii) Finally, the complex is difficult to isolate and to reconstitute functionally in larger amounts, making in vitro assays unfavorable.

Although a lot of structural information has been published for the Sec61 complex in the past decade, no attempts have been made to define inhibitors by in silico screening. The main obstacle for in silico screening is the Sec61α architecture itself: it forms a huge and highly dynamic aqueous pore with diameters ranging from 12 to 22 Angstrom. The fact that the channel handles a multitude of different proteins implies that it has slightly different interaction sites for protein substrates. It is consequently very difficult to define specific inhibitor binding sites by in silico approaches.

Despite all of these potential obstacles, a whole-cell screening approach was recently published for inhibitors of the Sec61 complex using two succeeding screening steps [67]. In a primary screen, inhibitors for transcription, translation, and the SRP-Sec61 targeting/translocation pathway were selected (Figure 4a). To this end, a heptahelical G protein-coupled receptor was used as the target, which was C-terminally tagged with GPP (CRF1R.GFP; Figure 4a). The CRF1R.GFP possesses a cleavable signal peptide and thus uses the SRP-Sec61 targeting/translocation pathway [134]. SRP binding to the signal peptide of this construct should decelerate or even arrest translation. The idea is that inhibitors of the SRP-Sec61 targeting/translocation pathway from a compound library should decrease or even prevent CRF1R.GFP biosynthesis, and consequently expression of the C-terminal GFP tag, which could be measured fluorimetrically [77]. A decreased GFP fluorescence, however, may also be observed when inhibitors of the transcription/translation machinery are present. Taking all hit compounds of the primary screen, the latter were deselected with a secondary screen using unfused, cytosolic GFP protein as a target (Figure 4b) [77]. GFP alone does not use the SRP-Sec61 targeting/translocation pathway and its expression depends only on transcription and translation. Compounds were considered as inhibitors of the SRP-Sec61 targeting/translocation pathway when they behaved as hits in the primary screen, but not in the secondary screen [77].

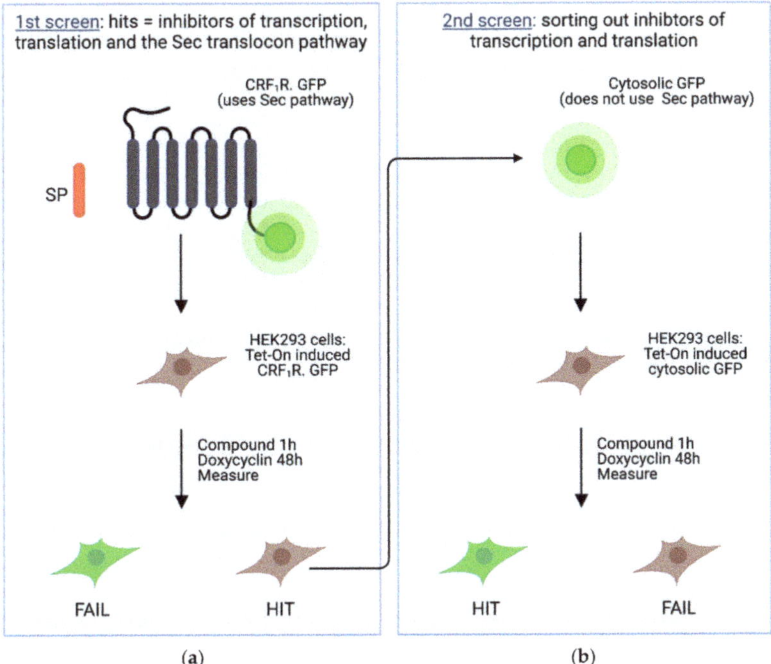

Figure 4. Scheme of the primary and the secondary screen in stably transfected HEK 293 cells. In the primary screen (**a**), Tet-On-controlled CRF1R.GFP was used, a GFP-tagged GPCR possessing a cleavable signal peptide that uses the SRP-Sec61 targeting/translocation pathway. The secondary screen (**b**) was performed with Tet-On-controlled, unfused, soluble GFP, which does not use the SRP-Sec61 targeting/translocation pathway. Hits of the primary screen were used in the secondary screen to deselect inhibitors of the transcription/translation machinery. Figure modified from [67].

Using this screening setup and a library of 37,312 substances, 1052 compounds were identified in the primary screen [77]. This number was reduced to 28 compounds following the secondary screen. Following an in vitro biosynthesis assay in live cells, five compounds were considered to represent real hits with a potential to inhibit the SRP-Sec61 targeting/translocation pathway. For one of them, namely FMP-401319-3, it could be shown by an in vitro transcription/translation/translocation assay that it acts indeed in a post targeting step at the level of the Sec61 complex. The potency of compound FMP-401319-3, however, was only in the low micromolar range. It remains to be determined whether it could be optimized by medicinal chemistry methods in the future. Using a much larger library might also be helpful to identify compounds with higher potency.

Of note, a slightly modified methodology may be used in the future to screen for substances affecting SecYEG, the bacterial ortholog of the Sec61 complex in order to derive novel antibiotic drugs. No such specific inhibitor for the prokaryotic SecYEG complex has been reported thus far, except for decatransin [120] and eeyarestatin [184], which both inhibit the Sec61 and SecYEG complexes.

5. Summary

Protein translocation is by far the most crucial process for the overall protein biogenesis and correct functioning of proteins in cellular processes, and homeostasis in general. This is evidenced by the fact that incorrect protein translocation is linked to numerous metabolic and protein folding diseases.

Today, different inhibitors of the Sec61 dependent protein translocation pathway have been identified. The chemical structure, compound concentration, and substrate targeting sequence are factors that ultimately contribute to the substrate specificity and selectivity of these compounds. As many inhibitors share binding regions within the Sec61α subunit, the translocon shows great potential as a molecular target in different therapeutic areas such as anticancer, immunosuppressive, and antiviral treatment. An intriguing fact of the inhibitors discussed in this review, is that they share a certain level of structural resemblance: they belong to the family of macrocyclic depsipeptides. The macrocyclic nature of the compounds, however, is associated with challenges regarding the synthesis, plasma stability, and/or stereochemical complexity.

By means of two-step whole cell screening approaches, researchers therefore aim to discover novel inhibitors specific to the SRP-Sec61 translocation pathway. A methodology that might also be expanded to screen for molecules that affect SecYEG, the bacterial ortholog of the Sec61 complex, in order to discover new antibiotic drugs. Undoubtedly, inhibitors of protein translocation will find their way into the clinic as promising therapeutics to treat various diseases.

Author Contributions: E.P., R.S. and K.V. jointly wrote the manuscript. All authors have read and agreed to the published version of the manuscript.

Funding: This research received no external funding.

Conflicts of Interest: The authors declare no conflict of interest.

References

1. Siekevitz, P.; Palade, G.E. A cytochemical study on the pancreas of the guinea pig. 5. In vivo incorporation of leucine-1-C14 into the chymotrypsinogen of various cell fractions. *J. Cell Biol.* **1958**, *7*, 619–630. [CrossRef] [PubMed]

2. Caro, L.G.; Palade, G.E. Protein synthesis, storage, and discharge in the pancreatic exocrine cell. An autoradiographic study. *J. Cell Biol.* **1964**, *20*, 473–495. [PubMed]

3. Jamieson, J.D.; Palade, G.E. Intracellular transport of secretory proteins in the pancreatic endocrine cell: II. Transport to Condensing Vacuoles and Zymogen Granules. *J. Cell Biol.* **1967**, *34*, 597–615. [CrossRef]

4. Jamieson, J.D.; Palade, G.E. Intracellular transport of secretory proteins in the pancreatic endocrine cell: I. Role of the Peripheral Elements of the Golgi Complex. *J. Cell Biol.* **1967**, *34*, 577–596. [CrossRef]

5. Matlin, K.S.; Caplan, M.J. The secretory pathway at 50: A golden anniversary for some momentous grains of silver. *Mol. Biol. Cell* **2017**, *28*, 229–232. [CrossRef]

6.   Walter, P.; Gilmore, R.; Müller, M.; Blobel, G. The protein translocation machinery of the endoplasmic reticulum. *Philos. Trans. R. Soc. Lond. Ser. B Biol. Sci.* **1982**, *300*, 225–228. [CrossRef]

7.   Walter, P.; Gilmore, R.; Blobel, G. Protein translocation across the endoplasmic reticulum. *Cell* **1984**, *38*, 5–8. [CrossRef]

8.   Von Heijne, G. Signal sequences. The limits of variation. *J. Mol. Biol.* **1985**, *184*, 99–105. [CrossRef]

9.   Cross, B.C.; Sinning, I.; Luirink, J.; High, S. Delivering proteins for export from the cytosol. *Nat. Rev. Mol. Cell Biol.* **2009**, *10*, 255–264. [CrossRef]

10.   Mateja, A.; Keenan, R.J. A structural perspective on tail-anchored protein biogenesis by the GET pathway. *Curr. Opin. Struct. Biol.* **2018**, *51*, 195–202. [CrossRef]

11.   Ast, T.; Cohen, G.; Schuldiner, M. A network of cytosolic factors targets SRP-independent proteins to the endoplasmic reticulum. *Cell* **2013**, *152*, 1134–1145. [CrossRef]

12.   Aviram, N.; Ast, T.; Costa, E.A.; Arakel, E.C.; Chuartzman, S.G.; Jan, C.H.; Hassdenteufel, S.; Dudek, J.; Jung, M.; Schorr, S.; et al. The SND proteins constitute an alternative targeting route to the endoplasmic reticulum. *Nature* **2016**, *540*, 134–138. [CrossRef] [PubMed]

13.   Greenfield, J.J.; High, S. The Sec61 complex is located in both the ER and the ER-Golgi intermediate compartment. *J. Cell Sci.* **1999**, *112 Pt 10*, 1477–1486. [PubMed]

14.   Kim, S.J.; Mitra, D.; Salerno, J.R.; Hegde, R.S. Signal sequences control gating of the protein translocation channel in a substrate-specific manner. *Dev. Cell* **2002**, *2*, 207–217. [CrossRef]

15.   Kraut-Cohen, J.; Afanasieva, E.; Haim-Vilmovsky, L.; Slobodin, B.; Yosef, I.; Bibi, E.; Gerst, J.E. Translation- and SRP-independent mRNA targeting to the endoplasmic reticulum in the yeast Saccharomyces cerevisiae. *Mol. Biol. Cell* **2013**, *24*, 3069–3084. [CrossRef] [PubMed]

16.   Johnson, N.; Powis, K.; High, S. Post-translational translocation into the endoplasmic reticulum. *Biochim. Biophys. Acta* **2013**, *1833*, 2403–2409. [CrossRef]

17.   Mandon, E.C.; Trueman, S.F.; Gilmore, R. Translocation of proteins through the Sec61 and SecYEG channels. *Curr. Opin. Cell Biol.* **2009**, *21*, 501–507. [CrossRef] [PubMed]

18.   Borgese, N.; Coy-Vergara, J.; Colombo, S.F.; Schwappach, B. The Ways of Tails: The GET Pathway and more. *Protein J.* **2019**. [CrossRef] [PubMed]

19.   Denic, V. A portrait of the GET pathway as a surprisingly complicated young man. *Trends Biochem. Sci.* **2012**, *37*, 411–417. [CrossRef] [PubMed]

20.   Chartron, J.W.; Clemons, W.M., Jr.; Suloway, C.J. The complex process of GETting tail-anchored membrane proteins to the ER. *Curr. Opin. Struct. Biol.* **2012**, *22*, 217–224. [CrossRef]

21.   Hassdenteufel, S.; Nguyen, D.; Helms, V.; Lang, S.; Zimmermann, R. ER import of small human presecretory proteins: Components and mechanisms. *FEBS Lett.* **2019**, *593*, 2506–2524. [CrossRef]

22.   Jadhav, B.; McKenna, M.; Johnson, N.; High, S.; Sinning, I.; Pool, M.R. Mammalian SRP receptor switches the Sec61 translocase from Sec62 to SRP-dependent translocation. *Nat. Commun.* **2015**, *6*, 10133. [CrossRef] [PubMed]

23.   Rapoport, T.A. Protein translocation across the eukaryotic endoplasmic reticulum and bacterial plasma membranes. *Nature* **2007**, *450*, 663–669. [CrossRef]

24.   Aviram, N.; Schuldiner, M. Embracing the void—How much do we really know about targeting and translocation to the endoplasmic reticulum? *Curr. Opin. Cell Biol.* **2014**, *29*, 8–17. [CrossRef]

25.   Johnson, N.; Hassdenteufel, S.; Theis, M.; Paton, A.W.; Paton, J.C.; Zimmermann, R.; High, S. The signal sequence influences post-translational ER translocation at distinct stages. *PLoS ONE* **2013**, *8*, e75394. [CrossRef]

26.   Kriegler, T.; Magoulopoulou, A.; Amate Marchal, R.; Hessa, T. Measuring Endoplasmic Reticulum Signal Sequences Translocation Efficiency Using the Xbp1 Arrest Peptide. *Cell Chem. Biol.* **2018**, *25*, 880–890. [CrossRef] [PubMed]

27.   Voorhees, R.M.; Hegde, R.S. Structure of the Sec61 channel opened by a signal sequence. *Science* **2016**, *351*, 88–91. [CrossRef]

28.   Hassdenteufel, S.; Schauble, N.; Cassella, P.; Leznicki, P.; Muller, A.; High, S.; Jung, M.; Zimmermann, R. Ca^{2+}-calmodulin inhibits tail-anchored protein insertion into the mammalian endoplasmic reticulum membrane. *FEBS Lett.* **2011**, *585*, 3485–3490. [CrossRef]

29.   Casson, J.; McKenna, M.; Hassdenteufel, S.; Aviram, N.; Zimmerman, R.; High, S. Multiple pathways facilitate the biogenesis of mammalian tail-anchored proteins. *J. Cell Sci.* **2017**, *130*, 3851–3861. [CrossRef] [PubMed]

30.   Colombo, S.F.; Cardani, S.; Maroli, A.; Vitiello, A.; Soffientini, P.; Crespi, A.; Bram, R.F.; Benfante, R.; Borgese, N. Tail-anchored Protein Insertion in Mammals: Function and reciprocal interactions of the two subunits of the TRC40 receptor. *J. Biol. Chem.* **2016**, *291*, 15292–15306. [CrossRef] [PubMed]

31.   Coy-Vergara, J.; Rivera-Monroy, J.; Urlaub, H.; Lenz, C.; Schwappach, B. A trap mutant reveals the physiological client spectrum of TRC40. *J. Cell Sci.* **2019**, *132*, jcs230094. [CrossRef]

32.   Borgese, N.; Fasana, E. Targeting pathways of C-tail-anchored proteins. *Biochim. Biophys. Acta (BBA) Biomembr.* **2011**, *1808*, 937–946. [CrossRef]

33.   Yamamoto, Y.; Sakisaka, T. Molecular Machinery for Insertion of Tail-Anchored Membrane Proteins into the Endoplasmic Reticulum Membrane in Mammalian Cells. *Mol. Cell* **2012**, *48*, 387–397. [CrossRef] [PubMed]

34.   Ott, M.; Marques, D.; Funk, C.; Bailer, S.M. Asna1/TRC40 that mediates membrane insertion of tail-anchored proteins is required for efficient release of Herpes simplex virus 1 virions. *Virol. J.* **2016**, *13*, 175. [CrossRef]

35. Lang, S.; Benedix, J.; Fedeles, S.V.; Schorr, S.; Schirra, C.; Schauble, N.; Jalal, C.; Greiner, M.; Hassdenteufel, S.; Tatzelt, J.; et al. Different effects of Sec61alpha, Sec62 and Sec63 depletion on transport of polypeptides into the endoplasmic reticulum of mammalian cells. *J. Cell Sci.* **2012**, *125*, 1958–1969. [CrossRef] [PubMed]

36. Lang, S.; Pfeffer, S.; Lee, P.H.; Cavalie, A.; Helms, V.; Forster, F.; Zimmermann, R. An Update on Sec61 Channel Functions, Mechanisms, and Related Diseases. *Front. Physiol.* **2017**, *8*, 887. [CrossRef] [PubMed]

37. Linxweiler, M.; Schick, B.; Zimmermann, R. Let's talk about Secs: Sec61, Sec62 and Sec63 in signal transduction, oncology and personalized medicine. *Signal Transduct. Target. Ther.* **2017**, *2*, 17002. [CrossRef] [PubMed]

38. Kalies, K.U.; Romisch, K. Inhibitors of Protein Translocation across the ER Membrane. *Traffic* **2015**, *16*, 1027–1038. [CrossRef]

39. Van Puyenbroeck, V.; Vermeire, K. Inhibitors of protein translocation across membranes of the secretory pathway: Novel antimicrobial and anticancer agents. *Cell. Mol. Life Sci. CMLS* **2018**, *75*, 1541–1558. [CrossRef]

40. Zimmermann, R.; Eyrisch, S.; Ahmad, M.; Helms, V. Protein translocation across the ER membrane. *Biochim. Biophys. Acta* **2011**, *1808*, 912–924. [CrossRef]

41. Matlack, K.E.S.; Mothes, W.; Rapoport, T.A. Protein Translocation: Tunnel Vision. *Cell* **1998**, *92*, 381–390. [CrossRef]

42. Luesch, H.; Paavilainen, V.O. Natural products as modulators of eukaryotic protein secretion. *Nat. Prod. Rep.* **2020**, *37*, 717–736. [CrossRef] [PubMed]

43. Lakkaraju, A.K.; Thankappan, R.; Mary, C.; Garrison, J.L.; Taunton, J.; Strub, K. Efficient secretion of small proteins in mammalian cells relies on Sec62-dependent posttranslational translocation. *Mol. Biol. Cell* **2012**, *23*, 2712–2722. [CrossRef] [PubMed]

44. Denks, K.; Vogt, A.; Sachelaru, I.; Petriman, N.A.; Kudva, R.; Koch, H.G. The Sec translocon mediated protein transport in prokaryotes and eukaryotes. *Mol. Membr. Biol.* **2014**, *31*, 58–84. [CrossRef] [PubMed]

45. Pfeffer, S.; Dudek, J.; Zimmermann, R.; Forster, F. Organization of the native ribosome-translocon complex at the mammalian endoplasmic reticulum membrane. *Biochim. Biophys. Acta* **2016**, *1860*, 2122–2129. [CrossRef]

46. Luirink, J.; Sinning, I. SRP-mediated protein targeting: Structure and function revisited. *Biochim. Biophys. Acta* **2004**, *1694*, 17–35. [CrossRef] [PubMed]

47. Lakkaraju, A.K.; Mary, C.; Scherrer, A.; Johnson, A.E.; Strub, K. SRP keeps polypeptides translocation-competent by slowing translation to match limiting ER-targeting sites. *Cell* **2008**, *133*, 440–451. [CrossRef]

48. Wild, K.; Juaire, K.D.; Soni, K.; Shanmuganathan, V.; Hendricks, A.; Segnitz, B.; Beckmann, R.; Sinning, I. Reconstitution of the human SRP system and quantitative and systematic analysis of its ribosome interactions. *Nucleic Acids Res.* **2019**, *47*, 3184–3196. [CrossRef]

49. Dudek, J.; Pfeffer, S.; Lee, P.H.; Jung, M.; Cavalie, A.; Helms, V.; Forster, F.; Zimmermann, R. Protein transport into the human endoplasmic reticulum. *J. Mol. Biol.* **2015**, *427*, 1159–1175. [CrossRef]

50. Nyathi, Y.; Wilkinson, B.M.; Pool, M.R. Co-translational targeting and translocation of proteins to the endoplasmic reticulum. *Biochim. Biophys. Acta* **2013**, *1833*, 2392–2402. [CrossRef]

51. Voorhees, R.M.; Hegde, R.S. Structures of the scanning and engaged states of the mammalian SRP-ribosome complex. *eLife* **2015**, *4*, e07975. [CrossRef] [PubMed]

52. Voorhees, R.M.; Fernandez, I.S.; Scheres, S.H.; Hegde, R.S. Structure of the mammalian ribosome-Sec61 complex to 3.4 A resolution. *Cell* **2014**, *157*, 1632–1643. [CrossRef]

53. Voorhees, R.M.; Hegde, R.S. Toward a structural understanding of co-translational protein translocation. *Curr. Opin. Cell Biol.* **2016**, *41*, 91–99. [CrossRef] [PubMed]

54. Pfeffer, S.; Dudek, J.; Schaffer, M.; Ng, B.G.; Albert, S.; Plitzko, J.M.; Baumeister, W.; Zimmermann, R.; Freeze, H.H.; Engel, B.D.; et al. Dissecting the molecular organization of the translocon-associated protein complex. *Nat. Commun.* **2017**, *8*, 14516. [CrossRef]

55. Braunger, K.; Pfeffer, S.; Shrimal, S.; Gilmore, R.; Berninghausen, O.; Mandon, E.C.; Becker, T.; Forster, F.; Beckmann, R. Structural basis for coupling protein transport and N-glycosylation at the mammalian endoplasmic reticulum. *Science* **2018**, *360*, 215–219. [CrossRef] [PubMed]

56. Bai, L.; Wang, T.; Zhao, G.; Kovach, A.; Li, H. The atomic structure of a eukaryotic oligosaccharyltransferase complex. *Nature* **2018**, *555*, 328–333. [CrossRef] [PubMed]

57. Pfeffer, S.; Dudek, J.; Gogala, M.; Schorr, S.; Linxweiler, J.; Lang, S.; Becker, T.; Beckmann, R.; Zimmermann, R.; Forster, F. Structure of the mammalian oligosaccharyl-transferase complex in the native ER protein translocon. *Nat. Commun.* **2014**, *5*, 3072. [CrossRef] [PubMed]

58. Liaci, A.M.; Steigenberger, B.; Telles de Souza, P.C.; Tamara, S.; Grollers-Mulderij, M.; Ogrissek, P.; Marrink, S.J.; Scheltema, R.A.; Forster, F. Structure of the human signal peptidase complex reveals the determinants for signal peptide cleavage. *Mol. Cell* **2021**, *81*, 3934–3948. [CrossRef] [PubMed]

59. Auclair, S.M.; Bhanu, M.K.; Kendall, D.A. Signal peptidase I: Cleaving the way to mature proteins. *Protein Sci.* **2012**, *21*, 13–25. [CrossRef] [PubMed]

60. Wang, S.; Jomaa, A.; Jaskolowski, M.; Yang, C.I.; Ban, N.; Shan, S.O. The molecular mechanism of cotranslational membrane protein recognition and targeting by SecA. *Nat. Struct. Mol. Biol.* **2019**, *26*, 919–929. [CrossRef]

61. Hassdenteufel, S.; Sicking, M.; Schorr, S.; Aviram, N.; Fecher-Trost, C.; Schuldiner, M.; Jung, M.; Zimmermann, R.; Lang, S. hSnd2 protein represents an alternative targeting factor to the endoplasmic reticulum in human cells. *FEBS Lett.* **2017**, *591*, 3211–3224. [CrossRef] [PubMed]

62. Strzyz, P. The third route to the ER. *Nat. Rev. Mol. Cell Biol.* **2017**, *18*, 3. [CrossRef]
63. Ast, T.; Schuldiner, M. All roads lead to Rome (but some may be harder to travel): SRP-independent translocation into the endoplasmic reticulum. *Crit. Rev. Biochem. Mol. Biol.* **2013**, *48*, 273–288. [CrossRef]
64. Egea, P.F.; Stroud, R.M.; Walter, P. Targeting proteins to membranes: Structure of the signal recognition particle. *Curr. Opin. Struct. Biol.* **2005**, *15*, 213–220. [CrossRef] [PubMed]
65. Halic, M.; Beckmann, R. The signal recognition particle and its interactions during protein targeting. *Curr. Opin. Struct. Biol.* **2005**, *15*, 116–125. [CrossRef]
66. Hwang Fu, Y.H.; Chandrasekar, S.; Lee, J.H.; Shan, S.O. A molecular recognition feature mediates ribosome-induced SRP-receptor assembly during protein targeting. *J. Cell Biol.* **2019**, *218*, 3307–3319. [CrossRef]
67. Kellogg, M.K.; Miller, S.C.; Tikhonova, E.B.; Karamyshev, A.L. SRPassing Co-translational Targeting: The Role of the Signal Recognition Particle in Protein Targeting and mRNA Protection. *Int. J. Mol. Sci.* **2021**, *22*, 6284. [CrossRef]
68. Saraogi, I.; Shan, S.O. Molecular mechanism of co-translational protein targeting by the signal recognition particle. *Traffic* **2011**, *12*, 535–542. [CrossRef]
69. Pfeffer, S.; Burbaum, L.; Unverdorben, P.; Pech, M.; Chen, Y.; Zimmermann, R.; Beckmann, R.; Forster, F. Structure of the native Sec61 protein-conducting channel. *Nat. Commun.* **2015**, *6*, 8403. [CrossRef]
70. Pfeffer, S.; Brandt, F.; Hrabe, T.; Lang, S.; Eibauer, M.; Zimmermann, R.; Forster, F. Structure and 3D arrangement of endoplasmic reticulum membrane-associated ribosomes. *Structure* **2012**, *20*, 1508–1518. [CrossRef] [PubMed]
71. Conti, B.J.; Devaraneni, P.K.; Yang, Z.; David, L.L.; Skach, W.R. Cotranslational stabilization of Sec62/63 within the ER Sec61 translocon is controlled by distinct substrate-driven translocation events. *Mol. Cell* **2015**, *58*, 269–283. [CrossRef] [PubMed]
72. Nguyen, D.; Stutz, R.; Schorr, S.; Lang, S.; Pfeffer, S.; Freeze, H.H.; Forster, F.; Helms, V.; Dudek, J.; Zimmermann, R. Proteomics reveals signal peptide features determining the client specificity in human TRAP-dependent ER protein import. *Nat. Commun.* **2018**, *9*, 3765. [CrossRef] [PubMed]
73. Fons, R.D.; Bogert, B.A.; Hegde, R.S. Substrate-specific function of the translocon-associated protein complex during translocation across the ER membrane. *J. Cell Biol.* **2003**, *160*, 529–539. [CrossRef]
74. Klein, M.C.; Lerner, M.; Nguyen, D.; Pfeffer, S.; Dudek, J.; Forster, F.; Helms, V.; Lang, S.; Zimmermann, R. TRAM1 protein may support ER protein import by modulating the phospholipid bilayer near the lateral gate of the Sec61-channel. *Channels* **2020**, *14*, 28–44. [CrossRef] [PubMed]
75. Tamborero, S.; Vilar, M.; Martinez-Gil, L.; Johnson, A.E.; Mingarro, I. Membrane insertion and topology of the translocating chain-associating membrane protein (TRAM). *J. Mol. Biol.* **2011**, *406*, 571–582. [CrossRef]
76. Russo, A. Understanding the mammalian TRAP complex function(s). *Open Biol.* **2020**, *10*, 190244. [CrossRef]
77. Klein, W.; Rutz, C.; Eckhard, J.; Provinciael, B.; Specker, E.; Neuenschwander, M.; Kleinau, G.; Scheerer, P.; von Kries, J.P.; Nazare, M.; et al. Use of a sequential high throughput screening assay to identify novel inhibitors of the eukaryotic SRP-Sec61 targeting/translocation pathway. *PLoS ONE* **2018**, *13*, e0208641. [CrossRef]
78. Hegde, R.S.; Voigt, S.; Rapoport, T.A.; Lingappa, V.R. TRAM Regulates the Exposure of Nascent Secretory Proteins to the Cytosol during Translocation into the Endoplasmic Reticulum. *Cell* **1998**, *92*, 621–631. [CrossRef]
79. Voigt, S.; Jungnickel, B.; Hartmann, E.; Rapoport, T.A. Signal sequence-dependent function of the TRAM protein during early phases of protein transport across the endoplasmic reticulum membrane. *J. Cell Biol.* **1996**, *134*, 25–35. [CrossRef]
80. Alder, N.N.; Shen, Y.; Brodsky, J.L.; Hendershot, L.M.; Johnson, A.E. The molecular mechanisms underlying BiP-mediated gating of the Sec61 translocon of the endoplasmic reticulum. *J. Cell Biol.* **2005**, *168*, 389–399. [CrossRef]
81. Alfaro-Valdés, H.; Burgos-Bravo, F.; Casanova-Morales, N.; Quiroga-Roger, D.; Wilson, C. *Mechanical Properties of Chaperone BiP, the Master Regulator of the Endoplasmic Reticulum*; IntechOpen: London, UK, 2018. [CrossRef]
82. Nguyen, T.H.; Law, D.T.; Williams, D.B. Binding protein BiP is required for translocation of secretory proteins into the endoplasmic reticulum in Saccharomyces cerevisiae. *Proc. Natl. Acad. Sci. USA* **1991**, *88*, 1565–1569. [CrossRef] [PubMed]
83. Matlack, K.E.; Misselwitz, B.; Plath, K.; Rapoport, T.A. BiP acts as a molecular ratchet during posttranslational transport of prepro-alpha factor across the ER membrane. *Cell* **1999**, *97*, 553–564. [CrossRef]
84. Araki, K.; Nagata, K. Protein folding and quality control in the ER. *Cold Spring Harb. Perspect. Biol.* **2011**, *3*, a007526. [CrossRef]
85. Comyn, S.A.; Chan, G.T.; Mayor, T. False start: Cotranslational protein ubiquitination and cytosolic protein quality control. *J. Proteom.* **2014**, *100*, 92–101. [CrossRef] [PubMed]
86. Ellgaard, L.; Helenius, A. Quality control in the endoplasmic reticulum. *Nat. Rev. Mol. Cell Biol.* **2003**, *4*, 181–191. [CrossRef] [PubMed]
87. Hwang, J.; Qi, L. Quality Control in the Endoplasmic Reticulum: Crosstalk between ERAD and UPR pathways. *Trends Biochem. Sci.* **2018**, *43*, 593–605. [CrossRef] [PubMed]
88. Kannan, K.; Vazquez-Laslop, N.; Mankin, A.S. Selective protein synthesis by ribosomes with a drug-obstructed exit tunnel. *Cell* **2012**, *151*, 508–520. [CrossRef] [PubMed]
89. Verchot, J. The ER quality control and ER associated degradation machineries are vital for viral pathogenesis. *Front. Plant Sci.* **2014**, *5*, 66. [CrossRef]
90. Sicking, M.; Lang, S.; Bochen, F.; Roos, A.; Drenth, J.P.H.; Zakaria, M.; Zimmermann, R.; Linxweiler, M. Complexity and Specificity of Sec61-Channelopathies: Human Diseases Affecting Gating of the Sec61 Complex. *Cells* **2021**, *10*, 1036. [CrossRef]

91. Lu, Z.; Zhou, L.; Killela, P.; Rasheed, A.B.; Di, C.; Poe, W.E.; McLendon, R.E.; Bigner, D.D.; Nicchitta, C.; Yan, H. Glioblastoma proto-oncogene SEC61gamma is required for tumor cell survival and response to endoplasmic reticulum stress. *Cancer Res.* **2009**, *69*, 9105–9111. [CrossRef]

92. Jung, V.; Kindich, R.; Kamradt, J.; Jung, M.; Müller, M.; Schulz, W.A.; Engers, R.; Unteregger, G.; Stöckle, M.; Zimmermann, R.; et al. Genomic and expression analysis of the 3q25-q26 amplification unit reveals TLOC1/SEC62 as a probable target gene in prostate cancer. *Mol. Cancer Res.* **2006**, *4*, 169–176. [CrossRef]

93. Greiner, M.; Kreutzer, B.; Jung, V.; Grobholz, R.; Hasenfus, A.; Stöhr, R.F.; Tornillo, L.; Dudek, J.; Stöckle, M.; Unteregger, G.; et al. Silencing of the SEC62 gene inhibits migratory and invasive potential of various tumor cells. *Int. J. Cancer* **2011**, *128*, 2284–2295. [CrossRef]

94. Linxweiler, M.; Linxweiler, J.; Barth, M.; Benedix, J.; Jung, V.; Kim, Y.J.; Bohle, R.M.; Zimmermann, R.; Greiner, M. Sec62 bridges the gap from 3q amplification to molecular cell biology in non-small cell lung cancer. *Am. J. Pathol.* **2012**, *180*, 473–483. [CrossRef] [PubMed]

95. Muller, C.S.L.; Pfohler, C.; Wahl, M.; Bochen, F.; Korner, S.; Kuhn, J.P.; Bozzato, A.; Schick, B.; Linxweiler, M. Expression of SEC62 Oncogene in Benign, Malignant and Borderline Melanocytic Tumors-Unmasking the Wolf in Sheep's Clothing? *Cancers* **2021**, *13*, 1645. [CrossRef] [PubMed]

96. Meng, H.; Jiang, X.; Wang, J.; Sang, Z.; Guo, L.; Yin, G.; Wang, Y. SEC61G is upregulated and required for tumor progression in human kidney cancer. *Mol. Med. Rep.* **2021**, *23*, 427. [CrossRef] [PubMed]

97. Witham, C.M.; Paxman, A.L.; Baklous, L.; Steuart, R.F.L.; Schulz, B.L.; Mousley, C.J. Cancer associated mutations in Sec61γ alter the permeability of the ER translocase. *PLoS Genet.* **2021**, *17*, e1009780. [CrossRef] [PubMed]

98. Ravindran, M.S.; Bagchi, P.; Cunningham, C.N.; Tsai, B. Opportunistic intruders: How viruses orchestrate ER functions to infect cells. *Nat. Rev. Microbiol.* **2016**, *14*, 407–420. [CrossRef]

99. Vermeire, K.; Schols, D. Specific CD4 down-modulating compounds with potent anti-HIV activity. *J. Leukoc. Biol.* **2003**, *74*, 667–675. [CrossRef] [PubMed]

100. Heaton, N.S.; Moshkina, N.; Fenouil, R.; Gardner, T.J.; Aguirre, S.; Shah, P.S.; Zhao, N.; Manganaro, L.; Hultquist, J.F.; Noel, J.; et al. Targeting Viral Proteostasis Limits Influenza Virus, HIV, and Dengue Virus Infection. *Immunity* **2016**, *44*, 46–58. [CrossRef]

101. Gillece, P.; Pilon, M.; Römisch, K. The protein translocation channel mediates glycopeptide export across the endoplasmic reticulum membrane. *Proc. Natl. Acad. Sci. USA* **2000**, *97*, 4609–4614. [CrossRef]

102. Vermeire, K.; Bell, T.W.; Choi, H.-J.; Jin, Q.; Samala, M.F.; Sodoma, A.; De Clercq, E.; Schols, D. The Anti-HIV Potency of Cyclotriazadisulfonamide Analogs Is Directly Correlated with Their Ability to Down-Modulate the CD4 Receptor. *Mol. Pharmacol.* **2003**, *63*, 203. [CrossRef] [PubMed]

103. Vermeire, K.; Zhang, Y.; Princen, K.; Hatse, S.; Samala, M.F.; Dey, K.; Choi, H.J.; Ahn, Y.; Sodoma, A.; Snoeck, R.; et al. CADA inhibits human immunodeficiency virus and human herpesvirus 7 replication by down-modulation of the cellular CD4 receptor. *Virology* **2002**, *302*, 342–353. [CrossRef]

104. Foster, C.A.; Dreyfuss, M.; Mandak, B.; Meingassner, J.G.; Naegeli, H.U.; Nussbaumer, A.; Oberer, L.; Scheel, G.; Swoboda, E.-M. Pharmacological Modulation of Endothelial Cell-associated Adhesion Molecule Expression: Implications for Future Treatment of Dermatological Diseases. *J. Dermatol.* **1994**, *21*, 847–854. [CrossRef] [PubMed]

105. Hommel, U.; Weber, H.P.; Oberer, L.; Naegeli, H.U.; Oberhauser, B.; Foster, C.A. The 3D-structure of a natural inhibitor of cell adhesion molecule expression. *FEBS Lett.* **1996**, *379*, 69–73. [CrossRef]

106. Boger, D.L.; Chen, Y.; Foster, C.A. Synthesis and evaluation of aza HUN-7293. *Bioorg. Med. Chem. Lett.* **2000**, *10*, 1741–1744. [CrossRef]

107. Chen, Y.; Bilban, M.; Foster, C.A.; Boger, D.L. Solution-phase parallel synthesis of a pharmacophore library of HUN-7293 analogues: A general chemical mutagenesis approach to defining structure-function properties of naturally occurring cyclic (depsi)peptides. *J. Am. Chem. Soc.* **2002**, *124*, 5431–5440. [CrossRef]

108. Harant, H.; Lettner, N.; Hofer, L.; Oberhauser, B.; de Vries, J.E.; Lindley, I.J. The translocation inhibitor CAM741 interferes with vascular cell adhesion molecule 1 signal peptide insertion at the translocon. *J. Biol. Chem.* **2006**, *281*, 30492–30502. [CrossRef] [PubMed]

109. Garrison, J.L.; Kunkel, E.J.; Hegde, R.S.; Taunton, J. A substrate-specific inhibitor of protein translocation into the endoplasmic reticulum. *Nature* **2005**, *436*, 285–289. [CrossRef]

110. Klein, W.; Westendorf, C.; Schmidt, A.; Conill-Cortes, M.; Rutz, C.; Blohs, M.; Beyermann, M.; Protze, J.; Krause, G.; Krause, E.; et al. Defining a conformational consensus motif in cotransin-sensitive signal sequences: A proteomic and site-directed mutagenesis study. *PLoS ONE* **2015**, *10*, e0120886. [CrossRef]

111. Schreiner, E.P.; Kern, M.; Steck, A.; Foster, C.A. Synthesis of ether analogues derived from HUN-7293 and evaluation as inhibitors of VCAM-1 expression. *Bioorg. Med. Chem. Lett.* **2004**, *14*, 5003–5006. [CrossRef]

112. Besemer, J.; Harant, H.; Wang, S.; Oberhauser, B.; Marquardt, K.; Foster, C.A.; Schreiner, E.P.; de Vries, J.E.; Dascher-Nadel, C.; Lindley, I.J. Selective inhibition of cotranslational translocation of vascular cell adhesion molecule 1. *Nature* **2005**, *436*, 290–293. [CrossRef]

113. Harant, H.; Wolff, B.; Schreiner, E.P.; Oberhauser, B.; Hofer, L.; Lettner, N.; Maier, S.; de Vries, J.E.; Lindley, I.J. Inhibition of vascular endothelial growth factor cotranslational translocation by the cyclopeptolide CAM741. *Mol. Pharm.* **2007**, *71*, 1657–1665. [CrossRef] [PubMed]

114. Westendorf, C.; Schmidt, A.; Coin, I.; Furkert, J.; Ridelis, I.; Zampatis, D.; Rutz, C.; Wiesner, B.; Rosenthal, W.; Beyermann, M.; et al. Inhibition of biosynthesis of human endothelin B receptor by the cyclodepsipeptide cotransin. *J. Biol. Chem.* **2011**, *286*, 35588–35600. [CrossRef] [PubMed]
115. Maifeld, S.V.; MacKinnon, A.L.; Garrison, J.L.; Sharma, A.; Kunkel, E.J.; Hegde, R.S.; Taunton, J. Secretory protein profiling reveals TNF-alpha inactivation by selective and promiscuous Sec61 modulators. *Chem. Biol.* **2011**, *18*, 1082–1088. [CrossRef]
116. MacKinnon, A.L.; Paavilainen, V.O.; Sharma, A.; Hegde, R.S.; Taunton, J. An allosteric Sec61 inhibitor traps nascent transmembrane helices at the lateral gate. *eLife* **2014**, *3*, e01483. [CrossRef]
117. MacKinnon, A.L.; Garrison, J.L.; Hegde, R.S.; Taunton, J. Photo-leucine incorporation reveals the target of a cyclodepsipeptide inhibitor of cotranslational translocation. *J. Am. Chem. Soc.* **2007**, *129*, 14560–14561. [CrossRef] [PubMed]
118. Palladino, M.A.; Bahjat, F.R.; Theodorakis, E.A.; Moldawer, L.L. Anti-TNF-alpha therapies: The next generation. *Nat. Rev. Drug Discov.* **2003**, *2*, 736–746. [CrossRef] [PubMed]
119. Ruiz-Saenz, A.; Sandhu, M.; Carrasco, Y.; Maglathlin, R.L.; Taunton, J.; Moasser, M.M. Targeting HER3 by interfering with its Sec61-mediated cotranslational insertion into the endoplasmic reticulum. *Oncogene* **2015**, *34*, 5288–5294. [CrossRef]
120. Junne, T.; Wong, J.; Studer, C.; Aust, T.; Bauer, B.W.; Beibel, M.; Bhullar, B.; Bruccoleri, R.; Eichenberger, J.; Estoppey, D.; et al. Decatransin, a new natural product inhibiting protein translocation at the Sec61/SecYEG translocon. *J. Cell Sci.* **2015**, *128*, 1217–1229. [CrossRef]
121. Huang, K.C.; Chen, Z.; Jiang, Y.; Akare, S.; Kolber-Simonds, D.; Condon, K.; Agoulnik, S.; Tendyke, K.; Shen, Y.; Wu, K.M.; et al. Apratoxin A Shows Novel Pancreas-Targeting Activity through the Binding of Sec 61. *Mol. Cancer Ther.* **2016**, *15*, 1208–1216. [CrossRef] [PubMed]
122. Paatero, A.O.; Kellosalo, J.; Dunyak, B.M.; Almaliti, J.; Gestwicki, J.E.; Gerwick, W.H.; Taunton, J.; Paavilainen, V.O. Apratoxin Kills Cells by Direct Blockade of the Sec61 Protein Translocation Channel. *Cell Chem. Biol.* **2016**, *23*, 561–566. [CrossRef] [PubMed]
123. Hau, A.M.; Greenwood, J.A.; Lohr, C.V.; Serrill, J.D.; Proteau, P.J.; Ganley, I.G.; McPhail, K.L.; Ishmael, J.E. Coibamide A induces mTOR-independent autophagy and cell death in human glioblastoma cells. *PLoS ONE* **2013**, *8*, e65250. [CrossRef]
124. Luesch, H.; Chanda, S.K.; Raya, R.M.; DeJesus, P.D.; Orth, A.P.; Walker, J.R.; Izpisúa Belmonte, J.C.; Schultz, P.G. A functional genomics approach to the mode of action of apratoxin A. *Nat. Chem. Biol.* **2006**, *2*, 158–167. [CrossRef]
125. Liu, Y.; Law, B.K.; Luesch, H. Apratoxin a reversibly inhibits the secretory pathway by preventing cotranslational translocation. *Mol. Pharm.* **2009**, *76*, 91–104. [CrossRef]
126. Wan, X.; Serrill, J.D.; Humphreys, I.R.; Tan, M.; McPhail, K.L.; Ganley, I.G.; Ishmael, J.E. ATG5 Promotes Death Signaling in Response to the Cyclic Depsipeptides Coibamide A and Apratoxin A. *Mar. Drugs* **2018**, *16*, 77. [CrossRef]
127. Medina, R.A.; Goeger, D.E.; Hills, P.; Mooberry, S.L.; Huang, N.; Romero, L.I.; Ortega-Barria, E.; Gerwick, W.H.; McPhail, K.L. Coibamide A, a potent antiproliferative cyclic depsipeptide from the Panamanian marine cyanobacterium *Leptolyngbya* sp. *J. Am. Chem. Soc.* **2008**, *130*, 6324–6325. [CrossRef]
128. Serrill, J.D.; Wan, X.; Hau, A.M.; Jang, H.S.; Coleman, D.J.; Indra, A.K.; Alani, A.W.; McPhail, K.L.; Ishmael, J.E. Coibamide A, a natural lariat depsipeptide, inhibits VEGFA/VEGFR2 expression and suppresses tumor growth in glioblastoma xenografts. *Investig. New Drugs* **2016**, *34*, 24–40. [CrossRef]
129. Molinski, T.F.; Dalisay, D.S.; Lievens, S.L.; Saludes, J.P. Drug development from marine natural products. *Nat. Rev. Drug Discov.* **2009**, *8*, 69–85. [CrossRef]
130. Gerwick, W.H.; Tan, L.T.; Sitachitta, N. Nitrogen-containing metabolites from marine cyanobacteria. *Alkaloids. Chem. Biol.* **2001**, *57*, 75–184. [CrossRef] [PubMed]
131. Tan, L.T. Bioactive natural products from marine cyanobacteria for drug discovery. *Phytochemistry* **2007**, *68*, 954–979. [CrossRef] [PubMed]
132. Tripathi, A.; Puddick, J.; Prinsep, M.R.; Rottmann, M.; Tan, L.T. Lagunamides A and B: Cytotoxic and antimalarial cyclodepsipeptides from the marine cyanobacterium Lyngbya majuscula. *J. Nat. Prod.* **2010**, *73*, 1810–1814. [CrossRef]
133. Luesch, H.; Yoshida, W.Y.; Moore, R.E.; Paul, V.J.; Corbett, T.H. Total structure determination of apratoxin A, a potent novel cytotoxin from the marine cyanobacterium Lyngbya majuscula. *J. Am. Chem. Soc.* **2001**, *123*, 5418–5423. [CrossRef] [PubMed]
134. Naidoo, D.; Roy, A.; Kar, P.; Mutanda, T.; Anandraj, A. Cyanobacterial metabolites as promising drug leads against the M(pro) and PL(pro) of SARS-CoV-2: An in silico analysis. *J. Biomol. Struct. Dyn.* **2021**, *39*, 6218–6230. [CrossRef] [PubMed]
135. Tranter, D.; Paatero, A.O.; Kawaguchi, S.; Kazemi, S.; Serrill, J.D.; Kellosalo, J.; Vogel, W.K.; Richter, U.; Mattos, D.R.; Wan, X.; et al. Coibamide A Targets Sec61 to Prevent Biogenesis of Secretory and Membrane Proteins. *ACS Chem. Biol.* **2020**, *15*, 2125–2136. [CrossRef]
136. Shi, W.; Lu, D.; Wu, C.; Li, M.; Ding, Z.; Li, Y.; Chen, B.; Lin, X.; Su, W.; Shao, X.; et al. Coibamide A kills cancer cells through inhibiting autophagy. *Biochem. Biophys. Res. Commun.* **2021**, *547*, 52–58. [CrossRef] [PubMed]
137. Hall, B.; Simmonds, R. Pleiotropic molecular effects of the Mycobacterium ulcerans virulence factor mycolactone underlying the cell death and immunosuppression seen in Buruli ulcer. *Biochem. Soc. Trans.* **2014**, *42*, 177–183. [CrossRef] [PubMed]
138. Morel, J.D.; Paatero, A.O.; Wei, J.; Yewdell, J.W.; Guenin-Mace, L.; Van Haver, D.; Impens, F.; Pietrosemoli, N.; Paavilainen, V.O.; Demangel, C. Proteomics Reveals Scope of Mycolactone-mediated Sec61 Blockade and Distinctive Stress Signature. *Mol. Cell. Proteom. MCP* **2018**, *17*, 1750–1765. [CrossRef]
139. McKenna, M.; Simmonds, R.E.; High, S. Mechanistic insights into the inhibition of Sec61-dependent co- and post-translational translocation by mycolactone. *J. Cell Sci.* **2016**, *129*, 1404–1415. [CrossRef] [PubMed]

140. Demangel, C.; High, S. Sec61 blockade by mycolactone: A central mechanism in Buruli ulcer disease. *Biol. Cell* **2018**, *110*, 237–248. [CrossRef]

141. Mve-Obiang, A.; Lee, R.E.; Umstot, E.S.; Trott, K.A.; Grammer, T.C.; Parker, J.M.; Ranger, B.S.; Grainger, R.; Mahrous, E.A.; Small, P.L.C. A newly discovered mycobacterial pathogen isolated from laboratory colonies of Xenopus species with lethal infections produces a novel form of mycolactone, the Mycobacterium ulcerans macrolide toxin. *Infect. Immun.* **2005**, *73*, 3307–3312. [CrossRef] [PubMed]

142. Guenin-Macé, L.; Ruf, M.-T.; Pluschke, G.; Demangel, C. Mycolactone: More than Just a Cytotoxin. In *Buruli Ulcer: Mycobacterium Ulcerans Disease*; Pluschke, G., Röltgen, K., Eds.; Springer International Publishing: Cham, Switzerland, 2019; pp. 117–134. [CrossRef]

143. George, K.M.; Chatterjee, D.; Gunawardana, G.; Welty, D.; Hayman, J.; Lee, R.; Small, P.L. Mycolactone: A polyketide toxin from Mycobacterium ulcerans required for virulence. *Science* **1999**, *283*, 854–857. [CrossRef]

144. Guenin-Macé, L.; Baron, L.; Chany, A.C.; Tresse, C.; Saint-Auret, S.; Jönsson, F.; Le Chevalier, F.; Bruhns, P.; Bismuth, G.; Hidalgo-Lucas, S.; et al. Shaping mycolactone for therapeutic use against inflammatory disorders. *Sci. Transl. Med.* **2015**, *7*, 289ra285. [CrossRef] [PubMed]

145. Baron, L.; Paatero, A.O.; Morel, J.D.; Impens, F.; Guenin-Mace, L.; Saint-Auret, S.; Blanchard, N.; Dillmann, R.; Niang, F.; Pellegrini, S.; et al. Mycolactone subverts immunity by selectively blocking the Sec61 translocon. *J. Exp. Med.* **2016**, *213*, 2885–2896. [CrossRef] [PubMed]

146. Ogbechi, J.; Hall, B.S.; Sbarrato, T.; Taunton, J.; Willis, A.E.; Wek, R.C.; Simmonds, R.E. Inhibition of Sec61-dependent translocation by mycolactone uncouples the integrated stress response from ER stress, driving cytotoxicity via translational activation of ATF4. *Cell Death Dis.* **2018**, *9*, 397. [CrossRef]

147. McKenna, M.; Simmonds, R.E.; High, S. Mycolactone reveals the substrate-driven complexity of Sec61-dependent transmembrane protein biogenesis. *J. Cell Sci.* **2017**, *130*, 1307–1320. [CrossRef] [PubMed]

148. Gerard, S.F.; Hall, B.S.; Zaki, A.M.; Corfield, K.A.; Mayerhofer, P.U.; Costa, C.; Whelligan, D.K.; Biggin, P.C.; Simmonds, R.E.; Higgins, M.K. Structure of the Inhibited State of the Sec Translocon. *Mol. Cell* **2020**, *79*, 406–415.e7. [CrossRef] [PubMed]

149. Zong, G.; Barber, E.; Aljewari, H.; Zhou, J.; Hu, Z.; Du, Y.; Shi, W.Q. Total Synthesis and Biological Evaluation of Ipomoeassin F and Its Unnatural 11R-Epimer. *J. Org. Chem.* **2015**, *80*, 9279–9291. [CrossRef]

150. Cao, S.; Norris, A.; Wisse, J.H.; Miller, J.S.; Evans, R.; Kingston, D.G. Ipomoeassin F, a new cytotoxic macrocyclic glycoresin from the leaves of Ipomoea squamosa from the Suriname rainforest. *Nat. Prod. Res.* **2007**, *21*, 872–876. [CrossRef] [PubMed]

151. Zong, G.; Hu, Z.; O'Keefe, S.; Tranter, D.; Iannotti, M.J.; Baron, L.; Hall, B.; Corfield, K.; Paatero, A.O.; Henderson, M.J.; et al. Ipomoeassin F Binds Sec61alpha to Inhibit Protein Translocation. *J. Am. Chem. Soc.* **2019**, *141*, 8450–8461. [CrossRef] [PubMed]

152. O'Keefe, S.; Roboti, P.; Duah, K.B.; Zong, G.; Schneider, H.; Shi, W.Q.; High, S. Ipomoeassin-F inhibits the in vitro biogenesis of the SARS-CoV-2 spike protein and its host cell membrane receptor. *J. Cell Sci.* **2021**, *134*, jcs257758. [CrossRef]

153. Roboti, P.; O'Keefe, S.; Duah, K.B.; Shi, W.Q.; High, S. Ipomoeassin-F disrupts multiple aspects of secretory protein biogenesis. *Sci. Rep.* **2021**, *11*, 11562. [CrossRef]

154. Zong, G.; Hu, Z.; Duah, K.B.; Andrews, L.E.; Zhou, J.; O'Keefe, S.; Whisenhunt, L.; Shim, J.S.; Du, Y.; High, S.; et al. Ring Expansion Leads to a More Potent Analogue of Ipomoeassin F. *J. Org. Chem.* **2020**, *85*, 16226–16235. [CrossRef] [PubMed]

155. Kazemi, S.; Kawaguchi, S.; Badr, C.E.; Mattos, D.R.; Ruiz-Saenz, A.; Serrill, J.D.; Moasser, M.M.; Dolan, B.P.; Paavilainen, V.O.; Oishi, S.; et al. Targeting of HER/ErbB family proteins using broad spectrum Sec61 inhibitors coibamide A and apratoxin A. *Biochem. Pharm.* **2021**, *183*, 114317. [CrossRef]

156. Shoemaker, R.H. The NCI60 human tumour cell line anticancer drug screen. *Nat. Rev. Cancer* **2006**, *6*, 813–823. [CrossRef]

157. Pauwels, E.; Rutz, C.; Provinciael, B.; Stroobants, J.; Schols, D.; Hartmann, E.; Krause, E.; Stephanowitz, H.; Schulein, R.; Vermeire, K. A Proteomic Study on the Membrane Protein Fraction of T Cells Confirms High Substrate Selectivity for the ER Translocation Inhibitor Cyclotriazadisulfonamide. *Mol. Cell. Proteom. MCP* **2021**, *20*, 100144. [CrossRef]

158. Claeys, E.; Pauwels, E.; Humblet-Baron, S.; Provinciael, B.; Schols, D.; Waer, M.; Sprangers, B.; Vermeire, K. Small Molecule Cyclotriazadisulfonamide Abrogates the Upregulation of the Human Receptors CD4 and 4-1BB and Suppresses In Vitro Activation and Proliferation of T Lymphocytes. *Front. Immunol.* **2021**, *12*, 1340. [CrossRef] [PubMed]

159. Van Puyenbroeck, V.; Claeys, E.; Schols, D.; Bell, T.W.; Vermeire, K. A Proteomic Survey Indicates Sortilin as a Secondary Substrate of the ER Translocation Inhibitor Cyclotriazadisulfonamide (CADA). *Mol. Cell. Proteom. MCP* **2017**, *16*, 157–167. [CrossRef]

160. Cross, B.C.; McKibbin, C.; Callan, A.C.; Roboti, P.; Piacenti, M.; Rabu, C.; Wilson, C.M.; Whitehead, R.; Flitsch, S.L.; Pool, M.R.; et al. Eeyarestatin I inhibits Sec61-mediated protein translocation at the endoplasmic reticulum. *J. Cell Sci.* **2009**, *122*, 4393–4400. [CrossRef]

161. Lowe, E.; Fan, R.; Jiang, J.; Johnson, H.; Kirk, C.; McMinn, D.; Millare, B.; Muchamuel, T.; Qian, Y.; Tuch, B.; et al. Preclinical evaluation of KZR-261, a novel small molecule inhibitor of Sec61. *J. Clin. Oncol.* **2020**, *38*, 3582. [CrossRef]

162. Vermeire, K.; Princen, K.; Hatse, S.; De Clercq, E.; Dey, K.; Bell, T.W.; Schols, D. CADA, a novel CD4-targeted HIV inhibitor, is synergistic with various anti-HIV drugs in vitro. *AIDS* **2004**, *18*, 2115–2125. [CrossRef]

163. Berger, K.; Pauwels, E.; Parkinson, G.; Landberg, G.; Le, T.; Demillo, V.G.; Lumangtad, L.A.; Jones, D.E.; Islam, M.A.; Olsen, R.; et al. Reduction of Progranulin-Induced Breast Cancer Stem Cell Propagation by Sortilin-Targeting Cyclotriazadisulfonamide (CADA) Compounds. *J. Med. Chem.* **2021**, *64*, 12865–12876. [CrossRef] [PubMed]

164. Bell, T.W.; Anugu, S.; Bailey, P.; Catalano, V.J.; Dey, K.; Drew, M.G.; Duffy, N.H.; Jin, Q.; Samala, M.F.; Sodoma, A.; et al. Synthesis and structure-activity relationship studies of CD4 down-modulating cyclotriazadisulfonamide (CADA) analogues. *J. Med. Chem.* **2006**, *49*, 1291–1312. [CrossRef] [PubMed]

165. Chawla, R.; Van Puyenbroeck, V.; Pflug, N.C.; Sama, A.; Ali, R.; Schols, D.; Vermeire, K.; Bell, T.W. Tuning Side Arm Electronics in Unsymmetrical Cyclotriazadisulfonamide (CADA) Endoplasmic Reticulum (ER) Translocation Inhibitors to Improve their Human Cluster of Differentiation 4 (CD4) Receptor Down-Modulating Potencies. *J. Med. Chem.* **2016**, *59*, 2633–2647. [CrossRef] [PubMed]

166. Vermeire, K.; Lisco, A.; Grivel, J.C.; Scarbrough, E.; Dey, K.; Duffy, N.; Margolis, L.; Bell, T.W.; Schols, D. Design and cellular kinetics of dansyl-labeled CADA derivatives with anti-HIV and CD4 receptor down-modulating activity. *Biochem. Pharm.* **2007**, *74*, 566–578. [CrossRef] [PubMed]

167. Lumangtad, L.A.; Claeys, E.; Hamal, S.; Intasiri, A.; Basrai, C.; Yen-Pon, E.; Beenfeldt, D.; Vermeire, K.; Bell, T.W. Syntheses and anti-HIV and human cluster of differentiation 4 (CD4) down-modulating potencies of pyridine-fused cyclotriazadisulfonamide (CADA) compounds. *Bioorg. Med. Chem.* **2020**, *28*, 115816. [CrossRef] [PubMed]

168. Demillo, V.G.; Goulinet-Mateo, F.; Kim, J.; Schols, D.; Vermeire, K.; Bell, T.W. Unsymmetrical cyclotriazadisulfonamide (CADA) compounds as human CD4 receptor down-modulating agents. *J. Med. Chem.* **2011**, *54*, 5712–5721. [CrossRef] [PubMed]

169. Bell, T.W.; Demillo, V.G.; Schols, D.; Vermeire, K. Improving potencies and properties of CD4 down-modulating CADA analogs. *Expert Opin. Drug Discov.* **2012**, *7*, 39–48. [CrossRef]

170. Vermeire, K.; Bell, T.W.; Van Puyenbroeck, V.; Giraut, A.; Noppen, S.; Liekens, S.; Schols, D.; Hartmann, E.; Kalies, K.U.; Marsh, M. Signal peptide-binding drug as a selective inhibitor of co-translational protein translocation. *PLoS Biol.* **2014**, *12*, e1002011. [CrossRef]

171. Van Puyenbroeck, V.; Pauwels, E.; Provinciael, B.; Bell, T.W.; Schols, D.; Kalies, K.U.; Hartmann, E.; Vermeire, K. Preprotein signature for full susceptibility to the co-translational translocation inhibitor cyclotriazadisulfonamide. *Traffic* **2020**, *21*, 250–264. [CrossRef]

172. Fiebiger, E.; Hirsch, C.; Vyas, J.M.; Gordon, E.; Ploegh, H.L.; Tortorella, D. Dissection of the dislocation pathway for type I membrane proteins with a new small molecule inhibitor, eeyarestatin. *Mol. Biol. Cell* **2004**, *15*, 1635–1646. [CrossRef] [PubMed]

173. Wang, Q.; Li, L.; Ye, Y. Inhibition of p97-dependent protein degradation by Eeyarestatin I. *J. Biol. Chem.* **2008**, *283*, 7445–7454. [CrossRef] [PubMed]

174. Smith, D.M. Could a Common Mechanism of Protein Degradation Impairment Underlie Many Neurodegenerative Diseases? *J. Exp. Neurosci.* **2018**, *12*, 1179069518794675. [CrossRef] [PubMed]

175. Wang, Q.; Shinkre, B.A.; Lee, J.G.; Weniger, M.A.; Liu, Y.; Chen, W.; Wiestner, A.; Trenkle, W.C.; Ye, Y. The ERAD inhibitor Eeyarestatin I is a bifunctional compound with a membrane-binding domain and a p97/VCP inhibitory group. *PLoS ONE* **2010**, *5*, e15479. [CrossRef] [PubMed]

176. Wang, Q.; Mora-Jensen, H.; Weniger, M.A.; Perez-Galan, P.; Wolford, C.; Hai, T.; Ron, D.; Chen, W.; Trenkle, W.; Wiestner, A.; et al. ERAD inhibitors integrate ER stress with an epigenetic mechanism to activate BH3-only protein NOXA in cancer cells. *Proc. Natl. Acad. Sci. USA* **2009**, *106*, 2200–2205. [CrossRef] [PubMed]

177. Lang, S.; Erdmann, F.; Jung, M.; Wagner, R.; Cavalie, A.; Zimmermann, R. Sec61 complexes form ubiquitous ER Ca2+ leak channels. *Channels* **2011**, *5*, 228–235. [CrossRef] [PubMed]

178. Gamayun, I.; O'Keefe, S.; Pick, T.; Klein, M.C.; Nguyen, D.; McKibbin, C.; Piacenti, M.; Williams, H.M.; Flitsch, S.L.; Whitehead, R.C.; et al. Eeyarestatin Compounds Selectively Enhance Sec61-Mediated Ca(2+) Leakage from the Endoplasmic Reticulum. *Cell Chem. Biol.* **2019**, *26*, 571–583.e6. [CrossRef]

179. Ding, R.; Zhang, T.; Xie, J.; Williams, J.; Ye, Y.; Chen, L. Eeyarestatin I derivatives with improved aqueous solubility. *Bioorg. Med. Chem. Lett.* **2016**, *26*, 5177–5181. [CrossRef]

180. Du, R.; Sullivan, D.K.; Azizian, N.G.; Liu, Y.; Li, Y. Inhibition of ERAD synergizes with FTS to eradicate pancreatic cancer cells. *BMC Cancer* **2021**, *21*, 237. [CrossRef] [PubMed]

181. Aletrari, M.O.; McKibbin, C.; Williams, H.; Pawar, V.; Pietroni, P.; Lord, J.M.; Flitsch, S.L.; Whitehead, R.; Swanton, E.; High, S.; et al. Eeyarestatin 1 interferes with both retrograde and anterograde intracellular trafficking pathways. *PLoS ONE* **2011**, *6*, e22713. [CrossRef]

182. McKibbin, C.; Mares, A.; Piacenti, M.; Williams, H.; Roboti, P.; Puumalainen, M.; Callan, A.C.; Lesiak-Mieczkowska, K.; Linder, S.; Harant, H.; et al. Inhibition of protein translocation at the endoplasmic reticulum promotes activation of the unfolded protein response. *Biochem. J.* **2012**, *442*, 639–648. [CrossRef]

183. Auner, H.W.; Moody, A.M.; Ward, T.H.; Kraus, M.; Milan, E.; May, P.; Chaidos, A.; Driessen, C.; Cenci, S.; Dazzi, F.; et al. Combined inhibition of p97 and the proteasome causes lethal disruption of the secretory apparatus in multiple myeloma cells. *PLoS ONE* **2013**, *8*, e74415. [CrossRef] [PubMed]

184. Steenhuis, M.; Koningstein, G.M.; Oswald, J.; Pick, T.; O'Keefe, S.; Koch, H.G.; Cavalie, A.; Whitehead, R.C.; Swanton, E.; High, S.; et al. Eeyarestatin 24 impairs SecYEG-dependent protein trafficking and inhibits growth of clinically relevant pathogens. *Mol. Microbiol.* **2021**, *115*, 28–40. [CrossRef] [PubMed]

International Journal of
Molecular Sciences

MDPI

Article

Lights, Camera, Interaction: Studying Protein–Protein Interactions of the ER Protein Translocase in Living Cells

Mark Sicking, Martin Jung and Sven Lang *

Department of Medical Biochemistry and Molecular Biology, Saarland University, 66421 Homburg, Germany;
mark.sicking@uni-saarland.de (M.S.); martin.jung@uks.eu (M.J.)
* Correspondence: sven.lang@uni-saarland.de

Abstract: Various landmark studies have revealed structures and functions of the Sec61/SecY complex in all domains of live demonstrating the conserved nature of this ancestral protein translocase. While the bacterial homolog of the Sec61 complex resides in the plasma membrane, the eukaryotic counterpart manages the transfer of precursor proteins into or across the membrane of the endoplasmic reticulum (ER). Sec61 complexes are accompanied by a set of dynamically recruited auxiliary proteins assisting the transport of certain precursor polypeptides. TRAP and Sec62/Sec63 are two auxiliary protein complexes in mammalian cells that have been characterized by structural and biochemical methods. Using these ER membrane protein complexes for our proof-of-concept study, we aimed to detect interactions of membrane proteins in living mammalian cells under physiological conditions. Bimolecular luminescence complementation and competition was used to demonstrate multiple protein–protein interactions of different topological layouts. In addition to the interaction of the soluble catalytic and regulatory subunits of the cytosolic protein kinase A, we detected interactions of ER membrane proteins that either belong to the same multimeric protein complex (intra-complex interactions: Sec61α–Sec61β, TRAPα–TRAPβ) or protein complexes in juxtaposition (inter-complex interactions: Sec61α–TRAPα, Sec61α–Sec63, and Sec61β–Sec63). In the process, we established further control elements like synthetic peptide complementation for expression profiling of fusion constructs and protease-mediated reporter degradation demonstrating the cytosolic localization of a reporter complementation. Ease of use and flexibility of the approach presented here will spur further research regarding the dynamics of protein–protein interactions in response to changing cellular conditions in living cells.

Keywords: bimolecular luminescence complementation; competition; split luciferase; membrane proteins; protein–protein interactions; Sec61 complex; Sec63; synthetic peptide complementation; TRAP complex; ER protein translocase

Citation: Sicking, M.; Jung, M.; Lang, S. Lights, Camera, Interaction: Studying Protein–Protein Interactions of the ER Protein Translocase in Living Cells. *Int. J. Mol. Sci.* **2021**, *22*, 10358. https://doi.org/10.3390/ijms221910358

Academic Editor: Toshiyuki Kaji

Received: 3 August 2021
Accepted: 22 September 2021
Published: 26 September 2021

Publisher's Note: MDPI stays neutral with regard to jurisdictional claims in published maps and institutional affiliations.

1. Introduction

As the fundamental unit of life, cells endow biological systems with tremendous powers and astonishing features. In the case of differentiated eukaryotic cells, these are often compartmentalized and different membrane-surrounded or membrane-less organelles help to shape cellular fitness, metabolism, and signaling. One of the organelles that supports both intracellular signaling, for example calcium (Ca^{2+}) signaling or the unfolded protein response, and secretion of proteohormones is the endoplasmic reticulum (ER) [1–4]. A major membrane protein at the crossroad of ER signaling and protein transport is the heterotrimeric Sec61 complex acting as the pore-forming component of the ER protein translocase [5–7].

Many different membrane-spanning (~5000) and soluble (~3000) proteins of the human proteome can be guided by the Sec61 complex to enter the secretory pathway [8]. To meet the demands for the transport of such topologically diverse precursor proteins crosslinking and native gel electrophoresis studies have demonstrated the dynamic association of the Sec61 complex with accessory factors to form the active holo-translocon [9–12].

For example, structural and proteomic data as well as pulling force studies verified the hetero-tetrameric translocon-associated protein (TRAP) complex as an important player supporting the Sec61 complex during co-translational transport of precursor proteins with an above-average glycine-plus-proline content in the signal peptide [13–15]. Similarly, the proteins Sec62/Sec63 were identified to support post-translational as well as substrate-specific co-translational opening of the Sec61 complex [9,16–21]. In analogy to enzyme-catalyzed reactions, auxiliary protein complexes like TRAP or Sec62/Sec63 act as allosteric effectors supporting the gating of the Sec61 complex [22].

As exemplified above for the Sec61 complex, protein translocation machineries of other organelles like the PIM complex of peroxisomes [23], the TIM/TOM complexes of mitochondria [24], or the TIC/TOC equivalents in chloroplasts [25] all represent large, heteromultimeric assemblies and their activity depends on protein–protein interactions (PPI) [26,27]. The necessity of protein multimerization for proper functionality is by no means restricted to protein translocation machineries and governs pivotal cellular processes like DNA replication and transcription, mRNA translation, transmembrane signaling mechanisms or enzymatic catalysis, and ATP production with estimates of up to 80% of proteins operating in complexes [28,29]. Typically, PPI control the assembly of protein complexes via non-covalent contacts between the side chains of amino acids from different polypeptide strands and can form large-scale networks, often referred to as the interactome [30,31]. Many methods have been developed to identify interactions of proteins and define the interactome of cells. With regard to the experimental setup, those methods can be classified as in vitro, in vivo, or in silico and include popular examples like coimmunoprecipitation, the yeast-two-hybrid system, and structure-based prediction via algorithms, respectively. However, method-specific strengths and weaknesses depending on the scientific question asked is a feature they all share [29,32–34]. One of the more challenging aspects when studying PPI is their dynamic nature that can trigger the transient or stable formation of homo- or heterooligomers whose association can be influenced by various cellular cues. A textbook example of a dynamic PPI of a heterooligomer regulated by cellular cues is the protein kinase A (PKA). PKA represents a soluble tetramer of two catalytic subunits harboring kinase function and two regulatory subunits that act as cAMP sensors. Separation of the catalytic from the regulatory subunits activates the kinase function and is influenced by intracellular (cAMP) and extracellular (hormones, e.g., glucagon) cues, which together with the corresponding hormone receptor and adenylate cyclase represent critical elements of a signal transduction pathway [35]. Bimolecular protein-fragment complementation assays are well suited to detect the spatiotemporal dynamics of binary PPI like that of PKA in intact living cells. These assays are based on proteinaceous split reporters with the complementary reporter fragments genetically fused to proteins of interest, whose interaction can be directly visualized [36].

Here we used bimolecular luminescence complementation (BiLC) as live cell assay to further complement the studies that so elegantly demonstrate the association of the Sec61 complex with the TRAP complex or Sec62/Sec63 either in silico, in vitro, or after reconstitution or vitrification. In the case of BiLC, the interaction of tagged protein pairs reassembles a functional, split luciferase and can be visualized upon addition of a luciferase substrate [37,38]. After verifying the dynamic PPI of the catalytic and regulatory subunits of PKA, our proof-of-concept shows "intra-complex" interactions between subunits of the Sec61 complex (Sec61α–Sec61β) as well as of the TRAP complex (TRAPα–TRAPβ). Furthermore, we demonstrate "inter-complex" interactions between membrane proteins of different complexes such as Sec61α–TRAPα, Sec61α–Sec63, and Sec61β–Sec63. Both types of PPI, intra- and inter-complex ones, can be perturbed by competitive over-expression of untagged variants. Alternatively, interactions of membrane or soluble protein pairs that rely on reassembly of the split luciferase components in the cytosol are abolished by the combination of plasma membrane permeabilization and trypsin-mediated reporter degradation.

2. Results

To reliably verify the PPI of ER membrane proteins in living cells, we established a microplate reader–based bimolecular protein-fragment complementation assay widely used for investigations of protein interactions [33]. The split reporter used here is an optimized variant of the catalytic subunit of the luciferase derived from deep sea luminous shrimp called NanoLuc [38]. The 19 kDa polypeptide consisting of a ten-stranded β-barrel topology is separated after the ninth β-barrel providing an 18 kDa and 1 kDa fragment called LgBiT and SmBiT, respectively. Although the two fragments can reconstruct a functional luciferase that provides a bright luminescence in the presence of its substrate furimazine, the engineered LgBiT and SmBiT show a low intrinsic affinity for each other. The association constant of LgBiT and SmBiT has a k_D of 190 µM, which is above the k_D of most physiological relevant PPI [37]. Thus, when LgBiT and SmBiT are added as fusion tags to proteins of interest, luminescence will occur if two prerequisites are met: (i) the presence of furimazine and (ii) direct interaction, i.e., close proximity of the proteins of interest (Figure 1A).

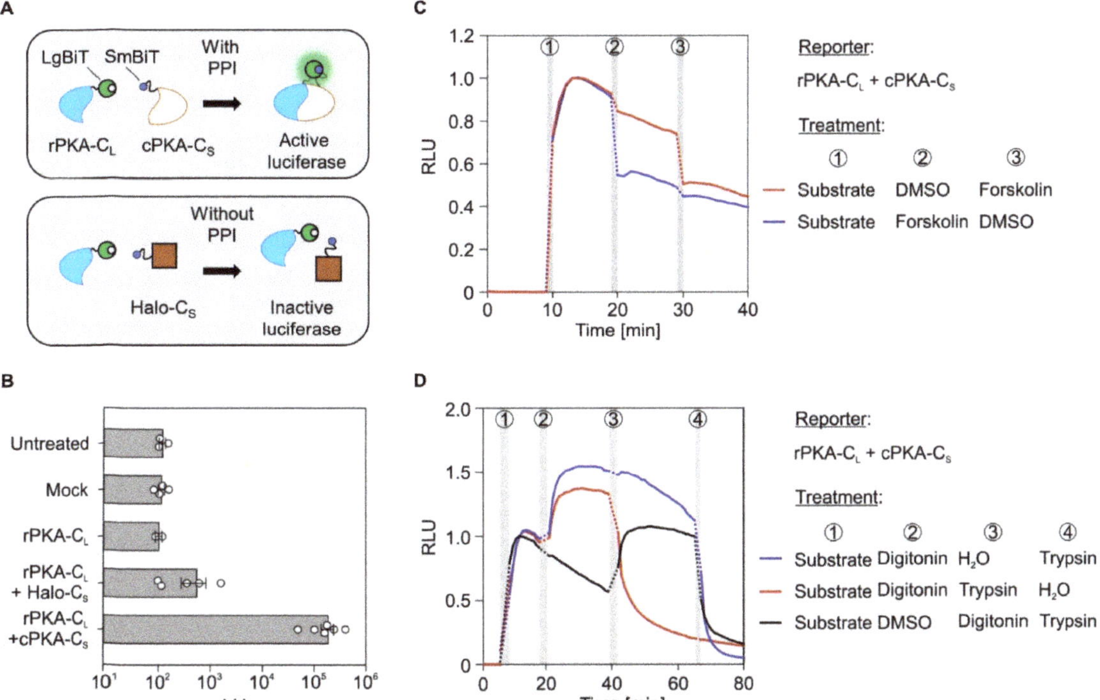

Figure 1. Establishing a bimolecular luminescence complementation assay based on protein kinase A and cell permeabilization. (**A**) Principle of the bimolecular luminescence complementation (BiLC) assay. The upper cell shows the expression of two interacting proteins, the regulatory and catalytic subunit of protein kinase A (rPKA and cPKA), C-terminally (**C**) tagged with the LgBiT (L) and SmBiT (S) fragment of the luciferase, respectively. Protein–protein interaction reassembles a functional luciferase (glowing green/blue sphere) that emits light upon addition of the substrate furimazine (not shown). In contrast, expression of non-interacting proteins (lower cell) prevents functional reassembly of the luciferase and luminescence. (**B**) In addition to untreated and mock treated cells, total luminescence (LU) was also measured from single or double transfected cells using the rPKA-C_L, cPKA-C_S, or Halo-C_S fusion constructs. For each condition and the corresponding biological replicates,

the signals recorded 4 min after addition of furimazine (this time point usually depicts the peak intensity after furimazine addition, cf. Figure 1C) were plotted. (**C**) Twenty-four hours post transfection with the rPKA-C$_L$ and cPKA-C$_S$ reporter pair, cells were subjected to the indicated treatment regimens (circled numbers). Nine minutes after starting the measurement, the addition of furimazine stimulated luminescence. After 19 min, cells were treated with forskolin (15 µM) to activate adenylate cyclases and cAMP production, causing disassembly of the PKA subunits. DMSO served as vehicle control. After 29 min, cells were treated in a reciprocal fashion; the ones that received DMSO first now received forskolin and vice versa. Measurements were normalized to the signal intensity recorded 4 min after furimazine application and plotted as relative luminescence units (RLU). (**D**) As in (**C**), but after substrate application, cells were treated with digitonin (0.002%) to permeabilize the plasma membrane and trypsin (50 µg/mL) to digest the reporter and other cytosolic proteins. DMSO and water treatments served as vehicle controls for digitonin and trypsin, respectively. Vertical gray bars in the line diagrams represent manual 1 min application periods without luminescence readings. The dotted lines are extrapolated based on the last and first data points before and after application. C$_L$, C-terminally located LgBiT tag; C$_S$, C-terminally located SmBiT tag.

2.1. Establishing the NanoBiT Assay Based on the cAMP-Dependent Protein Kinase A, Forskolin, and Semi-Permeabilization

The interacting regulatory and catalytic subunits of protein kinase A (rPKA, cPKA) were C-terminally tagged with the LgBiT ($_L$) and SmBiT ($_S$) fragments, respectively. Corresponding fusion constructs were called rPKA-C$_L$ and cPKA-C$_S$. The non-interacting HaloTag-based fusion construct Halo-C$_S$ served as negative control (Figure 1A). HeLa cells were seeded in a 96-well format and left untreated or were transfected for 24 h. Besides mock transfection, cells were transfected with the plasmid encoding for rPKA-C$_L$ alone or in combination with either Halo-C$_S$ or cPKA-C$_S$. Four minutes after addition of furimazine, the luminescence units (LU) of the five conditions were measured. As expected, strong luminescence was detected only for the bona fide interacting fusion proteins rPKA-C$_L$ and cPKA-C$_S$ providing a more than 600-fold brighter luminescence than the rPKA-C$_L$ and Halo-C$_S$ pair (Figure 1B). The reversibility of the rPKA and cPKA interaction was demonstrated using forskolin, a non-selective activator of most adenylate cyclase isoforms [39]. Conversion of ATP to cAMP by forskolin-activated adenylate cyclases caused the rapid disassembly of rPKA from cPKA. Accordingly, forskolin treatment 10 or 20 min after addition of the NanoLuc substrate furimazine induced a rapid decrease of luminescence (Figure 1C). As depicted by the vertical gray bars in Figure 1C the addition of furimazine and all other substances required their manual application and caused a short measurement gap that was kept constant using a time window of one minute. The same applies to other measurements shown later. We also tweaked the system for the use of larger, membrane-impermeable effectors including soluble enzymes such as trypsin. Cells expressing the rPKA-C$_L$ and cPKA-C$_S$ reporter pair as representation of a verified PPI were first subjected to permeabilization of the plasma membrane by digitonin and subsequently treated with trypsin for proteolytic cleavage of the reporter proteins to eliminate luminescence. As shown in the line graphs in Figure 1D, addition of the substrate caused a sharp increase in luminescence within 4 to 5 min (treatment 1). Somewhat surprisingly, the application of digitonin for permeabilization of the plasma membrane (treatment 2, red or blue line) further increased signal intensity. We attributed this effect to improved access of the bulky furimazine substrate to the re-assembled luciferase. The subsequent addition of the protease trypsin after 40 min (treatment 3, red line) or 70 min (treatment 4, blue or black line) eliminated luminescence almost entirely due to protease-mediated digestion of the fusion proteins (Figure 1D). Of note, trypsin treatment of intact cells prior to permeabilization did not cause a significant drop in luminescence due to reporter degradation or eventual detachment of adherent cells from the microplate surface (cf. Figures 5B and 6B, blue lines). Correspondingly, using a non-proteolytic enzyme like RNase A further corroborated the specific elimination of luminescence by trypsin-mediated reporter degradation (Figure S1).

This line of experiments showed the applicability and dynamics of the BiLC system for cytosolic interaction partners in living cells. Permeabilization of the plasma membrane by digitonin also granted access of otherwise membrane-impermeable agents including large biological effectors such as trypsin.

2.2. Expression Profiling via Synthetic Complementation Using Chemically Synthesized Low and High Affinity Oligopeptides

An often-ignored issue with protein-fragment complementation assays is the sufficient expression of fusion constructs and synthesis of the encoded fusion proteins, which is equally relevant for constructs encoding soluble and membrane proteins. Therefore, before applying BiLC to ER membrane proteins, we deconstructed the system and designed an assay for measuring expression of the LgBiT fusion constructs, when the LgBit is located in the cytosol (Figure 2A). Instead of expressing the 11 amino acids (aa) long SmBiT peptide as a fusion construct in parallel with a LgBiT fusion protein, we chemically synthesized the SmBiT peptide and dissolved it in DMSO (Figure S2A). During synthesis, the SmBiT peptide was N-terminally extended by a single cysteine for further downstream processes (see below). Like the 11 + 1 aa SmBiT peptide, we also synthesized a scrambled version thereof and an independent actin peptide (16 aa) as two negative controls that should not reconstitute a functional luciferase in presence of a genetically expressed LgBiT fusion protein. Using the semi-permeabilization protocol with digitonin (cf. Figure 1D), cells expressing the rPKA-C$_L$ construct were provided 100 μM of a negative control peptide (scrambled or actin) or the SmBiT. As expected, only upon the addition of the SmBiT peptide did a complementation with rPKA-C$_L$ occur (Figure 2B). Three points are worth emphasizing. One, taking into consideration the inefficient association between LgBiT and SmBiT (k_D = 190 μM) the use of a high peptide concentration like 100 μM was anticipated. Two, titrating the concentrations of the SmBiT peptide (1, 10, 100 μM) showed a clear dose-dependence of complementation and luminescence. Three, even at a concentration of 100 μM, the synthetic SmBiT provided a 10-fold lower signal intensity than the genetically expressed PPI pair rPKA-C$_L$ and cPKA-C$_S$ (Figure 2B, dark blue versus gray line).

During the initial mutational screens for low affinity versions of the SmBiT, Dixon and colleagues also reported variants with exceptionally high affinities for the LgBiT in the nanomolar range [37]. For one of the variants the glutamates in position 8 and 9 of the original SmBiT undecapeptide were exchanged by lysines. We also synthesized this peptide and based on its high affinity called it HaBiT (Figure S2A). As shown in Figure 2C, after semi-permeabilization, HaBiT was also suited for expression profiling of the rPKA-C$_L$ construct in a dose-dependent manner. Due to the high affinity, HaBiT interacted with rPKA-C$_L$ in a concentration range from 10 to 100 nM providing efficient luminescence (Figures 2C and S2C). At 100 nM the HaBiT signal exceeded that of the genetically encoded reporter pair rPKA-C$_L$ and cPKA-C$_S$. Combining the data of dose-dependence and tested concentrations for the synthetic SmBiT and HaBiT peptide, the latter showed an approximately 10,000-fold higher activity in the expression profiling assay (Figure 2D). Taken together, both peptides SmBiT and HaBiT can be used to verify and compare steady-state expression of generated fusion constructs.

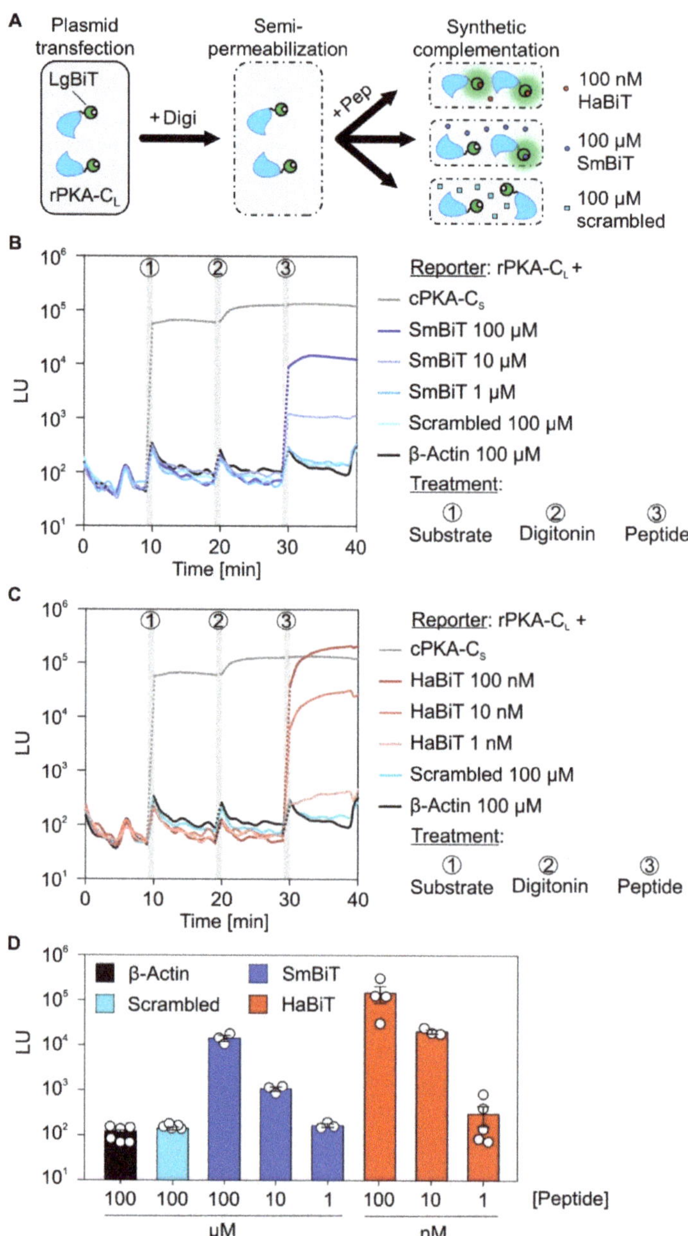

Figure 2. Expression profiling via synthetic complementation using a low (SmBiT) and high affinity (HaBiT) peptide. (**A**) Cartoon of the synthetic complementation approach. Twenty-four hours after transfection of cells with the LgBiT fusion construct (here rPKA-C$_L$), the plasma membrane was permeabilized using digitonin and synthetic complementation was achieved by adding one of the chemically synthesized peptides (Figure S2A). The scrambled (100 µM) peptide served as negative control. SmBiT (100 µM) and HaBiT (100 nM) peptides were used as low and high affinity positive

controls able to reconstitute the functional luciferase. (**B,C**) Twenty-four hours post transfection with rPKA-C_L, the measurements of luminescence units (LU) were started. At the indicated times, the luciferase substrate was added (treatment 1) and followed by 0.002% digitonin (treatment 2) and the addition of the peptide with the indicated final concentration (treatment 3). The scrambled and unrelated actin peptide were used as negative controls. Cells transfected with the genetically encoded rPKA-C_L plus cPKA-C_S reporter pair were used for comparison of signal intensity and treated with DMSO instead of peptide as last treatment. Vertical gray bars in the line diagrams represent manual 1 min application periods without luminescence readings and dotted lines are extrapolated based on the last and first data points before and after indicated treatments. Each trace is the average of at least three replicates. Note that traces for the reporter pair as well as the scrambled and actin peptide are the same in (**B**) and (**C**). (**D**) Summary of the recorded signals 4 min after peptide supplementation for each condition and repeat. C_L, C-terminally located LgBiT tag; Digi, digitonin; Pep, peptide.

2.3. Verifying Expression of LgBit Fusion Constructs of ER Protein Translocase Subunits via Synthetic Peptide Complementation

Before testing the PPI of ER membrane proteins in living cells, the synthetic complementation approach was used to verify synthesis of the LgBiT fusion proteins 24 h post transfection. In addition to the cytosolic positive control rPKA-C_L, five N- or C-terminally LgBiT-tagged ER membrane proteins were tested including Sec61α, Sec61β, TRAPα, and TRAPβ (Figure 3A). The latter two are classical type I membrane proteins with one N_{exo}-C_{cyto} transmembrane helix (TMH) and a cleavable signal peptide [40]. Sec61β is a type IV or tail-anchored membrane protein with a single TMH at the very C-terminus inserted in an N_{cyto}-C_{exo} orientation [41]. Sec61α has 10 TMH with its N- and C-terminus located in the cytosol [42]. All six of the LgBiT fusion proteins were tested with three different peptides (cf. Figure 2A). While the scrambled peptide served as negative control and for normalization, the SmBiT and HaBiT peptide served as positive controls that allowed reconstitution of a functional luciferase in case the LgBiT fusion construct was properly expressed. Taking the different affinities of SmBiT and HaBiT for the LgBiT into consideration, different peptide concentrations had to be used. Scrambled and SmBiT were added to a final concentration of 100 μM, whereas HaBiT was used at 100 nM (Figure 3B). Despite the three orders of magnitude lower concentration of HaBiT, it provided a 5–10-times stronger signal in all six tested synthetic complementations. The strongest signal was achieved in combination with the cytosolic rPKA-C_L. Compared to the scrambled peptide, four out of five LgBiT-tagged membrane protein fusions (Sec61α-N_L, Sec61β-C_L, TRAPα-C_L, TRAPβ-C_L) provided a more than 10-fold stronger luminescence when complemented with SmBiT peptide. Only the Sec61α-C_L protein showed lower complementation efficiency of ~2 relative luminescent units (RLU), likely caused by its lower abundance (Figure 3B). When the HaBiT peptide was used in combination with the ER membrane protein fusions, the RLU were at least 50-fold higher in comparison to the scrambled peptide, with the exception of the Sec61α-C_L protein peaking at 15 RLU. Thus, the synthetic complementation approach substantiated the plasmid-driven production of the LgBiT-tagged ER membrane proteins with the Sec61α-C_L protein providing a low complementation likely reflecting the limited synthesis or stability of this fusion protein.

Despite its ER luminally located LgBiT tag, the Sec61β-C_L protein also responded strongly using the SmBiT or HaBiT peptide after semi-permeabilization (Figure 3). While an inverted topology of Sec61β-C_L could be one explanation, other reasons might entail unintentional permeabilization of the ER membrane by digitonin or the active transport of the synthetic oligopeptides into the ER by the transporter associated with antigen processing, TAP [43]. As soon as more sensors with luminally located LgBiT tags are available, further experiments will test the impact of TAP, permeabilization, and orientation more systematically.

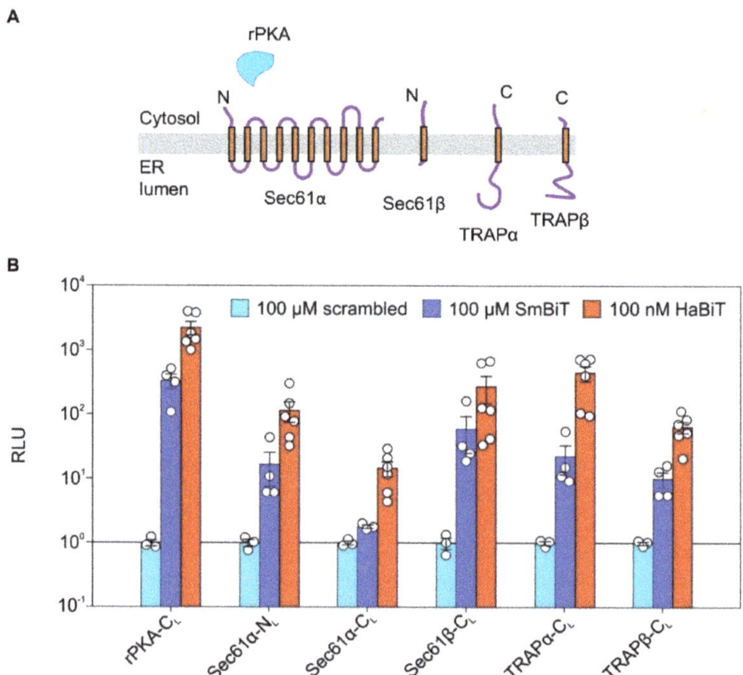

Figure 3. Expression profiles of LgBit-tagged components of the ER protein translocase. (**A**) Topology of the four ER membrane proteins (Sec61α/β, TRAPα/β) used for generating five LgBiT fusions. (**B**) Twenty-four hours post transfection, the expression of the indicated LgBiT fusion constructs tagged at the N- or C-terminus (N_L or C_L) was verified using the synthetic complementation approach with the scrambled, SmBiT, and HaBiT peptide, as demonstrated in Figure 2. The light intensity 4 min after peptide addition was plotted as relative luminescent units (RLU). For each experiment and tested fusion protein, the luminescence measured 4 min after complementation with the scrambled peptide was used for normalization and set to 1. To depict the variation of the individual scrambled peptide readings the average luminescence of this condition was used for its normalization. C_L, C-terminally located LgBiT tag; N_L, N-terminally located LgBiT tag.

2.4. Validating Expression of SmBit Fusion Constructs via Western Blotting Using a Polyclonal Antibody Raised against the SmBiT

The BiLC approach relies on the presence of both types of fusion proteins allowing the 1:1 stoichiometric interaction of the LgBiT and SmBiT tags. After confirming expression of the LgBiT-tagged constructs via synthetic complementation (Figure 3), we also tested for the sufficient expression of the SmBiT-tagged fusion constructs via Western blotting. Therefore, we raised a polyclonal rabbit antibody against the SmBiT undecapeptide, which required the aforementioned N-terminal cysteine for coupling of the SmBiT to the immunogenic keyhole limpet hemocyanin. As the α-SmBiT antibody showed unspecific binding and the occurrence of background bands, mock transfected cells served as negative control and helped to verify the correct signal of the individual SmBiT-tagged fusion proteins. First, the cytosolic control fusions cPKA-N_S and Halo-C_S were probed for abundance on protein level. Both proteins provided an additional band in comparison to the mock transfected cells and confirmed proper expression of the constructs (Figure 4). Likewise, the SmBiT fusion proteins TRAPα-C_S, TRAPβ-C_S, Sec61α-C_S, and Sec61β-N_S could be detected using the generated α-SmBiT antibody (Figure 4). We also generated two SmBiT fusions of the much larger Sec63 protein, Sec63-C_S and Sec63-N_S. Neither of those two constructs could be detected with the α-SmBiT antibody. We attributed the lack of detection to the more pronounced occurrence of unspecific cross-rations of the α-SmBiT antibody in the molecular weight range above 40 kDa which might obscure efficient detection of the Sec63 fusion protein at approximately 100 kDa. Further, the use of polyclonal antibodies raised against the native portion of the Sec63 fusion proteins did not provide an additional band of higher molecular weight likely due to a combination of the small size of the SmBiT tag, the intended low expression of fusion constructs, and partial masking of the antibody-epitope by addition of the tag. The next alternative for testing expression of the SmBiT fusion construct was to test their co-expression together with a LgBiT as part of a functional PPI experiment. Luminescence arising from a combination of a LgBit and a SmBiT fusion protein would indicate synthesis of both components as shown in the following section.

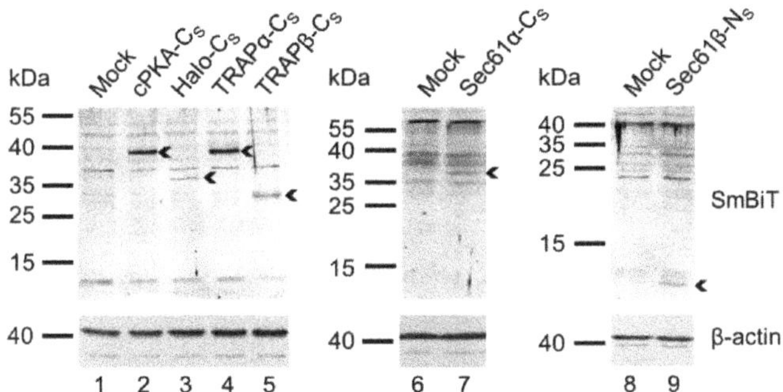

Figure 4. Protein abundance levels of SmBit-tagged components using a tag-specific antibody. Representative Western blot panels for different SmBiT fusion proteins twenty-four hours after transfection of cells are shown. As indicated, lower panels were incubated with anti-β-actin antibody as loading control and upper panels with the in-house generated anti-SmBiT antibody. Despite multiple cross-reactions of the anti-SmBiT antibody (see mock transfected cells in lanes 1, 6, and 8), signals for SmBiT-tagged fusion proteins at the expected molecular weight can be identified (arrowheads). Expected molecular weights: cPKA-C$_S$ (42 kDa), Halo-C$_S$ (36 kDa), TRAPα-C$_S$ (38 kDa), TRAPβ-C$_S$ (28 kDa), Sec61α-C$_S$ (40 kDa), Sec61β-N$_S$ (13 kDa). C$_S$, C-terminally located SmBiT tag; N$_S$, N-terminally located SmBiT tag.

2.5. α- and β-Subunit Interactions of the Sec61 or TRAP Complex

Next, we addressed the direct PPI of ER membrane protein pairs tagged with SmBiT and LgBiT. As proof of concept, we tested the intra-complex interaction of the heterotrimeric Sec61 complex. In its native environment, the Sec61complex consists of the channel-forming Sec61α subunit flanked by the two tail-anchored subunits β and γ [44,45]. As depicted in Figure 5A, we first tested if the cytosol facing N-termini of Sec61α and Sec61β are juxtaposed and do interact. After transfection of cells with the corresponding reporter constructs (Sec61α-N$_L$, Sec61β-N$_S$) and addition of the NanoLuc substrate luminescence was detected (Figure 5B). As seen before for the cytosolic PKA constructs (Figure 1C), light emission for the Sec61α-N$_L$ and Sec61β-N$_S$ pair was peaking 4 min past furimazine addition. To exclude drastic mislocalization of the constructs and verify that the Sec61α–Sec61β interaction occurs intracellularly, cells were subjected to trypsin-mediated reporter degradation. Indeed, only after permeabilization of the plasma membrane by digitonin, the luminescent signal was eliminated by addition of trypsin digesting the intracellular reporters. Instead, when cells were treated with DMSO prior to trypsin, no sharp decline of luminescence could be detected upon addition of the membrane impermeable protease (Figure 5B). Similar to what was observed before, the mild digitonin treatment amplified luminescence likely by providing easier access of furimazine to the cytosol (cf. Figures 1D, 2B and 5B). To further substantiate the validity of the PPI between Sec61α and Sec61β, a strategy based on competition was used. In addition to transfection with the reporter pair, cells received a third plasmid encoding an untagged, wild typic variant of either interaction partner. In contrast to the reporter pair plasmids carrying the HSV-TK promotor, the plasmids used for competition with untagged variants harbored a CMV promotor generally allowing stronger expression of a downstream gene [46]. An empty vector (EV) transfection was run in parallel and served as negative control. Contrary to the non-interacting Sec61α-N$_L$ and Halo-C$_S$ pair, a strong luminescent signal was detected for both the Sec61α-N$_L$ and Sec61β-N$_S$ pair as well as the Sec61α-N$_L$ and Sec61β-N$_S$ plus EV transfection (Figure 5C). Notably, introducing either untagged Sec61α or Sec61β into the system luminescence of the Sec61α-N$_L$ and Sec61β-N$_S$ reporter pair was strongly reduced and remained almost at the background level of the negative control pair Sec61α-N$_L$ and Halo-C$_S$ (Figure 5C). Thus, increasing the protein level of one of the interaction partners as untagged variant significantly competed with the interaction of the tagged proteins and underlines the authenticity of the tested Sec61α–Sec61β interaction in living cells at the ER (Figure 5D).

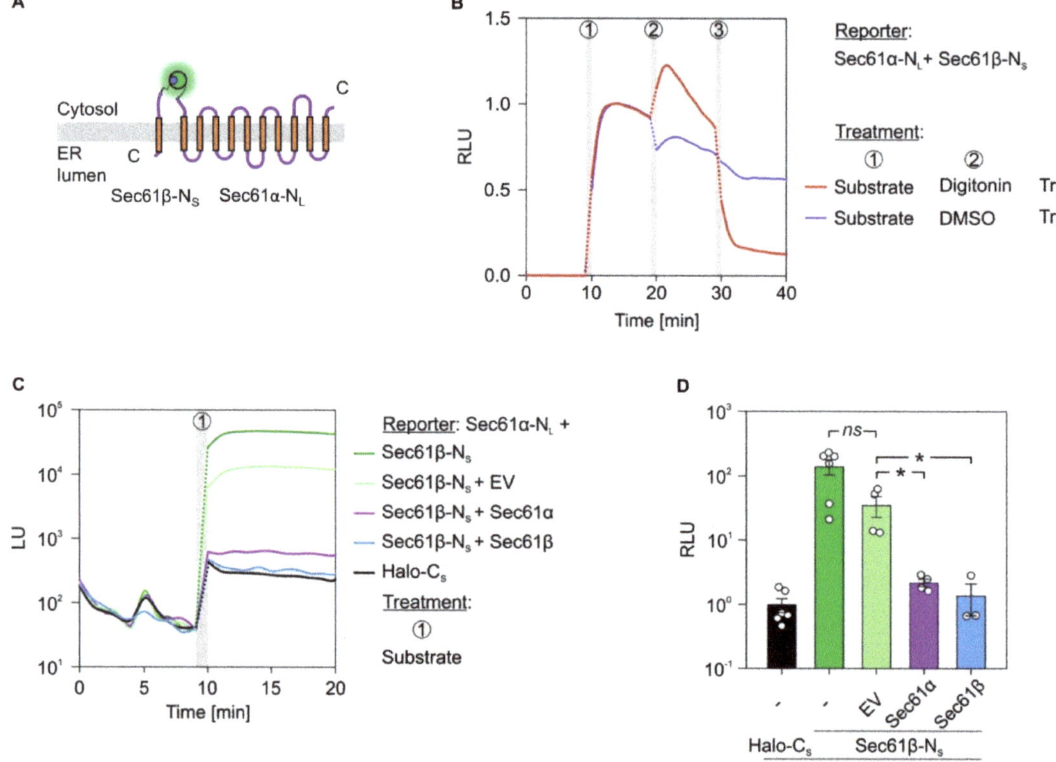

Figure 5. Protein–protein interactions within subunits of the Sec61 complex. (**A**) Topology cartoon of the presumed interaction for the Sec61α-N$_L$ and Sec61β-N$_S$ pair. As before, PPI reassembles the active luciferase (glowing green/blue sphere) that emits light upon addition of the substrate furimazine (not shown). (**B**) Twenty-four hours post transfection with the Sec61α-N$_L$ and Sec61β-N$_S$ reporter pair, cells were subjected to the indicated treatments (circled numbers). The furimazine treatment to activate luminescence (after 9 min) was followed by permeabilization using digitonin (0.002%, after 19 min) and trypsin mediated reporter digestion (50 µg/mL, after 29 min). DMSO served as vehicle control for digitonin. Measurements were normalized to the signal intensity recorded 4 min after furimazine application. (**C**) BiLC was combined with competition. Twenty-four hours after double or triple transfection with the indicated constructs, total luminescent units (LU) were measured. Protein names without a suffix do not carry a tag and were designed to compete with the interaction of the tagged interaction partners. (**D**) Measurements from (**C**) were used for quantification. For each condition

and the corresponding biological replicates, the signals 4 min after furimazine addition were evaluated and plotted as relative luminescent units (RLU). To do so, the luminescence of the negative control pair (Sec61α-N$_L$ plus Halo-C$_S$) that was run in parallel for each experiment was used for normalization and set to 1. To depict the variation of negative control pair readings the average LU of this condition was used for its normalization. Statistical comparison of two conditions was based on a student's *t*-test (indicated by italic labels) and comparison of multiple conditions was done using ANOVA. Vertical gray bars in the line diagrams represent manual 1 min application periods without luminescence readings. The dotted lines are extrapolated based on the last and first data points before and after application. C$_S$, C-terminally located SmBiT tag; EV, empty vector; N$_L$, N-terminally located LgBiT tag; ns, not significant; N$_S$, N-terminally located SmBiT tag. *p*-values are indicated by asterisks with $p < 0.05$ (*) or as non-significant (ns) if $p \geq 0.05$.

We also tested the intra-complex PPI of the heterotetrameric TRAP complex. Three of the four TRAP subunits (α, β, δ) traverse the ER membrane as single-pass type I membrane protein. Together with the tetra-spanning γ subunit these four membrane proteins tightly associate and form the native TRAP complex that supports as allosteric effector gating of the Sec61 complex [14,15,40,47]. Regarding TRAP, we focused on the interaction between the α- and β-subunit by generating the C-terminally tagged variants TRAPα-C$_L$ and TRAPβ-C$_S$ (Figure 6A).

Consistent with the Sec61 complex, we subjected the TRAPα/β reporter pair to the same luminescence measurements including trypsin-mediated reporter degradation and competition with the untagged counterparts (Figure 6). When cells transfected with TRAPα-C$_L$ and TRAPβ-C$_S$ were subjected to the furimazine–digitonin–trypsin, protocol we saw (i) the direct interaction between TRAPα and TRAPβ, (ii) the amplified luminescence after permeabilization of the plasma membrane, and (iii) rapid loss of the signal and PPI due to the proteolytic digestion Figures 6B and S1B). Like the Sec61α/β reporter pair, also TRAPα-C$_L$ and TRAPβ-C$_S$ showed a strong interaction in comparison to the non-interacting control pair TRAPα-C$_L$ and Halo-C$_S$ (cf. Figures 5C and 6C). This interaction was further tested using the competition setup based on plasmid-driven expression of the untagged *TRAPA* or *TRAPB* gene. Increasing the protein level of TRAPα or TRAPβ caused a significant reduction of the PPI between TRAPα-C$_L$ and TRAPβ-C$_S$ (Figure 6C,D). Furthermore, we tested if the expression efficiency of the LgBiT reporter construct TRAPα-C$_L$ was impaired by the additional expression of the untagged counterpart TRAPα or the presence of the EV in the transfection mix. The underlying reason was that the addition of a third plasmid during the transfection procedure might reduce transfection and/or expression efficiency of the plasmids. As the exogenously added high-affinity HaBiT peptide was able to substitute for and replace the genetically encoded SmBiT fusion construct of a given reporter pair, the synthetic peptide complementation setup was used to test for expression of the TRAPα-C$_L$ during transfections with two or three plasmids. As shown in Figure S3, the presence of an additional plasmid in the mix was not affecting expression of the TRAPα-C$_L$ construct, which still showed efficient complementation with HaBiT even in the presence of the EV or of the plasmid encoding the untagged TRAPα protein. Compared to the negative control pair TRAPα-C$_L$ and Halo-C$_S$, all other three conditions combining TRAPα-C$_L$ with TRAPβ-C$_S$ alone or in combination with either EV or TRAPα provided the same level of luminescence upon addition of HaBiT (Figure S3B).

As Figures 5 and 6 showed the broad applicability of the BiLC approach for ER membrane proteins, two questions arose. First, is the assay suited to detect inter-complex interactions, for example between Sec61 and TRAP? Second, can the assay detect PPI arising from reconstitution of the split luciferase either in the cytosol or the ER lumen? Detection of luminescence emanating from the ER lumen would further prove the applicability of the system to study PPI between membrane and/or soluble proteins occurring in the ER lumen.

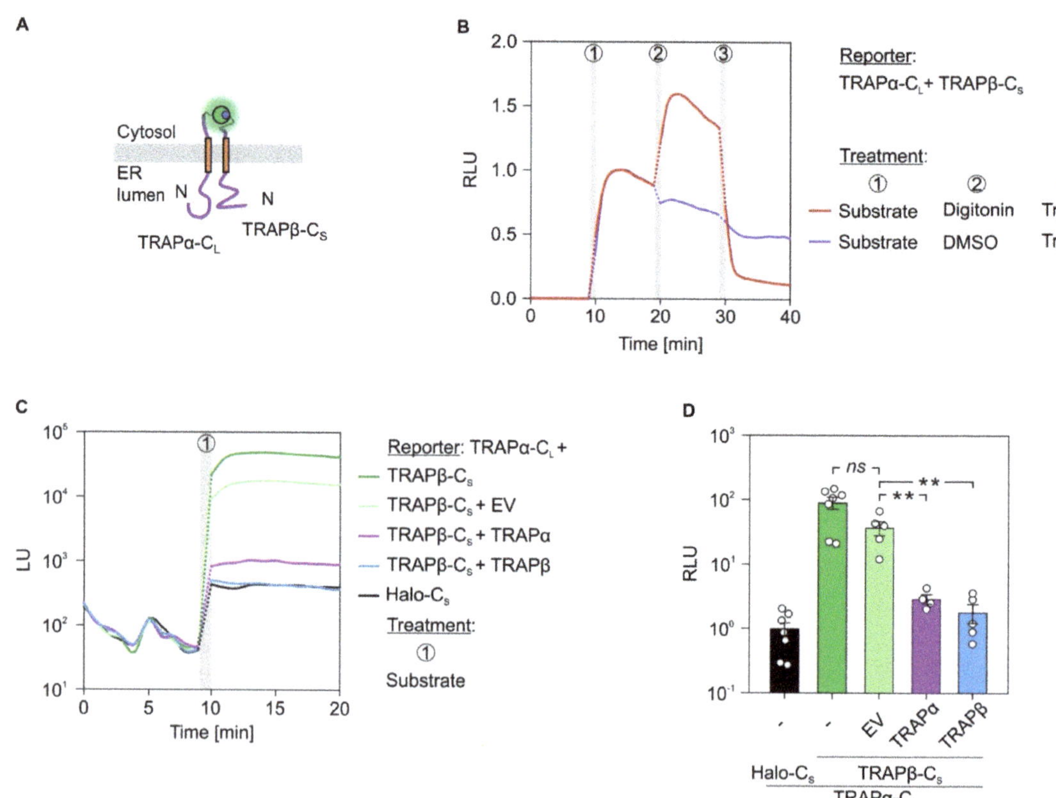

Figure 6. Protein–protein interactions within subunits of the TRAP complex. (**A**) Topology cartoon of the presumed interaction for the TRAPα-C$_L$ and TRAPβ-C$_S$ pair. Active luciferase is symbolized as glowing green/blue sphere. (**B**) Twenty-four hours post transfection with the TRAPα-C$_L$ and TRAPβ-C$_S$ reporter pair, cells were subjected to the indicated treatments (circled numbers). The furimazine treatment to activate luminescence (after 9 min) was followed by permeabilization using digitonin (0.002%, after 19 min) and trypsin mediated reporter digestion (50 μg/mL, after 29 min). DMSO served as vehicle control for digitonin. Measurements were normalized to the signal intensity recorded 4 min after furimazine application. (**C**) BiLC was combined with competition. Twenty-four hours after double or triple transfection with the indicated constructs, total luminescent units (LU) were measured. Consistency of the expression efficiency for double and triple transfections was confirmed by synthetic peptide complementation (Figure S3). Protein names without a suffix do not carry a tag and were designed to compete with the interaction of the tagged interaction partners. (**D**) Measurements from (**C**) were used for quantification. For each condition and the corresponding biological replicates, the signals 4 min after furimazine addition were evaluated and plotted as relative luminescent units (RLU). To do so, luminescence of the negative control pair (TRAPα-C$_L$ plus Halo-C$_S$) that was run in parallel for each experiment was used for normalization and set to 1. To depict the variation of negative control pair readings, the average LU of this condition was used for its normalization. Statistical comparison of two conditions was based on a student's t-test (indicated by italic labels) and comparison of multiple conditions was performed using ANOVA. Vertical gray bars in the line diagrams represent manual 1 min application periods without luminescence readings. The dotted lines are extrapolated based on the last and first data points before and after application. C$_L$, C-terminally located LgBiT tag; C$_S$, C-terminally located SmBiT tag; EV, empty vector; ns, not significant. *p*-values are indicated by asterisks with $p < 0.01$ (**) or as non-significant (ns) if $p \geq 0.05$.

2.6. Inter-Complex Interactions between Sec61 and TRAP as well as Sec61 and Sec63

The generated set of fusion constructs was suited to address the questions of inter-complex interaction as well as detection of ER luminal emitted luminescence. Based on the vicinity of the TRAP and Sec61 complex in native membranes of different organisms and specimens, we first tested the inter-complex interaction between the α-subunits of TRAP and Sec61 [15]. As before, cells were co-transfected with the reporter pair Sec61α-N_L and TRAPα-C_S (Figure 7A). Co-transfection of cells with Sec61α-N_L and Halo-C_S served as negative control. After addition of the luciferase substrate furimazine, strong luminescence could be detected for the inter-complex interaction pair between Sec61α and TRAPα, but not for the negative control (Figure 7B). To verify the specificity of the Sec61α–TRAPα interaction in the native, cellular context additional complementation experiments were carried out. In addition to the Sec61α-N_L and TRAPα-C_S reporter pair cells were transfected with an EV or *SEC61A1* encoding plasmid. In comparison to the additional EV transfection the expression of the untagged *SEC61A1* caused a significant reduction of luminescence (Figure 7C). Thus, increasing the amount of untagged Sec61α competed with the Sec61α-N_L and TRAPα-C_S interaction. This competition by the untagged Sec61α protein substantiated the PPI of Sec61α and TRAPα in living cells.

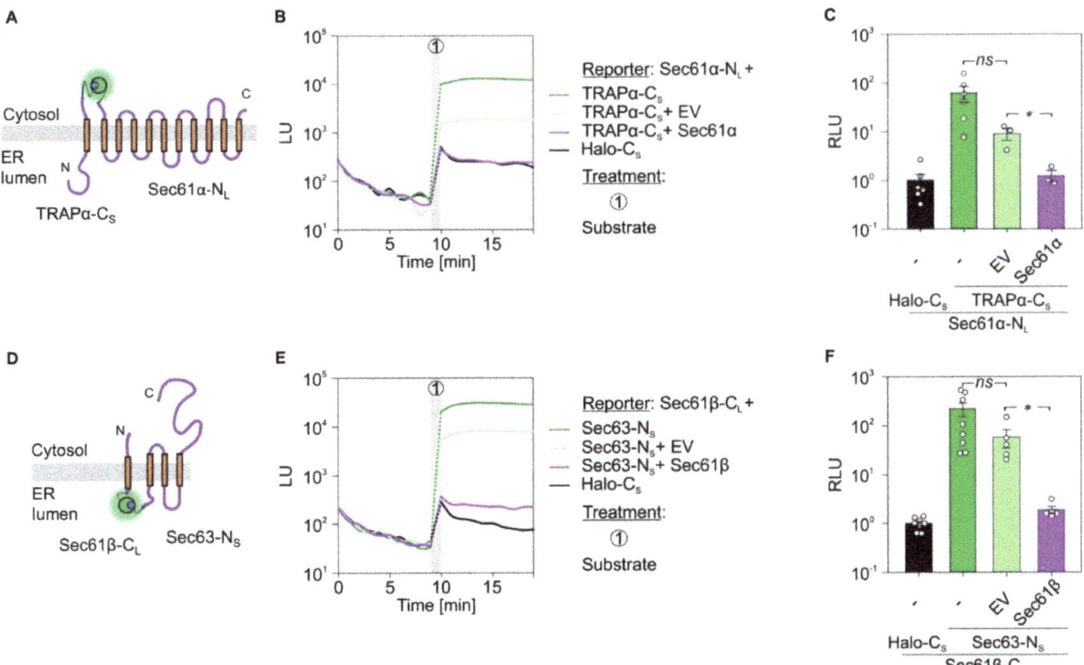

Figure 7. Protein–protein interactions within the holo-translocon: Sec61-TRAP and Sec61-Sec63. (**A,D**) Topology cartoon of the presumed interaction pairs Sec61α-N_L and TRAPα-C_S as well as Sec61β-C_L and Sec63-N_S. Active luciferase is symbolized as glowing sphere. (**B,E**) Twenty-four hours post double or triple transfection with the indicated constructs, total luminescent units (LU) were measured. Protein names without a suffix do not carry a tag. Nine minutes into the measurements, furimazine was added to activate luminescence. (**C,F**) Quantification and statistical analysis of luminescent readings from (**B,E**). For each condition and the corresponding biological replicates, the signals 4 min after furimazine addition were evaluated and plotted as relative luminescent units (RLU). Luminescence of the corresponding negative control pair (Sec61α-N_L plus Halo-C_S in **C**; Sec61β-C_L plus Halo-C_S in (**F**)) that was run in parallel for each experiment was used for normalization and set to 1. To depict the variation of negative control pair readings, the average LU of this condition was used for its normalization, respectively. Statistical comparison of two conditions was based on a student's

t-test. Vertical gray bars in the line diagrams represent manual 1 min application periods without luminescence readings. The dotted lines are extrapolated based on the last and first data points before and after application. C_L, C-terminally located LgBiT tag; C_S, C-terminally located SmBiT tag; EV, empty vector; N_L, N-terminally located LgBiT tag; ns, not significant; N_S, N-terminally located SmBiT tag. *p*-values are indicated by asterisks with $p < 0.05$ (*) or as non-significant (ns) if $p \geq 0.05$.

Next, we used a set of proteins whose LgBiT and SmBiT tags were both located in the ER lumen and belong to different translocon subcomplexes, Sec61β-C_L and Sec63-N_S (Figure 7D). When cells were co-transfected with the corresponding plasmids application of furimazine indicated (i) the functionality of the ER luminally assembled luciferase converting furimazine to furimamide plus light, and (ii) the inter-complex interaction between Sec61β and Sec63 (Figure 7E). Analogous to the TRAPα–Sec61α interaction, the Sec61β–Sec63 PPI was also susceptible to competition by expression of the untagged interaction partner. When cells carrying the interaction pair Sec61β-C_L and Sec63-N_S were transfected with a plasmid encoding wild type *SEC61B*, luminescence was much reduced compared to the corresponding EV transfection (Figure 7F).

2.7. Sterically Impossible Complementations Set the Threshold for Authentic Protein Interactions

Excluding the positive and negative control based on the PKA reporter pair (Figure 1), 35 combinations of LgBiT- and SmBiT-tagged proteins were tested and the signal intensity of interactions is summarized as heatmap in Figure 8A. As proof of concept, we mostly focused on well-established interactions between proteins of soluble and membrane-standing multimeric complexes like PKA, TRAP, and Sec61. Other than the interaction of the cPKA and rPKA subunits, nine PPI of ER membrane proteins were identified (Figure 8A). Undoubtedly, difficult to control methodological constraints such as transfection and expression efficiency of individual cells as well as transfection uniformity across a cell population can cloud deeper interpretation of the luminescence signal intensity of a given PPI. Yet, it is tempting to consider the emitted light intensity as surrogate marker and as first approximation for the strength of a tested PPI. The three strongest, or at least most light-intense, interactions of ER membrane proteins included the intra-complex interaction between Sec61α–Sec61β (141 RLU) and TRAPα–TRAPβ (91 RLU) as well as the inter-complex interaction Sec61β–Sec63 (202 RLU). Compared to the 613 RLU measured for the soluble rPKA–cPKA pair, those RLU are 3–6 times lower and might be explained by the limited structural flexibility and/or local mobility of membrane proteins (Figure 8B). Including those three, the validity of the five strongest PPI of ER membrane proteins was further confirmed by competition experiments with untagged proteins and is highlighted in the heatmap by check marks (Figures 5–7, 8A and S3A).

As addendum to the previous data with the LgBiT and SmBiT tags of fusion proteins being located in the same compartment, the heatmap also features the sterically impossible interactions pairs that were tested (Figure 8A, hachures). By definition, physical separation of the reporter tags by a lipid bilayer should prevent their complementation. As assumed, all ten tested pairs with LgBiT and SmBiT tags placed on opposite sides of the ER membrane provided only background luminescence and, thus, served as reliable negative controls to set the threshold value for a meaningful PPI above 10 RLU (Figure 8A hachures and Figure 8B). In light of a potential degradation, mislocalization, or inverted orientation of tagged membrane proteins, luminescence below threshold can provide useful information, too. For example, the Sec61β-C_L reporter represents a C-terminally extended tail-anchored protein (cf. Figure 3A). Adding 176 aa (LgBiT plus linker) shifts the single TMH of Sec61β from the C-terminus to the center of the protein and might affect topology [48–50]. On the one hand, Sec61β-C_L did not show interaction with the five tested cytosolic SmBiT variants of proteins (Sec61α/β-C_S, TRAPα/β-C_S, Sec63-C_S), most of which are known to associate in close proximity to Sec61β in native membranes and, thus, speaking against an inverted topology of Sec61β-C_L [7,19,51]. Instead, steady-state synthesis of the Sec61β-C_L protein (Figure 4) as well as strong interaction with the luminally located Sec63-N_S (Figure 7) were detected. On the other hand, of the sterically impossible interactions the Sec61β-C_L plus Sec61α-C_S reporter pair provided the highest RLU with 8.2, which might indicate a negligible fraction of Sec61β-C_L proteins oriented in an inverted topology (Figure 8B). Based on these data, we considered 10 RLU as a reasonable threshold to differentiate between the luminescence of an unspecific interaction (RLU < 10) and that of valid PPI of ER membrane proteins (RLU > 10).

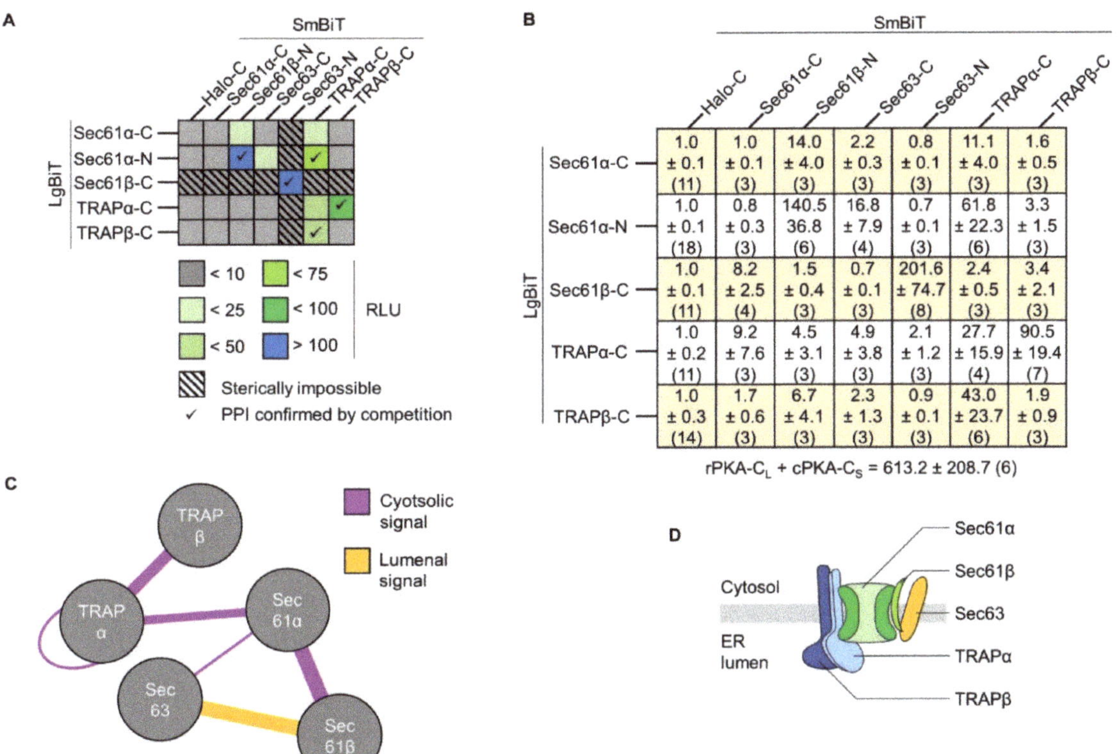

Figure 8. Summary of protein–protein interactions measured in living mammalian cells. (**A**) Heatmap summarizing the tested PPI. Interaction strength is based on relative luminescence units (RLU) normalized to the negative control Halo-C_S set as 1 and shown in the first column. RLU below 10 were considered as background (gray color). In addition to the RLU-based interaction strength, the heatmap encodes (i) if a given PPI could be confirmed by competition experiments (checkmark symbols, cf. Figures 5–7) and (ii) if the interaction of the tested reporter pair is sterically impossible, i.e., the tags are oriented on different sides of the ER membrane (hachured squares). (**B**) Data summary of tested PPI pairs as average ± standard error of the mean from (n) tested replicates. For better comparison, layout of the grid and heatmap in panel A are identical. (**C**) Interaction network based on the five proteins of interest tested here. While thickness of connecting lines indicates the detected RLU (strength) of the interaction, purple and orange color implies origin of the luminescent signal in the cytosol or ER lumen, respectively. (**D**) Cartoon view of the verified PPI. The suffix -N or -C following protein names indicates N- or C-terminal position of the LgBiT or SmBiT tag.

Taken together, BiLC provided a robust and versatile method to detect topology-dependent PPI of various ER membrane proteins including associations within heteromultimeric complexes such as TRAP or Sec61 as well as inter-complex subunit connections between Sec61 and TRAP or Sec61 and Sec63/Sec62. Authenticity of the described interactions was further corroborated by competition experiments. Increasing the protein level of one of the interaction partners as untagged protein reduced luminescence, i.e., PPI.

3. Discussion

Using a split luciferase and the principle of BiLC, multiple PPI of ER membrane proteins were probed in living cells. The luciferase used here represents an optimized variant of the catalytic subunit of the deep sea luminous shrimp (*Oplophorus gracilirostris*) enzyme and was called NanoLuc. The original 27 aa long signal peptide was substituted by the MV dipeptide. While the wild type luciferase had one cysteine residue close to the C-terminus, mutational analysis of the SmBit tag replaced this cysteine by phenylalanine [37,38]. Thus, despite the presence of protein disulfide isomerases and the oxidative environment found in the ER, the lack of cysteine residues in the NanoLuc should allow proper assembly of the functional monooxygenase also in this compartment. Indeed, luminescence of the re-assembled luciferase could be detected in the cytosol as well as the ER lumen. The latter was demonstrated by the Sec61β–Sec63 interaction. This interaction demonstrated (i) the efficient reconstitution of the functional luciferase in the ER lumen as well as (ii) that the oxygen- and furimazine-dependent luminescence is not blocked by the presence of an additional membrane or the milieu of the mammalian ER lumen (Figure 7D–F). However, future experiments will test more PPI for which the topological layout of the reporter pair requires reconstitution of the split luciferase in the ER lumen. In addition, considering the N-terminal signal peptide of the original shrimp luciferase, at least the wild typic version of the enzyme was adapted to the secretory pathway.

Dynamic adaptability and reversibility of PPI in living cells was demonstrated for assembly of the PKA subunits using forskolin as an activator of the adenylate cyclase (Figure 1). Thus, the system will be well suited to address the dynamics of the detected translocon interactions under different cellular conditions or perturbations. For instance, Snapp and colleagues designed a FRET assay that was used to detect ribosome associations of translocon proteins via antibody accessibility before and after RNase A treatment [52,53]. Instead of trypsin-mediated reporter degradation (Figures 1D, 5B and 6B, RNase A can be used upon semi-permeabilization to test either alterations in the PPI patterns of intra- and inter-complex associations or the efficiency of the synthetic complementation with the SmBiT and HaBiT peptide. Similarly, but without the necessity of semi-permeabilization, the effect of specific ribosomal or translocon inhibitors mimicking different stages of the protein transport process can be tested [54–57].

Luminescence data from the heatmap (Figure 8A) are also displayed as interaction network summarizing the intra- and inter-complex PPI of the five ER proteins tested here (Figure 8C). Even on this small scale some central nodes and connections within the holo-translocon become evident. As expected, Sec61α represents a major node with connections to Sec61β, TRAPα, and Sec63. All of those Sec61α-related connections have previously been observed by X-ray crystallography, cryo-electron microscopy/tomography, native gel electrophoresis, co-immunoprecipitation, or cross-linking studies [10,12,15,58–63]. However, none did so in living mammalian cells. Intriguingly, using different types of substrates the existence, and eventually dynamic assembly, of substrate-matched translocon sub-complexes was demonstrated [9]. As proposed earlier, the RLU measured in living cells might be indicative of the strength and permanency of a given PPI. This idea fits reasonably well with both cross-linking data and in situ structures. While the Sec61α–Sec61β (141 RLU) pair is consistently observed being associated, the Sec61α–TRAPα (62 RLU) association was seen to dissipate when substrates like prion protein or ERj3 engaged the translocon. In turn, those substrates, which probably present a minor fraction of the totality of transported cargos, rather stimulate the Sec61–Sec63 complex formation in mammalian settings [9]. Accordingly, we found Sec61α–Sec63 as weak or transiently interacting pair with only 17 RLU. Based on the recent structural data from the yeast translocon, the Sec61 complex and Sec63/Sec62 module assemble to form a translocon for the post-translational protein transport, a mode of transport less frequently encountered in mammalian cells [16,64,65]. Yet another recent structure of the fully closed yeast Sec complex from the Park lab shows juxtaposition of the Sec61β C-terminus and Sec63 N-terminus in the ER lumen [66]. While we measured only a weak association between Sec61α and Sec63, it was the reporter pair of

Sec61β C-terminus and Sec63 N-terminus that provided strong luminescence of 202 RLU (Figure 8B). Maybe Sec61β acts as the mediator that keeps Sec63 in vicinity of the Sec61 complex during transport events not relying on the Sec63/Sec62 module. Once substrates with special requirements start the gating process, Sec61β allows rapid integration of Sec63 into the active translocon. If so, Sec61β and Sec63 eventually will show a partially overlapping substrate-spectrum, which remains to be seen [6,18,67]. This scenario is reminiscent of a previous report showing the recruitment of the signal peptidase complex to the active, ribosome-engaged translocon by Sec61β [68].

TRAPα emerged as a second node in the interaction network. Aside from the interaction with Sec61α discussed before, TRAPα interacted strongly with its complex partner TRAPβ (91 RLU). Biochemical purifications identified the TRAP complex as a heterotetramer with the four subunits α-δ in a 1:1:1:1 stoichiometry [40,69,70]. Interestingly, using an appropriate pair of constructs the BiLC approach can also identify homo-oligomers, which we found for TRAPα, but not TRAPβ or Sec61α (Figure 8A–C). Based on the bimolecular design of the assay, it is unclear if two TRAPα molecules from adjacent TRAP complexes interact or if TRAPα forms a dimer or even a multimer within the same TRAP complex. Consequently, it will be intriguing to check if the TRAPα/β volume in the ER lumen defined by in situ structures of the mammalian TRAP complex could accommodate for more than one TRAPα protein [14,15,62].

In sum, luminescence complementation provided multiple interactions of subunits of the ER protein translocase in living cells (Figure 8D). Ease of use and expandability of the set of constructs plus the combination of the system with biological and synthetic agents will help to examine the dynamics of ER membrane proteins under different cellular conditions or in different cell lines (Figure S4).

4. Materials and Methods

4.1. Creation of a Plasmid Library

To clone the different reporter constructs four different backbone plasmids (X-N_S, X-N_L, X-C_S, X-C_L) for the insertion of cDNA were used. Backbone plasmids as well as the three control plasmids encoding for rPKA-C_L, cPKa-C_S, and Halo-C_S were provided as part of the NanoBiT PPI Starter System (Promega, Madison, WI, USA). To insert cDNA encoding for a protein of interest a standardized workflow was used. Forward and reverse primers were designed that anneal to the 5′- and 3′-end of the cDNA (Table S1). Primer overhangs included restriction sites for subcloning of cDNA into the multiple cloning site of the backbone vectors. After purification of the PCR product with QIAquick PCR purification kit (Qiagen, Hilden, Germany) the cleaned PCR product and backbone plasmid were double digested for directed insertion via sticky ends. Fragments were separated on agarose gels and purified via the QIAquick gel extraction kit (Qiagen). Insert and plasmid were ligated in a 3:1 ratio for 1 h at room temperature using T4 ligase (Thermo Fisher, Waltham, MA, USA). The ligation product was used for heat-shock transformation (42 °C for 90 s) of JM101 *E. coli* cells, which were plated on a 100 μg/mL ampicillin/LB-agar plate. After 16 h single colonies were picked and grown in a 2 mL liquid culture using 100 μg/mL ampicillin/TB medium. Cultures were harvested by centrifugation in a table-top centrifuge (3000 rpm, 5 min) and plasmid DNA was purified using the PureYield Plasmid Miniprep System (Promega). All plasmids were verified by sequencing (LGC Genomics, Berlin, Germany) and subsequently amplified in the DH5α *E. coli* strain after heat shock transformation (42 °C for 45 s). Plasmids from a 100 mL ampicillin/LB medium culture were purified using the Plasmid Midi Kit (Qiagen).

4.2. Cell Culture and Western Blot

Cell culture experiments were performed using HeLa ATCC no. CCL2 cells. The cells were cultivated in standard DMEM + GlutaMAX media with 10% FCS and 1% Pen/Strep (Thermo Fisher) in a humid environment at 37 °C and 5% CO_2. To prepare samples for Western blot 6×10^5 cells were seeded in a 6 cm dish (GreinerBioOne, Frickenhausen, Germany) and cultivated for 24 h before plasmid transfection. Amounts of 190 µL Opti-MEM (Thermo Fisher), 2 µL DNA (1 µg/µL), and 8 µL FuGeneHD (Promega) were mixed and incubated for 10 min at room temperature. The transfection mixture was added dropwise to cells, which were refreshed by 4 mL media beforehand. After 24 h cells were washed with PBS (Thermo Fisher), trypsinized (Trypsin-EDTA, Thermo Fisher) and harvested in 2 mL of KHM buffer (110 mM potassium acetate, 2 mM magnesium acetate, 20 mM HEPES/KOH, pH 7.2) plus 125 µg/mL trypsin inhibitor (MP Biomedicals, Illkirch-Graffenstaden, France). Cells were counted with the Countess Automated Cell Counter (Invitrogen, Darmstadt, Germany) and semi-permeabilized as described before [71]. Samples were mixed with Laemmli buffer (60 mM Tris/HCl pH 6.8, 10% (*v/v*) glycerol, 2% (*w/v*) SDS, 5% (*v/v*) 2-mercaptoethanol, 0.01% (*w/v*) bromophenol blue), and denatured at 56 °C for 10 min. A total of 2×10^5 cells were loaded on an SDS-PAGE gel, followed by transfer on a PVDF membrane (Merck Millipore, Billerica, MA, USA). Membranes were blocked for 30 min with 3% (*w/v*) BSA (Roth, Karlsruhe, Germany) dissolved in TBS (150 mM NaCl, 10 mM Tris/HCl pH 7.4). Afterwards, the membrane was incubated for 90 min each with primary and secondary antibodies diluted in blocking solution. After every antibody incubation the membrane was washed four times for 5 min using TBS as first and last washing step and TBS-T (TBS plus 0.005% Tween-20 (Sigma-Aldrich, Steinheim, Germany)) in between. The membrane was dried and scanned by the Typhoon Trio (GE Healthcare, Uppsala, Sweden). The following peptide was used to raise the primary antibody against the SmBiT tag (CVTGYRLFEEIL). The β-actin antibody (Sigma-Aldrich) was used as loading control. Visualization of primary antibodies was done using ECL Plex goat anti-rabbit IgG-Cy5 (VWR, Radnor, PA, USA) or ECL Plex goat anti-mouse IgG-Cy3 (GE Healthcare) and the Typhoon-Trio imaging system in combination with the ImageQuant TL software version 7.0 (GE Healthcare, Uppsala, Sweden).

4.3. NanoBiT Assay

Per cavity of a white, flat-bottom 96-well plate (GreinerBioOne) 2×10^4 cells were seeded in 100 µL media and cultivated for 24 h in the incubator. Transfection with the reporter constructs was performed without previous media exchange according to Table 1. The transfection mix was incubated at room temperature for 10 min and added to the wells. Per well per plasmid 50 ng of the DNA construct was used. After incubation for 24 h and 5 min before start of the measurement, the medium was replaced with prewarmed Opti-MEM without phenol red (Thermo Fisher) and the 96-well plate was placed in the preheated microplate reader (Tecan Infinite M200). Luminescence was recorded with the following settings: interval 1 min; shaking before every interval; integration time 1000 ms; settle time 50 ms. To activate the luciferase, 20 µL of the 1:20 diluted Nano-Glo Live Cell Assay System (Promega) were added according to manufacturer's protocol. Substances were added at the indicated timepoints in the figures with the following final concentrations: Digitonin (Merck, Darmstadt, Germany), 0.002%; Forskolin (Sigma-Aldrich), 15 µM; Peptides (see below), 10 pM–100 µM; RNase A (Roche, Basel, Switzerland), 80 µg/mL; Trypsin (Roche), 50 µg/mL. Stocks of digitonin, forskolin, and peptides were dissolved in DMSO (Sigma-Aldrich) whereas RNase A and trypsin were dissolved in H_2O. Stocks and solvents as negative control were prediluted in Opti-MEM without phenol red.

4.4. Peptide Synthesis

Peptides were synthesized with an automated ResPepSL synthesizer (Intavis, Cologne, Germany) using amide Rink resin as the solid phase and Fmoc-protected amino acids (Carbolution, St. Ingbert, Germany) for coupling according to the method of Merrifield [72]. The resin-coupled peptides were cleaved and deprotected with trifluoroacetic acid (Sigma-Aldrich) followed by precipitation with tert-butyl methyl ether (Thermo Fisher) and further analysis and purification by preparative HPLC (Merck, Darmstadt, Germany). Lyophilized peptides (Lyovac GT2, Finn-Aqua, Tuusula, Finland) were stored at 4 °C. Stock solutions of the peptides were prepared fresh and used the same day.

Table 1. Composition and intended purpose of different plasmid transfection mixtures. Stock concentrations of the plasmids were 100 ng/μL. All volumes are provided in microliters.

Transfection Mix	One Plasmid Mix	Two Plasmid Mix	Three Plasmid Mix
Purpose	Synthetic complementation	PPI experiment	PPI competition experiment
Opti-MEM w/o phenol red	7.36	6.72	6.08
LgBit plasmid	0.50	0.50	0.50
SmBiT plasmid	-	0.50	0.50
Competition plasmid	-	-	0.50
FuGENE HD [1]	0.14	0.28	0.42

[1] FuGENE HD (Promega).

4.5. Antibody Generation

Synthesized peptides were coupled to maleimide-activated keyhole limpet haemocyanin via an N- or C-terminal cysteine for 16 h at room temperature. After size exclusion chromatography, the protein fraction was dialyzed excessively with physiological buffer solution (137 mM NaCl, 2.7 mM KCl, 10 mM Na_2HPO_4, 1.8 MM KH_2PO_4, pH 7.4). Rabbits were immunized subcutaneously in 250 μL doses (100 μg antigen/Freud's Adjuvant) at 14-day intervals. Blood was collected at 8-day intervals after each immunization. The serum was prepared after agglutination by several 10-min centrifugation steps at 4 °C in a table-top centrifuge.

4.6. Statistics and Graphical Representation

Graphs were visualized using Sigma Plot 14.0 (Systat Software GmbH, Erkrath, Germany) and Corel Draw Graphics Suite 2018 software (Coral Cooperation, Ottawa, Canada). Statistical comparison of multiple groups was performed using ANOVA in combination with Dunnett's multiple comparison test whereas two groups were compared based on an unpaired, two-tailed t-test. p-values are indicated by asterisks with $p < 0.001$ (***) < 0.01 (**) < 0.05 (*) or as non-significant (ns) if $p \geq 0.05$. For better differentiation, the p-value symbols emanating from a t-test are written in italic in bar graphs. The number of repeats for each experiment are indicated by the individual data points shown in the bar graphs.

Supplementary Materials: The following are available online at https://www.mdpi.com/article/10.3390/ijms221910358/s1.

Author Contributions: M.S. and S.L. conceptualized the work; M.S., M.J. and S.L. established the methodology and resources; M.J. and S.L. acquired funding; M.S. performed the experiments under the supervision of S.L. and analyzed the data together with S.L.; M.S. and S.L. visualized the data; S.L. prepared the original draft; all authors reviewed and edited the manuscript. All authors have read and agreed to the published version of the manuscript.

Funding: This work was supported by the Deutsche Forschungsgemeinschaft (DFG, German Research Foundation) grants SFB 894 (S.L. and M.J.) and IRTG 1830 (S.L.).

Acknowledgments: We are grateful for the excellent technical assistance provided by M. Lerner (Homburg, Germany) and the fruitful discussions and inputs provided by our colleagues R. Zimmermann, G. Schlenstedt, A. Cavalié, and M. van der Laan. We also like to thank B. Schrul for kindly providing HeLa-Kyoto cells.

Conflicts of Interest: The authors declare no conflict of interest.

References

1. Palade, G.E. The endoplasmic reticulum. *J. Biophys. Biochem. Cytol.* **1956**, *2*, 85–98. [CrossRef]
2. Schwarz, D.S.; Blower, M.D. The endoplasmic reticulum: Structure, function and response to cellular signaling. *Cell. Mol. Life Sci.* **2016**, *73*, 79–94. [CrossRef]
3. Bravo, R.; Parra, V.; Gatica, D.; Rodriguez, A.E.; Torrealba, N.; Paredes, F.; Wang, Z.V.; Zorzano, A.; Hill, J.A.; Jaimovich, E.; et al. Chapter Five—Endoplasmic Reticulum and the Unfolded Protein Response: Dynamics and Metabolic Integration. In *International Review of Cell and Molecular Biology*; Jeon, K.W., Ed.; Academic Press: Cambridge, MA, USA, 2013; Volume 301, pp. 215–290.
4. Raffaello, A.; Mammucari, C.; Gherardi, G.; Rizzuto, R. Calcium at the Center of Cell Signaling: Interplay between Endoplasmic Reticulum, Mitochondria, and Lysosomes. *Trends Biochem. Sci.* **2016**, *41*, 1035–1049. [CrossRef]
5. Lang, S.; Nguyen, D.; Pfeffer, S.; Förster, F.; Helms, V.; Zimmermann, R. Functions and Mechanisms of the Human Ribosome-Translocon Complex. In *Macromolecular Protein Complexes II: Structure and Function*; Harris, J.R., Marles-Wright, J., Eds.; Springer International Publishing: Cham, Switzerland, 2019; pp. 83–141.
6. O'Keefe, S.; High, S. Membrane translocation at the ER: With a little help from my friends. *FEBS J.* **2020**, *287*, 4607–4611. [CrossRef] [PubMed]
7. Gemmer, M.; Förster, F. A clearer picture of the ER translocon complex. *J. Cell Sci.* **2020**, *133*, jcs231340. [CrossRef] [PubMed]
8. Thul, P.J.; Åkesson, L.; Wiking, M.; Mahdessian, D.; Geladaki, A.; Aitblal, H.; Alm, T.; Asplund, A.; Björk, L.; Breckels, L.M.; et al. A subcellular map of the human proteome. *Science* **2017**, *26*, 356. [CrossRef] [PubMed]
9. Conti, B.J.; Devaraneni, P.K.; Yang, Z.; David, L.L.; Skach, W.R. Cotranslational Stabilization of Sec62/63 within the ER Sec61 Translocon Is Controlled by Distinct Substrate-Driven Translocation Events. *Mol. Cell* **2015**, *58*, 269–283. [CrossRef] [PubMed]
10. Dejgaard, K.; Theberge, J.-F.; Heath-Engel, H.; Chevet, E.; Tremblay, M.L.; Thomas, D.Y. Organization of the Sec61 Translocon, Studied by High Resolution Native Electrophoresis. *J. Proteome Res.* **2010**, *9*, 1763–1771. [CrossRef] [PubMed]
11. Jadhav, B.; McKenna, M.; Johnson, N.; High, S.; Sinning, I.; Pool, M.R. Mammalian SRP receptor switches the Sec61 translocase from Sec62 to SRP-dependent translocation. *Nat. Commun.* **2015**, *6*, 133. [CrossRef]
12. Wang, L.; Dobberstein, B. Oligomeric complexes involved in translocation of proteins across the membrane of the endoplasmic reticulum. *FEBS Lett.* **1999**, *457*, 316–322. [CrossRef]
13. Kriegler, T.; Lang, S.; Notari, L.; Hessa, T. Prion Protein Translocation Mechanism Revealed by Pulling Force Studies. *J. Mol. Biol.* **2020**, *432*, 4447–4465. [CrossRef] [PubMed]
14. Nguyen, D.; Stutz, R.; Schorr, S.; Lang, S.; Pfeffer, S.; Freeze, H.H.; Förster, F.; Helms, V.; Dudek, J.; Zimmermann, R. Proteomics reveals signal peptide features determining the client specificity in human TRAP-dependent ER protein import. *Nat. Commun.* **2018**, *9*, 65. [CrossRef]
15. Pfeffer, S.; Dudek, J.; Schaffer, M.; Ng, B.G.; Albert, S.; Plitzko, J.M.; Baumeister, W.; Zimmermann, R.; Freeze, H.H.; Engel, B.D.; et al. Dissecting the molecular organization of the translocon-associated protein complex. *Nat. Commun.* **2017**, *8*, 72. [CrossRef]
16. Weng, T.-H.; Steinchen, W.; Beatrix, B.; Berninghausen, O.; Becker, T.; Bange, G.; Cheng, J.; Beckmann, R. Architecture of the active post-translational Sec translocon. *EMBO J.* **2021**, *40*, e105643. [CrossRef]
17. Bhadra, P.; Yadhanapudi, L.; Römisch, K.; Helms, V. How does Sec63 affect the conformation of Sec61 in yeast? *PLOS Comput. Biol.* **2021**, *17*, e1008855. [CrossRef]
18. Schorr, S.; Nguyen, D.; Haßdenteufel, S.; Nagaraj, N.; Cavalié, A.; Greiner, M.; Weissgerber, P.; Loi, M.; Paton, A.W.; Paton, J.C.; et al. Identification of signal peptide features for substrate specificity in human Sec62/Sec63-dependent ER protein import. *FEBS J.* **2020**, *287*, 4612–4640. [CrossRef]
19. Ziska, A.; Tatzelt, J.; Dudek, J.; Paton, A.W.; Paton, J.C.; Zimmermann, R.; Haßdenteufel, S. The signal peptide plus a cluster of positive charges in prion protein dictate chaperone-mediated Sec61 channel gating. *Biol. Open* **2019**, *4*, 691. [CrossRef]
20. Schuren, A.B.C.; Boer, I.G.J.; Bouma, E.; Lebbink, R.J.; Wiertz, E.J.H.J. Genetic editing of SEC61, SEC62, and SEC63 abrogates human cytomegalovirus US2 expression in a signal peptide-dependent manner. *bioRxiv* **2019**, *9*, 653857. [CrossRef]
21. Lang, S.; Benedix, J.; Fedeles, S.V.; Schorr, S.; Schirra, C.; Schäuble, N.; Jalal, C.; Greiner, M.; Haßdenteufel, S.; Tatzelt, J.; et al. Different effects of Sec61α, Sec62 and Sec63 depletion on transport of polypeptides into the endoplasmic reticulum of mammalian cells. *J. Cell Sci.* **2012**, *125*, 1958–1969. [CrossRef]
22. Lang, S.; Pfeffer, S.; Lee, P.-H.; Cavalié, A.; Helms, V.; Förster, F.; Zimmermann, R. An Update on Sec61 Channel Functions, Mechanisms, and Related Diseases. *Front. Physiol.* **2017**, *8*, 887. [CrossRef]
23. Walter, T.; Erdmann, R. Current Advances in Protein Import into Peroxisomes. *Protein J.* **2019**, *38*, 351–362. [CrossRef] [PubMed]
24. Wiedemann, N.; Pfanner, N. Mitochondrial Machineries for Protein Import and Assembly. *Annu. Rev. Biochem.* **2017**, *86*, 685–714. [CrossRef]
25. Li, H.-m.; Chiu, C.-C. Protein Transport into Chloroplasts. *Annu. Rev. Plant Biol.* **2010**, *61*, 157–180. [CrossRef]
26. Wickner, W.; Schekman, R. Protein Translocation Across Biological Membranes. *Science* **2005**, *310*, 1452–1456. [CrossRef] [PubMed]
27. Kunze, M.; Berger, J. The similarity between N-terminal targeting signals for protein import into different organelles and its evolutionary relevance. *Front. Physiol.* **2015**, *6*, 259. [CrossRef]
28. Berggård, T.; Linse, S.; James, P. Methods for the detection and analysis of protein–protein interactions. *Proteomics* **2007**, *7*, 2833–2842. [CrossRef] [PubMed]
29. Rao, V.S.; Srinivas, K.; Sujini, G.N.; Kumar, G.N.S. Protein–protein interaction detection: Methods and analysis. *Int. J. Proteom.* **2014**, *2014*, 147648. [CrossRef]

30. Lu, H.; Zhou, Q.; He, J.; Jiang, Z.; Peng, C.; Tong, R.; Shi, J. Recent advances in the development of protein–protein interactions modulators: Mechanisms and clinical trials. *Signal Transduct. Target. Ther.* **2020**, *5*, 213. [CrossRef]

31. Koh, G.C.K.W.; Porras, P.; Aranda, B.; Hermjakob, H.; Orchard, S.E. Analyzing Protein–Protein Interaction Networks. *J. Proteome Res.* **2012**, *11*, 2014–2031. [CrossRef]

32. Nero, T.L.; Morton, C.J.; Holien, J.K.; Wielens, J.; Parker, M.W. Oncogenic protein interfaces: Small molecules, big challenges. *Nat. Rev. Cancer* **2014**, *14*, 248–262. [CrossRef]

33. Braun, P. Interactome mapping for analysis of complex phenotypes: Insights from benchmarking binary interaction assays. *Proteomics* **2012**, *12*, 1499–1518. [CrossRef]

34. Xing, S.; Wallmeroth, N.; Berendzen, K.W.; Grefen, C. Techniques for the Analysis of Protein–Protein Interactions In Vivo. *Plant Physiol.* **2016**, *171*, 727–758. [CrossRef]

35. Milroy, L.-G.; Grossmann, T.N.; Hennig, S.; Brunsveld, L.; Ottmann, C. Modulators of Protein–Protein Interactions. *Chem. Rev.* **2014**, *114*, 4695–4748. [CrossRef]

36. Michnick, S.W.; Ear, P.H.; Manderson, E.N.; Remy, I.; Stefan, E. Universal strategies in research and drug discovery based on protein-fragment complementation assays. *Nat. Rev. Drug Discov.* **2007**, *6*, 569–582. [CrossRef]

37. Dixon, A.S.; Schwinn, M.K.; Hall, M.P.; Zimmerman, K.; Otto, P.; Lubben, T.H.; Butler, B.L.; Binkowski, B.F.; Machleidt, T.; Kirkland, T.A.; et al. NanoLuc Complementation Reporter Optimized for Accurate Measurement of Protein Interactions in Cells. *ACS Chem. Biol.* **2016**, *11*, 400–408. [CrossRef]

38. Hall, M.P.; Unch, J.; Binkowski, B.F.; Valley, M.P.; Butler, B.L.; Wood, M.G.; Otto, P.; Zimmerman, K.; Vidugiris, G.; Machleidt, T.; et al. Engineered Luciferase Reporter from a Deep Sea Shrimp Utilizing a Novel Imidazopyrazinone Substrate. *ACS Chem. Biol.* **2012**, *7*, 1848–1857. [CrossRef]

39. Alasbahi, R.H.; Melzig, M.F. Forskolin and derivatives as tools for studying the role of cAMP. *Die Pharm. Int. J. Pharm. Sci.* **2012**, *67*, 5–13. [CrossRef]

40. Hartmann, E.; Görlich, D.; Kostka, S.; Otto, A.; Kraft, R.; Knespel, S.; Bürger, E.; Rapoport, T.A.; Prehn, S. A tetrameric complex of membrane proteins in the endoplasmic reticulum. *Eur. J. Biochem.* **1993**, *214*, 375–381. [CrossRef]

41. Borgese, N.; Colombo, S.; Pedrazzini, E. The tale of tail-anchored proteins. *J. Cell Biol.* **2003**, *161*, 1013–1019. [CrossRef]

42. Zimmermann, R.; Eyrisch, S.; Ahmad, M.; Helms, V. Protein translocation across the ER membrane. *Biochim. Biophys. Acta* **2011**, *1808*, 912–924. [CrossRef]

43. Lehnert, E.; Tampé, R. Structure and Dynamics of Antigenic Peptides in Complex with TAP. *Front. Immunol.* **2017**, *8*, 10. [CrossRef]

44. Pfeffer, S.; Burbaum, L.; Unverdorben, P.; Pech, M.; Chen, Y.; Zimmermann, R.; Beckmann, R.; Förster, F. Structure of the native Sec61 protein-conducting channel. *Nat. Commun.* **2015**, *6*, 9403. [CrossRef]

45. Pfeffer, S.; Dudek, J.; Zimmermann, R.; Förster, F. Organization of the native ribosome–translocon complex at the mammalian endoplasmic reticulum membrane. *Biochim. Et Biophys. Acta (BBA) Gen. Subj.* **2016**, *1860*, 2122–2129. [CrossRef]

46. Ali, R.; Ramadurai, S.; Barry, F.; Nasheuer, H.P. Optimizing fluorescent protein expression for quantitative fluorescence microscopy and spectroscopy using herpes simplex thymidine kinase promoter sequences. *FEBS Open Bio* **2018**, *8*, 1043–1060. [CrossRef]

47. Bañó-Polo, M.; Martínez-Garay, C.A.; Grau, B.; Martínez-Gil, L.; Mingarro, I. Membrane insertion and topology of the translocon-associated protein (TRAP) gamma subunit. *Biochim. Biophys. Acta (BBA) Biomembr.* **2017**, *1859*, 903–909. [CrossRef]

48. Spiess, M.; Junne, T.; Janoschke, M. Membrane Protein Integration and Topogenesis at the ER. *Protein J.* **2019**, *38*, 306–316. [CrossRef]

49. Junne, T.; Spiess, M. Integration of transmembrane domains is regulated by their downstream sequences. *J. Cell Sci.* **2017**, *130*, 372–381. [CrossRef]

50. Brambillasca, S.; Yabal, M.; Makarow, M.; Borgese, N. Unassisted translocation of large polypeptide domains across phospholipid bilayers. *J. Cell Biol.* **2006**, *175*, 767–777. [CrossRef]

51. Pfeffer, S.; Förster, F. Sec61: A static framework for membrane-protein insertion. *Channels* **2016**, *10*, 167–169. [CrossRef]

52. Snapp, E.L.; Reinhart, G.A.; Bogert, B.A.; Lippincott-Schwartz, J.; Hegde, R.S. The organization of engaged and quiescent translocons in the endoplasmic reticulum of mammalian cells. *J. Cell Biol.* **2004**, *164*, 997–1007. [CrossRef]

53. Benedix, J.; Lajoie, P.; Jaiswal, H.; Burgard, C.; Greiner, M.; Zimmermann, R.; Rospert, S.; Snapp, E.L.; Dudek, J. BiP modulates the affinity of its co-chaperone ERj1 for ribosomes. *J. Biol. Chem.* **2010**, *285*, 36427–36433. [CrossRef]

54. Van Puyenbroeck, V.; Vermeire, K. Inhibitors of protein translocation across membranes of the secretory pathway: Novel antimicrobial and anticancer agents. *Cell. Mol. Life Sci.* **2018**, *75*, 1541–1558. [CrossRef]

55. Vermeire, K.; Bell, T.W.; Van Puyenbroeck, V.; Giraut, A.; Noppen, S.; Liekens, S.; Schols, D.; Hartmann, E.; Kalies, K.-U.; Marsh, M. Signal Peptide-Binding Drug as a Selective Inhibitor of Co-Translational Protein Translocation. *PLoS Biol.* **2014**, *12*, e1002011. [CrossRef]

56. Gilles, A.; Frechin, L.; Natchiar, K.; Biondani, G.; Loeffelholz, O.v.; Holvec, S.; Malaval, J.-L.; Winum, J.-Y.; Klaholz, B.P.; Peyron, J.-F. Targeting the Human 80S Ribosome in Cancer: From Structure to Function and Drug Design for Innovative Adjuvant Therapeutic Strategies. *Cells* **2020**, *9*, 629. [CrossRef]

57. Dmitriev, S.E.; Vladimirov, D.O.; Lashkevich, K.A. A Quick Guide to Small-Molecule Inhibitors of Eukaryotic Protein Synthesis. *Biochemistry* **2020**, *85*, 1389–1421. [CrossRef] [PubMed]

58. Van den Berg, B.; Clemons, W.M., Jr.; Collinson, I.; Modis, Y.; Hartmann, E.; Harrison, S.C.; Rapoport, T.A. X-ray structure of a protein-conducting channel. *Nature* **2004**, *427*, 36–44. [CrossRef]

59. Voorhees, R.M.; Hegde, R.S. Structure of the Sec61 channel opened by a signal sequence. *Science* **2016**, *351*, 88–91. [CrossRef]
60. Mothes, W.; Prehn, S.; Rapoport, T.A. Systematic probing of the environment of a translocating secretory protein during translocation through the ER membrane. *EMBO J.* **1994**, *13*, 3973–3982. [CrossRef]
61. Fons, R.D.; Bogert, B.A.; Hegde, R.S. Substrate-specific function of the translocon-associated protein complex during translocation across the ER membrane. *J. Cell Biol.* **2003**, *160*, 529–539. [CrossRef]
62. Ménétret, J.F.; Hegde, R.S.; Aguiar, M.; Gygi, S.P.; Park, E.; Rapoport, T.A.; Akey, C.W. Single copies of Sec61 and TRAP associate with a nontranslating mammalian ribosome. *Structure* **2008**, *16*, 1126–1137. [CrossRef]
63. Meyer, H.-A.; Grau, H.; Kraft, R.; Kostka, S.; Prehn, S.; Kalies, K.-U.; Hartmann, E. Mammalian Sec61 Is Associated with Sec62 and Sec63. *J. Biol. Chem.* **2000**, *275*, 14550–14557. [CrossRef] [PubMed]
64. Wu, X.; Cabanos, C.; Rapoport, T.A. Structure of the post-translational protein translocation machinery of the ER membrane. *Nature* **2019**, *566*, 136–139. [CrossRef]
65. Itskanov, S.; Park, E. Structure of the posttranslational Sec protein-translocation channel complex from yeast. *Science* **2018**, *32*, 6740. [CrossRef] [PubMed]
66. Itskanov, S.; Kuo, K.M.; Gumbart, J.C.; Park, E. Stepwise gating of the Sec61 protein-conducting channel by Sec63 and Sec62. *Nat. Struct. Mol. Biol.* **2021**, *28*, 162–172. [CrossRef]
67. O'Keefe, S.; Pool, M.R.; High, S. Membrane protein biogenesis at the ER: The highways and byways. *FEBS J.* **2021**, *11*, 51. [CrossRef]
68. Kalies, K.-U.; Rapoport, T.A.; Hartmann, E. The beta-Subunit of the Sec61 Complex Facilitates Cotranslational Protein Transport and Interacts with the Signal Peptidase during Translocation. *J. Cell Biol.* **1998**, *141*, 887–894. [CrossRef] [PubMed]
69. Migliaccio, G.; Nicchitta, C.V.; Blobel, G. The signal sequence receptor, unlike the signal recognition particle receptor, is not essential for protein translocation. *J. Cell Biol.* **1992**, *117*, 15–25. [CrossRef] [PubMed]
70. Wiedmann, M.; Kurzchalia, T.V.; Hartmann, E.; Rapoport, T.A. A signal sequence receptor in the endoplasmic reticulum membrane. *Nature* **1987**, *328*, 830–833. [CrossRef]
71. Dudek, J.; Lang, S.; Schorr, S.; Linxweiler, J.; Greiner, M.; Zimmermann, R. Analysis of Protein Translocation into the Endoplasmic Reticulum of Human Cells. In *Membrane Biogenesis*; Rapaport, D., Herrmann, J.M., Eds.; Methods in Molecular Biology; Humana Press: Totowa, NJ, USA, 2013; Volume 1033, pp. 285–299.
72. Merrifield, R.B. Solid Phase Peptide Synthesis. I. *The Synthesis of a Tetrapeptide*. *J. Am. Chem. Soc.* **1963**, *85*, 2149–2154. [CrossRef]

International Journal of
Molecular Sciences

MDPI

Article

Reduced DNAJC3 Expression Affects Protein Translocation across the ER Membrane and Attenuates the Down-Modulating Effect of the Translocation Inhibitor Cyclotriazadisulfonamide

Eva Pauwels [1], Becky Provinciael [1], Anita Camps [1], Enno Hartmann [2] and Kurt Vermeire [1,*]

[1] KU Leuven, Department of Microbiology, Immunology and Transplantation, Rega Institute for Medical Research, Laboratory of Virology and Chemotherapy, B-3000 Leuven, Belgium; eva.pauwels@kuleuven.be (E.P.); becky.provinciael@kuleuven.be (B.P.); anita.camps@kuleuven.be (A.C.)

[2] Centre for Structural and Cell Biology in Medicine, Institute of Biology, University of Lübeck, 23562 Lübeck, Germany; enno.hartmann@uni-luebeck.de

*   Correspondence: kurt.vermeire@kuleuven.be

Abstract: One of the reported substrates for the endoplasmic reticulum (ER) translocation inhibitor cyclotriazadisulfonamide (CADA) is DNAJC3, a chaperone of the unfolded protein response during ER stress. In this study, we investigated the impact of altered DNAJC3 protein levels on the inhibitory activity of CADA. By comparing WT DNAJC3 with a CADA-resistant DNAJC3 mutant, we observed the enhanced sensitivity of human CD4, PTK7 and ERLEC1 for CADA when DNAJC3 was expressed at high levels. Combined treatment of CADA with a proteasome inhibitor resulted in synergistic inhibition of protein translocation and in the rescue of a small preprotein fraction, which presumably corresponds to the CADA affected protein fraction that is stalled at the Sec61 translocon. We demonstrate that DNAJC3 enhances the protein translation of a reporter protein that is expressed downstream of the CADA-stalled substrate, suggesting that DNAJC3 promotes the clearance of the clogged translocon. We propose a model in which a reduced DNAJC3 level by CADA slows down the clearance of CADA-stalled substrates. This results in higher residual translocation into the ER lumen due to the longer dwelling time of the temporarily stalled substrates in the translocon. Thus, by directly reducing DNAJC3 protein levels, CADA attenuates its net down-modulating effect on its substrates.

Keywords: co-translational translocation; endoplasmic reticulum; cyclotriazadisulfonamide; ER quality control; DNAJC3; signal peptide; preprotein; Sec61 translocon; ribosome stalling

Citation: Pauwels, E.; Provinciael, B.; Camps, A.; Hartmann, E.; Vermeire, K. Reduced DNAJC3 Expression Affects Protein Translocation across the ER Membrane and Attenuates the Down-Modulating Effect of the Translocation Inhibitor Cyclotriazadisulfonamide. *Int. J. Mol. Sci.* **2022**, *23*, 584. https://doi.org/10.3390/ijms23020584

Academic Editors: Richard Zimmermann and Sven Lang

Received: 15 December 2021
Accepted: 5 January 2022
Published: 6 January 2022

Publisher's Note: MDPI stays neutral with regard to jurisdictional claims in published maps and institutional affiliations.

1. Introduction

In eukaryotic cells, protein translocation into the endoplasmic reticulum (ER) is the first and decisive step in the biogenesis of secretory and integral membrane proteins [1–6]. This translocation process is typically guided by protein-specific signal peptides (SPs) and their interplay with the different components of the translocation machinery that are present in the cytosol, the ER membrane and the ER lumen [1–6].

Eukaryotic ER protein translocation is mediated by the heterotrimeric Sec61 translocon. The Sec61 translocon consists of α, β and γ monomers that together form an aqueous pore that spans the ER membrane [7–10]. Depending on the overall size of the precursor protein and its hydrophobicity and/or amino acid content, ER protein translocation occurs post- or co-translationally [11–14]. In higher eukaryotes, co-translational protein translocation is the most common and couples protein translation directly to its translocation over the ER membrane [11,14,15]. The interaction of the SP with the Sec61 translocon results in conformational changes of the translocon that eventually lead to the translocation of the preprotein into the ER lumen [7,9,16,17]. In the case of integral membrane proteins, the hydrophobic transmembrane domain (TMD) is integrated into the ER membrane via lateral movement from the translocon, mediated by the lateral gate of the Sec61 translocon [2,7]. Once preproteins are co-translationally translocated into the ER lumen, they are post-translationally modified (i.e., SP cleavage by the signal peptidase complex and protein glycosylation by the oligosaccharyl transferase complex), folded and transported to the Golgi apparatus for further maturation [2,3,17–20].

The accumulation of mislocalised, misfolded and/or malfunctioning proteins due to dysfunctional protein processing (i.e., protein post-translational modifications, folding and/or assembly in the ER lumen), triggers ER stress, which is detrimental for overall cellular functions. Cells have, therefore, acquired sophisticated ER quality control processes such as the unfolded protein response (UPR) [21–24]. Three ER stress sensors, i.e., activating transcription factor 6 (ATF6), inositol requiring enzyme 1 (IRE1) and PKR-like endoplasmic reticulum kinase (PERK) activate the UPR via the transcriptional upregulation of molecular chaperones to refold misfolded proteins in the ER lumen and inhibit global protein synthesis to reduce the load of client proteins, while apoptotic pathways are activated to eliminate severely damaged cells [21–23,25–31].

During ER stress, DnaJ Homolog Subfamily C member 3 (DNAJC3, also known as ERdj6 or p58^{IPK}) plays a role in the UPR to restore ER homeostasis [32–34]. After its transcriptional upregulation via ATF6 and IRE1, DNAJC3 is co-translationally translocated into the ER lumen where it functions as the co-chaperone alongside the binding immunoglobulin protein (BiP) to refold misfolded proteins in the ER lumen [32,33,35]. Once the lumenal proteins are properly folded and ER stress is (at least partly) relieved, DNAJC3 is involved in the activation of protein synthesis via the inhibition of PERK, thus, indirectly enhancing the initiation of protein translation [31,36,37]. Hence, DNAJC3 helps to restore protein levels and, therefore, reassures cellular processes post ER stress.

As stated earlier, ER stress is undoubtedly detrimental for cells, and, in general, contributes to the pathogenesis of many, mostly neurological, diseases [38–41]. Mutations in BiP co-chaperones and proteins involved in UPR have been associated with nervous system abnormalities [39,42–49]. Furthermore, DNAJC3 loss-of-function mutations have been identified in patients that were diagnosed with early-onset diabetes mellitus and suffered from multisystemic neurodegeneration (including ataxia, sensorimotor neuropathy and sensorineural hearing loss), which appear to be systematic features of the DNAJC3-diseased phenotype [50–52]. Studies highlight the importance of cholesterol homeostasis and lipid metabolism at the ER to underly the pathophysiology of DNAJC3 loss-of-function mutations. DNAJC3-deficient cells were presented with accumulated levels of cholesterol and lipids at the ER membrane [39,53,54], which disrupted its composition and overall structure and, thus, triggered ER stress [39,55–59]. Altogether, it is clear that DNAJC3 provides a multifunctional role in protecting cells during ER stress.

As well as folding-related ER stress, evidence has shown that erroneously halted protein translocation also induces considerable levels of ER stress, as the partially translocated polypeptide clogs the translocon [24,60,61]. From this perspective, the retarding of protein translocation into the ER by specific Sec61 translocon inhibitors might also induce ER stress and activate cellular pathways to clear the clogged translocons. How cells cope with clogged translocons has since emerged as an intriguing question; however, little is known about the translocon-associated quality control processes to safeguard protein translocation [24].

CADA is a small synthetic macrocycle (Figure 1A) that was discovered during an anti-HIV screening program [62–65]. Previous work by our laboratory showed that CADA inhibits the co-translational translocation process of human CD4 (huCD4), a type I membrane protein, by direct interaction with the receptor's SP, resulting in the downmodulation of huCD4 on the surface of cells and a significant reduced entry and replication of HIV in human cells [62–66]. Later, sortilin and 4-1BB were also identified as targets for CADA. A recent proteomic study conducted on the membranome of T-cells revealed only three additional substrates, i.e., ERLEC1, PTK7 and DNAJC3, evidencing high substrate specificity for the ER protein translocation inhibitor CADA [67–70].

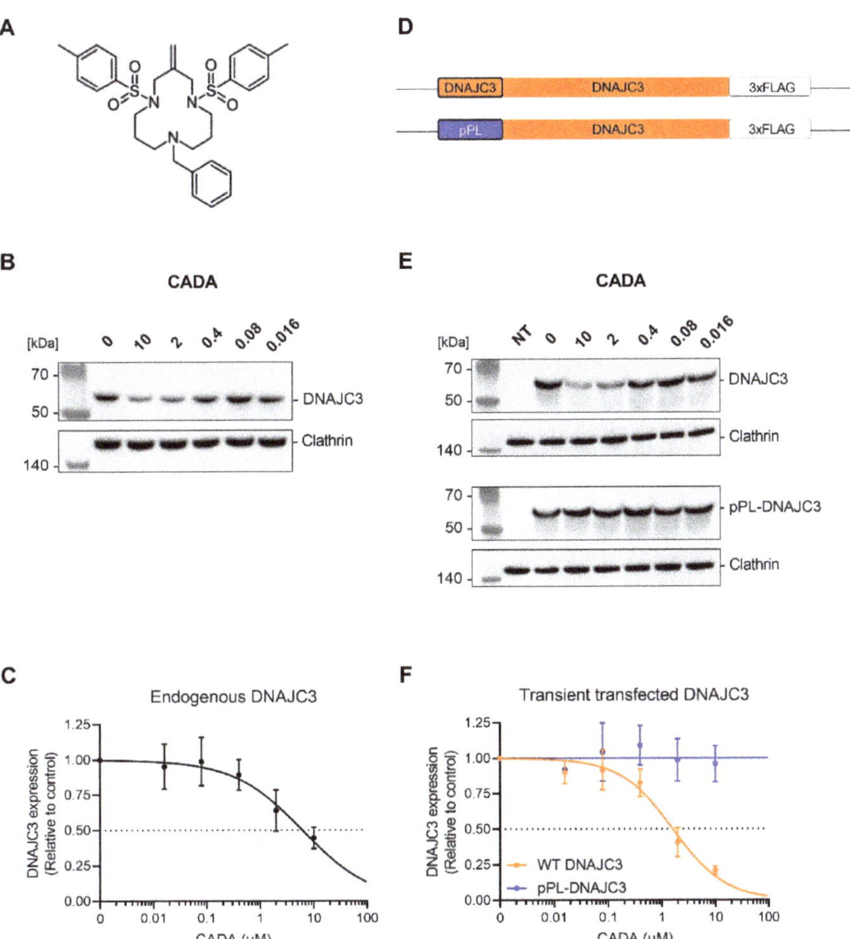

Figure 1. CADA sensitivity of endogenous and transfected DNAJC3 in HEK293T cells. (**A**) Chemical structure of CADA. (**B**) Western blot images of cell lysates from HEK293T cells treated for 24 h with different CADA concentrations. Protein bands were visualized with an antibody against DNAJC3 and an anti-clathrin antibody was used for the cell loading control. One representative experiment out of four is shown. (**C**) Concentration–response curves of CADA for endogenous DNAJC3 expression in HEK293T cells. Samples from (**B**) were quantified and normalised to the clathrin internal control. A four-parameter concentration–response curve was fitted to the data from four replicate experiments. Values are mean ± SD; n = 4. (**D**) Schematic representation of the DNAJC3 variants, with different signal peptides. (**E**) Western blot images of cell lysates from non-transfected (NT), DNAJC3-FLAG- or pPL-DNAJC3-FLAG-transfected HEK293T cells treated for 24 h with different CADA concentrations. Protein bands were visualized with an antibody against the FLAG tag and an anti-clathrin antibody was used for the cell loading control. One representative experiment out of five is shown. (**F**) Concentration–response curves of CADA for DNAJC3- and pPL-DNAJC3-transfected HEK293T cells. Samples from (**E**) were quantified and normalised to the clathrin internal control. A four-parameter concentration–response curve was fitted to the data from at least four replicate experiments. Values are mean ± SD; $n \geq 4$. CADA: cyclotriazadisulfonamide; HEK293T: human embryonic kidney 293T cells; DNAJC3: DnaJ homolog subfamily C member 3; pPL: pre-prolactin.

Thus, (i) CADA inhibits ER translocation of a specific set of substrates, presumably resulting in (temporarily) clogged translocons that might cause ER stress, and (ii) DNAJC3 is a CADA substrate that should be normally upregulated during ER stress. Therefore, we questioned if restored DNAJC3 levels under CADA pressure could influence the inhibitory effect of CADA. We compared the sensitivity of substrates to CADA in the presence of WT DNAJC3 or a CADA-resistant DNAJC3 mutant protein and observed the enhanced sensitivity of huCD4, PTK7 and ERLEC1 for CADA when DNAJC3 was expressed at high levels.

2. Results

2.1. CADA Reduces Cellular Expression of DNAJC3

In a recent proteomics survey, DNAJC3 has been validated as a target protein for CADA [69]. The treatment of HEK293T cells with increasing concentrations of CADA results in a concentration-dependent reduction in endogenous DNAJC3 expression (Figure 1B,C), with profoundly reduced DNAJC3 levels at 2 and 10 μM of CADA (IC_{50} = 6.2 μM). Due to the way DNAJC3 functions in the unfolded protein response during ER stress, we questioned if the reduced expression of DNAJC3 by CADA could affect the ER translocation of other CADA substrates. Previous work has shown that the sensitivity of targets to CADA, such as huCD4, 4-1BB, SORT, ERLEC1, PTK7 and DNAJC3, is intrinsic to the cleavable SP and the N-terminal region of the mature protein [64,68,69]. By inserting the huCD4 SP and the first 7 amino acid residues of the mature huCD4 protein into the CADA-resistant preprolactin (pPL) protein, we can introduce sensitivity of pPL to CADA as determined by immunoblotting (Supplementary Figure S1A,B), as similarly described for the CADA-resistant murine CD4 (mCD4) [68]. Vice versa, exchanging the wild-type SP of CADA-sensitive proteins by that of pPL or mCD4 results in the generation of a CADA-resistant protein [64]. Thus, we generated a CADA-resistant DNAJC3 protein (designated as pPL-DNAJC3) by replacing the CADA-sensitive SP of DNAJC3 with the CADA-resistant SP of pPL (see Figure 1D). We also introduced a triple-FLAG sequence at the C-terminus of DNAJC3 for detection purposes. As expected, the transfection of HEK293T cells with the DNAJC3 plasmids resulted in high protein levels of DNAJC3 that outranged the endogenous expression of DNAJC3 (Supplementary Figure S1C,D). HEK293T cells transfected with WT DNAJC3 showed a concentration-dependent sensitivity of DNAJC3 to CADA (IC_{50} of 1.6 μM) (Figure 1E,F), which was slightly higher than that of endogenous DNAJC3 (Figure 1C). In contrast, the transfection of cells with the pPL-DNAJC3 construct resulted in a CADA-resistant phenotype with a constant expression of DNAJC3, as well as at the highest CADA concentrations tested (Figure 1F). Thus, the replacement of the DNAJC3 SP with the SP of pPL resulted in the complete loss of sensitivity of DNAJC3 to CADA.

2.2. CADA Inhibits the Signal Peptide Dependent Co-Translational Translocation of the ER Lumenal DNAJC3 Protein

As the SP was cleaved from the DNAJC3 preprotein during ER translocation, we expected that both WT DNAJC3 and the chimaeric pPL-DNAJC3 constructs expressed the same mature DNAJC3 protein. To verify that DNAJC3 is translocated into the ER lumen, cell-free in vitro translation experiments were performed (Figure 2). As initial attempts to translate the full-length WT DNAJC3 protein were not successful (data not shown), we switched to an alternative DNAJC3/huCD4 chimaeric protein. This truncated soluble huCD4 construct contained the SP of DNAJC3 and the first 62 residues of mature DNAJC3 (see scheme in Figure 2A). In vitro translation of DNAJC3/huCD4 resulted in substantial levels of translated preprotein and detectable SP cleavage (Figure 2B, lane 2). Of note, more than one-third of the DNAJC3 species retained the SP after translocation as determined by proteinase K (PK) treatment (Figure 2C, right panel), an observation that is in line with another report [71] (Supplementary Figure S2). The translocated SP-containing DNAJC3 species were fully protected from PK treatment (Figure 2B, lanes 8–14), indicating that these proteins were localised in the lumen of the ER. Of course, for the translocated SP-containing DNAJC3 protein, the PK protection assay could not distinguish between membrane-anchored (by the SP) and luminal free-floating DNAJC3 species. Notably, CADA inhibited the translocation of both non-cleaved and cleaved DNAJC3 species in an equal manner (Figure 2B, lanes 9–14 and Figure 2D, right panel), whereas the preprotein levels in the control samples seemed to be unaffected by CADA (Figure 2B, lanes 2–7 and Figure 2D, left panel). In addition, DNAJC3 has also been proposed as an uncleaved type I transmembrane protein that inserts in the ER membrane in a head-on formation with the SP as the membrane anchor and with the majority of the protein exposed to the cytosol [71]. The relatively high amount of DNAJC3 preprotein as compared to the SP-cleaved species in the non-PK samples (Figure 2B, lanes 2–7 and Figure 2C, left panel), might suggest the appearance of such a membrane-anchored cytosolic protein. However, from these radioblots, we could not make a distinction between a free non-targeted preprotein and ER membrane-anchored DNAJC3 preprotein. Finally, DNAJC3 did not contain potential N-glycosylation sites, thus, excluding the possibility of tracking the translocated protein fraction by ER lumenal N-glycosylation.

2.3. Presence of Cellular DNAJC3 Enhances the Sensitivity of Substrates to CADA

To analyze if, under CADA pressure, the reduced DNAJC3 levels can affect the sensitivity of proteins to CADA, we co-transfected HEK293T cells with a V5-tagged huCD4 construct and either the FLAG-tagged WT DNAJC3 plasmid or the CADA-resistant pPL-DNAJC3 plasmid. The transiently transfected cells were then treated with increasing concentrations of CADA and subjected to immunoblotting. Similar co-transfections were performed with the V5-tagged ERLEC1 or PTK7 constructs. Both huCD4 and PTK7 are type I integral membrane proteins mainly expressed at the cell surface, whereas ERLEC1 is a soluble protein that resides in the ER lumen. In accordance with our previous report, the expression of the three V5-tagged proteins was differentially affected by CADA [69] (Supplementary Figure S3A,B). As summarized in Figure 3A, in the non-treated control cells, transfection with the different plasmids resulted in the high expression level of the V5-tagged target proteins and the co-transfected FLAG-tagged DNAJC3 protein. Treatment with CADA concentration dependently reduced the protein expression of huCD4, ERLEC1 and PTK7 when WT DNAJC3 was expressed (Figure 3A,C). Interestingly, the co-expression of the target proteins with pPL-DNAJC3 resulted in a stronger CADA effect as compared to WT DNAJC3 co-expression. This was evidenced by the significantly lower protein levels detected (Figure 3B versus Figure 3A) and by the comparison of the corresponding calculated IC$_{50}$ values (Figure 3C). Altogether, these data indicate that enhanced levels of DNAJC3 result in the sensitization of selective substrates to CADA.

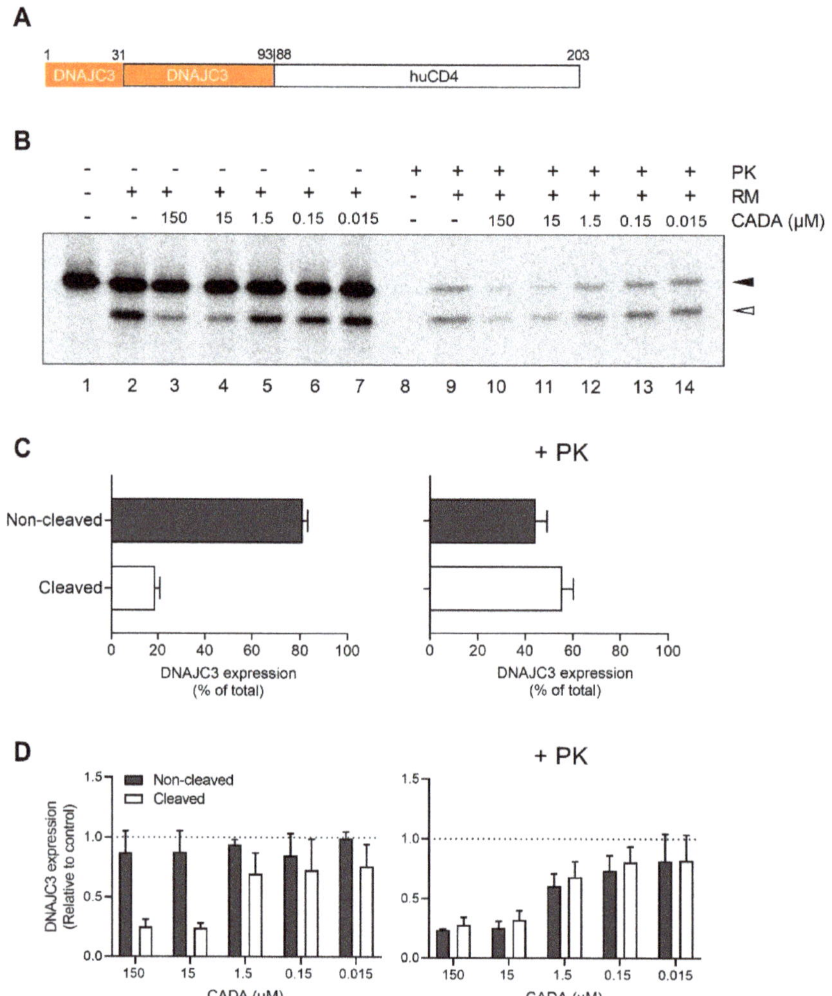

Figure 2. CADA inhibits the signal peptide-dependent co-translational translocation of the ER lumenal DNAJC3 protein. (**A**) Representation of the construct used for cell-free in vitro translation and translocation assay. (**B**) Cell-free in vitro translation and translocation in rabbit reticulocyte lysate supplemented with ovine microsomes, CADA and proteinase K (PK). Autoradiogram of the in vitro translated and translocated DNAJC3-huCD4 chimaeric protein. In the presence of rough microsomes (RM), the preprotein (black arrowhead) was translocated into the ER lumen and the SP was cleaved, resulting in a faster migrating mature protein (open arrowhead). One representative experiment out of three is shown. (**C**) Percentage of non-cleaved (black arrowhead, preprotein) and cleaved (open arrowhead, mature protein) DNAJC3 protein in the DMSO-treated control sample of untreated and PK-treated samples. Samples from (A, lane 2 and lane 9) were quantified. Bars are mean ± SE; $n = 3$. (**D**) CADA sensitivity of the non-cleaved and cleaved DNAJC3 protein fraction in control samples and PK-treated samples. Samples from A (lanes 2–7 and 9–14) were quantified and normalised to the respective protein fraction in the DMSO control (lane 2 or 9). Bars are mean ± SE; $n = 3$. CADA: cyclytriazadisulfonamide; DNAJC3: DnaJ homolog subfamily C member 3; PK: Proteinase K; RM: rough microsomes; ER: endoplasmic reticulum; SP: signal peptide; DMSO: dimethyl sulfoxide.

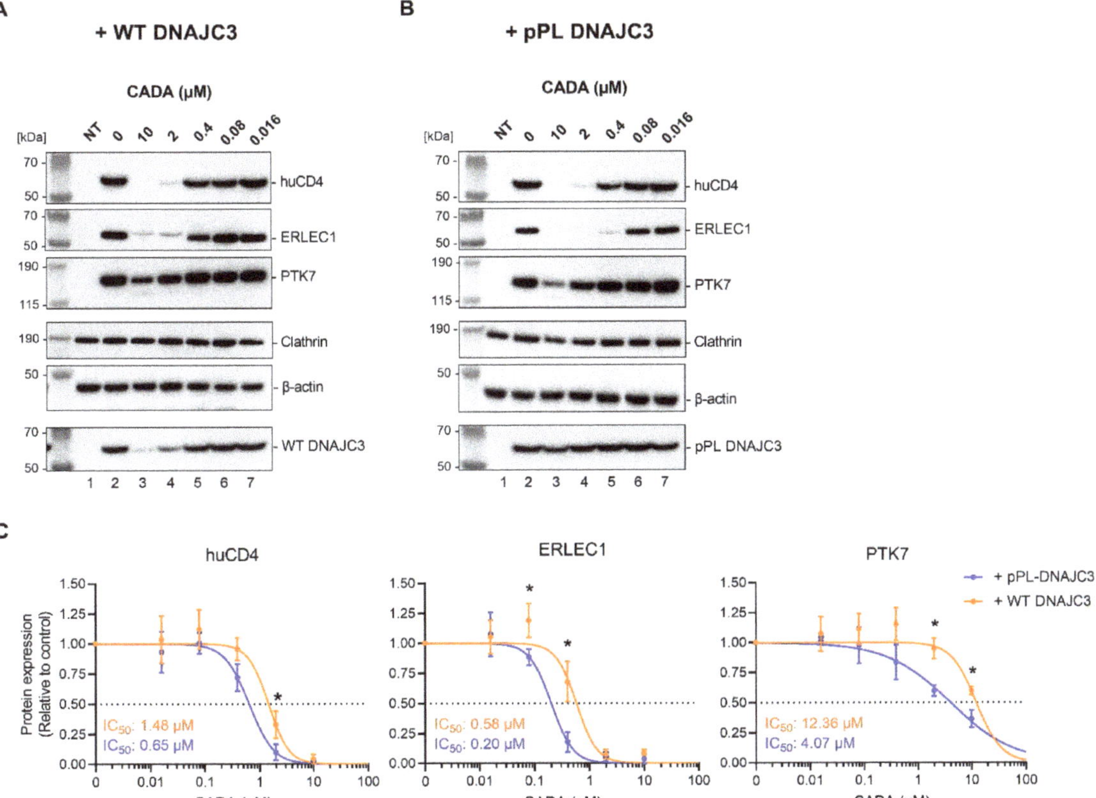

Figure 3. DNAJC3 sensitizes targets to CADA. (**A**) Western blot images of cell lysates from non-transfected (NT) HEK293T cells, and huCD4-V5, ERLEC1-V5 or PTK7-V5, co-transfected with WT DNAJC3. Cells were treated for 24 h with different CADA concentrations and subjected to immunoblotting. Protein bands were visualized with an antibody against the V5 tag, and an antibody against clathrin (huCD4 and ERLEC1) or β-actin (PTK7) was used for the cell loading controls. An antibody against the FLAG tag was used to detect the co-transfected WT DNAJC3 in the samples. One representative experiment out of three to six is shown. The clathrin loading control shown is that of the ERLEC1 sample. The respective loading control for huCD4 and WT DNAJC3 is presented in Supplementary Figure S3D. (**B**) Same as in (**A**) but for co-transfection with pPL-DNAJC3. One representative experiment out of three to six is shown. The clathrin loading control shown is that of the huCD4 sample. The respective loading control for ERLEC1 and pPL-DNAJC3 is presented in Supplementary Figure S3D. (**C**) Concentration–response curves of CADA for huCD4, ERLEC1 and PTK7 in transfected HEK293T cells. Samples from (**A,B**) were quantified and normalised to the clathrin (huCD4 and ERLEC1) or β-actin (PTK7) internal control. A four-parameter concentration–response curve was fitted to data from at least three replicate experiments. Values are mean ± SD; $n \geq 3$. Statistical analysis (multiple unpaired *t*-tests) showed significantly decreased expression of huCD4, ERLEC1 and PTK7 when co-expressed with pPL-DNAJC3 as compared to WT DNAJC3 (* = $p < 0.05$). HEK293T: human embryonic kidney 293T cells; huCD4: human CD4; ERLEC1: endoplasmic reticulum lectin 1; PTK7: inactive tyrosine-protein kinase 7; pPL: pre-prolactin; DNAJC3: DnaJ homolog subfamily C member 3; CADA: cyclotriazadisulfonamide.

2.4. DNAJC3 Enhances Proteasomal Degradation of CADA-Stalled PTK7 Preprotein

Previous experiments clearly indicate that the amount of translocated (mature) protein is more reduced by CADA when DNAJC3 is highly expressed. The lower levels of translocated protein induced by CADA treatment can be the net result of the stronger inhibition of protein translocation (lower entry efficiency into ER) or faster clearance of putative mis-targeted proteins from the ER by DNAJC3 (higher exit efficiency from ER). However, the role of DNAJC3 in this process is not entirely understood.

First, we addressed if the CADA inhibition of protein translocation is linked to the proteasomal degradation of the CADA substrates. Thus, we analyzed the effect of the proteasome inhibitor MG132 on the CADA samples. Briefly, we co-transfected HEK293T cells with V5-tagged huCD4, ERLEC1 or PTK7 and either the FLAG-tagged WT DNAJC3 plasmid or the CADA-resistant pPL-DNAJC3 plasmid. Cells were then treated with increasing concentrations of CADA in combination with a fixed dose of MG132 (200 nM) for 24 h and subjected to immunoblotting. As shown in Figure 4A, in the non-treated control cells, the expression of the substrates was not affected by treatment with the proteasome inhibitor MG132 only (Figure 4A, lanes 2 vs. 3). However, for PTK7, a faint lower band on the gel could be detected for the MG132-treated control sample (Figure 4A, lane 3) that corresponded to the non-glycosylated preprotein fraction as determined by Endo H treatment (Supplementary Figure S3C). The combined treatment of CADA with MG132 did result in the rescue of the preprotein, which was most evident for PTK7 (Figure 4A lanes 5, 7 and 9). Remarkably, the combination of 10 or 2 µM of CADA with MG132 completely abolished the expression of mature PTK7, demonstrating that MG132 has a synergistic effect on CADA in blocking the translocation of PTK7. Additionally, for huCD4, 2 µM CADA treatment in combination with MG132 further reduced the low level of mature (glycosylated) protein, as compared to CADA treatment only, and resulted in the rescue of the huCD4 preprotein, although at a nearly detectable and very low amount (Figure 4C). For ERLEC1, the complete inhibition of protein translocation was achieved with the higher concentrations of CADA. Here, the only fraction of ERLEC1 that could be (weakly) visualized was the slower migrating protein band that corresponds to the preprotein, given that ERLEC1 is an ER-resident protein that is not glycosylated (Supplementary Figure S3C), but of which the SP is cleaved upon translocation into the ER lumen. The addition of MG132 did not change the outcome of CADA treatment. This also suggests that, in contrast to huCD4 and PTK7, ERLEC1 has a more stable preprotein that is already visualised without the inhibition of the proteasome. In line with the data described above, the treatment of CADA alone resulted in the reduced expression of the mature protein fraction in a concentration-dependent manner that was more pronounced when the proteins were co-transfected with pPL-DNAJC3 as compared to co-transfection with WT DNAJC3 (Figure 4A,B, respectively). For ERLEC1, the absolute amount of rescued preprotein with MG132 remained little and somehow constant over the different CADA concentrations, irrespective of enhanced DNAJC3 expression (Figure 4C). Interestingly, when DNAJC3 was highly expressed (Figure 4B), the combined treatment of CADA with MG132 rescued proportionally more preprotein of PTK7 that reached significance for the 10 and 2 µM CADA samples ($p = 0.034$ and 0.047, respectively) (Figure 4C), indicating that DNAJC3 enhanced the proteasomal degradation of CADA-stalled PTK7 preprotein.

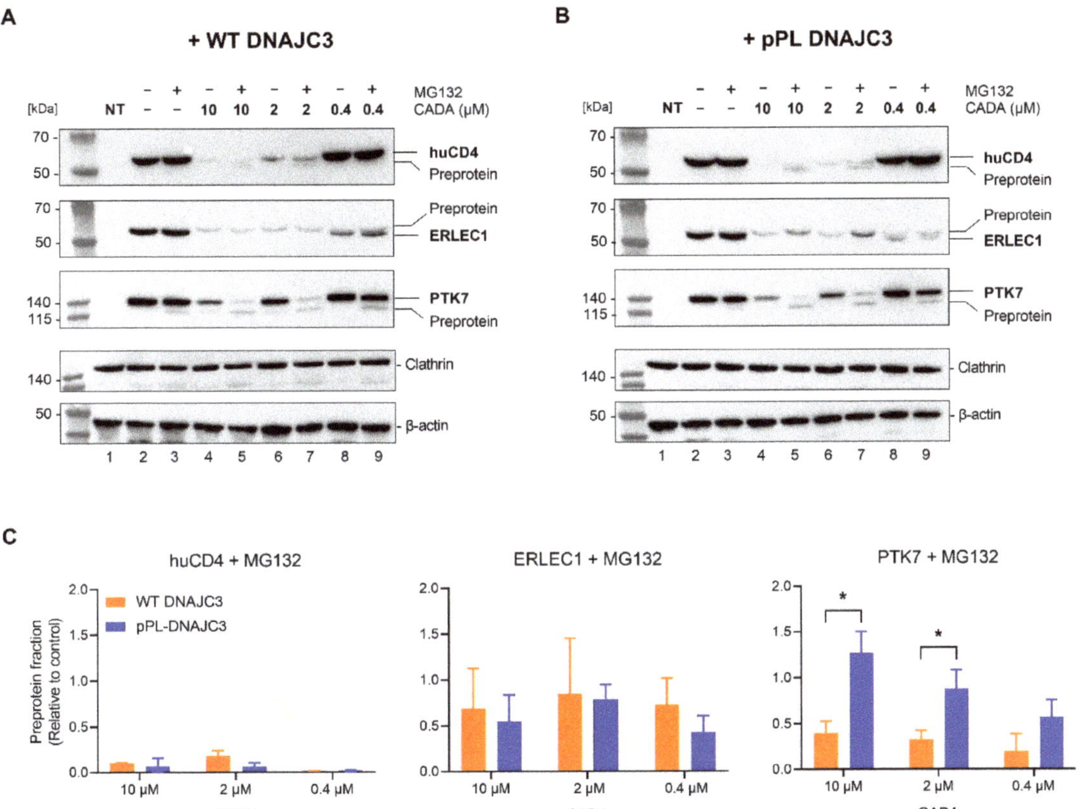

Figure 4. Inhibition of the proteasome in the presence of CADA rescues a preprotein fraction which is DNAJC3-dependent for PTK7. (**A**) Western blot images of cell lysates from non-transfected (NT) HEK293T cells, and huCD4-V5, ERLEC1-V5 or PTK7-V5, co-transfected with WT DNAJC3. Cells were treated for 24 h with different CADA concentrations and a constant dose of MG132 (200 nM). Protein bands were visualized with an antibody against the V5 tag, and an antibody against clathrin (huCD4 and ERLEC1) or β-actin (PTK7) was used for the cell loading controls. One representative experiment out of two to four is shown. The clathrin loading control shown is that of the ERLEC1 sample. The respective loading control for huCD4 is presented in Supplementary Figure S3E. (**B**) Same as in (**A**) but for co-transfection with pPL-DNAJC3. One representative experiment out of two to four is shown. (**C**) Preprotein fraction of MG132-treated samples quantified from (**A,B**) and normalised to the internal loading control of the respective sample. Bars are mean ± SE; $n = 2$ for huCD4 and PTK7; $n = 4$ for ERLEC1. Statistical analysis (multiple unpaired *t*-tests) showed increased detection of the PTK7 preprotein when co-expressed with pPL-DNAJC3 as compared to WT DNAJC3 (* = $p < 0.05$). HEK293T: human embryonic kidney 293T cells; huCD4: human CD4; ERLEC1: endoplasmic reticulum lectin 1; PTK7: inactive tyrosine-protein kinase 7; pPL: pre-prolactin; DNAJC3: DnaJ homolog subfamily C member 3; CADA: cyclotriazadisulfonamide; DMSO: dimethyl sulfoxide.

2.5. DNAJC3 Differentially Affects the Expression of Cytosolic or ER Translocated Proteins in the Presence of CADA-Stalled Substrates

The previous observation that a preprotein fraction could be rescued when cells are treated with a combination of CADA and MG132 indicates that an early event in protein translocation is blocked by CADA at a stage when the SP is not yet cleaved from the preprotein, and the protein is not yet glycosylated. Additionally, the limited but constant amount of rescued ERLEC1 preprotein measured in the CADA samples suggests that a saturating level of preprotein is reached, presumably corresponding to the fraction of preprotein that is stalled at the protected ribosome-bound translocon. Thus, it seems that CADA induces stalling of the preprotein at the translocon with subsequent extraction and further degradation of the respective preprotein by the proteasome. To explore if DNAJC3 contributes to the clearance of the CADA-stalled preproteins, we made use of fluorescently labelled CADA substrates in a tGFP-P2A-BFP backbone as described in a recent report [69]. This reporter construct encodes BFP downstream of a viral P2A sequence (Figure 5A), resulting in the transcription of polycistronic mRNA that is translated into two separated proteins in equal amounts, with cytosolic BFP serving as an internal protein translation control. By means of flow cytometry, the amount of tGFP and BFP was quantified for cells co-transfected with substrate-tGFP-P2A-BFP and either WT DNAJC3 or pPL-DNAJC3. In line with our previous report [69], the three substrates (i.e., WT huCD4, PTK7/huCD4 and ERLEC1/huCD4 chimaeric proteins) showed sensitivity to CADA in a concentration-dependent way (Figure 5B) which was similar to the sensitivity of full-length WT protein, as determined for PTK7 (Supplementary Figure S4A,B). The effect of CADA was related to the presence of an SP, given that an ERLEC1 mutant without an SP remained resistant to CADA (Figure 5B). As shown in Figure 5C, a detectable concentration-dependent decrease in BFP expression was observed when the substrates were treated with CADA (Figure 5C, black lines). Interestingly, the CADA-induced decrease in BFP could be nearly completely restored by co-transfecting the cells with the WT DNAJC3 plasmid, with a significant effect for ERLEC1 at the highest CADA concentrations (Figure 5C, orange lines). Furthermore, the reduction in BFP could be prevented (for PTK7) or even reversed (for huCD4 and ERLEC1) by pPL-DNAJC3 co-transfection, resulting in significantly enhanced BFP expression for ERLEC1 and huCD4 as compared to the non-co-transfected cells (Figure 5C, blue lines). Comparable results were obtained with the full-length PTK7 protein (Supplementary Figure S4C). Of note, the enhanced BFP levels under CADA pressure when cells expressed high levels of DNAJC3 could only be evoked by a CADA-sensitive substrate, given that a CADA-resistant construct of ERLEC1 (missing the SP) expressed cytosolic BFP at constant levels, irrespective of the DNAJC3 expression (Figure 5D). This indicates that CADA-treatment has no direct inhibitory effect on protein translation as such (i.e., ER-translocation independent), an effect that was also observed in a cell-free in vitro translation system [64]. Additionally, our BFP data show that when co-translational translocation of a substrate is blocked by CADA, translation of a downstream cytosolic protein is enhanced when DNAJC3 is present, suggesting that multiple attempts of protein translation from the same polycistronic transcript can be made once initially stalled proteins have been removed and blocked translocons have been cleared to restart protein translation.

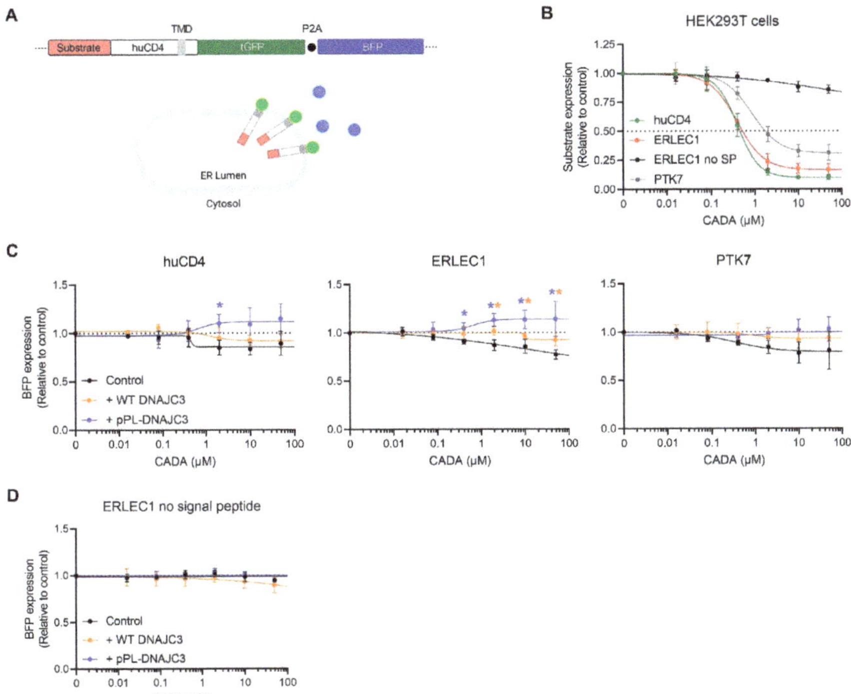

Figure 5. DNAJC3 differentially affected BFP expression in the presence of CADA-stalled proteins. (**A**) Representation of the tGFP-P2A-BFP construct. The construct expressed huCD4 that was anchored in the plasma membrane via its transmembrane domain (TMD) and with tGFP at the cytosolic tail. As the SP was cleaved by the ER lumenal signal peptidase during protein biogenesis, the mature huCD4 variants differed in only 62 amino acids at their N-terminus. (**B**) Four-parameter concentration–response curves for the CADA of huCD4, ERLEC1, ERLEC1 with no SP, and PTK7 cloned in the same tGFP-P2A-BFP plasmid backbone as shown in (**A**). HEK293T cells were transiently transfected with the tGFP-P2A-BFP constructs and incubated with different CADA concentrations for 24 h. Transfected tGFP-P2A-BFP plasmid DNA was equal to the transfected tGFP-P2A-BFP plasmid DNA in the co-transfected conditions of (**C,D**). Protein levels of tGFP (representing the level of substrate) in CADA-treated samples were normalised to the DMSO control (set at 1.00). Curves were fitted to data from three to five replicate experiments. Values are mean ± SD; $n \geq 3$. (**C**) Four-parameter concentration–response curves for CADA of the BFP signal of huCD4, ERLEC1 and PTK7 cloned in the same tGFP-P2A-BFP plasmid backbone as shown in (**A**). HEK293T cells were transiently transfected with the tGFP-P2A-BFP construct, or co-transfected with the tGFP-P2A-BFP construct and WT DNAJC3 or pPL-DNAJC3 and incubated with different CADA concentrations for 24 h. Transfected tGFP-P2A-BFP plasmid DNA was equal to the transfected tGFP-P2A-BFP plasmid DNA in the co-transfected conditions. BFP levels in CADA-treated samples were normalised to the DMSO control (set at 1.0). Curves were fitted to data from three to five replicate experiments. Values are mean ± SD; $n \geq 3$. Statistical analysis (multiple unpaired t-tests) showed significantly increased expression of BFP for huCD4 and ERLEC1 when co-expressed with pPL-DNAJC3 and for ERLEC1 when co-expressed with WT DNAJC3 as compared to the control (* = $p < 0.05$). (**D**) Same as in (**C**). Four-parameter concentration–response curve for the CADA of the BFP signal of ERLEC1 with no SP cloned in the tGFP-P2A-BFP plasmid backbone. Curves were fitted to data from three replicate experiments. Values are mean ± SD; $n = 3$. tGFP: turbo green fluorescent protein; BFP: blue fluorescent protein; huCD4: human CD4; SP: signal peptide; CADA: cyclotriazadisulfonamide; ERLEC1: endoplasmic reticulum lectin 1; PTK7: inactive tyrosine-protein kinase 7; HEK293T: human embryonic kidney 293T cells; DMSO: dimethyl sulfoxide; pPL: pre-prolactin; DNAJC3: DnaJ homolog subfamily C member 3; TMD, transmembrane domain.

Finally, we analyzed the combined effect of CADA with high DNAJC3 levels on the expression of a CADA-resistant type I transmembrane protein (i.e., mouse CD4) that is expressed in cis of a CADA-sensitive substrate (see representation in Figure 6A). In this construct, mCD4 is separated from ERLEC1 by a P2A sequence, similar to the construct with BFP. Interestingly, without co-expression of DNAJC3, levels of mCD4 were enhanced under CADA pressure (Figure 6B, black line), suggesting that targeting of the mCD4 protein to the ER membrane might be more efficient when the upstream ERLEC1 protein is stalled at the translocon by CADA. In accordance with the BFP data, high levels of DNAJC3 enhanced the expression of mCD4 (Figure 6B, blue line), which could be explained by the DNAJC3 mediated accelerated clearance of blocked translocons and/or enhanced re-initiation of protein translation. Remarkably, when the mCD4 sequence (containing its own SP) was fused directly to the ERLEC1 sequence without the separation by P2A, mCD4 protein expression became regulated by CADA (Figure 6C, black line). Blocking the translocation of ERLEC1 by CADA also slowed down the translocation of the C-terminal part of the fusion protein. However, the presence of the CADA-resistant mCD4 SP preserved some translocation autonomy for mCD4. Of note, in the control samples without CADA the translation of mCD4 when directly fused to ERLEC1 (Figure 6C) was clearly less productive as compared to the construct that contained the P2A sequence (Figure 6B), resulting in mCD4 levels that represent only 2% of the P2A counterpart as determined by the flow cytometric mean fluorescence intensity (MFI) values (Supplementary Figure S5A). Remarkably, low concentrations of CADA (400 nM) enhanced mCD4 expression tremendously, again pointing to a stabilizing effect of CADA on the insertion of the preprotein in the translocon with higher targeting and translocation effect as a result (Supplementary Figure S5B). Co-expression of DNAJC3 significantly reduced the level of translocated mCD4 protein, which was most evident when the cells were co-transfected with the CADA-resistant pPL-DNAJC3 (Figure 6C, blue line). This would suggest that DNAJC3 quickly removes the stalled mCD4 protein from the Sec61 translocon, thereby reducing the putative attempts of the stalled protein from being transported into the ER lumen, thus, preventing the protein from ultimately being translocated under CADA pressure.

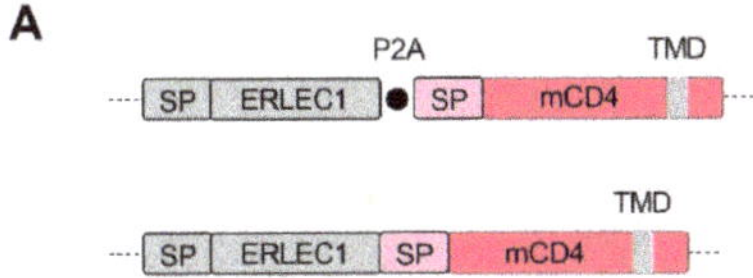

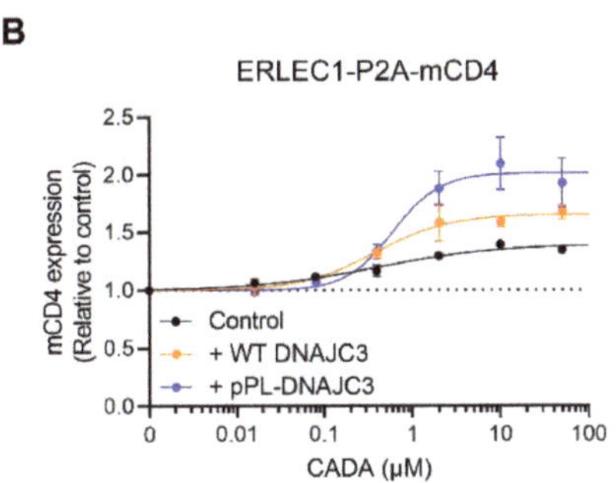

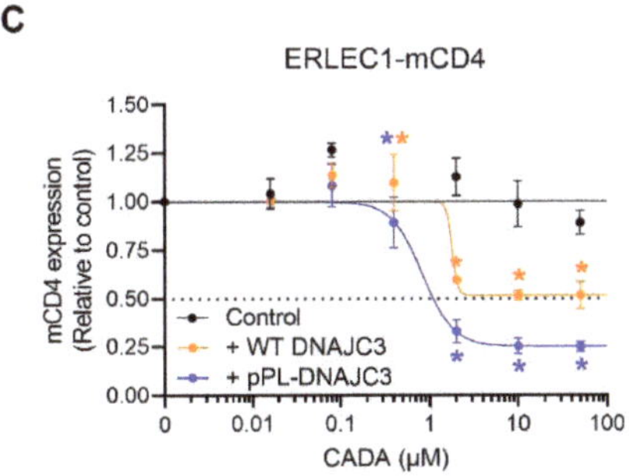

Figure 6. DNAJC3 affects the expression of a CADA-resistant protein differentially depending on the nature of the transcript. (**A**) Representation of the ERLEC1-mCD4 chimaeric constructs. The constructs expressed intracellular ERLEC1 and cell surface mCD4 that was anchored in the plasma membrane via its transmembrane domain. In the first construct, the two proteins were separated by a P2A sequence, whereas, in the second construct, the SP of mCD4 was directly fused to the C-terminus of ERLEC1. (**B**) Four-parameter concentration–response curves for the CADA of mCD4 that was cloned in the ERLEC1-P2A-mCD4 backbone as shown in (**A**). HEK293T cells were transiently transfected with ERLEC1-P2A-mCD4 plasmid DNA, or co-transfected with ERLEC1-P2A-mCD4 plasmid DNA and WT DNAJC3 or pPL-DNAJC3 and incubated with different CADA concentrations for 24 h. Transfected ERLEC1-P2A-mCD4 plasmid DNA was equal to the transfected ERLEC1-P2A-mCD4 plasmid DNA in the co-transfected conditions. Cells were fixed, permeabilized and stained with an anti-mCD4 antibody. Total levels of mCD4 in CADA-treated samples were normalised to the DMSO control (set at 1.0). Curves were fitted to data from two replicate experiments. Values are mean ± SD; $n = 2$. (**C**) Similar as in (**B**) but for the second construct without the P2A sequence. Curves were fitted to data from three replicate experiments. Values are mean ± SD; $n = 3$, except for the 0.08 µM CADA samples of control and WT DNAJC3 for which $n = 2$. Statistical analysis (multiple unpaired *t*-tests) showed significantly decreased expression of mCD4 when co-expressed with WT DNAJC3 and pPL-DNAJC3 as compared to the control. Of note, the values of 0.4 µM CADA for the ERLEC1-mCD4 construct only (black dots) have been removed to plot the black curve. These values are represented separately in Supplementary Figure S5. mCD4: mouse CD4; ERLEC1: endoplasmic reticulum lectin 1; CADA: cyclotriazadisulfonamide; HEK293T: human embryonic kidney 293T cells; pPL: pre-prolactin; DNAJC3: DnaJ homolog subfamily C member 3. (* = $p < 0.05$).

3. Discussion

Following up on a recent proteomics survey [69], in this study, we addressed the impact of altered DNAJC3 levels by small-molecule CADA on a few other CADA substrates, such as huCD4, ERLEC1 and PTK7. By comparison of a CADA-sensitive WT DNAJC3 with a CADA-resistant variant, we concluded that elevated DNAJC3 levels enhanced the inhibitory effect of CADA on the co-translational translocation of its substrates. Additional experiments with a proteasome inhibitor resulted in the rescue of a small fraction of preprotein under CADA pressure, pointing at a stalling effect of CADA on its substrates at the ribosome/Sec61 translocon complex. Finally, by means of flow cytometry, we demonstrated the DNAJC3-related enhanced protein translation of reporter proteins that were expressed downstream of a CADA-sensitive substrate. All these data together suggest a potential role of DNAJC3 in the clearance of clogged translocons by stalled ribosomes.

Although DNAJC3 has been validated as a substrate for CADA, our data confirm that DNAJC3 is not the most sensitive target for CADA. However, at high CADA concentrations, approximately 60% reduction in endogenous DNAJC3 and 75% decrease in transfected WT DNAJC3 was achieved that can clearly impact DNAJC3-related pathways. The inhibitory effect of CADA on the transfected WT DNAJC3 seemed to be stronger as compared to the endogenous protein (Figure 1F versus Figure 1C). This might be the result of the transfection protocol in which the compound is added as early as the DNAJC3 protein synthesis starts to prevent the biosynthesis of the protein, whereas, for endogenous DNAJC3, the net effect of CADA also relies on the natural turn-over (and degradation) of the existing cellular DNAJC3 source before the addition of CADA. Longer treatment (>24 h) with CADA generally enhances the protein down-modulating effect, as reported for huCD4 [62]. Regarding the cellular DNAJC3 protein level, given that (i) transfected cells express approximately 16 times more DNAJC3 as compared to the endogenous level (Supplementary Figure S1D), and (ii) high CADA-treatment still allows 25% expression of the protein (Figure 1F), different absolute expression levels of DNAJC3 were obtained under 10 µM CADA pressure, increasing gradually from a 0.4-fold, to 4-fold, to 16-fold increase for non-transfected, WT DNAJC3 and pPL-DNAJC3 transfected cells, respectively (as compared to the untreated non-transfected control). One should also interpret the obtained results (e.g., Figure 5) in light of these dose-response effects of CADA on DNAJC3.

One of the challenges of our study is the limited knowledge of DNAJC3, especially, about the different forms of the protein (with or without SP) and the cellular localisation (ER lumen or cytosolic). Our cell-free in vitro translation data clearly showed the existence of two different translocated species of DNAJC3 in the PK-protected ER lumen. The presence of a hydrophobic SP for one of the translocated DNAJC3 species suggests the anchoring of DNAJC3 in the ER membrane; however, additional experiments (e.g., alkaline flotation) are needed to verify this. The cell-free translation experiments were not performed with the WT DNAJC3 protein, though the DNAJC3/huCD4 chimaeric construct most likely represents a reliable alternative, given that it contains the N-terminal region of DNAJC3 sufficient to retain the targeting and gating features of WT DNAJC3 [72]. Accordingly, PK treatment of a DNAJC3/pPL chimaeric variant (used in a previous study [69]) also revealed a subfraction of uncleaved translocated DNAJC3 (Supplementary Figure S2). This indicates that the SP of DNAJC3 might contribute to the ultimate expression and subcellular localisation of the protein. Of course, most of our study is based on the comparison of WT DNAJC3 with a CADA-resistant pPL variant in which the SP of DNAJC3 has been exchanged. This could not only have an impact on sensitivity to CADA but also on the amount of ER membrane-anchored DNAJC3 species. From previous experiments with WT pPL (Supplementary Figure S2), we assume that SP cleavage of pPL-DNAJC3 should be complete and that solely SP-cleaved mature DNAJC3 proteins are present in the ER lumen, but this needs further investigation. Thus, depending on the role of the uncleaved WT DNAJC3 species, the comparison between CADA-sensitive and CADA-resistant DNAJC3 might not only be limited to the different levels of free DNAJC3 in the ER lumen under CADA pressure; therefore the presence or absence of the membrane-anchored DNAJC3 species should also be taken into account. However, in our immunoblot samples from HEK293T cells, we were not able to distinguish a putative uncleaved DNAJC3 variant, questioning the survival (and existence) of those species in cellulo. Furthermore, the significant differences in substrate expression between the WT DNAJC3 and pPL-DNAJC3 transfected cells (Figure 3) suggest that the SP-cleaved lumenal DNAJC3 protein is most likely the main driver in our study. Additionally, based on the consistent dose–response effects seen in Figures 5 and 6, the contribution of a membrane-anchored DNAJC3 species in CADA activity is questionable, given that the condition with transfected WT DNAJC3 (orange curves in Figures 5 and 6) would be the one with the highest amount of uncleaved translocated DNAJC3.

In the rested state, DNAJC3 is localised in the ER lumen where it has a multifunctional role in the protection of cells from the detrimental effects of ER stress. For instance, DNAJC3, alongside BiP, avoids protein misfolding in the early stages of ER stress after which it is involved in the upregulation of protein synthesis [32,33,35]. This way, DNAJC3 reassures the re-initiation of cellular processes post ER stress and, thus, contributes to the overall fitness of the cells [32–34]. The importance of DNAJC3 during ER stress has been evidenced in different independent clinical trial studies on the DNAJC3-related pathogenicity [51,52]. The DNAJC3-related pathogenicity involves a systematic phenotype, including the early onset of diabetes mellitus and different neurological disorders [39,50–52]. The latter is evidenced as a causal effect of the accumulation of cholesterol in the ER lumen of DNAJC3 deficient cells [39,53,54]. Interestingly, perturbed cholesterol metabolism is a reported pathophysiological observation in neurological diseases such as Alzheimer's disease and Niemann–Pick type C disease [73,74].

A clear and consistent effect of CADA treatment is the preservation of preprotein species that could be visualized by immunoblotting. For ERLEC1, these preproteins could be identified even without inhibition of the proteasome. As ERLEC1 is a soluble protein of the ER lumen, the protein has no TMD that can escape the translocon laterally to be inserted in the lipid bilayer as is the case for huCD4 and PTK7. Additionally, ERLEC1 has no cytosolic tail that can be sensed by other cytosolic chaperones for proteasomal degradation. It is plausible that the stalled ERLEC1 preprotein resides in the protected channel of the ribosome exit tunnel and Sec61 translocon to escape proteasomal degradation. In contrast, for huCD4, only very limited amounts of preprotein could be seen, and only when rescued by MG132 treatment. It is known that the half-life of huCD4 is significantly lower in non-lymphoid cells (such as HEK293T) as compared to T cells because of the lack of the tyrosine kinase p56^{lck} that normally stably interacts with the cytosolic tail of huCD4 to prevent lysosomal degradation [75]. Thus, the degradation of the huCD4 preprotein might be more directed towards the lysosomal pathway, explaining the barely detectable amount of rescued preprotein by proteasome inhibition. The most obvious results with MG132 treatment were obtained for PTK7. As this is the least sensitive substrate of the three tested CADA targets, with the partial inhibition of PTK7 expression at the highest CADA concentrations (Figure 4B and Supplementary Figure S3B), we were able to detect additional inhibitory effects by DNAJC3 co-transfection and/or MG132 treatment. The synergistic effect of MG132 with CADA also suggested that with CADA some residual translocation of PTK7 was still occurring, and that this could proceed if the few clogged translocons were continuously cleared. Complete proteasome inhibition shuts down the clearance of blocked translocons, thus, preventing new rounds of PTK7-targeting and translocation, thus, reducing the expression of the mature proteins. In contrast to the two other CADA substrates, only for PTK7 could significantly enhanced levels of preprotein be measured when the cells were expressing high levels of DNAJC3. This would indicate that the lumenal DNAJC3 is involved in diverting the stalled PTK7 preprotein to the proteasome. Of course, we have no evidence of direct interaction between DNAJC3 and the stalled protein PTK7, and DNAJC3 might not only enhance the clearance of clogged translocons but might also have a positive effect on restarting protein translation for the next round of PTK7-targeting and SP insertion in an available translocon.

The data from the P2A-BFP constructs are indicative of the reduced translation of the C-terminal part of a stalled CADA substrate, suggesting that because of stalling by CADA, the ribosome does not 'reach' the very end of the transcript and, thus, cannot translate the BFP part. In this stalled situation, the ribosome may detach more frequently from the transcript at the P2A sequence. However, reduced translation is not because of a direct effect of CADA on the translation process as such, as proven by the ERLEC1 construct without the SP. Considering PTK7, co-transfection with pPL-DNAJC3 can restore the reduction in BFP levels to normal control levels, both for the PTK7/huCD4 chimaeric protein (Figure 5C) as for the WT full-length PTK7 control (Supplementary Figure S4C). In this context, DNAJC3 can help to rapidly clear the clogged translocons and to restart protein translation to normal levels. Unexpectedly, for ERLEC1 (and to a lesser content also huCD4), high levels of DNAJC3 seem to enhance even the translation of BFP under CADA pressure. This might be related to the strong inhibition of CADA on the translocation of these substrates, as compared to the partial inhibition on PTK7 (5-fold less sensitive). The stronger inhibition of protein translocation by CADA might induce a stronger 'clogging signal' to clear the translocon more rapidly, resulting in enhanced re-initiation of BFP translation, whereas, for PTK7, CADA might have a more decelerating effect on translocation and subsequent BFP production.

The most exciting data were obtained with the ERLEC1-mCD4 constructs (Figure 6). Here, the presence of the P2A sequence clearly impacted the outcome of mCD4 expression in relation to DNAJC3 levels (Figure 7). In general, exchanging cytosolic BFP by an ER-targeting mCD4 did not change the relative order of the reporter expression level in relation to the different DNAJC3 constructs (with pPL-DNAJC3 co-transfection giving the highest upregulation). However, without exogenous DNAJC3, the presence of CADA enhanced even the expression of mCD4. Here, one of the most suitable explanations would be that of the enhanced targeting efficiency of mCD4 because of a stalled ribosome that had already docked onto the ER membrane and was slowed down in its translation. This would give mCD4 more time and higher chances to insert in a translocon quickly and successfully, thus, enhancing the expression levels of mCD4. The additional upregulating effects on mCD4 expression by DNAJC3 could then be because of more cleared translocons that would be available in the vicinity of the stalled N-terminal ERLEC1. In contrast, the expression of an mCD4 protein directly fused to ERLEC1 would become regulated and suppressed by CADA. This is not unexpected, given that the mCD4 protein is in fact an elongated ERLEC1 variant that is mainly controlled by the CADA-sensitive SP of ERLEC1. However, the rather limited CADA sensitivity of the mCD4 part in the control cells transfected with ERLEC1-mCD4 suggests that the expression of mCD4 was still under the control of its own (CADA-resistant) SP. Additional research is needed to further explore the functionality of the mCD4 SP in this context. The very low expression efficiency of this mCD4 fusion protein as compared to the P2A variant (Supplementary Figure S5A) might be related to the unsuccessful ER translocation of mCD4 and putative altered topology with a mainly cytosolic expression of the extracellular CD4 region that did not survive the cytosolic degradation machinery (the SP might have served as a kind of TMD in the head-on insertion without SP cleavage). Nevertheless, the low translocation speed of mCD4 (due to the stalled ERLEC1 part) might result in higher residual translocation into the ER lumen due to the longer dwelling time of the temporarily stalled substrate in the translocon. The low translocation efficiency of this construct magnified the impact of altered DNAJC3 levels in this process. Quickly removing the stalled targets by DNAJC3 considerably reduced the dwelling time of cargo in the translocon and prevented the residual 'slipping' of the preprotein into the ER lumen. Thus, by directly reducing DNAJC3 protein levels, CADA attenuates its net down-modulating effect on the expression of affected targets.

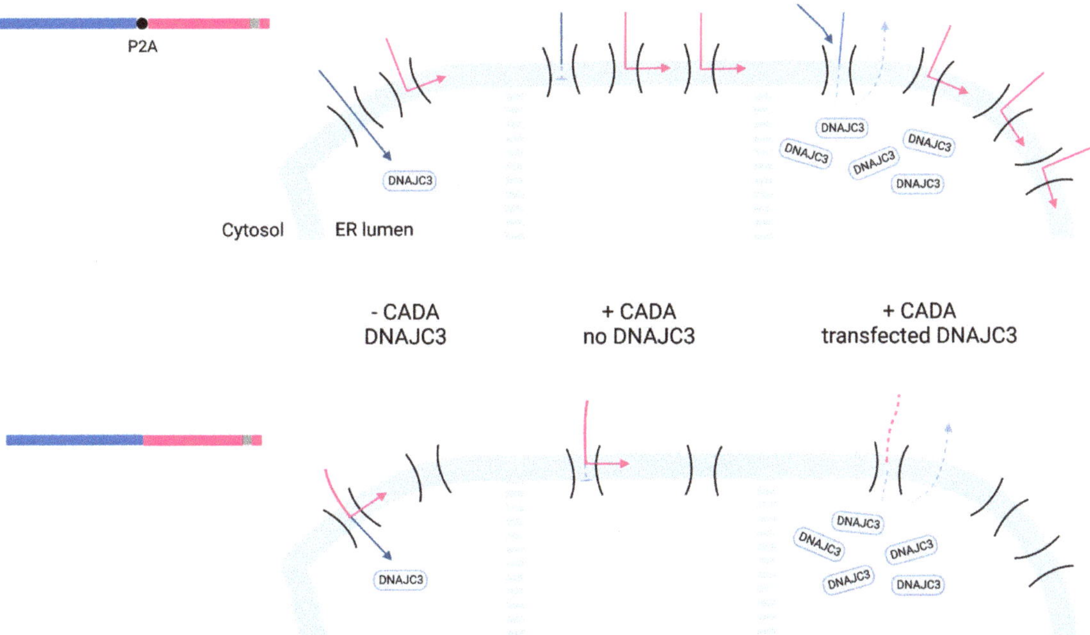

Figure 7. Proposed model of co-translational translocation events in the presence of CADA and DNAJC3. Top panel: The ERLEC1-P2A-mCD4 construct was translated by the ribosome and was targeted to the ER membrane with import into the ER lumen of the ERLEC1 protein (blue arrow) and insertion of mCD4 in the ER membrane (pink arrow). Under CADA pressure, ERLEC1 was prevented from being translocated (clogged translocon) while mCD4 could still get inserted into the ER membrane via its own SP, with even higher efficiency. When DNAJC3 was highly expressed, the ERLEC1 clogged translocon was rapidly cleared, allowing insertion of more mCD4 in the ER membrane while ERLEC1 translocation was consistently inhibited by CADA. Bottom panel: The ERLEC1-mCD4 construct was translated but the ER targeting and membrane insertion of mCD4 was at a very low level. Treatment with suboptimal levels of CADA stabilized the targeting complex and enhanced the targeting and membrane insertion of mCD4, whereas, at high CADA concentration, the translation and translocation of mCD4 were slightly down-regulated. The presence of high DNAJC3 levels rapidly removed (and degraded) the stalled ERLEC1 species (and downstream mCD4 protein) and prevented the stalled mCD4 from eventually getting inserted into the ER membrane, thus, intensifying the net down-modulating effect of CADA on its substrates.

In conclusion, the present study on the relation between CADA and DNAJC3 revealed that higher cellular levels of DNAJC3 enhanced the sensitivity of huCD4, PTK7 and ER-LEC1 for CADA. It also showed that CADA treatment resulted in the preservation of a small preprotein fraction, most likely corresponding with stalled CADA substrates at the ribosome/Sec61 translocon complex. DNAJC3 positively affected protein translation of a reporter protein expressed downstream of a CADA-stalled substrate, suggesting that DNAJC3 promotes the clearance of the clogged translocon. Altogether, we hypothesize that reduced DNAJC3 levels by CADA treatment retards the clearance of clogged Sec61 translocons filled with CADA-stalled substrates. This might allow more residual translocation of CADA substrates into the ER lumen due to the longer dwelling time of the temporarily stalled substrates in the translocon. Thus, by directly reducing DNAJC3 protein levels, CADA attenuates its net down-modulating effect on its substrates which can be related to the generally low impact of CADA on the proteome. However, further knowledge and research on DNAJC3-related pathways in ER stress would be of great help to improve our understanding of the cellular mechanism behind blocked co-translational translocation events, and the impact of small molecule translocation inhibitors at the molecular level.

4. Materials and Methods

4.1. Compounds and Antibodies

CADA hydrochloride was synthesized as described previously [76]. CADA was dissolved in dimethyl sulfoxide (DMSO) and stored at a stock concentration of 10 mM at room temperature. Western blot and flow cytometry antibodies were purchased from (i) Genscript (Piscataway, NJ, USA): anti-V5 (cat #A01724); (ii) BD Biosciences (Allschwil, Switzerland): anti-clathrin (cat. #610500); (iii) Thermo Fisher Scientific (Waltham, MA, USA): anti-β-actin (cat. #MA1-140) and allophycocyanin (APC)-labelled anti-mouse CD4 (clone GK1.5; cat. #47-0041-82); (iv) Sigma (Saint Louis, MO, USA): anti-FLAG (cat. #F1804); (v) Cell Signaling Technology (Danvers, MA, USA): anti-DNAJC3 (p58IPK) (cat. #2940S); (vi) Dako (Santa Clara, CA, USA): HRP-labelled goat anti-mouse immunoglobulins (cat. #P0447) and HRP-labelled swine anti-rabbit (cat. #P0399).

4.2. Plasmids and Mutagenesis

As described in a previous report [69], the ERLEC1 expression vector (pGEM-T backbone) was purchased from Sino biological (Beijing, China), the PTK7 expression vector (pDONR223 backbone) from Addgene (Watertown, MA, USA) and the DNAJC3 expression vector (pDONR223 backbone) from the DNASU plasmid repository (Tempe, AZ, USA). Constructs for western blot were designed to include the simian virus 5 (V5) epitope (GKPIPNPLLGLD) at the C-terminus of the protein of interest. Site-directed mutagenesis of all constructs was performed with the Q5 site-directed mutagenesis kit (New England Biolabs, Ipswich, MA, USA) or NEBuilder HiFi DNA assembly kit (New England Biolabs, Ipswich, MA, USA) following the manufacturer's instructions. Plasmid DNA was isolated using the Nucleospin Plasmid Transfection grade system (Macherey Nagel, Düren, Germany) supplemented with an endotoxin removal wash. The concentration of all constructs was determined with a NanoDrop 1000 spectrophotometer and sequences were confirmed by automated capillary Sanger sequencing (Macrogen Europe).

4.3. Transient Transfection

HEK293T cells were cultured in Dulbecco's modified eagle medium (DMEM), supplemented with 10% (v/v) fetal bovine serum (HyClone, Logan, UT, USA). Cells were seeded at 4×10^5 cells/mL and incubated overnight at 37 °C prior to transfection the next day. Lipofectamine LTX (Thermo Fisher Scientific, Waltham, MA, USA) was used for the transfection of plasmid DNA according to the manufacturer's protocol. CADA and/or MG132 (Sigma, Saint Louis, MO, USA) was added 6 h post transfection and cells were lysed for immunoblotting or fixed in paraformaldehyde for antibody staining and subsequent flow cytometric analysis.

4.4. Immunoblotting

After CADA treatment, cells were collected and lysed in Nonidet P-40 buffer (1%, supplemented with 50 mM Tris HCl, pH 8.0, 150 mM NaCl, protease inhibitor cocktail (Roche, Basel, Switzerland) and PMSF). Lysates were run on between 4 and 12% Criterion XT Bis-Tris gels in MES buffer (Bio-Rad, Hercules, CA, USA), transferred to PVDF or nitrocellulose membranes using the BioRad Trans-Blot Turbo transfer system (Bio-Rad, Hercules, CA, USA), blocked with 5% non-fat dried milk in TBST and incubated with a primary and secondary antibody. SuperSignal West Pico and Femto chemiluminescence reagent (Thermo Fisher scientific, Waltham, MA, USA) was used for detection with a ChemiDoc MP system (Bio-Rad, Hercules, CA, USA). Signal intensities were quantified with Image Lab software v5.0 (Bio-Rad, Hercules, CA, USA). Differences in protein concentration between each lane were compensated by normalisation to the clathrin heavy chain or β-actin signal. To compare the down-modulating activity of CADA, IC50 values were calculated with GraphPad Prism 8 software (San Diego, CA, USA) on a four-parameter concentration–response curve fitted to data from at least three replicate experiments with flow cytometry. The absolute IC50 value represented the compound concentration that resulted in the 50% reduction in the protein level.

4.5. Antibody Staining and Flow Cytometry

After CADA treatment, cells were collected, fixed and permeabilized with fixation/permeabilization solution (BD biosciences, Allschwil, Switzerland) and subsequently washed with perm/wash solution (BD biosciences, Allschwil, Switzerland). Next, cells were stained with antibodies and incubated at 4 °C for 45 min. Samples were then washed with perm/wash solution (BD biosciences) and fixed in PBS containing 1% formaldehyde before acquisition on BD FACS Celesta flow cytometer (Beckton Dickinson) with BD FACS-Diva 8.0.1 software (BD biosciences, Allschwil, Switzerland). All data were analyzed in FlowJo X v10 (BD biosciences, Allschwil, Switzerland).

4.6. Cell-Free In Vitro Translation and Translocation

The DNAJC3 SP and first 62AA of mature DNAJC3 were fused upstream of huCD4 with PCR. The Qiagen EasyXpress linear template kit was used to generate linear DNA fragments using PCR. The PCR products were purified and transcribed in vitro using T7 RNA polymerase (RiboMAX system; Promega, Madison, WI, USA). All transcripts were translated in the presence of rabbit reticulocyte lysate (Promega, Madison, WI, USA) in the presence of L-35S- methionine (PerkinElmer, Waltham, MA, USA). Translations were performed at 30 °C in the presence or absence of ovine pancreatic microsomes and CADA as described elsewhere. Samples were treated with proteinase K (Roche, Basel, Switzerland) on ice for 30 min. Protein digestion was stopped with phenylmethylsulfonyl fluoride (PMSF; Thermo Fisher Scientific, Waltham, MA, USA) after which samples were washed with a low-salt buffer (80 mM KOAc, 2 mM Mg(OAc)$_2$, 50 mM Hepes, pH 7.6), and radiolabelled proteins were isolated by centrifugation (10 min at 21.382 g, 4 °C). The proteins were then separated with SDS-PAGE and detected by phosphor imaging (Cyclone Plus storage phosphor system; PerkinElmer, Waltham, MA, USA).

Supplementary Materials: The following supporting information can be downloaded at: https://www.mdpi.com/article/10.3390/ijms23020584/s1.

Author Contributions: Conceptualization, K.V., E.P. and E.H.; methodology, E.P., B.P. and A.C.; writing—original draft preparation, E.P. and K.V.; writing—review and editing, E.P., K.V. and E.H.; visualization, E.P.; funding acquisition, K.V. All authors have read and agreed to the published version of the manuscript.

Funding: This research was funded by KU Leuven, Internal Funds KU Leuven, project 3M170314. The funders had no role in the study design, data collection and analysis, decision to publish or preparation of the manuscript.

Acknowledgments: We would like to thank Thomas W. Bell (UNR, Nevada, USA) for providing the CADA compound.

Conflicts of Interest: The authors declare no conflict of interest.

References

1. Wickner, W.; Schekman, R. Protein translocation across biological membranes. *Science* **2005**, *310*, 1452–1456. [CrossRef] [PubMed]
2. Zimmermann, R.; Eyrisch, S.; Ahmad, M.; Helms, V. Protein translocation across the ER membrane. *Biochim. Biophys. Acta* **2011**, *1808*, 912–924. [CrossRef]
3. Rapoport, T.A. Protein translocation across the eukaryotic endoplasmic reticulum and bacterial plasma membranes. *Nature* **2007**, *450*, 663–669. [CrossRef]
4. Dudek, J.; Pfeffer, S.; Lee, P.-H.; Jung, M.; Cavalié, A.; Helms, V.; Foerster, F.; Zimmermann, R. Protein Transport into the Human Endoplasmic Reticulum. *J. Mol. Biol.* **2015**, *427*, 1159–1175. [CrossRef] [PubMed]

5. Aviram, N.; Schuldiner, M. Targeting and translocation of proteins to the endoplasmic reticulum at a glance. *J. Cell Sci.* **2017**, *130*, 4079–4085. [CrossRef]
6. Walter, P.; Gilmore, R.; Blobel, G. Protein translocation across the endoplasmic reticulum. *Cell* **1984**, *38*, 5–8. [CrossRef]
7. Pfeffer, S.; Burbaum, L.; Unverdorben, P.; Pech, M.; Chen, Y.; Zimmermann, R.; Beckmann, R.; Foerster, F. Structure of the native Sec61 protein-conducting channel. *Nat. Commun.* **2015**, *6*, 8403. [CrossRef]
8. Voorhees, R.M.; Fernández, I.S.; Scheres, S.; Hegde, R.S. Structure of the Mammalian Ribosome-Sec61 Complex to 3.4 A Resolution. *Cell* **2014**, *157*, 1632–1643. [CrossRef] [PubMed]
9. Voorhees, R.M.; Hegde, R.S. Toward a structural understanding of co-translational protein translocation. *Curr. Opin. Cell Biol.* **2016**, *41*, 91–99. [CrossRef] [PubMed]
10. Pfeffer, S.; Brandt, F.; Hrabe, T.; Lang, S.; Eibauer, M.; Zimmermann, R.; Förster, F. Structure and 3D Arrangement of Endoplasmic Reticulum Membrane-Associated Ribosomes. *Structure* **2012**, *20*, 1508–1518. [CrossRef] [PubMed]
11. Hassdenteufel, S.; Nguyen, D.; Helms, V.; Lang, S.; Zimmermann, R. ER import of small human presecretory proteins: Components and mechanisms. *FEBS Lett.* **2019**, *593*, 2506–2524. [CrossRef] [PubMed]
12. Johnson, N.; Powis, K.; High, S. Post-translational translocation into the endoplasmic reticulum. *Biochim. Biophys. Acta* **2013**, *1833*, 2403–2409. [CrossRef]
13. Lakkaraju, A.K.; Thankappan, R.; Mary, C.; Garrison, J.L.; Taunton, J.; Strub, K. Efficient secretion of small proteins in mammalian cells relies on Sec62-dependent posttranslational translocation. *Mol. Biol. Cell* **2012**, *23*, 2712–2722. [CrossRef] [PubMed]
14. Denks, K.; Vogt, A.; Sachelaru, I.; Petriman, N.A.; Kudva, R.; Koch, H.-G. The Sec translocon mediated protein transport in prokaryotes and eukaryotes. *Mol. Membr. Biol.* **2014**, *31*, 58–84. [CrossRef] [PubMed]
15. Johnson, N.; Haßdenteufel, S.; Theis, M.; Paton, A.W.; Paton, J.C.; Zimmermann, R.; High, S. The Signal Sequence Influences Post-Translational ER Translocation at Distinct Stages. *PLoS ONE* **2013**, *8*, e75394. [CrossRef]
16. Voorhees, R.M.; Hegde, R.S. Structure of the Sec61 channel opened by a signal sequence. *Science* **2016**, *351*, 88–91. [CrossRef]
17. Gemmer, M.; Förster, F. A clearer picture of the ER translocon complex. *J. Cell Sci.* **2020**, *133*, jcs231340. [CrossRef]
18. Hegde, R.S.; Kang, S.W. The concept of translocational regulation. *J. Cell Biol.* **2008**, *182*, 225–232. [CrossRef]
19. Pfeffer, S.; Dudek, J.; Gogala, M.; Schorr, S.; Linxweiler, J.; Lang, S.; Becker, T.; Beckmann, R.; Zimmermann, R.; Foerster, F. Structure of the mammalian oligosaccharyl-transferase complex in the native ER protein translocon. *Nat. Commun.* **2014**, *5*, 3072. [CrossRef] [PubMed]
20. Braunger, K.; Pfeffer, S.; Shrimal, S.; Gilmore, R.; Berninghausen, O.; Mandon, E.C.; Becker, T.; Förster, F.; Beckmann, R. Structural basis for coupling protein transport and N-glycosylation at the mammalian endoplasmic reticulum. *Science* **2018**, *350*, 215–219. [CrossRef]
21. Kaufman, R.J. Stress signaling from the lumen of the endoplasmic reticulum: Coordination of gene transcriptional and translational controls. *Genes Dev.* **1999**, *13*, 1211–1233. [CrossRef]
22. Tirasophon, W.; Welihinda, A.A.; Kaufman, R.J. A stress response pathway from the endoplasmic reticulum to the nucleus requires a novel bifunctional protein kinase/endoribonuclease (Ire1p) in mammalian cells. *Genes Dev.* **1998**, *12*, 1812–1824. [CrossRef]
23. Mori, K. Tripartite Management of Unfolded Proteins in the Endoplasmic Reticulum. *Cell* **2000**, *101*, 451–454. [CrossRef]
24. Wang, L.; Ye, Y. Clearing Traffic Jams During Protein Translocation Across Membranes. *Front. Cell Dev. Biol.* **2020**, *8*, 610689. [CrossRef]
25. Bertolotti, A.; Zhang, Y.; Hendershot, L.M.; Harding, H.P.; Ron, D. Dynamic interaction of BiP and ER stress transducers in the unfolded-protein response. *Nat. Cell Biol.* **2000**, *2*, 326–332. [CrossRef]
26. Wang, X.Z.; Harding, H.P.; Zhang, Y.; Jolicoeur, E.M.; Kuroda, M.; Ron, D. Cloning of mammalian Ire1 reveals diversity in the ER stress responses. *EMBO J.* **1998**, *17*, 5708–5717. [CrossRef]
27. Yoshida, H.; Matsui, T.; Yamamoto, A.; Okada, T.; Mori, K. XBP1 mRNA Is Induced by ATF6 and Spliced by IRE1 in Response to ER Stress to Produce a Highly Active Transcription Factor. *Cell* **2001**, *107*, 881–891. [CrossRef]
28. Haze, K.; Yoshida, H.; Yanagi, H.; Yura, T.; Mori, K. Mammalian Transcription Factor ATF6 Is Synthesized as a Transmembrane Protein and Activated by Proteolysis in Response to Endoplasmic Reticulum Stress. *Mol. Biol. Cell* **1999**, *10*, 3787–3799. [CrossRef]
29. Fawcett, T.W.; Martindale, J.L.; Guyton, K.Z.; Hai, T.; Holbrook, N.J. Complexes containing activating transcription factor (ATF)/cAMP-responsive-element-binding protein (CREB) interact with the CCAAT/enhancer-binding protein (C/EBP)-ATF composite site to regulate Gadd153 expression during the stress response. *Biochem. J.* **1999**, *339 Pt 1*, 135–141. [CrossRef]
30. Harding, H.P.; Novoa, I.; Zhang, Y.; Zeng, H.; Wek, R.; Schapira, M.; Ron, D. Regulated Translation Initiation Controls Stress-Induced Gene Expression in Mammalian Cells. *Mol. Cell* **2000**, *6*, 1099–1108. [CrossRef]
31. Brostrom, C.O.; Brostrom, M.A. Regulation of Translational Initiation during Cellular Responses to Stress. *Prog. Nucleic Acid Res. Mol. Biol.* **1997**, *58*, 79–125.
32. Van Huizen, R.; Martindale, J.L.; Gorospe, M.; Holbrook, N.J. P58IPK, a novel endoplasmic reticulum stress-inducible protein and potential negative regulator of eIF2alpha signaling. *J. Biol. Chem.* **2003**, *278*, 15558–15564. [CrossRef] [PubMed]
33. Rutkowski, D.T.; Kang, S.-W.; Goodman, A.G.; Garrison, J.L.; Taunton, J.; Katze, M.G.; Kaufman, R.J.; Hegde, R.S. The Role of p58IPK in Protecting the Stressed Endoplasmic Reticulum. *Mol. Biol. Cell* **2007**, *18*, 3681–3691. [CrossRef] [PubMed]
34. Yan, W.; Frank, C.L.; Korth, M.J.; Sopher, B.L.; Novoa, I.; Ron, D.; Katze, M.G. Control of PERK eIF2 kinase activity by the endoplasmic reticulum stress-induced molecular chaperone P58IPK. *Proc. Natl. Acad. Sci. USA* **2002**, *99*, 15920–15925. [CrossRef]

35. Pobre, K.F.R.; Poet, G.J.; Hendershot, L.M. The endoplasmic reticulum (ER) chaperone BiP is a master regulator of ER functions: Getting by with a little help from ERdj friends. *J. Biol. Chem.* **2019**, *294*, 2098–2108. [CrossRef]
36. Harding, H.P.; Zhang, Y.; Ron, D. Protein translation and folding are coupled by an endoplasmic-reticulum-resident kinase. *Nature* **1999**, *397*, 271–274. [CrossRef]
37. Sood, R.; Porter, A.C.; Ma, K.; Quilliam, L.A.; Wek, R.C. Pancreatic eukaryotic initiation factor-2alpha kinase (PEK) homologues in humans, *Drosophila melanogaster* and *Caenorhabditis elegans* that mediate translational control in response to endoplasmic reticulum stress. *Biochem. J.* **2000**, *346 Pt 2*, 281–293. [CrossRef]
38. Lindholm, D.; Wootz, H.; Korhonen, L. ER stress and neurodegenerative diseases. *Cell Death Differ.* **2006**, *13*, 385–392. [CrossRef]
39. Jennings, M.J.; Hathazi, D.; Nguyen, C.D.L.; Munro, B.; Münchberg, U.; Ahrends, R.; Schenck, A.; Eidhof, I.; Freier, E.; Synofzik, M.; et al. Intracellular Lipid Accumulation and Mitochondrial Dysfunction Accompanies Endoplasmic Reticulum Stress Caused by Loss of the Co-chaperone DNAJC3. *Front. Cell Dev. Biol.* **2021**, *9*, 710247. [CrossRef]
40. Roussel, B.D.; Kruppa, A.; Miranda, E.; Crowther, D.; Lomas, D.; Marciniak, S. Endoplasmic reticulum dysfunction in neurological disease. *Lancet Neurol.* **2018**, *12*, 105–118. [CrossRef]
41. Colla, E. Linking the Endoplasmic Reticulum to Parkinson's Disease and Alpha-Synucleinopathy. *Front. Neurosci.* **2019**, *13*, 560. [CrossRef]
42. Anttonen, A.K.; Mahjneh, I.; Hämäläinen, R.H.; Lagier-Tourenne, C.; Kopra, O.; Waris, L.; Anttonen, M.; Joensuu, T.; Kalimo, H.; Paetau, A.; et al. The gene disrupted in Marinesco-Sjögren syndrome encodes SIL1, an HSPA5 cochaperone. *Nat. Genet.* **2005**, *37*, 1309–1311. [CrossRef]
43. Jauhari, P.; Sahu, J.K.; Roos, A.; Senderek, J.; Vyas, S.; Singhi, P. SIL1-negative Marinesco-Sjögren syndrome: First report of two sibs from India. *J. Pediatr. Neurosci.* **2014**, *9*, 291–292.
44. Senderek, J.; Krieger, M.; Stendel, C.; Bergmann, C.; Moser, M.; Breitbach-Faller, N.; Rudnik-Schöneborn, S.; Blaschek, A.; Wolf, N.; Harting, I.; et al. Mutations in SIL1 cause Marinesco-Sjögren syndrome, a cerebellar ataxia with cataract and myopathy. *Nat. Genet.* **2005**, *37*, 1312–1314. [CrossRef]
45. Buchkremer, S.; Coraspe, J.A.G.; Weis, J.; Roos, A. Sil1-Mutant Mice Elucidate Chaperone Function in Neurological Disorders. *J. Neuromuscul. Dis.* **2016**, *3*, 169–181. [CrossRef]
46. Zhao, L.; Rosales, C.; Seburn, K.; Ron, D.; Ackerman, S.L. Alteration of the unfolded protein response modifies neurodegeneration in a mouse model of Marinesco-Sjögren syndrome. *Hum. Mol. Genet.* **2010**, *19*, 25–35. [CrossRef]
47. Delépine, M.; Nicolino, M.; Barrett, T.; Golamaully, M.; Lathrop, G.M.; Julier, C. EIF2AK3, encoding translation initiation factor 2-alpha kinase 3, is mutated in patients with Wolcott-Rallison syndrome. *Nat. Genet.* **2000**, *25*, 406–409. [CrossRef] [PubMed]
48. Blumen, S.C.; Astord, S.; Robin, V.; Vignaud, L.; Toumi, N.; Cieslik, A.; Achiron, A.; Carasso, R.L.; Gurevich, M.; Braverman, I.; et al. A rare recessive distal hereditary motor neuropathy with HSJ1 chaperone mutation. *Ann. Neurol.* **2012**, *71*, 509–519. [CrossRef]
49. Khanim, F.; Kirk, J.; Latif, F.; Barrett, T.G. WFS1/wolframin mutations, Wolfram syndrome, and associated diseases. *Hum. Mutat.* **2001**, *17*, 357–367. [CrossRef] [PubMed]
50. Synofzik, M.; Haack, T.B.; Kopajtich, R.; Gorza, M.; Rapaport, D.; Greiner, M.; Schönfeld, C.; Freiberg, C.; Schorr, S.; Holl, R.W.; et al. Absence of BiP Co-chaperone DNAJC3 Causes Diabetes Mellitus and Multisystemic Neurodegeneration. *Am. J. Hum. Genet.* **2014**, *95*, 689–697. [CrossRef]
51. Bublitz, S.K.; Alhaddad, B.; Synofzik, M.; Kuhl, V.; Lindner, A.; Freiberg, C.; Schmidt, H.; Strom, T.; Haack, T.; Deschauer, M. Expanding the phenotype of DNAJC3 mutations: A case with hypothyroidism additionally to diabetes mellitus and multisystemic neurodegeneration. *Clin. Genet.* **2017**, *92*, 561–562. [CrossRef]
52. Lytrivi, M.; Senée, V.; Salpea, P.; Fantuzzi, F.; Philippi, A.; Abdulkarim, B.; Sawatani, T.; Marín-Cañas, S.; Pachera, N.; Degavre, A.; et al. DNAJC3 deficiency induces β-cell mitochondrial apoptosis and causes syndromic young-onset diabetes. *Eur. J. Endocrinol.* **2021**, *184*, 455–468. [CrossRef] [PubMed]
53. Tabas, I. Consequences of cellular cholesterol accumulation: Basic concepts and physiological implications. *J. Clin. Investig.* **2002**, *110*, 905–911. [CrossRef] [PubMed]
54. Luo, D.-X.; Cao, D.-L.; Xiong, Y.; Peng, X.-H.; Liao, D.-F. A novel model of cholesterol efflux from lipid-loaded cells. *Acta Pharmacol. Sin.* **2010**, *31*, 1243–1257. [CrossRef] [PubMed]
55. Feng, B.; Yao, P.M.; Li, Y.; Devlin, C.M.; Zhang, D.; Harding, H.P.; Sweeney, M.; Rong, J.X.; Kuriakose, G.; Fisher, E.A.; et al. The endoplasmic reticulum is the site of cholesterol-induced cytotoxicity in macrophages. *Nat. Cell Biol.* **2003**, *5*, 781–792. [CrossRef] [PubMed]
56. Shyu, P.; Ng, B.S.H.; Ho, N.; Chaw, R.; Seah, Y.L.; Marvalim, C.; Thibault, G. Membrane phospholipid alteration causes chronic ER stress through early degradation of homeostatic ER-resident proteins. *Sci. Rep.* **2019**, *9*, 8637. [CrossRef] [PubMed]
57. Volmer, R.; Ron, D. Lipid-dependent regulation of the unfolded protein response. *Curr. Opin. Cell Biol.* **2015**, *33*, 67–73. [CrossRef]
58. Boslem, E.; Weir, J.M.; MacIntosh, G.; Sue, N.; Cantley, J.; Meikle, P.; Biden, T.J. Alteration of Endoplasmic Reticulum Lipid Rafts Contributes to Lipotoxicity in Pancreatic β-Cells. *J. Biol. Chem.* **2013**, *288*, 26569–26582. [CrossRef]
59. Xu, J.; Taubert, S. Beyond Proteostasis: Lipid Metabolism as a New Player in ER Homeostasis. *Metabolites* **2021**, *11*, 52. [CrossRef]
60. Izawa, T.; Tsuboi, T.; Kuroha, K.; Inada, T.; Nishikawa, S.-I.; Endo, T. Roles of Dom34:Hbs1 in Nonstop Protein Clearance from Translocators for Normal Organelle Protein Influx. *Cell Rep.* **2012**, *2*, 447–453. [CrossRef]

61. Arakawa, S.; Yunoki, K.; Izawa, T.; Tamura, Y.; Nishikawa, S.-I.; Endo, T. Quality control of nonstop membrane proteins at the ER membrane and in the cytosol. *Sci. Rep.* **2016**, *6*, 30795. [CrossRef]

62. Vermeire, K.; Zhang, Y.; Princen, K.; Hatse, S.; Samala, M.F.; Dey, K.; Choi, H.-J.; Ahn, Y.; Sodoma, A.; Snoeck, R.; et al. CADA Inhibits Human Immunodeficiency Virus and Human Herpesvirus 7 Replication by Down-modulation of the Cellular CD4 Receptor. *Virology* **2002**, *302*, 342–353. [CrossRef]

63. Vermeire, K.; Schols, D. Cyclotriazadisulfonamides: Promising new CD4-targeted anti-HIV drugs. *J. Antimicrob. Chemother.* **2005**, *56*, 270–272. [CrossRef] [PubMed]

64. Vermeire, K.; Bell, T.W.; Van Puyenbroeck, V.; Giraut, A.; Noppen, S.; Liekens, S.; Schols, D.; Hartmann, E.; Kalies, K.-U.; Marsh, M. Signal Peptide-Binding Drug as a Selective Inhibitor of Co-Translational Protein Translocation. *PLoS Biol.* **2014**, *12*, e1002011. [CrossRef] [PubMed]

65. Vermeire, K.; Princen, K.; Hatse, S.; De Clercq, E.; Dey, K.; Bell, T.W.; Schols, D. CADA, a novel CD4-targeted HIV inhibitor, is synergistic with various anti-HIV drugs in vitro. *AIDS* **2004**, *18*, 2115–2125. [CrossRef] [PubMed]

66. Van Puyenbroeck, V.; Pauwels, E.; Provinciael, B.; Bell, T.W.; Schols, D.; Kalies, K.-U.; Hartmann, E.; Vermeire, K. Preprotein signature for full susceptibility to the co-translational translocation inhibitor cyclotriazadisulfonamide. *Traffic* **2020**, *21*, 250–264. [CrossRef]

67. Van Puyenbroeck, V.; Claeys, E.; Schols, D.; Bell, T.W.; Vermeire, K. A Proteomic Survey Indicates Sortilin as a Secondary Substrate of the ER Translocation Inhibitor Cyclotriazadisulfonamide (CADA). *Mol. Cell. Proteom.* **2017**, *16*, 157–167. [CrossRef]

68. Claeys, E.; Pauwels, E.; Humblet-Baron, S.; Provinciael, B.; Schols, D.; Waer, M.; Sprangers, B.; Vermeire, K. Small Molecule Cyclotriazadisulfonamide Abrogates the Upregulation of the Human Receptors CD4 and 4-1BB and Suppresses In Vitro Activation and Proliferation of T Lymphocytes. *Front. Immunol.* **2021**, *12*, 1340. [CrossRef] [PubMed]

69. Pauwels, E.; Rutz, C.; Provinciael, B.; Stroobants, J.; Schols, D.; Hartmann, E.; Krause, E.; Stephanowitz, H.; Schülein, R.; Vermeire, K. A Proteomic Study on the Membrane Protein Fraction of T Cells Confirms High Substrate Selectivity for the ER Translocation Inhibitor Cyclotriazadisulfonamide. *Mol. Cell. Proteom.* **2021**, *20*, 100144. [CrossRef]

70. Pauwels, E.; Schülein, R.; Vermeire, K. Inhibitors of the Sec61 Complex and Novel High Throughput Screening Strategies to Target the Protein Translocation Pathway. *Int. J. Mol. Sci.* **2021**, *22*, 12007. [CrossRef]

71. Daverkausen-Fischer, L.; Pröls, F. Dual topology of co-chaperones at the membrane of the endoplasmic reticulum. *Cell Death Discov.* **2021**, *7*, 203. [CrossRef] [PubMed]

72. Kim, S.J.; Mitra, D.; Salerno, J.R.; Hegde, R.S. Signal Sequences Control Gating of the Protein Translocation Channel in a Substrate-Specific Manner. *Dev. Cell* **2002**, *2*, 207–217. [CrossRef]

73. Fernández, A.; Llacuna, L.; Fernández-Checa, J.C.; Colell, A. Mitochondrial Cholesterol Loading Exacerbates Amyloid β Peptide-Induced Inflammation and Neurotoxicity. *J. Neurosci.* **2009**, *29*, 6394–6405. [CrossRef] [PubMed]

74. Marquer, C.; Laine, J.; Dauphinot, L.; Hanbouch, L.; Lemercier-Neuillet, C.; Pierrot, N.; Bossers, K.; Le, M.; Corlier, F.; Benstaali, C.; et al. Increasing membrane cholesterol of neurons in culture recapitulates Alzheimer's disease early phenotypes. *Mol. Neurodegener.* **2014**, *9*, 60. [CrossRef]

75. Pelchen-Matthews, A.; Boulet, I.; Littman, D.R.; Fagard, R.; Marsh, M. The protein tyrosine kinase p56lck inhibits CD4 endocytosis by preventing entry of CD4 into coated pits. *J. Cell Biol.* **1992**, *117*, 279–290. [CrossRef] [PubMed]

76. Bell, T.W.; Anugu, S.; Bailey, P.; Catalano, V.J.; Dey, K.; Drew, M.G.B.; Duffy, N.H.; Jin, Q.; Samala, M.F.; Sodoma, A.; et al. Synthesis and Structure—Activity Relationship Studies of CD4 Down-Modulating Cyclotriazadisulfonamide (CADA) Analogues. *J. Med. Chem.* **2006**, *49*, 1291–1312. [CrossRef] [PubMed]

International Journal of
Molecular Sciences

MDPI

Article

Quantitative Proteomics and Differential Protein Abundance Analysis after the Depletion of PEX3 from Human Cells Identifies Additional Aspects of Protein Targeting to the ER

Richard Zimmermann [1,*], Sven Lang [1], Monika Lerner [1], Friedrich Förster [2], Duy Nguyen [3], Volkhard Helms [3] and Bianca Schrul [1,*]

1   Medical Biochemistry and Molecular Biology, Saarland University, 66421 Homburg, Germany; sven.lang@uni-saarland.de (S.L.); monika.lerner@uks.eu (M.L.)
2   Bijvoet Center for Biomolecular Research, Utrecht University, 3584 CH Utrecht, The Netherlands; f.g.forster@uu.nl
3   Center for Bioinformatics, Saarland Informatics Campus, Saarland University, 66041 Saarbrücken, Germany; duy.nguyen@dkfz-heidelberg.de (D.N.); volkhard.helms@bioinformatik.uni-saarland.de (V.H.)
*   Correspondence: richard.zimmermann@uks.eu (R.Z.); Bianca.Schrul@uks.eu (B.S.)

Citation: Zimmermann, R.; Lang, S.; Lerner, M.; Förster, F.; Nguyen, D.; Helms, V.; Schrul, B. Quantitative Proteomics and Differential Protein Abundance Analysis after the Depletion of PEX3 from Human Cells Identifies Additional Aspects of Protein Targeting to the ER. *Int. J. Mol. Sci.* **2021**, *22*, 13028. https://doi.org/10.3390/ijms222313028

Academic Editor: Isao Ishii

Received: 29 September 2021
Accepted: 29 November 2021
Published: 1 December 2021

Publisher's Note: MDPI stays neutral with regard to jurisdictional claims in published maps and institutional affiliations.

Abstract: Protein import into the endoplasmic reticulum (ER) is the first step in the biogenesis of around 10,000 different soluble and membrane proteins in humans. It involves the co- or post-translational targeting of precursor polypeptides to the ER, and their subsequent membrane insertion or translocation. So far, three pathways for the ER targeting of precursor polypeptides and four pathways for the ER targeting of mRNAs have been described. Typically, these pathways deliver their substrates to the Sec61 polypeptide-conducting channel in the ER membrane. Next, the precursor polypeptides are inserted into the ER membrane or translocated into the ER lumen, which may involve auxiliary translocation components, such as the TRAP and Sec62/Sec63 complexes, or auxiliary membrane protein insertases, such as EMC and the TMCO1 complex. Recently, the PEX19/PEX3-dependent pathway, which has a well-known function in targeting and inserting various peroxisomal membrane proteins into pre-existent peroxisomal membranes, was also found to act in the targeting and, putatively, insertion of monotopic hairpin proteins into the ER. These either remain in the ER as resident ER membrane proteins, or are pinched off from the ER as components of new lipid droplets. Therefore, the question arose as to whether this pathway may play a more general role in ER protein targeting, i.e., whether it represents a fourth pathway for the ER targeting of precursor polypeptides. Thus, we addressed the client spectrum of the PEX19/PEX3-dependent pathway in both PEX3-depleted HeLa cells and PEX3-deficient Zellweger patient fibroblasts by an established approach which involved the label-free quantitative mass spectrometry of the total proteome of depleted or deficient cells, as well as differential protein abundance analysis. The negatively affected proteins included twelve peroxisomal proteins and two hairpin proteins of the ER, thus confirming two previously identified classes of putative PEX19/PEX3 clients in human cells. Interestingly, fourteen collagen-related proteins with signal peptides or N-terminal transmembrane helices belonging to the secretory pathway were also negatively affected by PEX3 deficiency, which may suggest compromised collagen biogenesis as a hitherto-unknown contributor to organ failures in the respective Zellweger patients.

Keywords: endoplasmic reticulum; lipid droplets; peroxisomes; PEX3; protein targeting; membrane protein insertion; protein translocation; label-free quantitative mass spectrometry; differential protein abundance analysis; Zellweger syndrome

1. Introduction

Analogously to the division of the human body into several organs, the nucleated human cell is divided into various different compartments, the cell organelles, which are surrounded—and thereby separated from the cytosol—by biological membranes. However, the vast majority of the approximately 30,000 types of the polypeptides—along with their isoforms—of a human cell are synthesized in the cytosol. Therefore, the proteins of the different organelles have to be sorted to the correct organelles and, subsequently, inserted into or translocated across the organellar membrane(s). The protein import into the endoplasmic reticulum (ER) is the first step in the biogenesis of about 10,000 different soluble and membrane proteins of human cells [1–4]. These were found to fulfill their functions in the membrane or lumen of the ER, along with the nuclear envelope, in one of the organelles belonging to the pathways of endo- and exocytosis (i.e., ERGIC, Golgi apparatus, endosome, lysosome, transport vesicles), or at the cell surface as plasma membrane- or secretory-proteins. Excluding resident proteins of the ER, most of the correctly folded and assembled proteins are delivered from the ER to their functional location by vesicular transport, which involves vesicles budding off from subdomains of the tubular ER, which are termed "exit sites" (ERES) [5,6]. In recent years, however, an increasing number of proteins destined to lipid droplets (LDs), peroxisomes or mitochondria were observed to be targeted to the ER as well, prior to their integration into budding LDs or peroxisomes, or prior to their delivery to mitochondria via the recently-identified ER–SURF pathway [7–11]. LDs and peroxisomes are ER-derived organelles, and their biogenesis occurs in specialized subdomains of the tubular ER [8,10].

Protein import into the ER involves ER membrane targeting as the initial step, and the insertion of nascent or fully-synthesized membrane proteins into—or the translocation of soluble precursor polypeptides across—the ER membrane as the second step [1–4]. Typically, both processes depend on N-terminal signal peptides (SPs) or transmembrane helices (TMHs) in the precursor polypeptides that serve as signals [4,12–14]. Generally, the Sec61 complex of the ER membrane represents the entry point for most of these precursor polypeptides into the organelle (Table 1) [1–4]. A variety of proteins rely on the Sec61 complex for their integration into the ER lumen or their translocation across the ER membrane (Figure 1a). These include SP-containing soluble or GPI-anchored proteins, bitopic type I (C-in N-out) and type II (N-in C-out) proteins, and multispanning membrane proteins. Sec61 is also implicated in the membrane insertion of proteins without N-terminal SPs, such as type III (C-in N-out) [4] and C-tail-anchored (TA) (N-in C-out) membrane proteins [15–17], and even monotopic hairpin proteins (C-in N-in) [18,19]. In these cases, however, the role of Sec61 is less clear, and auxiliary factors such as the TRAP complex, the Sec62/Sec63 complex, or insertases such as the ER membrane protein complex (EMC) or the TMCO1 complex are required in addition [4]. Especially the membrane insertion of TA-proteins predominantly relies on the insertase WRB/CAML [15–17]. In conclusion, several co- and post-translational protein targeting pathways merge at the Sec61 complex in the ER membrane, including the co-translational SRP/SR-pathway and the post-translational SRP-independent or SND pathways [2,4,20–26].

In addition, there is the targeting of mRNAs to the ER membrane, which involves mRNA receptors (such as KTN1), or receptors for ribosome nascent chain complexes with nascent chains, which are not yet long enough to be able to interact with SRP (such as RRBP1) [27–33]. However, one general lesson from the analysis of all these pathways is that they are not strictly separated from each other, and there are at least some precursor polypeptides which can be targeted to the ER by more than one pathway (such as some small presecretory proteins and some tail-anchored membrane proteins) [23,25,26]. Thus, the targeting pathways have overlapping substrate specificities, and can substitute for each other, at least to a certain extent. The characterization of all of these pathways and mechanisms is also of medical importance, as several of the components are linked to human hereditary or tumor diseases, or are hijacked by viral or bacterial agents [34].

Recent work identified the PEX19/PEX3-dependent pathway as a fourth pathway for the ER targeting of precursor polypeptides [35,36]. PEX3 (also termed peroxisomal biogenesis factor 3 or Peroxin-3) was first identified in yeast, and is a membrane protein with an N-terminal transmembrane domain and a large C-terminal domain, which faces the cytosol both in yeast and in humans [37–41]. Originally, it was characterized as a peroxisomal membrane protein, which cooperates with the cytosolic protein PEX19 in the targeting of peroxisomal membrane proteins to pre-existent peroxisomes and in the facilitation of their membrane insertion [38,40]. However, PEX3 is also present in discrete subdomains of ER membranes, and is involved in the targeting of an unknown number of precursor proteins to ER membranes, and possibly in their membrane insertion [35,36]. So far, these precursor proteins include membrane proteins, which either remain in the ER (the two-hairpin or reticulon-domain containing proteins ARL6IP1, RTN3A, and RTN4C) [35] or are pinched off in LDs (such as the hairpin protein UBXD8) [36]. At the ER, PEX3 cooperates with the farnesylated variant of PEX19 [36,39]. Thus, the farnesylation of PEX19 is most likely decisive in delivering precursor polypeptides to either pre-existent peroxisomes or the ER [7,8]. These observations raised the question of whether this pathway may play a more general role in ER protein targeting. Defects in the human PEX3 gene are linked to a particularly devastating form of Zellweger syndrome, which belongs to the peroxisome biogenesis disorders, and is also termed "cerebro-hepato-renal syndrome" to indicate the most important affected organs [41–44]. Infants with the disease typically die within their first year of life because of the complete absence of peroxisomes in all of the cells of the body.

Here, we address the client spectrum of PEX3 in ER protein targeting in human cells and, simultaneously, the question of whether the PEX19/PEX3-dependent pathway to the ER can also target precursor polypeptides to the Sec61 complex. The approach involves transiently PEX3-depleted HeLa cells or chronically PEX3-deficient Zellweger patient fibroblasts in combination with differential proteomic analysis by label-free quantitative mass spectrometry (MS) and differential protein abundance analysis (Figure 1b). Thus, we report on negatively and positively affected proteins after the partial depletion of PEX3 in HeLa cells and in PEX3-deficient Zellweger patient fibroblasts.

Table 1. Protein transport components and associated proteins in HeLa cells and linked diseases.

Component/Subunit for ER Targeting	Abundance [1]	Localization [2]	Linked Diseases
#p34 (LRC59, LRRC59) [3]	2480	ERM	
#p180 (RRBP1)	135	ERM	Hepatocellular Carcinoma, Colorectal Cancer
Kinectin 1 (KTN1)	263	ERM	
AEG-1 (LYRIC, MTDH)	575	ERM	
#SRP [4]		C	
— SRP72	355		Aplasia, Myelodysplasia
— SRP68	197		
— SRP54	228		Neutropenia, Pancreas Insufficiency
— SRP19	33		
— SRP14	4295		
— SRP9	3436		
— 7SL RNA			
SRP receptor		ERM	
— SRα (docking protein)	249		
— SRβ	173		
Calmodulin	9428	C	
for ER targeting and, possibly, for membrane integration			
hSnd1	unknown		
Snd receptor		ERM	
— hSnd2 (TMEM208)	81		
— hSnd3 [§]	49		
PEX19	80	C	Zellweger Syndrome
PEX3	103	ERM, PexM	Zellweger Syndrome

Table 1. *Cont.*

Component/Subunit for ER Targeting	Abundance [1]	Localization [2]	Linked Diseases
for ER targeting plus membrane integration			
#Bag6 complex		C	
— TRC35 (Get4)	171		
— Ubl4A	177		
— Bag6 (Bat3)	133		
SGTA	549	C	
TRC40 (Asna1, Get3)	381	C	
TA receptor		ERM	
— CAML (CAMLG, Get2)	5		
— WRB (CHD5, Get1)	4		Congenital Heart Disease
for ER membrane integration			
ERM protein complex			
— EMC1	124		
— EMC2	300		
— EMC3	270		
— EMC4	70		
— EMC5 (MMGT1)	35	ERM	
— EMC6 (TMEM93)	5		
— EMC7	247		
— EMC8	209		
— EMC9	1		
— EMC10	3		
#TMCO1 complex §			
— TMCO1	2013		
— Nicalin	99		Glaucoma, Cerebrofaciothoracic
— TMEM147	21	ERM	Dysplasia
— CCDC47 (Calumin)	193		
— NOMO	267		
PAT complex			
— PAT10 (Asterix)		ERM	
— CCDC47 (Calumin)	193		
for ER membrane integration plus translocation			
#Sec61 complex §		ERM	
— Sec61α1	139		Diabetes [5], CVID [6], TKD, Neutropenia
— Sec61β	456		PLD, Colorectal cancer
— Sec61γ	400		GBM, Hepatocellular carcinoma
#Sec62 (TLOC1)	26	ERM	Breast-, Prostate-, Cervix-, Lung-cancer
ER Chaperones			
— Sec63 (ERj2)	168	ERM	PLD, Colorectal cancer
— #ERj1 (DNAJC1)	8	ERM	
— BiP (Grp78, HSPA5)	8253	ERL	HUS
— Grp170 (HYOU1)	923	ERL	
— Sil1 (BAP)	149	ERL	MSS
*#Calnexin*_{palmitoylated}	7278	ERM	
#TRAM1	26	ERM	
TRAM2	40	ERM	
#TRAP complex		ERM	
— TRAPα ((SSR1)	568		
— TRAPβ (SSR2)			
— TRAPγ (SSR3)	1701		CDG, Hepatocellular Carcinoma
— TRAPδ (SSR4)	3212		CDG
#RAMP4 (SERP1)		ERM	

Table 1. *Cont.*

Component/Subunit for ER Targeting	Abundance [1]	Localization [2]	Linked Diseases
for covalent modification			
#Oligosaccharyltransferase (OST-A)		ERM	
— RibophorinI (Rpn1)	1956		
— RibophorinII (Rpn2)	527		
— OST48	273		CDG
— Dad1	464		
— OST4			
— TMEM258			
— Stt3A *	430		CDG
— DC2			
— Kcp2			
Oligosaccharyltransferase (OST-B)			
— RibophorinI (Rpn1)	1956		
— RibophorinII (Rpn2)	527		
— OST48	273		CDG
— Dad1	464		
— OST4			
— TMEM258			
— Stt3B *	150		CDG
— TUSC3			CDG
— MagT1	33		
Signal peptidase (SPC-A)		ERM	
— SPC12	2733		
— SPC18 * (SEC11A)			
— SPC22/23	334		
— SPC25	94		
Signal peptidase (SPC-C)		ERM	
— SPC12	2733		
— SPC21 * (SEC11C)			
— SPC22/23	334		
— SPC25	94		
GPI transamidase (GPI-T)			
— GPAA1	9		
— PIG-K	38	ERM	
— PIG-S	86		
— PIG-T	20		
— PIG-U	42		

[1] Abundance refers to the concentration (nM) of the respective protein in HeLa cells, as reported by Hein et al. [45]. [2] Localization refers to the functional intracellular localization(s) of the respective protein [1–4,35,36,41], i.e., C, Cytosol, ERL, ER lumen, ERM, ER membrane, PexM, and Peroxisome membrane. [3] Alternative protein names are given in parentheses. [4] Complexes are indicated by italics. Abbreviations for the protein names: EMC, ER membrane (protein) complex; GET, guided entry of tail-anchored proteins; SEC, (protein involved in) secretion; SND, SRP-independent; SR, SRP receptor; SRP, signal recognition particle; SSR, signal sequence receptor; TMEM, transmembrane (protein); TRAM, translocating chain-associating membrane (protein); TRAP, translocon-associated protein; TRC, transmembrane recognition complex. [5] Diabetes was linked to the particular protein in mice. [6] Abbreviation for diseases: CDG, congenital disorder of glycosylation; CVID, common variable immunodeficiency; GBM, glioblastoma multiforme; HUS, hemolytic-uremic syndrome; MSS, Marinesco-Sjögren syndrome; PLD, polycystic liver disease; TKD, tubulointerstitial kidney disease, as reported by Sicking et al. [34]. # indicates ribosome binding ability; § indicates ion channel activity; * indicates enzymatically active subunit.

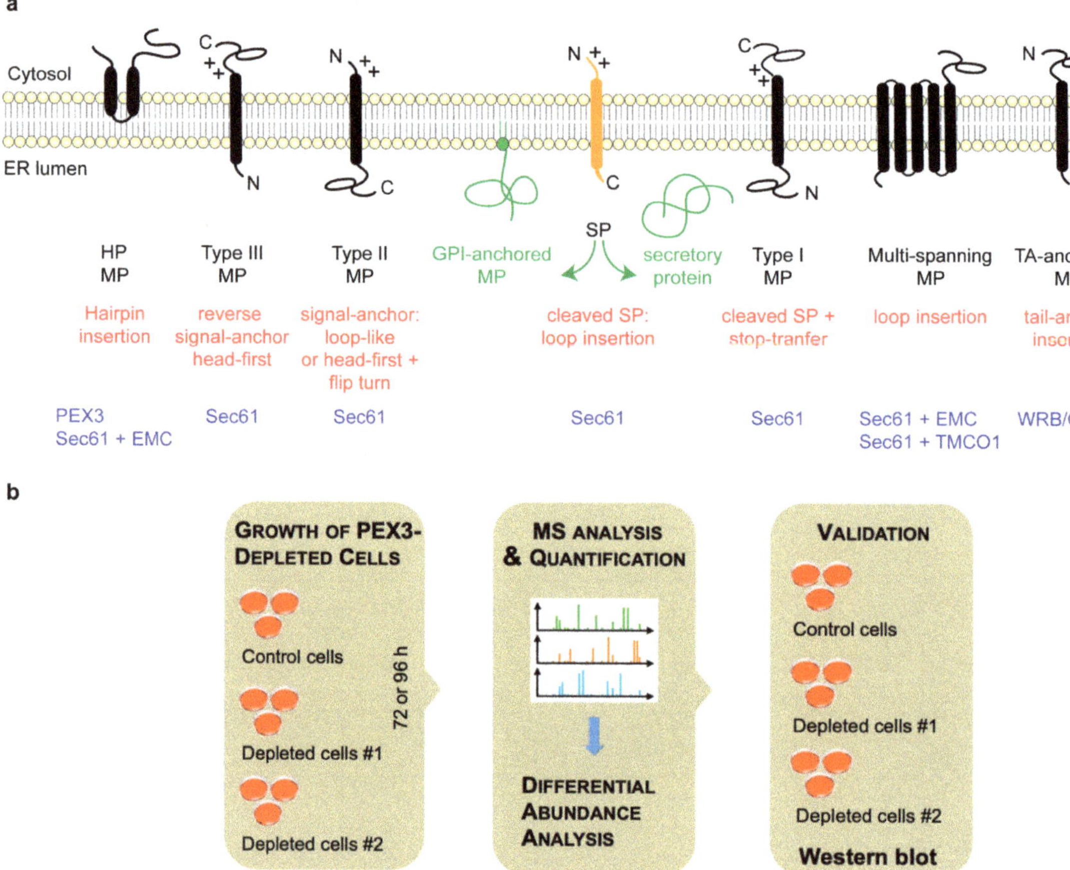

Figure 1. Types of ER membrane proteins and our experimental strategy to address their biogenesis. (**a**) The cartoon depicts a signal peptide (SP) (in yellow) and six types of ER membrane proteins (MP) (in black), together with their membrane protein type and the mechanism of membrane insertion (both indicated below the cartoon). Cleavable SPs (in yellow) can facilitate the ER import of secretory proteins (in green), glycosylphosphatidylinositol (GPI)-anchored membrane proteins (in green), and several types of membrane proteins, including single-spanning type I membrane proteins. Positively charged amino acid residues (+) play an important role in membrane protein and SP orientation, i.e., they typically follow the positive inside rule [14]. Amino-terminal transmembrane helices (TMHs) can serve as signal-anchor sequences to facilitate the membrane insertion of type II, type III, and many multi-spanning membrane proteins. In the case of membrane proteins with amino-terminal TMHs, membrane insertion typically involves the same components and mechanisms, which deliver secretory proteins (in green) and GPI-anchored membrane proteins (in green) to the ER lumen. The central component here is the Sec61 complex. In some cases, however, auxiliary membrane protein insertases, such as EMC or TMCO1 complex, play a role. These can also operate as stand-alone membrane protein insertases, an activity that they have in common with the WRB/CAML complex [4]. Hairpin (HP) proteins have a monotopic topology with N- and C-termini facing the cytosol, and some of them require PEX3 for membrane targeting. C, carboxy-terminus; N, amino-terminus. (**b**) The experimental strategy was as follows: siRNA-mediated gene silencing using two different siRNAs for each target and one non-targeting (control) siRNA, respectively, with three replicates for each siRNA for 96 h, followed by the label-free quantitative analysis of the total cellular proteome, and then differential protein abundance analysis to identify negatively affected proteins (i.e., putative clients of the target) and positively affected proteins (i.e., putative compensatory mechanisms), and finally validation by Western blot. In addition, PEX3-deficient Zellweger patient cells were analyzed in triplicates.

2. Results

2.1. Quantitative Proteomic Analysis of HeLa Cells after the Transient and Partial Depletion of PEX3 by siRNA

Our approach to the characterization of the client spectrum of PEX3 in ER protein targeting involves the gene silencing of the putative receptor PEX3 in HeLa Kyoto cells with two different targeting siRNAs in parallel to a non-targeting or control siRNA, and differential proteomic analysis by label-free quantitative MS analysis and differential protein abundance analysis (Figure 1b). This protocol was developed and previously used to characterize the client spectrum and client SP features of ER protein translocation components, including Sec61 complex (as a proof on concept), TRAP complex, Sec62/Sec63 complex, TRAM1-protein, ERj1, BiP, and the mRNA targeting components KTN1 and RRBP1 [33,46–48]. The approach is based on the assumption that polypeptide precursors, which have to be imported into the ER, are degraded by the proteasome in the cytosol upon interference with their ER targeting or translocation because their SPs or TMHs are not easily compatible with the aqueous character of the cytosol. Therefore, their cellular levels are decreased compared to those of the control cells, and this change is detected by quantitative MS and subsequent differential protein abundance analysis [46]. Typically, the decrease was observed to be accompanied by an increase of ubiquitin-conjugating enzymes [33,46–48]. Furthermore, a simultaneous increase in other ER import components was detected, which is consistent with the overlap of pathways and, additionally, may indicate a genetic interaction between different pathways and cellular compensation.

Here, we applied the established experimental strategy to identify precursor polypeptides that may depend on PEX3-dependent targeting to the ER [46]. They were expected among the negatively affected proteins in the label-free quantitative MS and subsequent differential protein abundance analysis. HeLa cells were treated in triplicates with two different *PEX3*-targeting siRNAs (*PEX3* #1 siRNA, *PEX3* #2 siRNA) in parallel to a non-targeting (control) siRNA for 96 h. Each MS experiment provided proteome-wide abundance data as LFQ intensities for three sample groups: one control (non-targeting siRNA treated) and two stimuli (down-regulation by two different targeting siRNAs directed against the same gene), which each having three data points (Figure 1b). In order to identify which proteins were affected by knock-down in siRNA-treated cells relative to the non-targeting (control) siRNA treated sample, we log2-transformed the ratio between the siRNA and control siRNA samples, and performed two separate unpaired t-tests for each siRNA against the control siRNA sample according to [46]. The p values obtained by the unpaired t-tests were corrected for multiple testing using a permutation-based false discovery rate (FDR) test. The proteins with an FDR-adjusted p value of below 5% were considered significantly affected by the knock-down of the targeted protein.

After PEX3 depletion, 6488 different proteins were quantitatively detected by MS in all of the samples (Figure S1, Tables 2, S1 and S2). The MS data were deposited to the ProteomeXchange Consortium via the PRIDE partner repository [49] with the dataset identifier PXD012005 (http://www.proteomexchange.org; last accessed on 2 May 2021). They included the expected representations of proteins with cleaved SP (7%), N-glycosylated proteins (9%), and membrane proteins (13%), which were comparable to the previously published Sec61 and TRAP experiments (Figure S1c, left small pies) [46]. Through the application of the established statistical analysis, we found that transient PEX3 depletion significantly affected the steady-state levels of 13 proteins: 13 negatively and none positively (with a permutation-based false discovery rate-adjusted p value < 0.05). As had to be expected, PEX3 itself was negatively affected (Figure S1a, volcano plots), which was confirmed by Western blot analysis (Figure S1b). The identified precursors included one protein with cleavable SP (the ER membrane protein AGPAT1), and two membrane proteins with TMH (not counting PEX3), i.e., the endo- and lysosomal membrane protein TMEM192 with four transmembrane domains, and the single-pass type II plasma membrane protein SGCD (Figure S1b, Table S2). Of these three negatively affected proteins, one was N-glycosylated (SGCD). There were no proteins with an annotated functional location in peroxisomes or LDs among the negatively affected proteins. There was no positively affected protein observed. These results raise the question of why PEX3 depletion from HeLa cells had hardly any effect on the cellular proteome. There are several possibilities. The simplest answer would be that the depletion efficiency of 85% may not have been high enough to cause the accumulation of precursor proteins. Another answer could be that PEX3 function in ER protein import in HeLa cells is not essential, i.e., it can be substituted by other proteins or pathways. Another possibility is that the accumulating precursors were not degraded but aggregated in the cytosol, or ended up in other organelles, where they were protected from degradation. These possibilities will we considered in further detail in the discussion.

Table 2. Statistics for the identification of putative PEX3 clients in comparison to the previously identified clients for ER membrane targeting and translocation components.

Proteins	PEX3	Z [1]	RRBP1 [2]	KTN1 [2]	SEC61 [2]	TRAP [2]	CDG [1,2]
Quantified proteins	8178	6328	4813	4947	7212	7670	5920
Statistically analyzed proteins	6488	6328	4813	4947	5129	5911	5920
representing the secretory pathway (%)	29	29	26	27	26	27	36
Proteins with SP (%)	7	7	6	6	6	7	nd [3]
N-Glycoproteins (%)	9	9	8	8	8	8	nd
Membrane proteins (%)	13	13	12	13	12	13	nd
Positively affected proteins	0	97	157	25	342	77	39
Negatively affected proteins	13	141	141	45	482	180	279
representing the secretory pathway (%)	54	39	37	41	61	40	36
Negatively affected proteins with SP (%)	8	19	18	7	41	22	12
Negatively affected N-glycoproteins (%)	8	21	17	18	45	23	17
Negatively affected membrane proteins (%)	31	18	18	22	36	26	23
Negatively affected proteins with SP	1	27	21	3	197	38	34
including N-glycoproteins	0	23	16	3	158	28	30
corresponding to %	0	85	76	100	80	74	88
including membrane proteins	1	6	6	1	77	19	16
corresponding to %	100	22	29	33	39	50	53
Negatively affected proteins with TMH	3	16	18	8	98	22	41
including N-glycoproteins	1	6	7	4	56	11	17
corresponding to %	33	38	39	50	57	50	41
Negatively affected peroxisomal proteins	1	12	0	1	1	0	1
corresponding to %	8	9	nd	2	0	nd	0
including membrane proteins	1	6	nd	0	1	0	1
corresponding to %	100	50	nd	nd	100	nd	100
Negatively affected mitochondrial proteins	0	14	6	1	29	14	21
corresponding to %	nd	10	4	2	1	1	1
including membrane proteins	nd	4	3	0	11	3	8
corresponding to %	nd	29	50	nd	38	21	38

[1] Z and CDG refer to immortalized fibroblasts from patients suffering from Zellweger syndrome or a congenital disorder of glycosylation. [2] Refers to siRNA-mediated knockdown HeLa cells, and was previously published [33,46]. [3] nd, not determined.

2.2. Quantitative Proteomic Analysis of PEX3-Deficient Zellweger Patient Fibroblasts

In the course of our previous analysis of Sec62- and Sec63 clients, we moved from the respective siRNA-treated and incompletely depleted HeLa cells with low client numbers on to CRISPR/Cas9-treated and deficient HEK293 cells, and indeed, we identified many more clients [47]. Therefore, we sought to test whether a complete PEX3 knockout in cells leads to the depletion of putative PEX3 client proteins. Therefore, we subjected control fibroblasts and immortalized Zellweger patient fibroblasts with PEX3 deficiency [41], which had been grown in triplicates, to label-free quantitative proteomic analysis and differential protein abundance analysis, and analyzed the data for negatively affected proteins, i.e., potential PEX3 clients (Figure 2, Tables 2, S3–S5). The MS data were deposited to the ProteomeXchange Consortium via the PRIDE partner repository [49] with the dataset identifier PXD012005 (http://www.proteomexchange.org; last accessed on 2 May 2021).

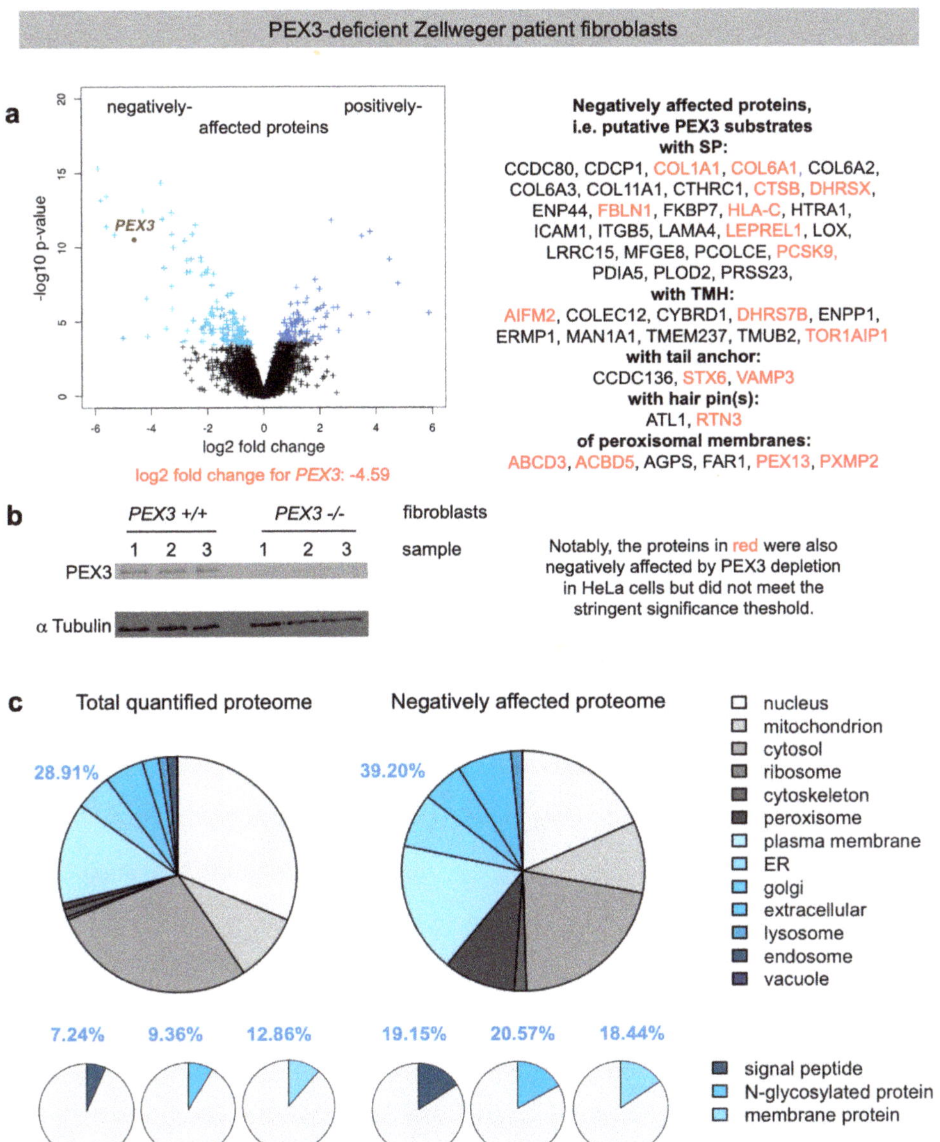

Figure 2. Volcano plots and Gene Ontology (GO) enrichment for PEX3-deficient Zellweger patient fibroblasts. (**a**) The differentially affected proteins were characterized by the mean difference of their intensities plotted against the respective permutation-based false discovery rate-adjusted *p*-values in the volcano plots; PEX3 is highlighted. In addition, the proteins, which were negatively affected by PEX3 deficiency are given in the right panel. (**b**) PEX3 deficiency was evaluated by Western blot. The molecular mass values are indicated in kilodaltons (KDa). Only the area of interest of the blot is shown; the original images are shown in the Supplementary Materials. (**c**) The classification of the putative PEX3 clients was based on GO enrichment factors where the results from the complete set of quantified proteins in the left panel are compared with the negatively affected proteome. The protein annotations of the SPs, membrane location, and N-glycosylation in humans were extracted from UniProtKB, and were used to determine the enrichment of the GO annotations among the negatively affected proteins.

We quantitatively identified a total of 6328 different proteins by MS, 141 of which were negatively affected by PEX3 deficiency in the patient fibroblasts versus the control fibroblasts. As had to be expected, PEX3 itself was negatively affected (Figure 2a, volcano plots), which was confirmed by Western blot (Figure 2b). Applying the established statistical analysis, we found that PEX3 deficiency significantly affected the steady-state levels of 238 proteins: 141 negatively and 97 positively (permutation-based false discovery rate-adjusted p value < 0.05). Of the negatively affected proteins, GO terms assigned 39.2% to organelles of the endocytic and exocytic pathways (Figure 2c, large pies), which corresponds to a 1.36-fold enrichment (Figure 2c, large pies, 39.2% divided by 28.91% = 1.36) and is below the average values of 1.46 and 1.94 observed after the depletion of the mRNA targeting components KTN1 (1.55) and RRBP1 (1.37), and the translocation components Sec61 (2.37) and TRAP (1.5), respectively (Table 2) [33,46]. In contrast to the PEX3-knockdown cells (Figure S1c), we also detected the enrichment of proteins with SP (2.65–fold), N-glycosylated proteins (2.2-fold), and membrane proteins (1.36–fold) (Figure 2c, small pies), which was lower compared to the Sec61 (6.51, 2.83, 2.51) and TRAP experiments (3.3, 2.7, 2.1), but higher compared to the KTN1 (1, 2.09, 1.76) and RRBP1 experiments (2.44, 2.12, 1.46) [33,46].

The negatively affected proteins of the secretory pathway included 27 proteins with cleavable SP (including six collagens or collagen-like proteins, three collagen-modifying proteins, three ER lumenal proteins—i.e., FKBP7, PCSK9, and PDIA5—the lysosomal cathepsin CTSB, eight secretory proteins, the ER membrane protein PLOD2, and the plasma membrane proteins ENPP4, HLA-C, ICAM1, ITGB5, and LRRC15 (Figure 2a).

Furthermore, 15 membrane proteins of the secretory pathway with TMH (not counting PEX3) were negatively affected, i.e., most notably the two ER-resident hairpin proteins ATL1 and RTN3; the three tail-anchored proteins CCDC136, STX6 and VAMP3; the ER membrane proteins DHRS7B, ERMP1, and TMUB2; the Golgi protein MAN1A1; the plasma membrane proteins AIFM2 (which has additional locations, see below), COLEC12, CYBRD1, ENPP1, and TMEM237; and the nuclear envelope protein TOR1AIP1 (Figure 2a, Tables 3 and S4). Of these 42 negatively affected proteins, 29 were N-glycosylated proteins (23 with SP and 6 with TMH).

Interestingly, there were 14 precursors of mitochondrial proteins negatively affected by PEX3 deficiency (Table S4), a phenomenon previously observed after the depletion of RRBP1 from HeLa cells and attributed to their physiological trafficking from the ER to mitochondria via the newly identified ER-SURF pathway [11,33]. Among these negatively affected mitochondrial proteins were three outer membrane proteins (AIFM2, RHOT1, VAT1), one inner membrane protein (NDUFV3), and ten matrix proteins, two of which had a dual localization in the mitochondria and peroxisomes (ACAD11, SCP2). Alternatively, our observation of mitochondrial proteins among the negatively affected ones may be due to the findings that PEX19, at least, associates with peroxisomes and mitochondria, and, for example, facilitates the protein biogenesis of the tail-anchored membrane proteins Fis1 and Gem1 for both organelles [50].

The positively affected proteins in the PEX3-deficient fibroblasts included cell adhesion molecules (such as ITGA1, L1CAM, and NCAM1), signal transduction components of the plasma membrane (ANXA3, PKD2), mitochondrial membrane proteins of the inner membrane with functions in protein import into the mitochondria (DNAJC15) or protein quality control (OMA1), and the cytosolic chaperone HSPB6.

2.3. Negatively Affected Precursor Proteins in Zellweger Patient Fibroblasts Are Specific for PEX3-Deficiency, and Are Partially Affected in PEX3-Depleted HeLa Cells

In order to validate the proteomic data on putative PEX3-substrates, we conducted independent Western blot experiments with the PEX3-deficient Zellweger and control fibroblasts for the SP-containing candidate PDIA5, the hairpin protein RTN3, the dual topology LD/peroxisome protein Far1, and the peroxisomal protein ACBD5 (Figure 3). All of these proteins were depleted in PEX3-deficient fibroblasts, fully confirming the proteomic analysis and verifying them as putative PEX3 clients. Interestingly, for RTN3, we observed that specifically the 100 kDa isoform was depleted in the PEX3-deficient cells, while the 25 kDa isoform remained largely unaltered (Figure 3, e versus f).

For PEX3-depleted HeLa cells, we observed that 17 of the 48 negative hits were not quantified, and that 18 out of 31 putative PEX3 clients in Zellweger patient fibroblasts (i.e., 58%) were also negatively affected by PEX3 depletion in HeLa cells (including two out of three tail-anchored proteins, one out of two hairpin proteins, and four out of six peroxisomal proteins), but did not meet the stringent significance threshold (Table S6 and proteins indicated in red in Figure 2). Notably, the PEX3 depletion in the HeLa cells was not as efficient as it was in the fibroblasts (log2-fold change: -3.5 versus -4.6) (Figures 2 and S1). Hence, it is understandable that the levels of the putative PEX3 clients are less perturbed in HeLa cells than in fibroblasts. Still, the majority of them (18 versus 13) are perturbed in the same negative direction as they are in fibroblasts (Table S1).

We consider these putative clients of PEX3 for organelles of the secretory pathway to be specific for additional reasons. First, the negatively affected proteins included, as expected, twelve precursors of proteins with a functional location in peroxisomes (not counting PEX3), including six peroxisomal membrane proteins (Figure 2a) (Table 3). The peroxisomal membrane proteins were ABCD3, ACBD5, AGPS, FAR1, PEX13, and PXMP2. Notably, the tail-anchored membrane protein FAR1 exhibits a dual topology, and can locate to peroxisomes as well as to LDs [51]. Second, with two hairpin proteins, ATL1 and RTN3, among the negatively affected proteins, we also confirmed a second class of already-known PEX3 clients in human cells under physiological conditions (Table 3). Third, only one of the negatively affected peroxisomal proteins, PEX13, had previously been observed for TRAP-deficient fibroblasts from patients who suffer from congenital disorders of glycosylation (CDG) and are either *SSR3-* (coding for TRAPγ) or *SSR4* (TRAPδ)-deficient [46]. Furthermore, there was no overlap in the positively affected proteins that accumulate in either CDG or Zellweger patient fibroblasts.

Table 3. Negatively affected proteins in PEX3-deficient cells, i.e., putative PEX3 substrates.

Gene	Subcellular Location	Membrane Protein Type	SS or TMH
ACBD5	Peroxisome membrane	Single-spanning membrane protein	
COLEC12	Membrane	Single-spanning type II membrane protein	TMH
LRRC15	Membrane	Single-spanning type I membrane protein	SP
PEX3	Peroxisome membrane	Single-spanning membrane protein	
TOR1AIP1	Nuclear envelope inner membrane	Single-spanning membrane protein	TMH
COL1A1	Secreted, Extracellular space, Extracellular matrix		SP
AGPS	Peroxisome membrane		
ACAD11	Peroxisome, Mitochondrion		
STX6	Golgi apparatus membrane	Tail-anchored membrane protein	tail anchor
CCDC136	Acrosome membrane, Secretory vesicle, Cytoplasmic vesicle	Tail-anchored membrane protein	tail anchor
FAR1	Peroxisome membrane	Tail-anchored membrane protein	tail anchor
PXMP2	Peroxisome membrane	Multi-spanning membrane protein	
ATL1	Cell projection, Golgi apparatus membrane, ER membrane,	Hairpin membrane protein with one HP	hairpin
COL6A2	Extracellular matrix, Membrane, Secreted, Extracellular space		SP
LOX	Extracellular space, Secreted		SP
ERMP1	ER membrane	Multi-spanning membrane protein	TMH
CYBRD1	Membrane	Multi-spanning membrane protein	TMH
TMUB2	Membrane	Multi-spanning membrane protein	TMH
ABCD3	Peroxisome membrane	Multi-spanning membrane protein	
SCP2	Peroxisome, Mitochondrion, Cytoplasm		
CDCP1	Secreted, Cell membrane		SP
COL6A3	Extracellular space, Secreted, Extracellular matrix		SP

Table 3. *Cont.*

Gene	Subcellular Location	Membrane Protein Type	SS or TMH
TMEM237	Cell projection, Membrane, Cilium	Multi-spanning membrane protein	TMH
ENPP4	Cell membrane	Single-spanning type I membrane protein	SP
HTRA1	Cell membrane, Secreted, Cytoplasm, Cytosol		SP
VAMP3	Synapse, Membrane, Cell junction, Synaptosome	Tail-anchored membrane protein	tail anchor
MFGE8	Membrane, Secreted		SP
PRSS23	Secreted		SP
DHRS4	Peroxisome, Nucleus		
ITGB5	Membrane	Single-pass type I membrane protein	SP
FBLN1	Extracellular space, Secreted, Extracellular matrix		SP
COL6A1	Extracellular space, Secreted, Extracellular matrix		SP
PCSK9	Endosome, Golgi apparatus, Cell surface, Secreted, ER, Lysosome		SP
CTHRC1	Extracellular space, Secreted, Extracellular matrix		SP
DHRSX	Secreted		SP
HLA-C	Membrane	Single-spanning type I membrane protein	SP
CCDC80	Secreted, Extracellular space, Extracellular matrix		SP
RTN3	ER membrane, Golgi apparatus membrane	Hairpin membrane protein with two HP	hairpin
ENPP1	Secreted, Basolateral cell membrane, Cell membrane	Single-spanning type II membrane protein	TMH
PLOD2	Rough ER membrane		SP
RHOT1	Mitochondrion outer membrane		
COL11A1	Extracellular matrix, Extracellular space, Secreted		SP
NDUFV3	Mitochondrion inner membrane		
PCOLCE	Secreted		SP
AIFM2	Membrane, Mitochondrion outer membrane, Lipid droplet	Single-spanning membrane protein	TMH
MAN1A1	Golgi apparatus membrane	Single-spanning type II membrane protein	TMH
ACBD7	Cytosol		
ICAM1	Membrane	Single-spanning type I membrane protein	SP
CTSB	Lysosome, Melanosome, Secreted, Extracellular space		SP
DHRS7B	ER membrane	Single-spanning type II membrane protein	TMH
LAMA4	Extracellular matrix, Extracellular space, Secreted		SP
LEPREL1	ER, Golgi apparatus		SP
PEX13	Peroxisome membrane	Single-spanning membrane protein	
PDIA5	ER lumen		SP
CTHRC1	Extracellular space, Secreted, Extracellular matrix		
FKBP7	ER lumen		SP

The proteins are listed according to the decreasing negative effects of PEX3 depletion. The colors refer to peroxisomal proteins (yellow), mitochondrial proteins (brown), and proteins of the secretory pathway with SP, TMH, tail anchors (green) or hairpins (orange). As compared to Table S4, the GO annotation for TOR1AIP1, the hairpin of RTN3, and the definitions of the membrane protein types were taken from GeneCards (https://www.genecards.org; last accessed on 1 September 2021). In addition, the term "TMH" is used here only for proteins of the secretory pathway. Red letters refer to incomplete annotations (see text for details). HP, hairpin.

As outlined in Table 2, and for comparison with the PEX3-deficient Zellweger patient fibroblasts, 5919 different proteins were previously quantified for CDG patient fibroblasts, 279 of which were negatively affected by TRAP absence, and 39 of which were positively affected. In total, 100 of the negatively affected proteins were assigned to the secretory pathway, including 34 precursor polypeptides with SP and 41 with TMH (including the subunits of the heterotetrameric TRAP complex) [46]. A total of 47 of the negatively affected proteins were N-glycoproteins (30 with SP and 17 with TMH). The peroxisomal membrane protein PEX13 was among the negatively affected proteins, which was also negatively affected in PEX3-deficient Zellweger patient fibroblasts (Table 2).

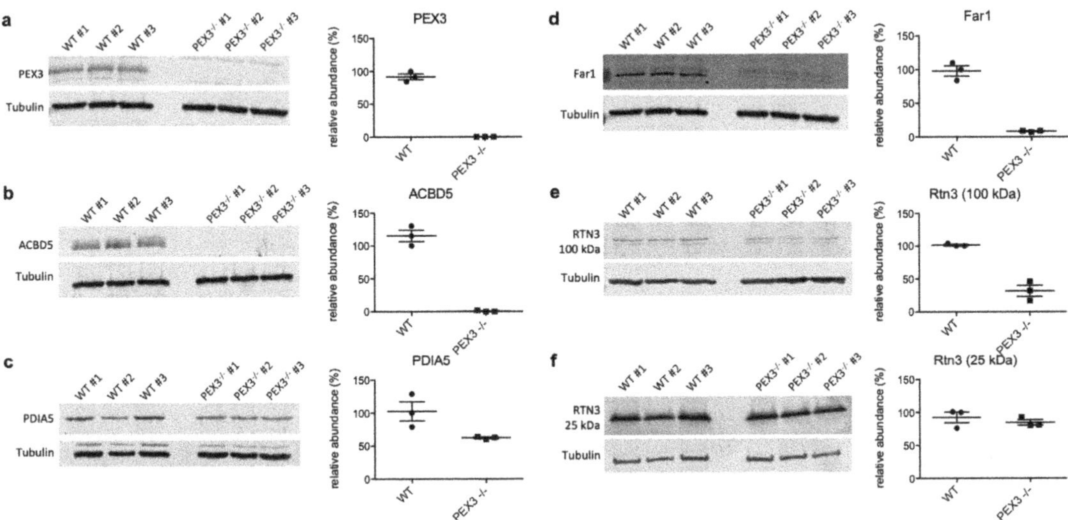

Figure 3. Validation of the PEX3 clients by Western blot analyses. (**a–f**) Three independent cell lysates from the control (WT) and PEX3-deficient fibroblasts (PEX3$^{-/-}$), respectively, were analyzed by Westen blotting using antibodies as indicated. Left panels: Relevant sections of the representative Western blots are shown; tubulin served as a loading control. We note that the full scans of all of the blots are shown in the supplement. Right panels: The scatter plots indicate the relative protein abundances in the control and PEX3-deficient fibroblasts, as derived from quantitative Western blots, as shown in the left panels. The signals were quantified by densitometry, and the relative abundances were calculated as the ratio of the signal of interest to the corresponding tubulin signal in the same lane, and were normalized against one control sample. The mean values with SEM from three independent lysates per cell line are indicated, as well as the individual data points for each replicate.

3. Discussion

Here, we addressed the question of which precursor polypeptides employ the PEX3-dependent pathway for the targeting of or insertion and translocation, respectively, into the ER of human cells. First, we employed our previously established approach of the siRNA-mediated depletion of PEX3 in HeLa cells, the label-free quantitative MS of the total cellular proteome, and differential protein abundance analysis. Next, we quantified the negatively and positively affected proteins under conditions of PEX3 deficiency in Zellweger patient fibroblasts.

On first sight, the result of the siRNA-mediated depletion of PEX3 in HeLa cells was not very informative (Figure S1). In addition to the depletion of PEX3, only three precursor polypeptides with SP or TMH were found among the negatively affected proteins and may be considered as PEX3 clients in ER protein targeting. Notably, the one with an SP, AGPAT1, is an important enzyme in lipid metabolism, and is required for the synthesis of phosphatidic acid and triacylglycerides [52]. Therefore, it may affect LD biogenesis, and LD localization would not be unexpected. The proteins with a TMH are the endo- and lysosomal membrane protein TMEM192 with four transmembrane domains, and the single-spanning type II plasma membrane protein SGCD. There was no peroxisomal protein negatively affected by PEX3 depletion, which has to be interpreted in light of the facts that PEX19 and PEX3 are essential for peroxisome formation [37,38], and that PEX3 deficiency causes the complete absence of peroxisomes [42–44]. Together, these results raise the question of why PEX3 depletion from HeLa cells had hardly any effect on the cellular proteome. There are several possibilities. (i) The simplest answer would be that the depletion efficiency and its duration may not have been sufficient to cause a significant accumulation of precursor proteins. According to the MS data, the log2 fold change was −3.4942, i.e., the depletion efficiency was higher than 90%, which is consistent with the Western blot analysis (Figure S1b). This residual amount of PEX3, however, may have been sufficient for physiological functions, and could explain the absence of an effect on peroxisomal proteins. (ii) Another answer may be that the PEX3 function in ER protein import in HeLa cells is not essential, i.e., it may also be provided by other proteins or pathways. Indeed, it was shown in cell-free ER protein import studies that certain peroxisomal membrane proteins can be targeted to the mammalian ER by SRP or TRC40 (including PEX3) [29,35,53–55]. Furthermore, some collagens, as well as some hairpin membrane proteins, were previously observed as RRBP1 clients (ATL2, ATL3, COL1A1, COL1A2, COL4A2) and SRP clients (ARL6IP1, RTN3), respectively, in cell biological or proteomic experiments [30,33,35]. (iii) Another possibility is that some accumulating precursors were not degraded, and either stayed soluble in the cytosol (as is known for catalase), aggregated, or ended up in other organelles (as is known for PEX14) [41], where they were protected from degradation. Indeed, we have previously observed the mis-targeting of certain precursors of secretory proteins into mitochondria in the absence of Sec61 function in HeLa cells [56]. Furthermore, the known PEX3 client, UBXD8, accumulates in mitochondria when PEX3 function is compromised [36]. (iv) Last but not least, all three possibilities may have contributed to the result we obtained in HeLa cells upon siRNA-mediated knockdown; we consider this the most likely explanation.

Under conditions of PEX3 deficiency in Zellweger patient fibroblasts, the results were more informative (Figure 2). First, the negatively affected proteins included twelve precursors of proteins with a functional location on or in peroxisomes, including six peroxisomal membrane proteins (Figure 4, Table 3). This was expected, and demonstrated once more the feasibility of the approach. However, this does not mean that all of these precursors of peroxisomal proteins are targeted to the ER. Rather, at least for the peroxisomal matrix proteins, it must be due to the complete absence of peroxisomes from the patient fibroblasts with PEX3 deficiency [42–44]. Likewise, we cannot rule out that the depletion of some precursors could derive from transcriptional effects. However, we consider this unlikely, as previous transcriptomics studies revealed that solely the absence of peroxisomes does not generally result in global mRNA expression changes [57].

Second, the identified precursors included the two ER-resident hairpin proteins ATL1 (with one hairpin) and RTN3 (with two hairpins), which is consistent with the previous finding that PEX3 is involved in the ER targeting of hairpin proteins [35,36]. It remains to be tested whether PEX3, possibly in cooperation with PEX16 [41], also facilitates the membrane insertion of these hairpin proteins, and whether additional membrane protein insertases—such as Sec61, EMC, TMCO1, and WRB/CAML—contribute to membrane insertion. Likewise, it is also conceivable that the members from the DHRS- and ACBD-protein families (ACBD5, DHRS4, DHRS7b, and DHRSX; Figure 2, Table 3) are first inserted into the ER membrane by PEX3 and additional insertion components of the ER. This will also have to be addressed in future research.

Furthermore, 14 α-helical membrane proteins, including four tail-anchored membrane proteins (CCDC136, FAR1, STX6 and VAMP3) and ten others (not counting PEX3)—i.e., DHRS7B (type II), ERMP1 (multi-pass), TMUB2 (multi-pass), MAN1A1 (type II), AIFM2, COLEC12 (type II), CYBRD1 (multi-pass), ENPP1 (type II), TOR1AIP1 (single-pass), and TMEM237 (multi-pass)—were negatively affected by PEX3 absence, and therefore can be considered as potential PEX3 clients in ER protein targeting (Table 3). Importantly, the annotated membrane topology of these potential clients should be considered with care, as they may be, in certain cases, incomplete. For example, it was recently shown that the tail-anchored membrane protein FAR1 exhibits a dual topology, including a monotopic hairpin topology, and can therefore locate to peroxisomes as well as LDs [51]. Because ERMP1, STX6, and TOR1AIP1 were previously found to depend on Sec61 for their membrane insertion in HeLa cells, it is tempting to speculate that PEX3 is able to target precursors to the Sec61 complex.

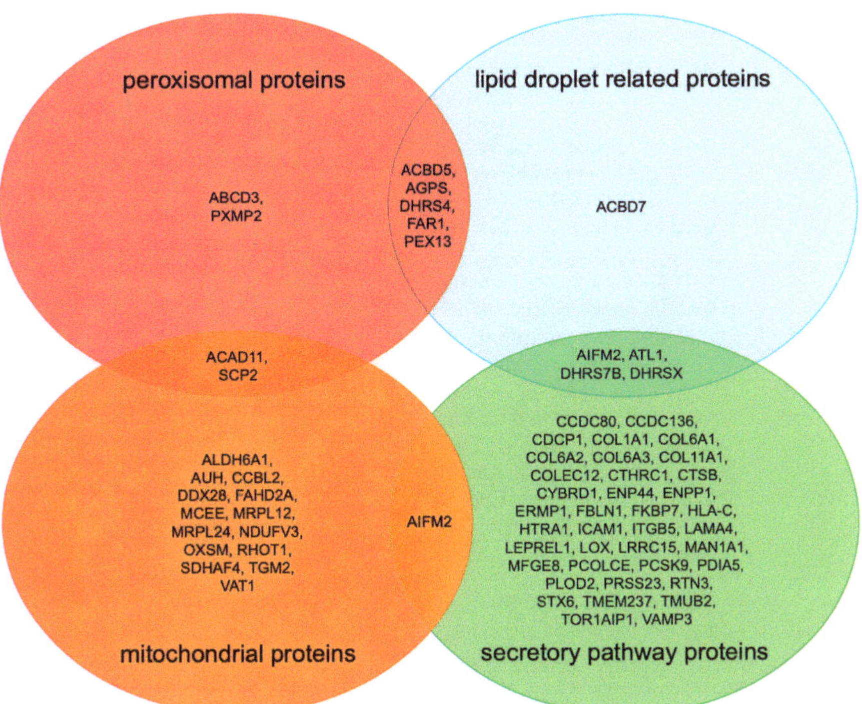

Figure 4. Venn diagram for negatively affected proteins in PEX3-deficient Zellweger patient fibroblasts. We note that Tables 3 and S4 served as the basis for this compilation.

In addition, the identified precursors included 27 proteins with cleavable SP, including five collagens and one collagen-like protein, four collagen-modifying proteins LE–REL1, three ER lumenal proteins, and the four plasma membrane proteins with a connection to the extracellular matrix or neighboring cells, i.e., CDCP1, ICAM1, ITGB5, LRRC15. Thus, between 14 and 17 of the total of 43 secretory pathway precursor proteins are collagens or related to collagens, which is a significant enrichment of 33–40%. This raises the question of how PEX3 could be involved in the targeting of these precursors, which are expected to involve a cotranslational targeting pathway to the Sec61 complex. We hypothesize that, in these particular cases, PEX3 may act in concert with mRNA targeting pathways, which would be consistent with the observation that RRBP1-mediated targeting pathways were found to be involved in the biogenesis of these proteins [30,33]. However, their degradation in the absence of PEX3 may have an alternative explanation. We speculate that the combination of farnesylated PEX19 and ER-membrane-resident PEX3 targets the membrane-shaping and hairpin(s)-containing membrane proteins (such as Atlastins, Reticulons and Spastin) to the PEX3-rich ER subdomain, and that this enrichment of hairpin proteins creates an environment which attracts collagens as well as some of their modifying enzymes and future interaction partners [6,58–61]. As is consistent with this scenario, the key player of the formation of large cargo secretory vesicles at ER exit sites, the membrane protein TANGO1, has two transmembrane domains, one of which is supposed to form a hairpin in the inner leaflet of the ER membrane [59,60]. Either way, a common ER subdomain may be conducive to the budding of both peroxisomal precursor vesicles and large cargo secretory vesicles, as well as the formation of LDs. Indeed, very recent evidence suggests that membrane bridges between ER exit sites and LDs allow the protein partitioning of hairpin proteins from the ER to LDs (doi: https://doi.org/10.1101/2021.09.

14.460330; accessed on 25 November 2021). In addition, this observation raises the question of whether defects in collagen biogenesis contribute to the devastating effects of PEX3 deficiency in Zellweger patients, which clearly warrants further work.

Furthermore, negatively affected proteins in PEX3-depleted or -deficient human cells need to be discussed with respect to LD biogenesis from a functional point of view (Figure 3, Table 3). DHRS- and ACBD-family proteins, which were negatively affected in PEX3-deficient fibroblasts (Table 3) and the dehydrogenase/reductase SDR family members 4, X and 7B (DHRS4, DHRSX, DHRS7B) play important roles in retinol biosynthesis, and may also localize to LDs [62]. The latter was suggested for DHRSX, and was actually shown for the family members DHRS3 and DHRS7B [62] (doi: https://doi.org/10.110 1/2021.09.14.460330). On the other hand, acyl-CoA-binding domain-containing proteins 5 and 7 (peroxisomal ACBD5, cytosolic ACBD7) are involved in organelle contacts and, therefore, may be important for lipid metabolism and/or organelle budding from the ER. These aspects, too, warrant further studies.

Taken together, our study revealed a putative client spectrum for PEX3-mediated protein targeting to the ER, including several unexpected protein classes, such as secreted collagens. Importantly, several PEX3 client candidates are involved in lipid metabolic pathways or membrane-shaping mechanisms, which affect the ER-derived organelle biogenesis of peroxisomes and/or LDs. As observed for other ER targeting pathways, PEX3 also likely shares some of its putative clients with other ER targeting components, including the Sec61 complex.

4. Materials and Methods

4.1. Cell Growth and Analysis

HeLa Kyoto cells [36] were cultivated at 37 °C in a humidified environment with 5% CO_2, in DMEM with 10% fetal bovine serum (FBS; Sigma-Aldrich, Taufkirchen, Germany). The cell growth and viability were monitored using the Countess® Automated Cell Counter (Invitrogen, Thermo Fisher Scientific, Darmstadt, Germany), following the manufacturer's instructions.

For gene silencing, 4×10^5 HeLa cells were seeded per 6-cm culture plate, followed by incubation under normal culture conditions. Next, the cells were transfected with either PEX3-targeting Silencer Select pre-designed siRNA (Life Technologies, Darmstadt, Germany, IDs s16154 and s16156) or with a scrambled siRNA control (Life Technologies ID 4390843) at a final concentration of 3.3 nM using Lipofectamine 2000 (Life Technologies), following the manufacturer's instructions. After 48 h, the cells were transfected a second time and grown for an additional 48 h. Thus, silencing was performed for a total of 96 h using two different siRNAs. We note that the cell viability was not affected by PEX3 depletion for 96h, i.e., the viability values were virtually identical for the two PEX3-targeting siRNAs (95.0% for RNA#1 and 96.0% for RNA#2, $n = 3$) versus the scrambled siRNA control (95.7%).

The silencing efficiencies were evaluated by Western blot analysis using PEX3-specific antibodies [63] (1:1000 dilution), which were kindly donated by Gabriele Dodt (University of Tübingen, Tübingen, Germany) and anti-tubulin antibodies (T6199, Sigma-Aldrich; 1:10,000 dilution). Donkey-derived, Cy3- and Alexa488-conjugated secondary antibodies (715-165-151, 711-545-152, Jackson Immunoresearch, Cambridgeshire, UK) were detected using the Typhoon-Trio imaging system combined with Image Quant TL software 7.0 (GE Healthcare, Freiburg, Germany).

Immortalized PEX3-deficient fibroblasts and control fibroblasts were obtained from Gabriele Dodt (University of Tübingen, Tübingen, Germany), and were previously characterized [41,64]. They were cultivated at 37 °C in a humidified environment with 5% CO_2, in DMEM/GlutaMAX with 10% fetal bovine serum (FBS; Sigma-Aldrich) for 72 h.

4.2. Label-Free Quantitative Proteomic Analysis

After growth for 96 h, 1×10^6 cells (corresponding to roughly 0.2 mg protein) were harvested, washed twice in PBS, and lysed in a buffer containing 6 M GnHCl, 20 mM tris(2-carboxyethyl)phosphine (TCEP; Pierce™, Thermo Fisher Scientific, Darmstadt, Germany), and 40 mM 2-chloroacetamide (CAA; Sigma-Aldrich) in 100 mM Tris, at pH 8.0 [33,46–48]. The lysate was heated to 95 °C for 2 min, and then sonicated in a Bioruptor sonicator (Diagenode, Seraing, Belgium) at the maximum power setting for 10 cycles of 30 s each. For a 10% aliquot of the sample, the entire process of heating and sonication was repeated once, and then the sample was diluted 10-fold with digestion buffer (25 mM Tris, pH 8, 10% acetonitrile). The protein extracts were digested for 4 h with Lysyl endoproteinase Lys-C (Wako Bioproducts, Fujifilm, Neuss, Germany, enzyme to protein ratio: 1:50), followed by the addition of trypsin (Promega, Heidelberg, Germany) for overnight digestion (at an enzyme to protein ratio of 1:100). The next day, a booster digestion was performed for 4 h using an additional dose of trypsin (enzyme to protein ratio: 1:100). After the digestion, a 10% aliquot of peptides (corresponding to about 2 μg of peptides) were purified via SDB-RPS StageTips [65], eluted as one fraction, and loaded for MS analysis. Purified samples were loaded onto a 50-cm column (inner diameter: 75 microns; packed in-house with ReproSil-Pur C18-AQ 1.9-micron beads, Dr. Maisch HPLC GmbH, Ammerbuch, Germany) via the autosampler of the Thermo Easy-nLC 1000 (Thermo Fisher Scientific) at 60 °C. Using the nanoelectrospray interface, the eluting peptides were directly sprayed onto the benchtop Orbitrap mass spectrometer Q Exactive HF (Thermo Fisher Scientific) [66]. The peptides were loaded in buffer A (0.1% (v/v) formic acid at 250 nL/min, and the percentage of buffer B was ramped to 30% over 180 min, followed by a ramp to 60% over 20 min, then 95% over the next 10 min, and maintained at 95% for another 5 min [33,47]. The mass spectrometer was operated in a data-dependent mode, with survey scans from 300 to 1700 m/z (resolution of 60,000 at m/z = 200). Up to 15 of the top precursors were selected and fragmented using higher energy collisional dissociation (HCD) with a normalized collision energy value of 28 [33,47]. The MS2 spectra were recorded at a resolution of 17,500 (at m/z = 200). The AGC targets for the MS and MS2 scans were set to 3E6 and 1E5, respectively, within a maximum injection time of 100 and 25 ms for the MS and MS2 scans, respectively. Dynamic exclusion was enabled in order to minimize the repeated sequencing of the same precursor ions, and was set to 30 s [33,47].

4.3. Data Analysis

The raw data were processed using the MaxQuant computational platform [67]. The peak list was searched against Human Uniprot databases, and the proteins were quantified across the samples using the label-free quantification algorithm in MaxQuant as the label-free quantification (LFQ) intensities [68]. We note that LFQ intensities do not reflect true copy numbers because they depend not only on the amounts of the peptides but also on their ionization efficiencies; thus, they only served to compare the abundances of the same protein in different samples [66–71]. Each MS experiment provided proteome-wide abundance data as LFQ intensities for three sample groups—one control (the non-targeting siRNA treated) and two stimuli (down-regulation by two different targeting siRNAs directed against the same gene)—with each having three data points. The missing data points were generated by imputation, whereby we distinguished two cases [46]. In order to identify which proteins were affected by PEX3 knock-down in siRNA-treated cells relative to the non-targeting (control) siRNA-treated sample, we log2-transformed the ratio between siRNA and the control siRNA samples, and performed two separate unpaired *t*-tests for each siRNA against the control siRNA sample [46]. The *p* values obtained by the unpaired *t*-tests were corrected for multiple testing using a permutation-based false discovery rate (FDR) test. The proteins with an FDR-adjusted *p* value of below 5% were considered to be significantly affected by the knockdown of the targeted protein. The results from the two unpaired *t*-tests were then intersected for further analysis, meaning that the abundance of all of the reported candidates was statistically significantly affected in both siRNA silencing experiments. For completely missing proteins lacking any valid data points, the imputed data points were randomly generated in the bottom tail of the whole proteomics distribution, following the strategy in the Perseus software (http://maxquant.net/perseus/; last accessed on 2 May 2021) [70]. For proteins with at least one valid MS data point, the missing data points were generated from the valid data points based on the local least squares (LLS) imputation method [71]. The validity of this approach was demonstrated [46]. Subsequent to the data imputation, gene-based quantile normalization was applied to homogenize the abundance distributions of each protein with respect to the statistical properties. All of the statistical analyses were performed using the R package of SAM (https://statweb.stanford.edu/~tibs/SAM/; last accessed on 2 May 2021) [72]. The protein annotations of the signal peptides, transmembrane regions, and N-glycosylation sites in humans and yeast were extracted from UniProtKB entries using custom scripts [46]. The enrichment of the functional Gene Ontology annotations (cellular components and biological processes) among the secondarily affected proteins was computed using the GOrilla package [73].

4.4. Validation of Putative PEX3 Substrates by Quantitative Western Blotting

Hits from the proteomics analysis were validated by quantitative Western blot analyses of Triton-X100 cell lysates (1% Triton X-100, 50 mM Hepes pH 7.5, 150 mM NaCl, 10% glycerol, 1 mM EDTA, 1 mM PMSF, Complete EDTA-free protease inhibitors (Roche)) using the following antibodies: anti-PEX3 (gift from G. Dodt; 1:1000 dilution), anti-ACBD5 (HPA012145 Merck, Taufkirchen, Germany; 1:1000 dilution), anti-RTN3 (12055-2-AP Proteintech, Manchester, UK; 1:1000 dilution), anti-PDIA5 (15545-1 Proteintech; 1:1000 dilution), anti-Far1 (ATA-HPA017322 Biozol, Eching, Germany; 1:1250 dilution), and anti-tubulin (T6199, Sigma-Aldrich; 1:10,000 dilution), which served as a loading control for normalization. The secondary antibodies were purchased from Licor Biosciences, Bad Homburg, Germany (926-68020, 926-68021, 926-32211, 926-32210, all in 1:20,000 dilution). The signals were detected using the Odyssey Clx system from Licor Biosciences, and were quantified by densitometry using the Image Studio software (Licor Biosciences). The relative protein abundance was calculated as the ratio of the signal of interest to the corresponding tubulin signal in the same lane, and was normalized against one control sample. The visualization of the quantification data was performed using Graphpad Prism software.

5. Conclusions

Recent studies characterized the PEX19/PEX3 pathway, which is best known for its role in the biogenesis of peroxisomal membrane proteins both at the peroxisomal and the ER membrane, as being involved in the biogenesis of hairpin membrane proteins of the ER as well as LDs. Therefore, the question arose as to whether this pathway may play a more general role in ER protein targeting, i.e., whether it may represent a fourth pathway for the ER targeting of precursor polypeptides next to SRP, SND, and TRC/GET. We have started to address this question by a novel approach which involves the label-free quantitative mass spectrometry of the total proteome of depleted or deficient cells, along with differential protein abundance analysis. Thus, we addressed the client spectrum of the PEX19/PEX3-dependent pathway in both PEX3 targeting siRNA-treated HeLa cells and PEX3-deficient Zellweger patient fibroblasts. The negatively affected proteins included six peroxisomal membrane proteins and two hairpin proteins of the ER, thus confirming the two previously identified classes of putative PEX19/PEX3 clients for ER targeting in human cells, as well as the validity of the experimental approach. In addition, 14 membrane proteins (including four tail-anchored proteins) and 27 proteins with SP (including 14 collagens and collagen-related proteins) belonging to the secretory pathway were also negatively affected by PEX3 deficiency. The latter findings are consistent with the idea that PEX3 represents a fourth pathway for the targeting of precursor polypeptides to the Sec61 complex. Furthermore, it may suggest a hitherto unknown spatial—or at least physical—relationship between the

ER subdomains that are involved in ER shaping and the budding of peroxisomal precursor vesicles, large cargo vesicles, and lipid droplets. In addition, these results may suggest compromised collagen biogenesis as a hitherto unknown contributor to organ failures in the respective Zellweger patients. All of these suggestions will have to be addressed in future research.

Supplementary Materials: The Supplementary Materials are available online at https://www.mdpi.com/article/10.3390/ijms222313028/s1.

Author Contributions: R.Z. and B.S. planned and supervised the sample generation for the MS analysis and validation experiments by M.L., D.N. performed the MS data analysis under supervision by V.H., F.F. and R.Z. designed the study. R.Z. and B.S. wrote the manuscript together with F.F., S.L. and V.H. All of the authors discussed the results. All authors have read and agreed to the published version of the manuscript.

Funding: F.F., V.H., S.L., B.S. and R.Z. were supported by the Deutsche Forschungsgemeinschaft (DFG, German Research Foundation), with grants FO716/4-1 to F.F., HE3875/15-1 to V.H., ZI234/13-1 to R.Z., IRTG1830 and SFB894 to S.L. and R.Z, and SFB1027 to B.S. Furthermore, the authors acknowledge support from the the Deutsche Forschungsgemeinschaft (DFG, German Research Foundation) and Saarland University within the funding programme Open Access Publishing.

Data Availability Statement: The novel MS proteomics data were deposited to the ProteomeXchange Consortium via the PRIDE partner repository, with the dataset identifier PXD012005 (http://www.proteomexchange.org, accessed on 18 October 2021). In addition, all of the data are available from the authors.

Acknowledgments: The authors thank Nagarjuna Nagaraj (Max-Planck Institute of Biochemistry, Biochemistry core facility, Martinsried, Germany) for the MS analyses, Silke Guthörl (Saarland University, Homburg, Germany) for the technical assistance, and Gabriele Dodt (University of Tübingen, Tübingen, Germany) for providing anti-PEX3 antibodies, as well as the immortalized PEX3-deficient cells and control fibroblasts.

Conflicts of Interest: The authors declare no conflict of interest.

Abbreviations

ATL	Atlastin
CDG	Congenital disorder of glycosylation
EMC	ER membrane complex
ER	Endoplasmic reticulum
GET	Guided entry of tail-anchored proteins
GO	Gene ontology
GPI	Glycosylphosphatidylinositol
LD	Lipid droplet
PEX	Peroxin
RAMP	Ribosome-associated membrane protein
RNC	Ribosome-nascent chain complex
RTN	Reticulon
SEC	(Protein involved in) secretion
SND	SRP-independent
SP	Signal peptide
SR	SRP receptor
SRP	Signal recognition particle
SSR	Signal sequence receptor
TMEM	Transmembrane (protein)
TMH	Transmembrane helix
TRAM	translocating chain-associating membrane (protein)
TRAP	Translocon-associated protein
TRC	Transmembrane recognition complex
Z	Zellweger (patient fibroblasts)

References

1. Dudek, J.; Pfeffer, S.; Lee, P.-H.; Jung, M.; Cavalié, A.; Helms, V.; Förster, F.; Zimmermann, R. Protein transport into the human endoplasmic reticulum. *J. Mol. Biol.* **2015**, *427*, 1159–1175. [CrossRef]
2. Aviram, N.; Schuldiner, M. Targeting and translocation of proteins to the endoplasmic reticulum at a glance. *J. Cell Sci.* **2017**, *130*, 4079–4085. [CrossRef] [PubMed]
3. Gemmer, M.; Förster, F. A clearer picture of the ER translocon complex. *J. Cell Sci.* **2020**, *133*, jcs231340. [CrossRef] [PubMed]
4. O´Keefe, S.; Pool, M.R.; High, S. Membrane protein biogenesis at the ER: The highways and byways. *FEBS J.* **2021**. [CrossRef]
5. Budnik, A.; Stephens, D.J. ER exit sites—Localization and control of COPII vesicle formation. *FEBS Lett.* **2009**, *583*, 3796–3803. [CrossRef] [PubMed]
6. Stephens, D.J. Collagen secretion explained. *Nature* **2012**, *482*, 474–475. [CrossRef]
7. Schrul, B.; Schliebs, W. Intracellular communication between lipid droplets and peroxisomes: The Janus face of PEX19. *Biol. Chem.* **2018**, *399*, 741–749. [CrossRef]
8. Jansen, R.L.M.; van der Klei, I.J. The peroxisome biogenesis factors Pex3 and Pex19: Multitasking proteins with disputed functions. *FEBS Lett.* **2019**, *593*, 457–474. [CrossRef]
9. Dhimann, R.; Caesar, S.; Thiam, A.R.; Schrul, B. Mechanisms of protein targeting to lipid droplets: A unified cell biological and biophysical perspective. *Sem. Cell Dev. Biol.* **2020**, *108*, 4–13. [CrossRef]
10. Goodman, J.M. Building the lipid droplet assembly complex. *J. Cell Biol.* **2020**, *219*, e202006025. [CrossRef] [PubMed]
11. Hansen, K.G.; Aviram, N.; Laborenz, J.; Bibi, C.; Meyer, M.; Spang, A.; Schuldiner, M.; Herrmann, J.M. An ER surface retrieval pathway safeguaerds the import of mitochondrial membrane proteins in yeast. *Science* **2018**, *361*, 1118–1122. [CrossRef]
12. Hegde, R.S.; Bernstein, H. The surprising complexity of signal peptides. *Trends Biochem. Sci.* **2006**, *31*, 563–571. [CrossRef]
13. Goder, V.; Spiess, M. Molecular mechanism of signal sequence orientation in the endoplasmic reticulum. *EMBO J.* **2003**, *22*, 3645–3653. [CrossRef] [PubMed]
14. Goder, V.; Junne, T.; Spiess, M. Sec61p contributes to signal sequence orientation according to the positive-inside rule. *Mol. Biol. Cell* **2004**, *15*, 1470–1478. [CrossRef]
15. Borgese, N.; Fasana, E. Targeting pathways of C-tail-anchored proteins. *Biochim. Biophys. Acta* **2011**, *1808*, 937–946. [CrossRef]
16. Yamamoto, Y.; Sakisaka, T. Molecular machinery for insertion of tail-anchored membrane proteins into the endoplasmic reticulum membrane in mammalian cells. *Mol. Cell* **2012**, *48*, 387–397. [CrossRef] [PubMed]
17. Borgese, N.; Coy-Vergara, J.; Colombo, S.F.; Schwappach, B. The ways of tails: The GET pathway and more. *Proteins* **2019**, *38*, 289–305. [CrossRef]
18. Abell, B.M.; High, S.; Moloney, M.M. Membrane protein topology of oleosin is constrained by its long hydrophobic domain. *J. Biol. Chem.* **2002**, *277*, 8602–8610. [CrossRef]
19. Leznicki, P.; Schneider, H.O.; Harvey, J.V.; Shi, W.Q.; High, S. Co-translational biogenesis of lipid droplet integral membrane proteins. *J. Cell Sci.* **2021**, *132*, jcs.259220. [CrossRef]
20. Gamerdinger, M.; Hanebuth, M.A.; Frickey, T.; Deuerling, E. The principle of antagonism ensures protein targeting specificity at the endoplasmic reticulum. *Science* **2015**, *348*, 201–207. [CrossRef]
21. Hsieh, H.-H.; Lee, J.H.; Chandrasekar, S.; Shan, S.-o. A ribosome-associated chaperone enables sustrate triage in a cotranslational protein targeting complex. *Nat. Commun.* **2020**, *11*, 5840. [CrossRef]
22. Aviram, N.; Ast, T.; Costa, E.A.; Arakel, E.; Chuartzman, S.G.; Jan, C.H.; Haßdenteufel, S.; Dudek, J.; Jung, M.; Schorr, S.; et al. The SND proteins constitute an alternative targeting route to the endoplasmic reticulum. *Nature* **2016**, *540*, 134–138. [CrossRef]
23. Casson, J.; McKenna, M.; Haßdenteufel, S.; Aviram, N.; Zimmermann, R.; High, S. Multiple pathways facilitate the biogenesis of mammalian tail-anchored proteins. *J. Cell Sci.* **2017**, *130*, 3851–3861. [CrossRef]
24. Haßdenteufel, S.; Sicking, M.; Schorr, S.; Aviram, N.; Fecher-Trost, C.; Schuldiner, M.; Jung, M.; Zimmermann, R.; Lang, S. hSnd2 protein represents an alternative targeting factor to the endoplasmic reticulum in human cells. *FEBS Lett.* **2017**, *591*, 3211–3224. [CrossRef]
25. Haßdenteufel, S.; Johnson, N.; Paton, A.W.; Paton, J.C.; High, S.; Zimmermann, R. Chaperone-mediated Sec61 channel gating during ER import of small precursor proteins overcomes Sec61 inhibitor-reinforced energy barrier. *Cell Rep.* **2018**, *23*, 1373–1386. [CrossRef]
26. Haßdenteufel, S.; Nguyen, D.; Helms, V.; Lang, S.; Zimmermann, R. Components and mechanisms for ER import of small human presecretory proteins. *FEBS Lett.* **2019**, *593*, 2506–2524. [CrossRef]
27. Cui, X.A.; Zhang, H.; Palazzo, A.F. p180 promotes the ribosome-independent localization of a subset of mRNA to the endoplasmic reticulum. *PLoS Biol.* **2012**, *10*, e1001336. [CrossRef]
28. Cui, X.A.; Zhang, Y.; Hong, S.J.; Palazzo, A.F. Identification of a region within the placental alkaline phosphatase mRNA that mediates p180-dependent targeting to the endoplasmic reticulum. *J. Biol. Chem.* **2013**, *288*, 29633–29641. [CrossRef] [PubMed]
29. Calvin, H.J.; Williams, C.C.; Weissman, J.S. Principles of ER coranslational translocation revealed by proximity-specific ribosome profiling. *Science* **2014**, *346*, 1257521. [CrossRef]
30. Ueno, T.; Tanaka, K.; Kaneko, K.; Taga, Y.; Sata, T.; Irie, S.; Shunji Hattori, S.; Ogawa-Goto, K. Enhancement of procollagen biosynthesis by p180 through augmented ribosome association on the endoplasmic reticulum in response to stimulated secretion. *J. Biol. Chem.* **2010**, *285*, 29942–29950. [CrossRef] [PubMed]

31. Hsu, J.C.-C.; Reid, D.W.; Hoffman, A.M.; Sarkar, D.; Nicchitta, C.V. Oncoprotein AEG-1 is an endoplasmic reticulum RNA-binding protein whose interactome is enriched in organelle resident protein-encoding mRNAs. *RNA* **2018**, *24*, 688–703. [CrossRef]

32. Hannigan, M.M.; Hoffman, A.M.; Thompson, J.W.; Zheng, T.; Nicchitta, C.V. Quantitative proteomics links the LRRC59 interactome to mRNA translation on the ER membrane. *Mol. Cell. Proteom.* **2020**, *19*, 1826–1849. [CrossRef]

33. Bhadra, P.; Schorr, S.; Lerner, M.; Nguyen, D.; Dudek, J.; Förster, F.; Helms, V.; Lang, S.; Zimmermann, R. Quantitative proteomics and differential protein abundance analysis after depletion of putative mRNA receptors in the ER membrane of human cells identifies novel aspects of mRNA targeting to the ER. *Molecules* **2021**, *26*, 3591. [CrossRef] [PubMed]

34. Sicking, M.; Lang, S.; Bochen, F.; Drenth, J.P.H.; Zacharia, M.; Zimmermann, R.; Roos, A.; Linxweiler, M. Complexity and specificity of sec61-channelopathies: Human diseases affecting gating of the Sec61 complex. *Cells* **2021**, *10*, 1036. [CrossRef]

35. Yamamoto, Y.; Sakisaka, T. The peroxisome biogenesis factors posttranslationally target reticulon homology-domain containing proteins to the endoplasmic reticulum membrane. *Sci. Rep.* **2018**, *8*, 2322. [CrossRef] [PubMed]

36. Schrul, B.; Kopito, R.R. Peroxin-dependent targeting of a lipid-droplet-destined membrane protein to ER subdomains. *Nat. Cell Biol.* **2016**, *18*, 740–751. [CrossRef] [PubMed]

37. Dahan, N.; Francisco, T.; Falter, C.; Rodrigues, T.; Kalel, V.; Kunze, M.; Hansen, T.; Schliebs, W.; Erdmann, R. Current advances in the function and biogenesis of peroxisomes and their roles I health and disease. *Hist. Cell Biol.* **2021**, *155*, 513–524. [CrossRef]

38. Erdmann, R.; Veenhuis, M.; Mertens, D.; Kunau, W.-H. Isolation of peroxisome-deficient mutants of Saccharomyces cerevisiae. *Proc. Natl. Acad. Sci. USA* **1989**, *86*, 5419–5423. [CrossRef]

39. Götte, K.; Girzalsky, W.; Linkert, M.; Baumgart, E.; Kammerer, S.; Hunau, W.-H.; Erdmann, R. Pex19p, a farnesylated protein essential for peroxisome biogenesis. *Mol. Cell. Biol.* **1998**, *18*, 616–628. [CrossRef] [PubMed]

40. Hettema, E.H.; Girzalsky, W.; van den Berg, M.; Erdmann, R.; Distel, B. Saccharomyces cerevisiae Pex3p and Pex19p are required for proper localization and stability of peroxisomal membrane proteins. *EMBO J.* **2000**, *19*, 223–233. [CrossRef] [PubMed]

41. Schmidt, F.; Dietrich, D.; Eylenstein, R.; Groemping, Y.; Stehle, T.; Dodt, G. The role of conserved PEX3 regions in PEX19-binding and peroxisome biogenesis. *Traffic* **2012**, *13*, 1244–1260. [CrossRef]

42. Muntau, A.C.; Maerhofer, P.U.; Paton, B.C.; Kammerer, S.; Roscher, A.A. Defective peroxisome membrane synthesis due to mutations in human PEX3 causes Zellweger syndrome, complementation group G. *Am. J. Hum. Genet.* **2000**, *67*, 967–975. [CrossRef]

43. Shimozawa, N.; Suzuki, Y.; Zhang, Z.; Imamura, A.; Ghaedi, K.; Fujiki, Y.; Kondo, N. Identification of PEX3 as the gene mutated in a Zellweger syndrome patient lacking peroxisomal remnant structures. *Hum. Mol. Genet.* **2000**, *9*, 1995–1999. [CrossRef] [PubMed]

44. Ghaedi, K.; Honsho, M.; Shimozawa, N.; Suzuki, Y.; Kondo, N.; Fujiki, Y. PEX3 is the causal gene responsible for peroxisome membrane assembly-defective Zellweger syndrome of complementation group G. *Am. J. Hum. Genet.* **2000**, *67*, 976–981. [CrossRef] [PubMed]

45. Hein, M.Y.; Hubner, N.C.; Poser, I.; Cox, J.; Nagaraj, N.; Toyoda, Y.; Gak, I.A.; Weisswange, I.; Mansfeld, J.; Buchholz, F.; et al. A human interactome in three quantitative dimensions organized by stoichiometries and abundances. *Cell* **2015**, *163*, 712–723. [CrossRef] [PubMed]

46. Nguyen, D.; Stutz, R.; Schorr, S.; Lang, S.; Pfeffer, S.; Freeze, H.F.; Förster, F.; Helms, V.; Dudek, J.; Zimmermann, R. Proteomics reveals signal peptide features determining the client specificity in human TRAP-dependent ER protein import. *Nat. Commun.* **2018**, *9*, 37639. [CrossRef]

47. Schorr, S.; Nguyen, D.; Haßdenteufel, S.; Nagaraj, N.; Cavalié, A.; Greiner, M.; Weissgerber, P.; Loi, M.; Paton, A.W.; Paton, J.C.; et al. Proteomics identifies signal peptide features determining the substrate specificity in human Sec62/Sec63-dependent ER protein import. *FEBS J.* **2020**, *287*, 4612–4640. [CrossRef]

48. Klein, M.-C.; Lerner, M.; Nguyen, D.; Pfeffer, S.; Dudek, J.; Förster, F.; Helms, V.; Lang, S.; Zimmermann, R. TRAM1 protein may support ER protein import by modulating the phospholipid bilayer near the lateral gate of the Sec61 channel. *Channels* **2020**, *14*, 28–44. [CrossRef] [PubMed]

49. Vizcaíno, J.A.; Csordas, A.; del-Toro, N.; Dianes, J.A.; Griss, J.; Lavidas, I.; Mayer, G.; Perez-Riverol, Y.; Reisinger, F.; Ternent, T.; et al. 2016 update of the PRIDE database and related tools. *Nucleic Acids Res.* **2016**, *44*, D447–D456. [CrossRef]

50. Cichocki, B.A.; Krumpe, K.; Vitali, D.G.; Rapaport, D. Pex19 is involved in importing dually targeted tail-anchored proteins to both mitochondria and peroxisomes. *Traffic* **2018**, *19*, 770–785. [CrossRef]

51. Exner, T.; Romero-Brey, I.; Yifrach, E.; Rivera-Monroy, J.; Schrul, B.; Zouboulis, C.C.; Stremmel, W.; Honsho, M.; Bartenschlager, R.; Zalckvar, E.; et al. An alternative membrane topology permits lipid droplet localization of peroxisomal fatty acyl-CoA reductase 1. *J. Cell Sci.* **2021**, *132*, jcs223016. [CrossRef]

52. Agarwal, A.K.; Sukumaran, S.; Cortés, V.A.; Tunison, K.; Mizrachi, D.; Sankella, S.; Gerard, R.D.; Horton, J.D.; Garg, A. Human 1-acylglycerol-3-phosphate O-acyltransferase isoforms 1 and 2: Biochemical characterization and inability to rescue hepatic steatosis in Agpat2(-/-) gene lipodystrophic mice. *J. Biol. Chem.* **2011**, *286*, 37676–37691. [CrossRef]

53. He, W.; Shi, Q.; Hu, X.; Yan, R. The membrane topology of RTN3 and its effect on binding of RTN3 to BACE1. *J. Biol. Chem.* **2007**, *282*, 29144–29151. [CrossRef]

54. Mayerhofer, P.U.; Bano-Polo, M.; Mingarro, I.; Johnson, A.E. Human peroxin PEX3 is co-translationally integrated into the ER and exits the ER in budding vesicles. *Traffic* **2016**, *17*, 117–130. [CrossRef] [PubMed]

55. Brar, G.A.; Weissman, J.S. Ribosome profiling reveals the what, when, where, and how of protein synthesis. *Nat. Rev. Mol. Cell Biol.* **2015**, *16*, 651–664. [CrossRef]
56. Pfeiffer, N.V.; Dirndorfer, D.; Lang, S.; Resenberger, U.K.; Restelli, L.M.; Hemion, C.; Miesbauer, M.; Frank, S.; Neutzner, A.; Zimmermann, R.; et al. Structural features within the nascent chain regulate alternative targeting of secretory proteins to mitochondria. *EMBO J.* **2013**, *32*, 1036–1051. [CrossRef] [PubMed]
57. Nuebel, E.; Morgan, J.T.; Fogarty, S.; Winter, J.M.; Lettlova, S.; Berg, J.A.; Chen, Y.-C.; Kidwell, C.U.; Maschek, J.A.; Clowers, K.J.; et al. The biochemical basis of mitochondrial dysfunction in Zellweger spectrum disorder. *EMBO Rep.* **2021**, *22*, e51991. [CrossRef] [PubMed]
58. Chen, C.; Li, J.; Qin, X.; Wang, W. Peroxisomal membrane contact sites in mammalian cells. *Front. Cell Dev. Biol.* **2020**, *8*, 512. [CrossRef]
59. Saito, K.M.; Chen, F.; Bard, S.; Chen, H.; Zhou, D.; Woodley, R.; Polischuk, R.; Schekman, R.; Malhotra, V. TANGO1 facilitates cargo loading at endoplasmic reticulum exit sites. *Cell* **2009**, *136*, 891–902. [CrossRef] [PubMed]
60. Raote, I.; Ortega-Bellido, M.; Santos, A.J.M.; Foresti, O.; Zhang, C.; Garcia-Parajo, M.F.; Campelo, F.; Malhotra, V. TANGO1 builds a machine for collagen export by recruiting and spatially organizing COPII, thethers and membranes. *eLife* **2018**, *7*, e32723. [CrossRef]
61. Raote, I.; Ernst, A.M.; Campelo, F.; Rothman, J.; Pincet, F.; Malhotra, V. TANGO1 membrane helices create a lipid diffusion barrier at curved membranes. *eLife* **2020**, *9*, e57822. [CrossRef]
62. Pataki, C.I.; Rodrigues, J.; Zhang, L.; Qian, J.; Efron, B.; Hastie, T.; Elias, J.E.; Levitt, M.; Kopito, R.R. Proteomic analysis of monolayer-integrated proteins of lipid droplets identifies amphipathic interfacial α-helical membrane anchors. *Proc. Natl. Acad. Sci. USA* **2018**, *115*, E8172–E8180. [CrossRef]
63. Pinto, M.P.; Grou, C.P.; Fransen, M.; Fransen, M.; Sa-Miranda, C. The cytosolic domain of PEX3, a protein inolved in the biogenesis of peroxisomes, binds membrane lipids. *Biochem. Biophys. Acta* **2009**, *1793*, 1669–1675. [CrossRef] [PubMed]
64. Poulos, A.; Christodoulou, J.; Chow, C.W.; Goldblatt, J.; Paton, B.C.; Orii, T.; Suzuki, Y.; Shimozawa, N. Peroxisomal assembly defects: Clinical, pathologic, and biochemical findings in two patients in a newly identified complementation group. *J. Pediatr.* **1995**, *127*, 596–599. [CrossRef]
65. Rappsilber, J.; Mann, M.; Ishihama, Y. Protocol for micro-purification, enrichment, pre-fractionation and storage of peptides for proteomics using StageTips. *Nat. Protoc.* **2007**, *2*, 1896–1906. [CrossRef]
66. Nagaraj, N.; Kulak, N.A.; Cox, J.; Neuhauser, N.; Mayr, K.; Hoerning, O.; Vorm, O.; Mann, M. System-wide perturbation analysis with nearly complete coverage of the yeast proteome by single-shot ultra HPLC runs on a bench top Orbitrap. *Mol. Cell. Proteom.* **2012**, *11*, M111.013722. [CrossRef]
67. Cox, J.; Mann, M. MaxQuant enables high peptide identification rates, individualized p.p.b.-range mass accuracies and proteome-wide protein quantification. *Nat. Biotechnol.* **2008**, *26*, 1367–1372. [CrossRef] [PubMed]
68. Cox, J.; Hein, M.Y.; Luber, C.A.; Paron, I.; Nagaraj, N.; Mann, M. Accurate proteome-wide label-free quantification by delayed normalization and maximal peptide ratio extraction, termed MaxLFQ. *Mol. Cell. Proteom.* **2014**, *13*, 2513–2526. [CrossRef]
69. Nagaraj, N.; Wisniewski, J.R.; Geiger, T.; Cox, J.; Kircher, M.; Kelso, J.; Pääbo, S.; Mann, M. Deep proteome and transcriptome mapping of a human cancer cell line. *Mol. Sys. Biol.* **2011**, *7*, 548. [CrossRef]
70. Tyanova, S.; Temu, T.; Sinitcyn, P.; Carlson, A.; Hein, M.Y.; Geiger, T.; Mann, M.; Coc, J. The Perseus computational platform for comprehensive analysis of (proteo)omics data. *Nature Meth.* **2016**, *13*, 731–740. [CrossRef] [PubMed]
71. Hyunsoo, K.; Golub, G.H.; Park, H. Missing value estimation for DNA microarray gene expression data: Local least squares imputation. *Bioinformatics* **2005**, *21*, 187–198. [CrossRef]
72. Tusher, V.G.; Tibshirani, R.; Chu, G. Significance analysis of microarrays applied to the ionizing radiation response. *Proc. Natl. Acad. Sci. USA* **2001**, *98*, 5116–5121. [CrossRef] [PubMed]
73. Eden, E.; Navon, R.; Steinfeld, I.; Lipson, D.; Yakhini, Z. Gorilla: A tool for discovery and visualization of enriched GO terms in ranked gene lists. *BMC Bioinform.* **2009**, *10*, 48. [CrossRef] [PubMed]

MDPI

St. Alban-Anlage 66

4052 Basel

Switzerland

Tel. +41 61 683 77 34

Fax +41 61 302 89 18

www.mdpi.com

International Journal of Molecular Sciences Editorial Office

E-mail: ijms@mdpi.com

www.mdpi.com/journal/ijms